P. Keating
UCLA '92

Speech Production and Speech Modelling

NATO ASI Series

Advanced Science Institutes Series

A Series presenting the results of activities sponsored by the NATO Science Committee, which aims at the dissemination of advanced scientific and technological knowledge, with a view to strengthening links between scientific communities.

The Series is published by an international board of publishers in conjunction with the NATO Scientific Affairs Division

A Life Sciences **B Physics**	Plenum Publishing Corporation London and New York
C Mathematical and Physical Sciences **D Behavioural and Social Sciences** **E Applied Sciences**	Kluwer Academic Publishers Dordrecht, Boston and London
F Computer and Systems Sciences **G Ecological Sciences** **H Cell Biology**	Springer-Verlag Berlin, Heidelberg, New York, London, Paris and Tokyo

Series D: Behavioural and Social Sciences - Vol. 55

Speech Production and Speech Modelling

edited by

William J. Hardcastle
Department of Linguistic Science,
University of Reading,
Reading, U.K.

and

Alain Marchal
C.N.R.S.,
Aix-en-Provence, France

Kluwer Academic Publishers

1990

Dordrecht / Boston / London

Published in cooperation with NATO Scientific Affairs Division

Proceedings of the NATO Advanced Study Institute on
Speech Production and Speech Modelling
Bonas, France
July 17–29, 1989

Library of Congress Cataloging in Publication Data

NATO Advanced Study Institute on Speech Production and Speech Modelling (1st : 1989 : Bonas, France)
Speech production and speech modelling / edited by William J. Hardcastle and Alain Marchal.
p. cm. -- (NATO ASI series. Series D, Behavioural and social sciences ; vol. 55)
"Proceedings of the NATO Advanced Study Institute on Speech Production and Speech Modelling, Bonas, France, July 17-29, 1989."
"Published in cooperation with NATO Scientific Affairs Division."
ISBN 0-7923-0746-1 (U.S. : alk. paper)
1. Speech--Congresses. 2. Linguistics--Congresses.
I. Hardcastle, William J. 1943- . II. Marchal, Alain. III. North Atlantic Treaty Organization. Scientific Affairs Division.
IV. Title. V. Series: NATO ASI series. Series D, Behavioural and social sciences ; no. 55.
QP399.N37 1989
612.7'8--dc20 90-34208

ISBN 0–7923–0746–1

Published by Kluwer Academic Publishers,
P.O. Box 17, 3300 AA Dordrecht, The Netherlands.

Kluwer Academic Publishers incorporates the publishing programmes of
D. Reidel, Martinus Nijhoff, Dr W. Junk and MTP Press.

Sold and distributed in the U.S.A. and Canada
by Kluwer Academic Publishers,
101 Philip Drive, Norwell, MA 02061, U.S.A.

In all other countries, sold and distributed
by Kluwer Academic Publishers Group,
P.O. Box 322, 3300 AH Dordrecht, The Netherlands.

Printed on acid-free paper

Printed in the Netherlands

CONTENTS

Speech Production and Speech Modelling

Preface

Speech sound production is one of the most complex human activities: it is also one of the least well understood. This is perhaps not altogether surprising as many of the complex neurological and physiological processes involved in the generation and execution of a speech utterance remain relatively inaccessible to direct investigation, and must be inferred from careful scrutiny of the output of the system - from details of the movements of the speech organs themselves and the acoustic consequences of such movements. Such investigation of the speech output have received considerable impetus during the last decade from major technological advancements in computer science and biological transducing, making it possible now to obtain large quantities of quantative data on many aspects of speech articulation and acoustics relatively easily. Keeping pace with these advancements in laboratory techniques have been developments in theoretical modelling of the speech production process. There are now a wide variety of different models available, reflecting the different disciplines involved - linguistics, speech science and technology, engineering and acoustics. The time seems ripe to attempt a synthesis of these different models and theories and thus provide a common forum for discussion of the complex problem of speech production. Such an activity would seem particularly timely also for those colleagues in speech technology seeking better, more accurate phonetic models as components in their speech synthesis and automatic speech recognition systems.

Such was the motivation behind the first NATO Advanced Study Institute on Speech Production and Speech Modelling held at Bonas, France from 17-29 July, 1989. Attended by 80 participants including senior research scientists from Universities and Industry, lecturers and graduate students, the ASI brought together a distinguished group from a wide variety of different backgrounds and orientations, all sharing a common interest in seeking a deeper understanding of the speech production process. This volume brings together the 16 invited lectures from that Institute and, in doing so, constitutes a unique collection of research findings and state-of-the-art reports from many of the leading international experts in this field.

There are four main sections. Section 1 provides a physiological framework for speech production processes. Neurophysiological aspects are outlined in Gentil's paper, while Ohala discusses underlying aerodynamic forces essential for speech sounds. From a different viewpoint, MacNeilage explains how many features of child language acquisition can be phylogenetically linked to underlying physiological behaviour in primates. In Section 2 there are two detailed discussions of connected speech processes: Farnetani outlines theories and experimental results on VCV coarticulation and, in a wide-ranging article on German, Kohler describes, and accounts for, various phonetic and phonological processes such as assimilations, weak forms, elisions and /r/ vocalizations that characterize normal connected speech.

The complex relationship between articulatory activities in the vocal tract and their acoustic consequences are addressed in Section 3. Maeda shows how, starting from a factor analysis of vocal tract profiles, he is able to propose a set of seven parameters for specifying vocal tract configurations. His articulatory model demonstrates how the jaw, lip and tongue can compensate acoustically for each other during vowel production. The modelling of more complex sounds such as fricatives is discussed in two papers; by Scully (on articulatory synthesis) and by Shadle (mechanical modelling). In Carré's paper, a new theoretical model of the vocal tract based on distinctive regions and modes is described. The section concludes with a report by Teager discussing complex non-linear aerodynamic features of the vocal tract.

Section 4 of the book covers a number of different models and theories of articulatory organization and timing. Perkell stresses the theoretical importance of variable articulatory data and develops his view of speech production as consisting of movement ensembles from one set of articulatory goals to another. In a detailed evaluation of Action Theory, Löfqvist discusses control and coordination of articulatory gestures. Next, writing as a physicist and a speech scientist, Fujimura proposes an alternative to the traditional linguistic view that speech is organized by concatenating phonemic segments. The subject of speech motor timing is treated at length by Keller, who evaluates the view that timing is subject to external neurocognitive control. Kent introduces various different theoretical frameworks for speech production and evaluates to what extent these frameworks can account for data from neurologic disorders. Finally Lindblom addresses the systematic nature of intra-speaker phonetic variation and develops a theoretical framework that accounts for the interplay between production-oriented control ("hypospeech") and output-oriented factors ("hyperspeech") during speech performance.

We are indebted to the NATO Scientific Affairs Division for providing the major funding and support to make this ASI possible. We are grateful also to the numerous governmental and research institutions that supported individual participants; the French Centre National de la Recherce Scientifique, the British Council, the British Science and Engineering Research Council, and the American National Science Foundation. Finally we thank all the participants who made this ASI a particularly productive and enjoyable experience.

Bill Hardcastle
Alain Marchal

List of Contributors to this Volume

Professor R. Carré
Institut de la Communication Parlée
INPG - ENSERG
46 Av. F. Viallet
38031 Grenoble
France

Dr Barbara L Davis
Dept of Linguistics
The University of Texas at Austin
Calhoun Hall 501
Austin Texas 78712
U.S.A.

Dr Edda Farnetani
Centro di Studio per le
Ricerche di Fonetica del CNR
Piazza Salvemini 13
35122 Padova
Italy

Professor Osamu Fujimura
The Ohio State University
Division of Speech and Hearing Science
Columbus, OH 43210
U.S.A.

Dr Michèle Gentil
CHU Pitié-Salpêtrière
Lab. de Physiologie de la Motricité
CNRS (URA 385)
91 Bd. de l'Hôpital
75634 Paris cédex 13
France

Dr Eric Keller
Université du Québec à Montréal
Linguistique
C.P. 8888
Succ. A
Montréal
QC H3C 3P8
Canada

Professor R.D. Kent
Dept of Communicative Disorders
University of Wisconsin-Madison
1975 Willow Drive
Madison WI 53706
U.S.A.

Prof.Dr. K.J. Kohler
Institut für Phonetik & Digitale
Sprachverarbeitung
Olshausenstrasse 40
D-2300 Kiel
F.R.G.

Professor Björn Lindblom
Dept of Linguistics
University of Texas at Austin
Calhoun Hall 501
Austin
Texas 78712 1196
U.S.A.

Dr Anders Löfqvist
Dept of Logopedics and Phoniatrics
University Hospital
S-22185 Lund
Sweden
and
Haskins Laboratories
New Haven
CT
U.S.A.

Professor Peter F. MacNeilage
Dept of Linguistics
The University of Texas at Austin
Calhoun Hall 501
Austin
Texas 58712
U.S.A.

Dr Shinji Maeda
Tour Sapporo Apt.203
70 rue du Javelot
75013 Paris
France

Dr M. Mrayati
Scientific Studies and Research Center
POB 4470
Damascus

Professor John Ohala
Phonology Laboratory
Dept of Linguistics
University of California
Berkeley CA 94720
U.S.A.

Dr Joseph S. Perkell
Research Lab of Electronics
Massachusetts Institute of Technology
Room 36-543
50 Vassar St
Cambridge
Mass 02139
U.S.A.

Ms Celia Scully
Dept of Linguistics & Phonetics
University of Leeds
Leeds
LS2 9JT
U.K.

Dr Christine H. Schadle
Dept of Electronics and Computer Science
University of Southampton
Southampton
SO9 5NH
U.K.

Dr H.M. Teager & Dr S.M. Teager
Boston University Hospital School of Medicine
Biomedical Engineering
75 East Newton Street
Boston
Mass 02118
U.S.A.

ORGANIZATION OF THE ARTICULATORY SYSTEM : PERIPHERAL MECHANISMS AND CENTRAL COORDINATION

Michèle GENTIL
INSERM, Paris

CHU Pitié-Salpêtrière
Laboratoire de Physiologie de la Motricité
CNRS (URA 385)
91, boulevard de l'Hôpital
75634 PARIS cédex 13 (France)

ABSTRACT. The physiological mechanisms underlying oral communication are of primary interest. Investigations concerning the motor acts of speech are limited to what can be inferred from observation of the peripheral mechanisms of speech production. The neuromuscular events of speech movement patterning are described. In addition, speech movement coordination involves substantial adjustments among the multiple muscles of the various articulatory organs. The cerebellum is an important regulatory center for the control of motor activity. It provides smoothness to the contraction of synergist and antagonist muscles. It uses sensory information to predict, judge and correct the motor act. Damage to a structure with this capability would be expected to have consequences on speech production . The role of the cerebellum in speech control is examined by assessing the effects of cerebellar lesions on speech through a short review of literature.

1. Introduction

Specification of the structure of the motor system is not an easy task. The last fifteen years have shown a considerable increase in research on motor control which has led to two main approaches : theories proposing a motor program in which the commands to the individual muscles are organized centrally by some sort of plan or programming, and action theories that reject such a computer analogy and instead describe motor control in terms of coordinated structures. Independently of these different views, a certain number of basic notions, at the peripheral level, can be simply defined, in particular, some elements relative to peripheral mechanisms underlying articulatory organ movement control, which are essential for the understanding of normal and disordered speech motor behavior.

The peripheral mechanisms include : the mechanical response of the peripheral structures to the neural impulses, and the contribution of afferent feedback mechanisms. Thus, nerves, synaptic transmissions, muscles and sensory receptors, which contribute significantly to speech system output, will be described. Contrary to the central mechanisms, these peripheral components are, in large part, observable. Moreover, the development of speech monitoring capabilities has facilitated observation of speech movements and muscular activities. From exploration of neuromuscular events, inferences to higher levels of control can be made.

Various systems contribute to speech production : a) orofacial system consisting of lips, jaw, tongue, b) velopharyngeal system, c) laryngeal system, d) respiratory system. Observations of a simple system or, a fortiori, a subsystem may not provide a clear picture of the complex adjustments underlying speech production. The nervous system has to

W. J. Hardcastle and A. Marchal (eds.), Speech Production and Speech Modelling, 1–22.

assume the coordination of multiple articulatory organ movements subordinated to an overall communication goal. Perhaps more can be gained in our understanding of the coordination if we consider the role of sensors as well as the importance of the cerebellum, a very interesting structure in the central nervous system. It is a highly organized center which exerts a regulatory influence on muscular activity. It receives a continuous stream of impulses from receptors in muscles, joints, tendons and skin, and from visual, auditory and vestibular end organs. These impulses supply the sensory cues essential to the control of movement. Signals from the cerebral cortex and other motor regions also reach the cerebellum which plays a crucial role in movement by adjusting the output of the major descending motor system of the brain. Through comparisons of external and internal feedback signals, the cerebellum is able to correct ongoing movements when they deviate from their intended course, and to modify central motor programs so that subsequent movements can fulfill their goal. If any damage occurs to the cerebellar system, the coordination of movements is disrupted.

The activity of speech is realized by the complex movements of the articulatory organs and the speech process can be viewed as a fine motor skill which must be regulated in terms of sequence and duration, with great accuracy, speed, ranging from 5 to 20 cm per second (Lund et al. 1982), and rhythmicity. Lesions affecting the cerebellum and/or its connection via the cerebellar peduncles result in a very distinctive-sounding dysarthria called "ataxic dysarthria" (Brown et al. 1970 ; Darley et al. 1975 ; Kent et al. 1975 ; 1979 ; Netsell 1982). This is characterized by excessive fluctuations in fundamental frequency and intensity, and irregularity in the velocity of articulatory organ movements bringing about timing errors (Hirose et al. 1978 ; Netsell et al. 1976). Thus, the cerebellar system is to some extent involved in the control of the vocal tract at various levels. The relationship between cerebellar lesions and speech disorders will be examined.

2. Peripheral Mechanisms Underlying Articulatory Organ Movement Control

An important research effort in physiology and biomechanics aims at elucidating the complex process of peripheral mechanisms. The validity of inferences made about higher levels of control depends on a sound knowledge of peripheral mechanisms. When one considers the neurological network involved in motor behavior, it can be found that a great part of it is implicated in the processing of sensory information. As a great deal of speech input is derived from proprioceptive sources, this suggests that the response is considerably influenced by peripheral mechanisms. First, fundamental physiological notions are discussed, which may give an insight into peripheral components of the speech control mechanism.

2.1. NERVE AND SYNAPTIC TRANSMISSION

Thanks to nerve cells and the connections between them that convey messages, the brain can issue motor commands for the coordinated movements and receive information from peripheral receptors in the skin, mucosa and muscles. Thus, the nerve fibres which send the neural command out from the central nervous system to muscle fibres belong to the efferent (motor) system and the nerve fibres which transmit the information derived from the receptors to the central nervous system belong to the afferent (sensory) system.

The muscles involved in speech articulation are innervated by the cranial nerve system which arises from the brain stem. The motor nucleus of the trigeminal nerve (V) lies in the pons. It contains motor neurons that innervate the muscles of the jaw. The motor nucleus of the facial nerve (VII) lies caudal to the motor nucleus of the trigeminal nerve in the pons and contains motor neurons that innervate the facial muscles. The motor neurons contributing to the glossopharyngeal (IX), and vagus (X) nerves lie most caudally in the medulla and are clustered in a group called the nucleus ambiguous. Neurons in this nucleus innervate striated muscles in the larynx and pharynx and are therefore critical for speech. The hypoglossal (XII) nucleus is in the medulla and innervates the tongue muscles (Ranson et al. 1953). The cranial nerve system consists mostly of mixed nerves which contain both efferent and afferent nerve fibres (Lazorthes 1967).

Although the nerve cells can be classified into different types, they share many features in common. Some of the structural and functional properties of nerve cells will be considered, followed by an examination of how they are interconnected.

2.1.1. *Structure Of The Nerve Cell.* The typical neuron has four morphologically defined regions : the cell body, the dendrites, the axon and the presynaptic terminals of the axon (Figure 1).

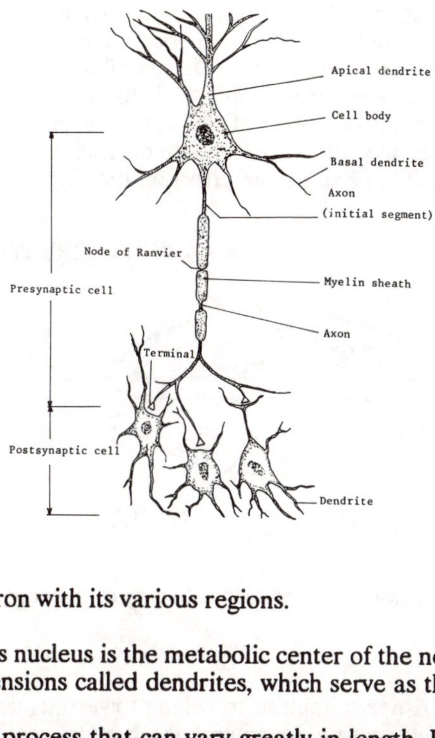

Figure 1. Typical neuron with its various regions.

The cell body with its nucleus is the metabolic center of the neuron. It gives rise to :
a) fine arborizing extensions called dendrites, which serve as the chief receptive apparatus for the neuron,
b) the axon, a tubular process that can vary greatly in length. It constitutes the conducting unit of the neuron. The axons of cranial nerves innervating the speech musculature are usually myelinated, that is an insulating sheath of myelin surrounds the axon. Myelin does not run continuously along the fibre. It is interrupted at regular intervals. These points of interruption are called the nodes of Ranvier which, in general, are more widely spaced in larger fibres than in smaller ones. Near its end, the axon divides into many branches which have specialized endings called presynaptic terminals. By means of its terminal, one neuron contacts and transmits neural impulse to the receptive surfaces of another neuron, a muscle or other kinds of effector cells.

2.1.2. *Propagation Of The Neural Impulse.* The membrane surrounding the axon is selectively permeable ; some chemical substances pass rather than others. In a resting condition, positive potassium ions pass out through the membrane and the inside of the axon is negatively charged with respect to the outside. During excitation, the membrane becomes much more permeable to positive sodium ions that pass inside the membrane and reverses the membrane potential. An impulse is produced actively only at a node of Ranvier. This region becomes depolarized i.e. the outside of the membrane is negatively charged. However, because of the presence of myelin which has high electric resistance, an eddy current cannot develop. The current generated in the depolarized region must travel along the inside of the fibre until it reaches the next node of Ranvier. Here, the current leaves the fibre and returns along the outside of the myelin sheath to the originally depolarized region. When current leaves the fibre at the second nodal region, this becomes depolarized and an action potential is developed. The latter generates a similar process and the next node is depolarized, etc. Thus, the impulse jumps from one node to the next according to the theory of saltatory conduction (Figure 2). Consequently, the impulse can travel along the fibre at great speed. Obviously, the transmission time of the impulse depends on the length of the nerve fibres (see Lenneberg, 1967, for the difference in length of nerves innervating the speech musculature). Large fibres, with more widely spaced nodes of Ranvier conduct impulses faster than small ones. In unmyelinated fibres, the velocity is slower. The conduction velocities of the motoneuron axons in the facial nerve range from 23 to 46 meters per second at the terminal portion of the nerve (Sawhney et al. 1971). The temperature of the fibre has also an effect on the conduction velocity.

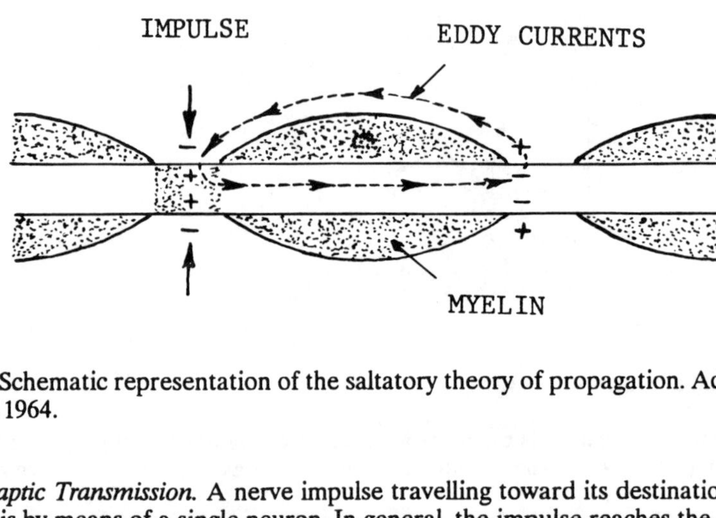

Figure 2. Schematic representation of the saltatory theory of propagation. Adapted from Hodgkin, 1964.

2.1.3. *Synaptic Transmission.* A nerve impulse travelling toward its destination usually does not get this by means of a single neuron. In general, the impulse reaches the end of a nerve that is in contact with another structure. Sherrington introduced the term synapse (Greek. synapsis, junction) at the turn of the century to refer to the specialized contact zone, described histologically by Ramon y Cajal where one neuron communicates with one another. The presynaptic element is the terminal portion of the nerve fibre that is in opposition with the following structure. The latter is called the postsynaptic element. Both are separated by a very small distance, called the synaptic cleft. The transfer of information across a synaptic junction is referred to as synaptic transmission which is either chemical or electric . Chemical synapses interconnect neurons for variable and complex behavior, such as speech. Chemical transmission is divided into two sets of processes. The presynaptic transmitting processes determine the release of the chemical substance called acetylcholine. The postsynaptic receptive processes determine the interaction between the acetylcholine and the reception molecule in the postsynaptic cell. The interaction gives rise to current flow that produces the synaptic potentials. The delay of transmission , called synaptic delay is less than 0.5 msec (Katz 1966).

Synapses can be either excitatory or inhibitory, that is either they facilitate the transmission of the neural impulse or they resist. Because of the ability of the central nervous system to regulate the firing of excitatory and inhibitory cells, movements and particularly speech movements can be continually modulated.

2.2. MUSCLE

The central nervous system can act on the outside world only by means of muscle. How facial muscles produce the forces necessary to move articulatory organs will be examined.

The muscles perform two kinds of functions : voluntary and involuntary. Histologically, they are of three types : striated, unstriated and cardiac. Speech muscles are striated muscles contracting at will. In a muscle, there are two groups of muscle fibres, extrafusal and intrafusal which will be considered separately.

2.2.1. *Extrafusal Muscle : Mechanisms Of Contraction.* Extrafusal muscle fibres have diameters of 10 to 100 and are surrounded by an excitable membrane, the sarcolemma. Each fibre consists of several myofibrils which are 1 to 2 in diameter. These myofibrils contain the muscle's contractile machinery in the form of thin and thick myofilaments. The interdigitating arrangement of myofilaments produces the characteristic pattern of transverse bands (hence the name "striated") (Figure 3).
The dark A band (anisotropic) corresponds to the zone where thin and thick filaments overlap. In its midsection there is a less dense strip, the H band or only thick filaments are present. The light isotopic bands or I bands consisting of thin filaments only, present also a dark strip, the Z line which limits the sarcomere (contractile unit). The most generally accepted explanation of mechanical events in muscular contraction is the sliding filament hypothesis. This asserts that the sarcomere length is reduced by about 30 % during shortening, while the length of both thin and thick filaments remains unchanged ; thus, thin and thick filaments slide past each other. Force is developed between the thick and thin filaments by the formation of cross-bridges composed of the head of the myosin molecule, the protein constituent of the thick filaments. The cross-bridges contact the thin filaments consisting mostly of another protein, actin. Thus, during contraction both structures become attached. Variations in the angle of attachment of the heads with the thin filaments permit the sliding filaments to move past each other (Huxley 1969). The development of mechanical force is the result of a sequence of structural changes associated with the hydrolysis of adenosine triphosphate (ATP) and its conversion to adenosine diphosphate (ADP).

Although sliding filament mechanisms operate in all muscles, individual muscle fibres can be subdivided into two distinct classes, fast and slow, each of which have different physiological and biochemical properties (Granit 1970). Slow muscle fibres contract and relax slowly, generate only low levels of force. Because these fibres have a high myoglobin content and are richly vascularized, they are red. They also have many mitochondria and utilize oxidative metabolism. They are therefore resistant to fatigue and are specialized for sustained contraction. Fast muscle fibres contract and relax rapidly when stimulated. These fibres are capable of generating great force. They have a relatively low myoglobin content and are poorly vascularized, which gives them a pale color ; they have few mitochondria.

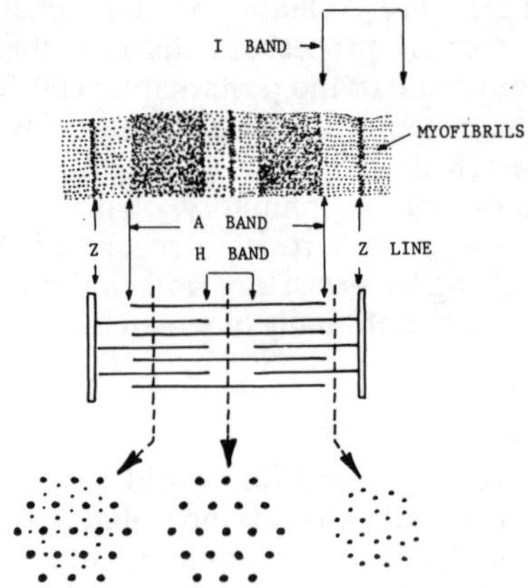

Figure 3. Structure of a muscle fibre. Adapted from Eyzaguirre and Fidone, 1977.

As a result, fast muscle fibres fatigue rapidly and are best suited for intense activity of short duration. Histochemical studies beginning with that of Denny-Brown (1929) have shown that slow fibres are mixed with fast fibres in most muscles. For example, mandibular muscles of humans contain different proportions of these two principal fibres : 50 % of masseter and temporalis muscle fibres are fast fibres and 50 % are slow fibres (Ringqvist 1971) .The anterior belly of digastric has 65 % of its fibres which are fast and 35 % slow (Erikson, 1982).

2.2.2. *Motor Unit.* A single nerve fibre innervates a certain number of muscle fibres. In 1925, Liddell and Sherrington introduced the term "motor unit" to describe the smallest functional unit that can be controlled by the nervous system. The motor unit consists of a single motor neuron, its axon and all the muscle fibres that it innervates. The number of muscle fibres in a motor unit varies from muscle to muscle according to the role they play in motor activity. Muscles requiring delicate adjustments have a small number of muscle fibres for each motor unit, whereas larger muscles requiring only gross control have many fibres for each motor unit. Thus, in the intrinsic tongue muscles there are may be as few as 7 muscular fibres in a motor unit (Hardcastle 1976), in the larynx muscles from 116 to 248 (Faaborg Andersen 1957), in masseter muscle 640 and in temporal 936 (Goodgold et al. 1977). For comparison, limb muscles have from 400 to 1700 fibres per motor unit (Buchtal 1961).

Moreover, motor neurons differ in their ability to fire rapidly and there is a close correspondance between the functional properties of the muscle and the motor neuron. Characteristic firing frequencies increase in the order : limb and trunk musculature, speech musculature, extraocular musculature. Higher firing frequencies appear to be assisted with a great number of possible states of contraction and an ability to change states more rapidly (MacNeilage 1973). Concerning lips and tongue, a high frequence of firing and few muscle fibres per motor unit allow fine movements with great precision (Fromkin et al. 1966).

Controlled precise gradation of contraction is necessary for speech articulation. The nervous system produces graded increase in muscle forces in two ways : by activation of previously silent motor units (recruitment), and by regulation of discharge frequency in already active motor units (rate coding) (for review see Burke 1981a and 1981b). Henneman et al. (1965) hypothesize an order of recruitment of motor units during a muscular contraction which is dependent upon the size of motoneurons. Smaller motoneurons are more excitable than larger. As the force in a muscle increases, motor units of increasing size are recruited resulting in a graded contraction (Goldberg et al. 1977). The other way for the nervous system to command greater muscular force is by increasing the firing rate of motor neurons. Thus, variation in the range, force or type of movement are determined by the pattern of recruitment and the frequency of firing of different motor units. The motor unit can therefore be considered the elementary unit of behavior in the motor system.

2.2.3. *Mechanical Properties.* A muscle can contract isotonically or isometrically. In isotonic contraction, the muscle shortens. The degree to which a muscle can change in length, in response to a given neural drive, depends on the initial length of the muscle and on the force opposing changes in length. The dependence on initial length arises because muscles behave mechanically like springs. In isomotric contraction, shortening of the muscle is prevented but tension develops. Concerning speech movement control, the evaluation of peripheral mechanical properties has been the subject of a considerable research effort (for a review, see Abbs et al. 1976). These complex processes cannot be examined in the limited scope of this chapter.

2.2.4. *Neuromuscular Junction And Propagation Of The Impulse.* The neuromuscular junction is the region where nerve and muscular fibres come into contact. A detailed account of the transmission of nerve impulses at the neuromuscular junction can be found in Eccles (1973c) and in Katz (1962 ; 1966). When the action potential reaches the nerve endings, a transmitter (acetylcholine) is released from the presynaptic element of the junction and floods the postsynaptic membrane of the junction (motor end-plate). This transmitter depolarizes this membrane and if the depolarization attains enough amplitude, it will trigger a muscle action potential. Thus neuromuscular transmission occurs in several stages : a) nerve impulse, b) nerve ending depolarization, c) transmitter release, d) end-plate potential, e) muscle action potential, f) muscle contraction.

The propagation of action potentials constitute the basic electrical phenomena from which electromyographic recordings are derived. EMG technique provides graphic information about the electrical activity which accompanies muscle contraction. EMG is the most powerful means for assessing speech control principles through direct measurements (Fujimura 1979 ; 1987). This technique has been used for studying muscular function (Alfonso et al. 1982b ; Bell-Berti 1983 ; Gentil et al. 1986 ; Hirano et al. 1969 ; 1970 ; Kakita et al. 1983 ; Smith 1971) and coordination of the various articulatory organs (Abbs et al. 1984 ; Alfonso et al. 1982a ; Gay 1977 ; 1979 ; Gentil et al. 1984 ; Tuller et al. 1982). Recently, EMG has been applied to the analysis of pathological kinesiology related to speech production (Hirose, 1986).

2.3. SENSORY RECEPTORS

Speech production, considered a fine motor skill (Hardy 1971 ; Netsell 1982), is hypothesized to be highly dependent on afferent information derived from receptors located within skin, muscles and joints (Abbs et al. 1982 ; 1983 ; Barlow et al. 1986a ; 1986b ; Evarts 1982 ; Lund et al. 1982 ; Mountcastle 1980a ; 1980b). These sensors transduce or convert mechanical energy into a neural signal encoded by the central nervous system (Barlow et al. in press).

2.3.1. *Mechanoreceptors.* One source of afferent information originates from mechanoreceptors which respond to different kinds of mechanical distortions by generating a depolarizing current in the sensory fibre (Gray 1959). The mucosae, although not part of the skin, have receptors with essentially the same characteristics as the cutaneous receptors. For example, sensations of touch, temperature, etc, can be evoked from the oral mucosa. The tongue surface is supplied with many different mechanoreceptors (Grossman et al. 1967). About the broad classification of receptors into "free" endings and "organized" endings (Winkelmann 1960), Hardcastle (1970) suggested a functional significance for the sensory control of speech articulation, the free endings subversing a general sensation of touch and the organized endings allowing a high degree of tactile acuity. The morphological structure and location in the oral mucosa could explain these functions. The free endings are diffuse overlapping terminal filaments which interweave with one another through the oral mucosa. The organized endings are well defined morphologically distinct structures consisting of fibrous tissue capsules. In the tongue, these organized receptors (Krause's end-bulb and Meissner's corpuscle) are situated in the deep layers of the lamina propria. They respond to a very slight deformation and stop immediately when the movement ceases. There are too a number of receptors present within the periodontal membranes of teeth. Reflex studies with electromyographic recording from jaw-closing muscle in human beings (Goldberg 1971 ; Sessle et al. 1972) have demonstrated short-latency excitatory responses in the masseter and temporal muscles evoked by mechanical stimulation of the periodontal receptors. All these various receptors play some important role in the control of speech.

Various studies (Abbs et al. 1984 ; Folkins et al. 1982 ; Gracco et al. 1985 ; Smith et al. 1985) have shown that small mechanical inputs can significantly affect orofacial muscle activity during speech. Afferent information is thought to originate from mechanoreceptors within perioral tissue (Barlow 1987). The type of these receptors is not yet known. The Pacinian corpuscles which may be capable of detecting very small movements as well as movement acceleration (Skoglund 1973) are absent in the face (Barlow 1987). In addition, there are no muscle spindles and Golgi tendon organs (cf below) in facial muscles (Folkins et al. 1978 ; Lovell et al. 1977). More histologic and physiologic data are needed to improve our knowledge of sensitive receptors in the human face.

2.3.2. *Muscle Receptors.* Among the different receptors, two have important and specific actions on motor neurons. These are the muscle spindles and the Golgi tendon organs. Muscle spindles, arranged in parallel with the extrafusal muscle fibres provide information about the length of the muscle. Golgi tendon organs, arranged in series with the extrafusal muscle fibres inform the nervous system of the tension exerted by the muscle.

2.3.2.1. Intrafusal Muscle (Muscle Spindle). The term "intrafusal" was coined by Sherrington (1894) to account for the muscles of the neuromuscular spindles. In the past 30 years, the structure and function of these have received almost continuous attention. The reader interested can find the subject developed in the following contributions : Barker (1962), Cooper (1960), Eldred (1965), Granit (1955 ; 1970) ; Houk et al. (1981) ; Matthews (1964 ; 1972 ; 1981), Stein (1974). Eldred and his colleagues (1967 ; 1977) published a remarkable bibliography on muscle receptors.

The morphological structure of the muscle spindle is essential to the understanding of its function. The muscle spindle is a fusiform structure 4 to 7 mm long, 80 to 200 wide located deep in the muscle mass. It consists of a group of thin, striated muscle fibres, the intrafusal bundle. This is enveloped by a capsule. In its center, the equatorial region, the capsule has a bulbous enlargment which gives the receptor its spindlelike appearance. The muscle spindle contains two types of intrafusal muscle fibres : long and thick nuclear bag

fibres which contain large aggregations of nuclei and shorter and thinner nuclear chain fibres which have fewer nuclei (Figure 4). Bag fibres produce slow contraction and chain fibres produce fast or twitch contractions.

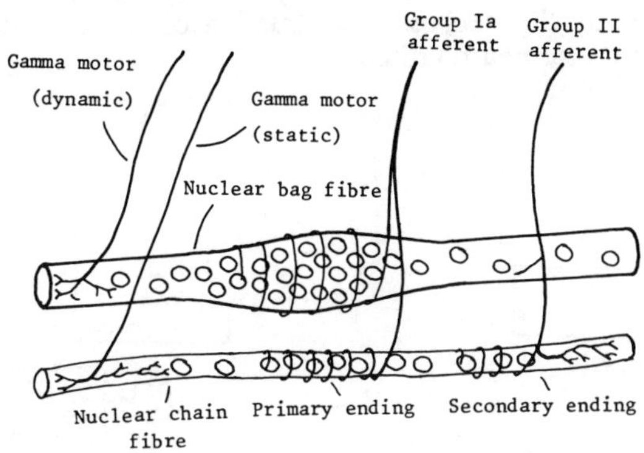

Figure 4. Nuclear bag and nuclear chain intrafusal fibres within a muscle spindle. Adapted from Matthews, 1964.

The sensory innervation of the muscle spindle arises from the groups Ia and II afferent fibres. The group Ia fibres coil about the central region of both the nuclear bag fibres and the nuclear chain fibres forming the primary ending and the group II fibres terminate primarily upon the nuclear chain fibres forming the secondary ending. When a muscle is stretched, the primary and secondary afferents in the spindles respond differently. Thus concerning the jaw stretch reflex (cf below), primary afferents fire phasically and are more sensitive to the velocity of muscle lengthening during opening, secondary afferents are mainly sensitive to length changes. As suggested by Barlow et al. (in press) the displacement information encoded by secondary endings may explain partially the ability of speech motor control system to reprogram lip and tongue movements for vowel production under jaw-blocked conditions. The intrafusal muscle fibres also receive a motor innervation from a group of efferent nerve fibres called gamma motor neurons. These produce contraction of the polar ends of the intrafusal fibres. They are separable into two groups according to their effects on the dynamic and static responses of the spindle afferents. Dynamic fibres increase the dynamic response of the afferents Ia and have no effect on spindle secondaries. Static fibres increase the static responses of both the primary and secondary spindle afferents.

In speech musculature, the presence of spindles is variable. They have been found in the lingual muscles (Cooper 1953 ; Kubota et al. 1975 ; Walker et al. 1959) and also in the laryngeal muscles (Barlow et al. in press). Concerning the jaw, the deep parts of the temporal and masseter muscles contain a high number of muscle spindles. In contrast, few spindles are observed in the lateral pterygoid muscle and only occasional ones in the anterior belly of digastric (Erikson, 1982). As previously mentioned, spindles do not exist in lip muscles.

2.3.2.2. Role Of Muscle Spindles. Activation of the muscles can be brought about by activation of gamma motor neurons from high centers. This is called activation through the gamma loop. Activation of gamma motor neurons produces shortening of the spindle giving rise to an increase in Ia fibre discharge which in turn increases the alpha motor neuron output, thereby producing contraction of the extrafusal muscle. Thus, spindles contribute to the stretch (myotatic) reflex. For example, it is the case of stretch reflex of masseter or temporal muscles (Hugelin and al. 1957). In fact, the stretch reflex could be made to function as a servomechanism regulating muscle length, the extrafusal muscle fibres following automatically the degree of contraction of the intrafusal fibre which is determined by the gamma motor neuron discharge.

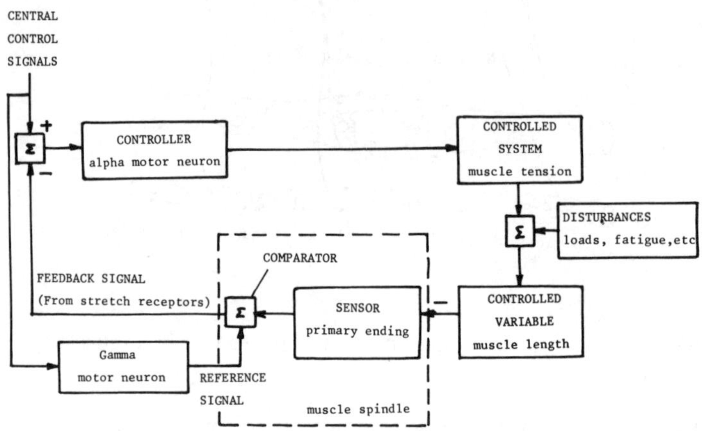

Figure 5. Diagram illustrating the role of spindles. From Carew, 1985.

However, the movement is not initiated by activation of the gamma motor neurons but by coactivation of both alpha and gamma motor neurons (Vallbo 1970). This activation has been observed in several motor systems including those involved in jaw contraction (Carew 1985). Then, the stretch reflex is a negative feedback mechanism. The descending control signal acts upon the alpha motor neurons where their action sums with feedback signals from the muscle spindle. This descending control is designed to maintain a set muscle length ; the controlled variable. This is achieved by controlling muscle tension which interacts and sums with external disturbances, such as load changes or muscle fatigue, to yield the actual length change. The difference between the actual and intended change in length is sensed by the spindle which is informed about the intended length by the gamma drive. If the muscle shortens more than intended, the spindle output will drop then reducing the Ia excitatory input to the motor neuron and it will fire less. This, in turn, will reduce muscle tension (Figure 5).

2.3.2.3. Golgi Tendon Organ. The Golgi tendon organ is a sensory receptor that sends information to the central nervous system, about the tension exerted by the muscle on its tendinous insertion to the bone. It is a slender capsule. Each organ is in series with some extrafusal muscle fibres that enter the capsule. These terminate in musculo-tendinous junctions and give rise to collagen fibre bundles which become braided and run the length of the capsule. Afferent fibres Ib enter the capsule and branch many times so that the axons of the afferent fibres become twisted between the collagen fibres. When the extrafusal fibres contract they cause the collagen bundle to straighten, this, in turn, compresses the axons of the afferent fibres, causing them to fire. Thus, the arrangement of fibres that compress the axons makes these very sensitive to small changes in muscle tension. Concerning the Golgi tendon organ, excellent reviews can be consulted (Matthews 1972, Hunt 1974). Tendon organs are present in the jaw system (Karlsson 1976) but, as already specified, they have not been found in lips (Folkins et al. 1978).

3 - Sensorimotor Processes

They are multiple sensorimotor actions underlying the generation of movements. The two subcortical structures most often implicated in movement programming are the basal ganglia and the cerebellum (Allen et al. 1974 ; Brooks et al. 1981 ; Delong et al. 1981 ; Paillard 1983). In particular, Schell et al. (1984) suggested that the orofacial representations of the basal ganglia and cerebellum project to two different nonprimary motor cortical regions : the supplementary motor and premotor areas, respectively. It is to be inferred that afferents from the basal ganglia and the cerebellum influence the primary motor cortex via non primary motor areas. Consideration of these processes offer hypotheses on the structures contributing to speech movement control. It is not intended to present here the different pathways of afferent input from the orofacial region to motor output (see Gracco et al. 1987 for a review of programming and execution processes of speech movement control). Only cerebellum-premotor area relations will be examined. Sensory information is an integral part of movement control and coordination. In speech, investigations are needed to determine the exact nature of the sensory contributions to the processes of speech motor control.

3.1. CEREBELLAR-PREMOTOR PATHWAY

It appears that the sensory projections to the cerebellum are involved in functions other than moment-to-moment control of motor execution. After a review of recent neuroanatomical and neurophysiological findings, Gracco et al. (1987) propose the following pathway for speech motor programming (Figure 6).

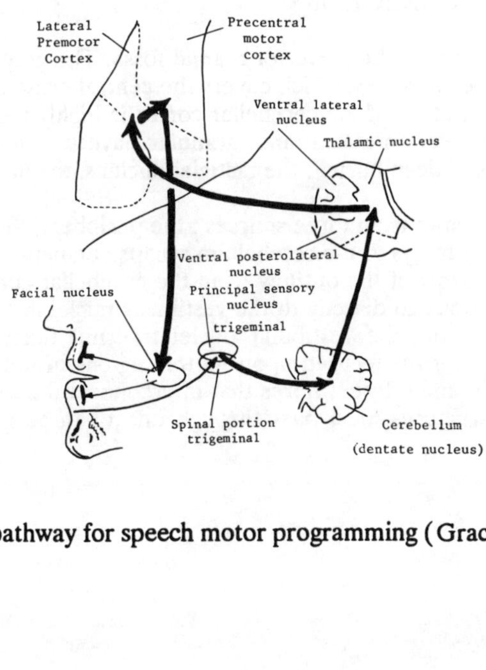

Figure 6. Proposed pathway for speech motor programming (Gracco et al. 1987).

The cerebellum receives orofacial sensory information from the periphery via trigeminal routes and from cortical projections (Allen et al. 1974). Orofacial afferents project directly to the dentate nucleus as well as indirectly to dentate and interpositus nuclei via the cerebellar cortex. The caudal portion of the dentate nucleus which contains a facial representation projects to the ventrolateral thalamus area (Asanuma et al. 1983a ; 1983b ; Brooks et al. 1981 ; Stanton 1980) that does not project directly to motor cortex but to the premotor area (Schell et al. 1984). Thus, sensory inputs to the cerebellum would influence motor cortex output to orofacial motoneurons indirectly via the premotor cortex. We shall see below speech aberrations resulting from cerebellar dysfunction. It has been suggested that cerebellar-premotor area damage may yield speech disorders (Kent et al. 1982).

4. Cerebellum

It is impossible here to review the extensive literature on the cerebellum, except in very broad outline. Important works that provide considerable information about cerebellar anatomy, physiology and circuitry include Dow et al. 1958 ; Eccles 1969 ; 1973a ; 1973b ; 1973c ; 1977 ; Eccles et al. 1967 ; Ito 1970 ; 1984 ; Ito et al. 1964 ; 1966a ; 1966b ; Jansen et al. 1954 ; Massion 1961 ; 1973 ; Massion et al. 1979. The cerebellum is classically considered to be involved in the coordination of complex movements. It serves to enrich the quality of motor performance by adding smoothness and harmony to the contraction of synergist and antagonist muscles. The cerebellum uses the gamut of sensory information to predict, judge and correct the motor act in order to achieve the proper force, direction and rate of muscle contraction and relaxation. Extrapolating from available data in the limbs (Meyer-Lohmann et al. 1977 ; Soechting et al. 1976 ; Terzuolo et al. 1974), it appears that the cerebellum may be involved in etablishing the necessary temporal and spatial adjustments among and within the various articulatory organs. The organization of the cerebellum : its structure, connections and divisions will be considered. Finally, the effects of cerebellar lesions on speech will be examined.

4.1. STRUCTURE AND CONNECTIONS

The cerebellum occupies most of the posterior cranial fossa. The gray matter forms a thin superficial layer, the cerebellar cortex, which covers the central white matter, or medullary body. The structural organization of the cerebellar cortex is highly regular and consists of three layers : molecular, Purkinje cell and granular layers. Lying deep within the cerebellum are three pairs of deep nuclei : the fastigial nucleus, the interposed nucleus and the dentate nucleus.

The cerebellum receives input from three sources : the periphery, the brain stem and the central cortex. The input pathways to the cerebellum synapse on neurons in the deep nuclei and the cerebellar cortex. Most of the outflow from the cerebellar cortex projects back to the deep nuclei, some portions go directly to the vestibular nuclei in the brain stem. Thus, the deep cerebellar nuclei and the vestibular nuclei transmit the entire output of the cerebellum. This output in turn is focused upon motor regions of the cerebral cortex and the brain stem. All afferent and efferent fibres that interconnect the cerebellum with other central nervous system structures must pass through one of three neural bridges called peduncles.

4.2. DIVISIONS OF THE CEREBELLUM

Two deep transverse fissures divide the cerebellum into three major lobes : anterior, posterior and flocculonodular lobes. Shallower fissures subdivide each lobe into several lobules. The surface of the cerebellum also has two longitudinal furrows. Thus, a thin longitudinal strip in the midline, known as the vermis is separated from the left and right cerebellar hemisphere on either side. Each hemisphere is composed of an intermediate and lateral zone (Figure 7).

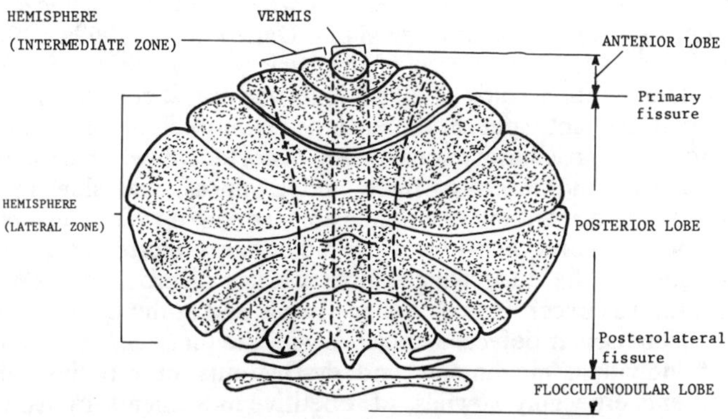

Figure 7. Schematic view showing the divisions of the cerebellar surface.

By virtue of its different connections, some authors (Ghez et al. 1985) divide the cerebellum into three distinct parts : the vestibulocerebellum, the spinocerebellum and the cerebrocerebellum. The vestibulocerebellum occupies the flocculonodular lobe. This region receives its input from the vestibular nuclei and projects back to them, hence its name. The spinocerebellum includes the vermis and the intermediate zone of the hemispheres. It contains topographical maps of the body that receive principal information from the spinal cord. Through the fastigial and interposed nuclei to which the spinocerebellum projects, this controls the medial and lateral descending systems, respectively. The cerebrocerebellum is the lateral zone of the cerebellum. Its information comes from the cerebral cortex via the pontine nuclei and its output is conveyed by the dentate nucleus to the motor and premotor cortices. Such a division has a functional perspective which is not yet well documented.

4.3. SPEECH DISORDERS ASSOCIATED WITH CEREBELLAR DISEASE

Careful observations of humans with lesions of the cerebellum or cerebellar pathways have documented a variety of motor deficits, analysis of which may lead to a better understanding of normal cerebellar physiology. These deficits include particularly disorders of voluntary movements which, as suggested by Holmes (1939), consist of abnormalities of a) rate, range, force (dysmetria), b) the timing, with irregularity of successive movements, (dysdiadochokinesia), c) the organization of a complex movement (incoordination). The activity of speech requiring perfect coordination of synergistic muscle groups of various articulatory organs ; the cerebellum seems concerned with speech motor control (Holmes 1922 ; Kornhuber 1977). In animals, lesions of the cerebellum and/or deep cerebellar nuclei alter call structure, including the vocalizations, intensity, fundamental frequency (Larson et al. 1978). In humans, lesions affecting the cerebellum or cerebellar system result in dysarthric speech which agrees with specific cerebellar motor symptoms : dysmetria, dysdiadochokinesia and incoordination.

Speech disorders associated with cerebellar disease have been described since the nineteenth century (Charcot 1877). Holmes (1917) reporting his experience in World War I with patients suffering posterior fossa wounds, described the speech of those as "drawling and monotonous", "staccato and scanning" and "remarkably irregular and jerky". Gremy et al. (1967) studied ataxic dysarthrias from an oscillographic and electroglottographic study. They found that the major dysarthrias were characterized by laryngeal irregularities, specific impairment of articulation and bradylalia. Darley et al. (1969a ; 1969b ; Brown et al. 1970) designed a classification system of the basic abnormalities of speech in various neurological disorders. From a study of 30 patients with cerebellar lesions they identified 10 major perceptual deviant speech dimensions : imprecise consonants, excess and equalized stress, irregular articulatory breakdown, distorted vowels, harsh voice, prolonged phonemes and intervals, monotony of pitch and loudness, and slow rate. Correlation analyses of the 10 major deviant speech dimensions revealed three clusters of disorders : cluster of articulatory inaccuracy, cluster of prosodic excess, cluster of phonatory-prosodic insufficiency. In addition, the underlying neuromuscular disorders for each cluster were suggested from literature concerning cerebellar disease and authors ' observations of ataxic dysarthria. The neuromuscular defects associated with the three clusters were respectively : "a) inaccuracy of individual movements and dysrhythmia of repetitive movements, b) slowness of movements especially slowness of repetitive movements, c) hypotonia". Kent et al. (1975) proposed some modifications to the conclusions of Brown et al. (1970). Their study, based not on perceptual but on cineradiographic and spectrographic observations revealed abnormalities in speaking rate, stress pattern, articulatory placements, velocities of articulatory movements and fundamental frequency contours. From these data the authors made suggestions concerning the nature of the underlying neuromuscular abnormalities. This study was limited to one subject. General acoustic characteristics of ataxic dysarthria were described in a further study (Kent et al. 1979). Dysprosodic aspects of ataxic dysarthria were particularly noticed and discussed in relation to cerebellar function in motor control. Other studies have been made which roughly confirm perceptual judgments or acoustic characteristics mentioned by the previous authors (Fujibayashi et al. 1977 ; Hirose 1973 ; Hirose et al. 1982 ; Gilman et Kluin 1985 ; Joanette et al. 1980).

The incoordination of speech movements observed in cerebellar disease may be revealed by electromyographic analysis of the primary muscles of the speech production system. A few EMG studies have been reported on the speech musculature in cerebellar disease (Hirose 1977 ; Hirose et al. 1978 ; Netsell 1972 ; Netsell et al. 1977).

The areas of cerebellar damage most commonly associated with dysarthria have not been strictly defined. The vermal origin of cerebellar speech (Mills et al. 1914 ; Holmes 1917) was questioned (Dow et al. 1958 ; Amici et al. 1976). Lechtenberg et al.'s study of patients with nondegenerative cerebellar disease (1978) implicated the left cerebellar hemisphere in the development of dysarthria. Further studies which use modern brain imaging techniques including magnetic resonance imaging, are necessary to determine precise localization in the human cerebellum.

5. Conclusion

The major challenge facing researchers of speech motor control lies in elucidating the manner in which central and peripheral systems interact in coordinating articulatory organ movements. In an approach oriented from a product to a process, underlying mechanisms are of primary concern and particularly peripheral mechanisms. On the understanding of these depend the validity of inferences. However, compared to our understanding of limb motor control, actual data concerning the neurophysiology of speech production are insufficient to know the complex organization of the vocal tract involved in speech.

In addition, it is well known that the cerebellum is responsible for the fine coordination of muscle tone and movement permitting a delicate and precise control of both posture and locomotion. But our knowledge of the real role of cerebellum in the regulation of speech movement is still limited. The description of what the speech motor control system can do in the face of damage to the cerebellar system, may permit us to deduce what this does in regulating speech. However, interpretations have to be made cautiously. Moreover, extensive studies are needed to throw light on mechanisms that contribute to the role of cerebellum in speech.

REFERENCES

Abbs, J.H. and Cole, K.L. (1982) 'Consideration of bulbar and suprabulbar afferent influences upon speech motor coordination and programming', in S. Grillner, B. Lindblom, J. Lubker and A. Persson (eds.), Speech motor control, Pergamon Press, Oxford, 159-186.

Abbs, J.H. and Eilenberg, G.R. (1976) 'Peripheral mechanisms of speech motor control', in N. Lass (ed.), Contemporary issues in experimental phonetics, Academic Press, New-York, 139-168.

Abbs, J.H. and Gracco, V.L. (1983) 'Sensorimotor actions in the control of multimovement speech gestures', Trends Neurosci. 6, 391-395.

Abbs,J.H. and Gracco, V.L. (1984) 'Control of complex motor gestures : orofacial muscle responses to load perturbations of the lip during speech', J.Neurophysiol. 51, 705-723.

Abbs, J.H., Gracco, V.L. and Cole, K.J. (1984) 'Control of multimovement coordination : sensorimotor mechanisms in speech motor programming, J. Mot. Behav. 16, 195-231.

Alfonso, P.E. and Baer, T. (1982a) 'Dynamic of vowel articulation' Language and speech 25, 151-173.

Alfonso, P.E., Honda, K., Baer, T. and Harris, K.S. (1982b) 'Multichannel study of tongue EMG during vowel production', JASA 71, S 54, (A).

Allen, G.I. and Tsukahara, N. (1974) 'Cerebrocerebellar communication systems', Physiol. Rev. 54 (4), 957-1006.

Amici, R., Avanzini, G., Pacini L. (1976) 'Cerebellar tumors', in monographs in Neural sciences, vol. 4, Karger, Basel.

Asanuma, C., Thach, W.T. and Jones, E.G. (1983a) 'Distribution of cerebellar terminations and their relations to other afferents terminations in the ventral lateral thalamic region of the monkey', Brain Res. Rev. 5, 237-265.

Asanuma, C., Thach, W.T. and Jones, E.G. (1983b) 'Cytoarchitectonic delineation of the ventral lateral thalamic region in the monkey', Brain Res. Rev. 5, 219-235.

Barker, D., Ip, M.C. and Adal, M.N. (1962) 'A correlation between the receptor population of the cat 's soleus muscle and the afferent fibre diameter spectrum of the nerve supplying it', in D. Barker (ed.), Symposium on muscle receptors, University Press, Hong-Kong, 257-261.

Barlow, S.M. (1987) 'Mechanical frequency detection thresholds in the human face', Exp. Neurol. 96, 253-261.

Barlow, S.M. and Farley, G.R. (in press) 'Speech neurophysiology', in D. Beukelman (ed.), Speech and swallowing disorders in traumatic brain injury.
Barlow, S.M. and Netsell, R. (1986a) 'Force biasing of the perioral reflex', Assoc. Res. Otolaryngol. 9, 135, (A).
Barlow, S.M. and Netsell R. (1986b) 'Mechanically evoked responses of perioral muscles during fine force control', Society for Neurosci. 12, 1539.
Basmajian, J.V. and Stecko, G.A. (1962) 'A new bipolar indwelling electrode for electromyography', J. Appl. Physiol. 17, 849.
Bell-Berti, F. (1973) 'The velopharyngeal mechanism : an electromyographic study', PhD diss. New-York.
Brooks, V.B. and Thach, W.T. (1981) 'Cerebellar control of posture and movement', in V.B. Brooks, (ed.), Handbook of Physiology, Sect. 1 : The nervous system, vol 2 : Motor control, Part 2, American Physiological Society, Bethesda MD, 877-946.
Brown, J., Darley F. and Aronson, A. (1970) 'Ataxic dysarthria', Int. J. Neurol. 7, 302-318.
Buchtal, F. (1961) 'The general concept of the motor unit', Res. Publ. Ass. Nerv. Ment. Dis. 38, 1-30.
Burke, R.E. (1981a) 'Motor units : Anatomy, physiology and functional organization', in V.B. Brooks (ed.), Handbook of Physiology, Sect I : The nervous system, vol 2 : Motor control, American Physiological Society, Washington, 345-422.
Burke, R.E. (1981b) 'Motor unit recruitment : what are the critical factors?', in J.E. Desmedt (ed.), Motor unit types, recruitment and plasticity in health and disease, Prog. Clin. Neurophysiol. vol 9, Karger, Basel, 61-84.
Carew, T.J. (1985) 'The control of reflex action', in E.R. Kandel and J.H. Schwartz (eds.), Principles of neural science, Elsevier Sciences Publishers, Amsterdam.
Charcot, J.M. (1877) 'Lectures on the diseases of the nervous system', vol. 1, The new Sydenham Society, London.
Cooper, S. (1953) 'Muscle spindles in the intrinsic muscles of the human tongue', J. Physiol. 122, 193-202.
Cooper, S. (1960) 'Muscle spindles and other muscle receptors', in G.H. Bourne (ed.), Structure and function of muscle, Academic Press, New-York, vol 1, 381-420.
Darley, F.L., Aronson, A.E. and Brown, J.R. (1969a) 'Differential diagnostic pattern of dysarthria', J. Speech Hear. Res. 12, 246-269.
Darley, F.L., Aronson A.E. and Brown, J.R. (1969b) 'Clusters of deviant speech dimensions in the dysarthrias', J. Speech Hear. Res. 12, 462-496.
Darley, F.L., Aronson, A.E. and Brown, J.R. (1975) 'Motor speech disorders', W.B. Saunders Co, Philadelphia.
Delong, M.R. and Georgopoulos, A.P. (1981) 'Motor functions of the basal ganglia', in V.B. Brooks (ed.), Handbook of Physiology, Sect 1 : The nervous system, vol 2 : Motor control, Part 2, American Physiological Society, Bethesda MD, 1017-1062.
Denny-Brown, D. (1929) 'The histological feature of striped muscle in relation to its functional activity, Proc. Roy. Soc. B. 104, 371-411.
Dow, R.S. and Moruzzi, G. (1958) 'The Physiology and Pathology of the cerebellum', University of Minnesota Press, Minneapolis.
Eccles, J.C. (1969) 'The dynamic loop hypothesis of movement control', in K.N. Leibovic (ed.), Information Processing in the central nervous system, Springer Verlag, New-York, Heidelberg, 245-269.
Eccles, J.C. (1973a) 'The cerebellum as a computer : patterns in space and time', J. Physiol. 228, 1-32.

Eccles, J.C. (1973b) 'A re-evaluation of cerebellar function in man', in J.E. Desmedt (ed.), New developments in electromyography and clinical neurophysiology, vol 3, Karger, Basel, 209-224.

Eccles, J.C. (1973c) 'The understanding of the brain', McGraw Hill, New-York.

Eccles, J.C. (1977) 'Cerebellar function in the control of movement', in F. Rose (ed.), The physiological aspect of clinical neurology, Blackwell, Oxford, 157-178.

Eccles, J.C., Ito, M. and Szentagothai, J. (1967) 'The cerebellum as a neuronal machine', Springer Verlag, Heidelberg, New-York, 1-335.

Eldred, E. (1965) 'The dual sensory role of muscle spindles', J. Amer. Phys. Ther. Ass. 45, 290-313.

Eldred, E., Yellin, H., Gabbois, L. and Sweeney, S. (1967) 'Bibliography on muscle receptors ; their morphology, pathology and physiology', Exp. Neurol., suppl. 3, 1-154.

Eldred, E., Yellin, H., Desantis, M. and Smith, C.M. (1977) 'Supplement to bibliography on muscle receptors : their morphology, pathology, physiology and pharmacology', Exp. Neurol. 55, 1-118.

Erikson, P.O. (1982) 'Muscle fibre composition of the human mandibular locomotor system. Enzyme-histochemical and morphological characteristics of functionally different parts' Swed. dent. J., suppl. 12, 1-44.

Evarts, E.V. (1982) 'Analogies between central motor programs for speech and for limb movements', in S. Grillner, B. Lindblom, J. Lubker and A.Persson (eds.), Speech motor control, Pergamon Press, Oxford, 19-41.

Eyzaguirre, C. and Fidone, S.J. (1977) 'Physiology of the nervous system', Year Book Medical Publishers, Chicago, London.

Faaborg-Andersen, K. (1957) 'Electromyographic investigation of intrinsic laryngeal muscles in humans', Acta Physiol. Scand. 41, 3-30.

Folkins, J. and Larson, C. (1978) 'In search of a tonic vibration reflex in the human lip', Brain Res. 151, 409-412.

Folkins, J. and Zimmerman, G. (1982) 'Lip and jaw interaction during speech responses to perturbation of lower lip movement prior to bilabial closure', JASA 71, 1225-1233.

Fromkin, V. and Ladefoged, P. (1966) 'Electromyography in speech research', Phonetica 15, 219-242.

Fujibayashi, M., Fukusako, Y., Monoi, H., Kobayashi, N. Tatsumi, I.F. and Hirose H. (1977) 'Characteristics of dysarthric speech due to cerebellar disorders, pseudo-bulbar palsy and amyotrophic lateral sclerosis' Jap. J. Logop. Phoniat. 18, 101-109.

Fujimura, O. (1979) 'Physiological functions of the larynx in phonetic control', in H. and P. Hollien (eds.), Current issues in the phonetic sciences, J. Benjamin, Amsterdam, 129-164.

Fujimura, O. (1987) 'Fundamentals and applications in speech production research', Proc. XIth IC Phon. Sciences, Tallinn, vol 6, 2.1-2.18.

Gay, T. (1977) 'Cinefluorographic and electromyographic studies of articulatory organization', in M. Sawashima and F. Cooper (eds.), Dynmic aspects of speech production, University of Tokyo Press, 85-105.

Gay, T. (1979) 'Coarticulation in some consonant-vowel and consonant cluster-vowel syllables', in B. Lindblom and S. Ohman (eds.), Frontiers of speech communication research, Academic Press, London, 69-77.

Gentil, M. and Gay, T. (1984) 'Temporal organization of interarticulator muscle activity in american english monosyllables' JASA 76 S 15 (A).

Gentil, M. and Gay, T. (1986) 'Neuromuscular specialization of the mandibular motor system : speech versus non-speech movements', Speech Com. 5, 69-82.

Ghez, C. and Fahn S. (1985) 'The cerebellum', in E.R. Kandel, and J.H. Schwartz (eds), Principles of neural sciences, Elsevier Science publishers, Amsterdam, 502-522.

Gilman, S. and Kluin, K. (1985) 'Perceptual analysis of speech disorders in Friedreich disease and olivopontocerebellar atrophy', in J.R. Bloedel, J. Dichgans, and W. Precht (eds.), Cerebellar functions', Springer Verlag, Berlin, 148-163.

Goldberg, L.J. (1971) 'Masseter muscle excitation induced by stimulation of periodontal and gingival receptors in man' Brain Res. 32, 369-381.

Goldberg, L.J. and Derfler, B. (1977) 'Relationship among recruitment order, spike, amplitude and twitch tension of single motor units in human masseter muscle' J. Neurophysiol. 40, 879-890.

Goodgold, M.D. and Eberstein, A. (1977) 'Electrodiagnosis of neuromuscular diseases' The William and Wilkins Co, Baltimore, 1-19.

Gracco, V.L. and Abbs, J.H. (1985) 'Dynamic control of the perioral system during speech : kinematic analyses of autogenic and nonautogenic sensorimotor processes', J. Neurophysiol. 54, 418-432.

Gracco, V.L. and Abbs, J.H. (1987) 'Programming and execution processes of speech movement control : potential neural correlates', in E. Keller and M. Gopnik (eds.), Motor and sensory processes of language, Lawrence Erlbaum Assoc. Inc., Hillsdale NJ, 163-201.

Granit, R. (1955) 'Receptors and sensory perception' University Press, New Haven, Yale.

Granit, R. (1970) 'The basis of motor control', Academic Press, London, New-York.

Gray, J.A.B. (1959) 'Mechanical into electrical energy in certain mechanoreceptors', Prog. in Biophysics and Biophysical Chemistry 9, 285-324.

Gremy, F., Chevrie-Muller, C. and Garde, E. (1967) 'Etude phoniatrique, clinique et instrumentale des dysarthries. I. Technique. Résultats chez les malades présentant un syndrome cérébelleux', Revue Neurologique 116, 401-426.

Grossman, R.C. and Hattis, B.F. (1967) 'Oral mucosal sensory innervation and sensory experiences : a review', in J.F. Bosma (ed.), Symposium on oral sensation and perception, Ch. C. Thomas, Springfield, Ill., 5-63.

Hardcastle, W.J. (1970) 'The role of tactile and proprioceptive feedback in speech production', Work in Progress 4, Dep. of linguist., Edinburg University, 100-112.

Hardcastle, W.J. (1976) 'Physiology of speech production : an introduction for speech scientists', Academic Press, London.

Hardy, J. (1971) 'Development of neuromuscular systems underlying speech production', in ASHA reports, Speech and the dentofacial complex : The state of the art, Washington DC, 49-68.

Henneman, E., Somjen, G. and Carpenter, D.O. (1965) 'Functional significance of cell size in spinal motoneurons', J. Neurophysiol. 28, 560-580.

Hirano, M., Ohala, J. and Vennard, W. (1969) 'The function of laryngeal muscles in regulating fundamental frequency and intensity of phonation, J. Speech Hear Res. 12, 616-628.

Hirano, M., Vennard, W. and Ohala, J. (1970) 'Regulation of register, pitch and intensity of voice. An electromyographic investigation of intrinsic laryngeal muscles, Folia phoniatrica 22, 1-20.

Hirose, H. (1973) 'Toward differential diagnosis of dysarthrias, in I. Kirikae (ed.), Approaches to the disorders of the central nervous system, Kanehara Publ. co, Tokyo, 214-232.

Hirose, H. (1977) 'Electromyography of the larynx and other speech organs', in M. Sawashima and F.S. Cooper (eds.), Dynamic aspects of speech production, University of Tokyo Press, 49-70.
Hirose, H. (1986) 'Pathophysiology of motor speech disorders (dysarthria)', Ann. Bull. RILP 20, 169-204.
Hirose, H., Kiritani, S. and Tatsumi, I. (1982) 'On the nature of bradylalia - a preliminary study of the acoustic characteristics of dysarthric speech', Ann. Bull RILP 16, 229-234.
Hirose, H., Kiritani, S., Ushijima, T. and Sawashisma, M. (1978) 'Analysis of abnormal articulatory dynamics in two dysarthric patients', J. Speech Hear. Dis. 43, 96-105.
Hodgkin, A.L. (1964) 'The conduction of the nervous impulse, University Press, Liverpool.
Holmes, G. (1917) 'The symptoms of acute cerebellar injuries due to gunshot injuries', Brain 40, 461-535.
Holmes, G. (1922) 'Clinical symptoms of cerebellar disease and their interpretation', The Croonian lectures, Lancet 1 : 1177-1182 ; 1231-1237 ; 2 : 59-65 ; 111-115.
Holmes, G. (1939) 'The cerebellum of man', Brain 62, 1-30.
Houk, J. and Rymer W. (1981) 'Neural control of muscle length and tension', in V.B. Brooks (ed.), Handbook of Physiology, The nervous system, vol 2, Sect 1, part 1, American Physiological Society, Bethesda, MD.
Hugelin, A. and Bonvallet, M. (1957) 'Etude oscillographique d'un réflexe monosynaptique cranien (réflexe masseterin)', J. Physiol. Paris 49, 210-211.
Hunt, C.C. (1974) 'Handbook of sensory physiology muscle receptors', vol. 3, Springer Verlag, Berlin.
Huxley, H.E. (1969) 'The mechanism of muscular contraction', Science 164, 1350.
Ito, M. (1970) 'Neurophysiological aspects of the cerebellar motor control system', Int. J. Neurol. 7, 162-176.
Ito, M. (1984) 'The cerebellum and neural control', Raven Press, New-York.
Ito, M., Obata, K. and Ochi, R. (1966b) 'The origin of cerebellar-evoked inhibition of Deiters neurons. II. Temporal correlation between transsynaptic activation of Purkinje cells and the inhibition of Deiters neurons, Exp. Brain Res. 2, 350-364.
Ito, M. and Yoshida, M. (1964) 'The cerebellar evoked monosynaptic inhibition of Deiters neurons', Experientia 20, 515-516.
Ito, M. and Yoshida, M. (1966a) 'The origin of cerebellar induced inhibition of Deiters neurons. I. Monosynaptic initiation of the inhibitory postsynaptic potentials', Exp. Brain Res. 2, 330.
Jansen, J. and Brodal, A. (1954) 'Aspects of cerebellar anatomy', J.G. Tanum, Oslo.
Joanette, Y. and Dudley, J.G. (1980) 'Dysarthric symptomatology of Friedreich ataxia', Brain and Lang. 10, 39-50.
Kakita, Y. and Fujimura, O. (1983) 'Computation of mapping from muscular contraction patterns to vowel space', JASA 74, S 117, (A).
Karlsson, U.L. (1976) 'The structure and function of muscle spindles and tendon organs in the masticatory muscles, in D.J. Anderson and B. Matthews (eds.), Mastication, Wright, Bristols.
Katz, B. (1962) 'The transmission of impulses from nerve to muscle, and the subcellular unit of synaptic action', Proc. R. Soc. Lond. (Biol.) 155, 455-477.
Katz, B. (1966) 'Nerve, muscle and synapse', Mc Graw Hill Book Co, New-York.
Kent, R. and Netsell, R. (1975) 'A case study of an ataxic dysarthric : cineradiographic and spectrographic observations', J. Speech Hear. Dis. 40, 115-134.

Kent R., Netsell R. and Abbs J.H. (1979) 'Acoustic characteristics of dysarthria associated with cerebellar disease', J. Speech Hear. Res. 22, 627-648.

Kent, R. and Rosenbeck, J.C. (1982) 'Prosodic disturbance and neurologic lesion', Brain and Lang. 15, 259-291.

Kobayashi, N., Fukusako, Y., Anno, M. and Hirose, H. (1976) 'Characteristic patterns of speech in cerebellar dysarthria', J. Speech Hear. Dis. 5, 63-68.

Kornhuber, H. (1977) 'A reconsideration of the cortical and subcortical mechanisms involved in speech and aphasia', in J. Desmedt (ed.), Language and hemispheric specialization in man : Cerebral ERPs Progress in Clinical Neurophysiology, vol. 3, 28-35.

Kubota, K., Negishi, T. and Hasegi, T. (1975) 'Topological distribution of muscle spindles in the human tongue and its significance in proprioception', Bull. of Tokyo Med. Dent. University 22, 235-242.

Larson, C.R., Sutton, D and Lindeman, R.C. (1978) 'Cerebellar regulation of phonation in rhesus monkey (macaca mulatta), Exp. Brain Res. 33, 1-18.

Lazorthes, G. (1967) 'Le système nerveux périphérique', Masson, Paris.

Lechtenberg, R. and Gilman, S. (1978) 'Speech disorders in cerebellar disease', Ann. Neurol. 3, 285-290.

Lenneberg, E.H. (1967) 'Biological foundations of language', J. Wiley and sons, New-York.

Liddell, E.G.T. and Sherrington, C.S. (1925) 'Recruitment and some other feature of reflex inhibition', Proc. R. Soc. Lond. (Biol.) 97, 488-518.

Lovell, M., Sutton, D. and Lindman, R. (1977) 'Muscle spindles in the nonhuman primate extrinsic auricular muscles', Anat. Rec. 189, 519-524.

Lund, J.P., Appenteng, K. and Seguin, J.J. (1982) 'Analogies and common features in the speech and masticatory control systems', in S. Grillner, B. Lindblom, J. Lubker and A. Persson (eds.), Speech Motor Control, Pergamon Press, Oxford, 231-245.

MacNeilage, P.F. (1973) 'Preliminaries to the study of single motor unit activity in speech musculature', J. Phonetics 1, 55-71.

Massion, J. (1961) 'Contribution à l'étude de la régulation cérebelleuse du système extrapyramidal', Masson and Cie, Paris.

Massion, J. (1973) 'Intervention des voies cerebello-corticales et cortico-cérebelleuses dans l'organisation et la régulation du mouvement', J. Physiol. Paris 67, 117-170.

Massion, J. and Sasaki, K. (1979) 'Cerebro-cerebellar interactions', Elsevier, North Holland Biomedical Press, Amsterdam, New-York.

Matthews, P.B.C. (1964) 'Muscle spindles and their motor control', Physiol. Rev.44, 219-288.

Matthews, P.B.C. (1972) 'Mammalian muscle receptors and their central action', Arnold, London.

Matthews, P.C.B. (1981) 'Muscle spindles their messages and their fusimotor supply', in V.B. Brooks (ed.), Handbook of Physiology, Sect. 1 : The nervous system, vol. 2, Motor Control, Part 1, American Physiological Society, Bethesda, MD.

Meyer-Lohmann, J., Hore, J. and Brooks, V.B. (1977) 'Cerebellar participation in generation of prompt arm movements', J. Neurophysiol.40, 1038-1050.

Mills, C.K. and Weisenburg, T.H. (1914) 'Cerebellar symptoms and cerebellar localization', JAMA 63, 1813-1818.

Mountcastle, V.B. (1980a) Neural mechanisms in somesthesis', in V.B. Mountcastle (ed.), Medical Physiology, vol. 1, C.V. Mosby Co, St-Louis, Mo.

Mountcastle, V.B. (1980b) 'Sensory receptors and neural encoding : introduction to sensory processes', in V.B. Mountcastle (ed.), Medical Physiology, vol. 1, C.V. Mosby Co, St-Louis, Mo.

Netsell, R. (1972) 'Lip electromyography in the dysarthrias', ASHA, San Francisco.

Netsell, R. (1982) 'Speech motor control and selected neurologic disorders', in S. Grillner, B. Lindblom, J. Lubker and A. Persson (eds.), Speech Motor Control, Pergamon Press, Oxford, 247-261.

Netsell, R. and Abbs, J.H. (1977) 'Some possible uses of neuromotor speech disturbances in understanding the normal mechanism, in M. Sawashima and F.S. Cooper (eds.), Dynamic aspects of speech production, University of Tokyo Press, 369-398.

Netsell, R. and Kent, R. (1976) 'Paroxysmal ataxic dysarthria', J. Speech Hear. Dis. 41, 93-109.

Paillard, J. (1983) 'Introductory lecture. The functional labelling of neural codes, in J. Massion, J. Paillard, W. Schultz and M. Wiesendanger (eds.), Neural coding of motor performance, Springer Verlag, Berlin, Heidelberg, 1-19.

Ranson, S.W. and Clark, S.L. (1953) 'The anatomy of the nervous system : its development and function', W.B. Saunders, Philadelphia.

Ringqvist, M. (1971) 'Histochemical fiber types and fiber sizes in human masticatory muscles', Scand. J. Dent. Res. 79, 366-368.

Sawhney, B.B. and Kayan, A. (1971) 'A study of the F wave from the facial muscles', Electroencephal. and Clinic Neurophysiol. 30, 255-274.

Schell, G.R. and Strick P.L. (1984) 'The origin of thalamic inputs to the arcuate premotor and supplementary motor areas', J. Neurosci. 4, 539-560.

Sessle, B.J. and Schmitt, A. (1972) 'Effects of controlled tooth stimulation on jaw muscle activity in man, Arch. Oral Biol. 17, 1597-1607.

Sherrington, C.S. (1894) 'On the anatomical constitution of nerves of skeletal muscles ; with remarks on recurrent fibres in the ventral spinal nerve root, J. Physiol. 17, 211-258.

Skoglund, S. (1973) 'Handbook of sensory physiology, somatosensory system' A. Iggo (ed.), Springer Verlag, Berlin, New-York, vol. 2, 111-136.

Smith, T.S. (1971) 'A phonetic study of the function of the extrinsic tongue muscles', Working Pap. in Phon. 18, UCLA 18, 1-131.

Smith, A., Moore, C., McFarland, D. and Weber, C. (1985) 'Reflex responses of human lip muscles to mechanical stimulation during speech', J. Mot. Behav. 17, 18-167.

Soechting, J.F., Ranish, N.A., Palminteri, R. and Terzuolo, C.A. (1976) 'Changes in a motor pattern following cerebellar and olivary lesions in the squirrel monkey', Brain Res. 105, 21-44.

Stanton, G.B. (1980) 'Topographical organization of ascending cerebellar projections from the dentate and interposed nuclei in macaca mulatta : An anterograde degeneration study, J. Comp. Neurol. 190, 699-731.

Stein, R.B. (1974) 'Peripheral control of movement', Physiol. Rev. 54, 215-243.

Strick, P.L. (1983) 'The influence of motor preparation on the response of cerebellar neurons to limb displacements', J. Neurosci. 3, 2007 - 2020.

Terzuolo, C.A. and Viviani, P. (1974) 'Parameters of motion and EMG activities during some simple motor tasks in normal subjects and cerebellar patients', in S. Cooper, M. Riklan and R.S. Snider (eds.), The cerebellum, epilepsy and behavior, Pergamon Press, Oxford, 173-215.

Tuller, B., Kelso, J.A.S. and Harris, K.S. (1982) 'Interarticulator phasing as an index of temporal regularity in speech', J. Exp. Psychol. H.P.P. 8, 460-472.

Vallbo, A.B. (1970) 'Discharge patterns in human muscle spindle afferents during isometric voluntary contraction', Acta Physiol. Scand. 80, 552-566.

Walker, L.B. and Rajagopal, M.D. (1959) 'Neuromuscular spindles in the human tongue', Anat. Rec. 133, 438.

Winkelmann, R.K. (1960) 'Similarities in cutaneous nerve end-organs', in W.Montagna (ed.), Advances in biology of skin : Cutaneous innervation, Pergamon Press, London, vol. 1, 48-62.

RESPIRATORY ACTIVITY IN SPEECH

John J. Ohala
Phonology Laboratory
Department of Linguistics
University of California
Berkeley, CA 94720

ABSTRACT. Functionally the subglottal respiratory system behaves for the most part like a piston in a piston chamber driven with a constant force (the sum of thoracic and abdominal muscular forces plus the elastic recoil force). Small, short term, variations in subglottal air pressure occur nevertheless due to (a) short-term variations in the resistance to the exiting airflow, i.e., from glottal and supraglottal articulations and (b) the inertia of the pulmonic system which creates delays in its passive response to such short-term variations in subglottal pressure. This casts doubt on claims of active contribution of the pulmonic system in the implementation of syllables, stress, certain sentence-final fundamental frequency contours, and certain segment types, e.g., aspirated stops.

1. INTRODUCTION

Speech consists primarily of the articulators transforming the dc pressure variations created during respiration into ac pressure variations or sound.[1] Speech is further superimposed almost exclusively on the expiratory phase of respiration[2] although this expiration is often prolonged far beyond what it would be in ordinary respiration. In this paper I will review research and claims on what the pulmonic system does during speech. Since a great many basic reference works cover the anatomy and physiology of the respiratory system in general and in speech

---

1 The glottalic and velaric speech sounds found in some languages, that is, clicks, ejectives, and implosives, are created out of dc pressure variations created by piston-like action of the tongue and/or the larynx (Catford 1939, Ladefoged 1962a).

2 The exceptions are ingressive speech used in vocal disguise, speaking when out of breath (e.g., when reporting emergencies), and interjections in various cultures, e.g., French "oui" or Scandinavian "ja" or "nej".

W. J. Hardcastle and A. Marchal (eds.), Speech Production and Speech Modelling, 23–53.

(e.g., Zemlin 1968, Luchsinger & Arnold 1965, Hixon 1973, di Cristo 1975, Catford 1977) I will not go into great detail on these matters but will instead focus on the less well resolved issues relating specifically to the involvement of the pulmonic system in speech.

2. ANATOMY

Respiration involves getting oxygen into and carbon dioxide out of the lungs which are two sponge-like lobes having great internal surface area in order to facilitate gas exchange. This is accomplished by varying the volume and thus the pressure in and around the lungs by muscular and passive elastic forces acting on the thoracic cavity. Although more than a dozen muscles can be involved in respiration under certain conditions, including any of the trunk muscles, the primary muscles used are the internal and external intercostals (which connect the ribs), the diaphragm (the thin dome-shaped muscular sheet separating the abdominal cavity and the thoracic cavity) and the various sheet-like muscles covering the abdominal cavity which are, in effect, antagonistic to the diaphragm. Of these, the external intercostals and the diaphragm are primarily associated with inspiration and the internal intercostals and the muscles covering the abdomen, primarily with expiration (although see below).

The lungs plus the thoracic and abdominal cavities also have an elasticity which gives rise to forces which oppose the volume changes induced by muscular action and which, when the muscular action ceases or diminishes, tend to restore the system to a certain resting or state.

The lungs of an adult male can hold maximally about 7 liters of air but have a resting capacity of about 4 liters. The latter volume, the so-called functional residual capacity (FRC), is the volume assumed by the lungs when the external airway (trachea, pharynx, plus mouth or nasal passage) is open and the net volume-changing forces acting on the thoracic cavity are zero.

The difference between the 7 l maximum volume and the 4 l at the FRC is the 3 l that can be taken in during a maximum inspiration; a maximum expiration can reduce the volume by 2 l, leaving 2 l of residual volume, the quantity of air that always remains in the lungs. The 5 l that can be exchanged (3 l under maximum inspiration plus 2 liters under maximum expiration) is called the vital capacity.

3. PHYSIOLOGY

There are two somewhat divergent accounts of the physiological activity of respiration during speech. Weismer (1985) refers to these as the 'classic' and 'contemporary' views. They differ primarily in the role assigned to the abdominal system and in the extent to which the accounts are justified by reference to biomechanical efficiency.

3.1. The 'Classic' Account of Respiration During Speech

According to the classic view, quiet breathing typically involves a respiratory cycle that is initiated by an inspiration consisting of an approximately 500 cc increase in lung volume (over the FRC) accomplished by contraction of external intercostals. Expiration is accomplished by the elastic recoil force which brings the lungs back to the FRC. Deep inspiration will also involve more forceful action of the external intercostals in addition to the diaphragm and associated muscles. Forced (rapid) expiration and deep expiration past the FRC also require the engagement of expiratory muscles (internal intercostals). In controlled expirations, which includes speech, the inspiratory muscles may be active when lung volume is above the FRC in order to oppose the elastic recoil force, that is, to prevent a too rapid decrease in lung volume. Short speech intervals (2 or 3 seconds or less) of low to average loudness may be accomplished entirely on the volume of air taken in during a normal inspiration typical of quiet breathing and, like quiet breathing, not require active involvement of expiratory musculature. Anything longer or louder requires a deviation from normal respiration, that is, either inspiring greater volumes of air and/or expiring volumes of air past the FRC.

3.2. The 'Contemporary' Account of Respiration During Speech

The contemporary view, while acknowledging that there are considerable idiosyncratic variations among speakers, emphasizes the crucial role of the abdominal system in respiration. Most speakers are found to make some use of the abdominal system at least to assist the thoracic system in the controlled expirations during speech and some speakers exploit the abdominal system more than the rib cage to change lung volume (Baglioni 1922, Gray 1936, Hixon, Goldman, & Mead 1973, Hixon, Mead, & Goldman 1976, Weismer 1985). Most speech utterances, especially those in conversational style, are said to be initiated on that level of lung volume which would generate the desired subglottal pressure (Ps) purely by the elastic recoil force (c. 60% of vital capacity). After initiation, however, the expiratory force must be supplemented by muscular action in order to maintain the Ps at this level. Most of the expirations on which speech is superimposed are said to be terminated near the FRC. Weismer offers speculations in terms of bioenergetics for the greater efficiency of this scheme over that depicted in the classic view.

3.3. What is Common to the 'Classic' and 'Contemporary' Accounts

The classic and contemporary accounts of speech respiration agree in attributing to the respiratory physiology the goal of providing a relatively constant Ps (for a given loudness level). They agree furthermore that this is accomplished by providing a nearly constant net force to the thoracic cavity and thus the lungs. They differ only in the specifics of how this is done: whether largely with the rib cage muscles, or assisted by the abdominal system, and in how the muscles interact with the elastic recoil force. From the point of view of the end-product, then, the respiratory system's function in speech is relatively simple: apply a constant force to the lungs in order to maintain Ps at a range useful for the supraglottal articulators to make sound. The complexity arises in implementing this task, given that the forces applied by separate components of the system, including the back pressure of the air inside the lungs themselves, may change radically during a single expiration.

3.1. The Principal Aerodynamic Respiratory Variables

3.1.1 Volume Velocity. The rate of airflow, the volume velocity, U, (units = cm^3/sec), is a function of both the pressure differential between lungs and atmosphere and the resistance offered to the airflow; resistance in turn is a function of the channel characteristics, primarily the area at its narrowest point (aperture). Theoretically the variations in flow as a function of aperture depend on whether the flow is laminar (smooth) or turbulent. The transition from laminar flow to turbulent flow, again theoretically, is supposed to take place suddenly when a certain combination of factors including the physical properties of the fluid (the warm moist air in the case of speech), the channel characteristics and the particle velocity[3] reach a certain value, the so-called Reynolds number. Practically, however, in non-uniform tubes--which is the case with the respiratory tract--there is turbulent flow at almost any rate of air flow and so one is actually dealing with cases where the volume velocity, U, varies with pressure drop, ΔP, and area of the aperture, a, as in (1).

$$U = c \cdot \Delta P^{\alpha} \cdot a^{\beta} \quad (1)$$

where c is a constant and the exponents α and β vary non-linearly as a function of particle velocity (Jaeger & Matthys 1970). In general, though, it is safe to take α as 0.5 and β as 1.

3.1.2. Pressure. Variations in air pressure in the respiratory tract are a function primarily of two factors: pressure varies inversely with the volume of the chamber in question and directly with the mass of air in the chamber. These functions are given in the equations in (2) and (3).

$$\frac{V1}{V2} = \frac{P2}{P1}, \quad \text{or} \quad P2 = \frac{V1}{V2} \cdot P1 \quad (2)$$

$$\frac{M1}{M2} = \frac{P1}{P2}, \quad \text{or} \quad P2 = \frac{M2}{M1} \cdot P1 \quad (3)$$

The influence of temperature variations are typically ignored under the assumption that pressure changes are essentially adiabatic.

---

3 Particle velocity, v, is a function of volume velocity and the cross-dimensional area of the channel, v = U/a.

(In addition a complete model would take into consideration inertial and elastic factors as well as compliance, the amount by which the surface of the respiratory tract yields to impinging pressure.)

3.2. Interactions Between Pulmonic and Oral Systems

Most of the issues surrounding respiratory activity during speech seek to determine the contribution of the pulmonic system independent of activity in the upper respiratory tract, that is, by the glottis plus supraglottal articulators (hereafter designated 'oral articulators' for convenience). Difficulties arise due to interactions between the two systems. Variation in the pressure in the lungs (for our purposes, equivalent to Ps) is a function both of the expiratory forces (muscular plus elastic recoil--by equation (2)) and of the resistance to air flow encountered downstream (by equations (1) and (3)). The same can be said of air flow. Variation in lung volume might be considered a definitive way to differentiate between activity initiated by the lungs and that due to oral articulators but even here complications arise. Variation in the rate of lung volume decrement could be a reflection of independent augmentation of the expiratory force or it could be a largely passive reaction to variation in the expiratory volume velocity. Only by sampling multiple parameters would it be possible to differentiate independent pulmonic activity from that dependent on upper tract activity.

To make this clear, consider a few simple cases. Assume, first, a constant muscular force acting to reduce lung volume--a situation that can be imagined as a weight resting on an inflated balloon. If no air leaves the lungs due to the exit aperture being zero, that is, during breath holding after equilibrium is achieved, then the forces inside the lungs exactly oppose the net force compressing the lungs (muscular force plus recoil force). In this static case there is no change in lung volume and lung pressure is at its maximum. If the exit aperture is non-zero, then there is some airflow and the lungs lose some mass of air--the amount determined by equation (1)--and has its pressure momentarily lowered as reflected in equation (3). When the pressure (force) inside the lungs is then momentarily less than the combined muscular and recoil force compressing the lungs, the latter will dominate and the lung volume will decrease. This lung volume decrement will in turn tend to increase lung pressure (by equation (2)). Given the inertia of the system, however, during a continuous venting of the air from the lungs there is some phase lag between the air mass decrement and the lung volume decrement; consequently the pressure in the lungs during expiration is less than that of the combined muscular and recoil force.

Evidence for this latter point is the fact that during intervocalic voiceless stops the pressure sampled just below the glottis (Ps) rises above the level of that for adjacent vowels. See Fig. 1. In such cases the variations in lung pressure and exit volume velocity are a function of "downstream" activity.

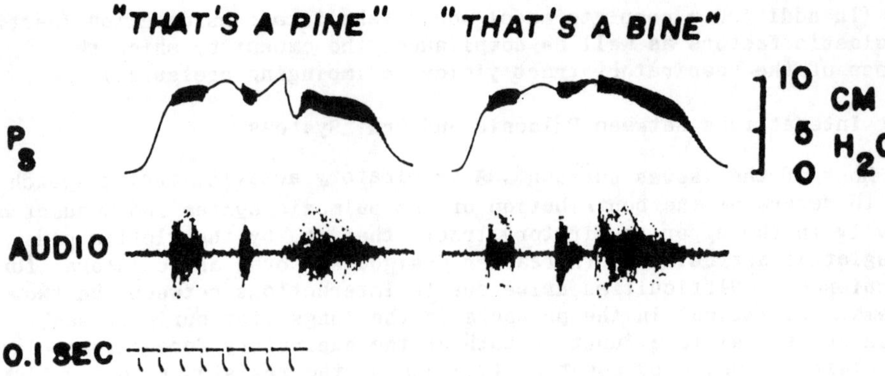

Figure 1. Subglottal pressure and microphone signal sampled during two utterances spoken by an adult male speaker of English.

On the other hand, consider the case of an unvarying exit aperture but a varying muscular force. Here the muscular variations will cause directly proportional variations in lung pressure (by eq. (2)) and in the airflow leaving the lungs (by eq. (1)). In this case variations in lung pressure and exit volume velocity are a function of "upstream" activity.

3.3. Measurement

A variety of physiological respiratory parameters may be sampled during speech.

3.3.1. Electromyography. The activity of various muscles may be sampled using electromyography (EMG) (Stetson 1951, Draper, Ladefoged, & Whitteridge 1959, Mead, Bouhuys, & Proctor 1968, Munro & Adams 1971, van Katwijk 1974). The main difficulty with this method is that since many of the relevant muscles are not near the surface or if they are, are overlaid or adjacent to other muscles, surface electromyographic electrodes offer problems. This leaves needle electrodes which (a) require considerable skill to use and (b) restrict the subject population.

3.3.2. Plethysmography. The variations in the volume of the lungs may be measured relatively easily and non-invasively with the use of a whole body plethysmograph (Mead 1960, Arutjunjan, Granstrem, & Kozhevnikov 1967, Hixon 1972, Ohala 1977a). When a subject is placed inside an airtight container slightly larger than his body and allowed to breathe air from outside the container (e.g., via a close-fitting face mask in the wall of the container), then the variations in the chest or abdominal volume during breathing and speaking will cause proportional changes in the pressure inside the container. Alternatively, one may measure the amount of air passing through a hole in the wall of the container due its displacement by the body volume changes. The former, a pressure plethysmograph has better time-resolution but is somewhat more difficult to calibrate (in terms of variations in the volume of the body); the latter, a flow plethysmograph tends to have less time resolution but is easy to calibrate. Other methods of measuring chest volume include belts that wrap around the chest, changes in the length of which reflect changes in the thoracic circumference (e.g., the accordion-like tube which is a familiar part of the polygraph or 'lie detector'; see also Scripture 1902 [1973], pp. 214ff. and Panconcelli-Calzia 1924: 5ff.) and newer devices which transduce chest or abdominal diameter by detecting change in the motion of a sensor placed on one part of the body relative to another reference device (e.g., the sensor detects strength of magnetic field and the reference device emits a magnetic field) (Hixon 1971).

3.3.3 Aerodynamic. Airflow from the mouth and nose and air pressure in the trachea are commonly used in studies addressing issues of the respiratory contribution in speech. These are familiar techniques in instrumental phonetics and will not be reviewed here (see Keller 1971, Ladefoged 1967.) It should be said, however, that most indirect ways of measuring Ps are far inferior to the most commonly used direct method which involves a tracheal puncture. However, as with needle electrodes in EMG, tracheal punctures can only be done by a skilled physician and are unacceptable to many subjects.

4. RESPIRATORY ACTIVITY IN SPEECH

The principal issues concerning the role of the pulmonic system during speech are: what role does it play in the production of stress, syllables, individual speech sounds (phonemes), intonation, including the gradual declination of voice fundamental frequency (F0) from beginning to end of a breath group (that uttered on a single expiration)?

4.1. Air Under Pressure.

Perhaps the only function of the respiratory system in speech which has never been in dispute is that the lungs provide the air under pressure which, by the action of the oral articulators, is turned into audible pressure variations, i.e., sound. Because the rise and fall of the thorax during speech is quite visible and since kinesthetic sensations relating to expiration are so evident, this has been known from ancient times. Some of the earliest genuinely scientific attempts at speech synthesis, that by Wolfgang von Kempelen (1791) and by Erasmus Darwin (1803, note XI), simulated the function of the pulmonic system quite adequately by means of a bellows. There was some confusion, however, over how the expired air was turned into sound until the matter was cleared up by Willis (1830, 1833).

4.2. Voice Quality

The voice teacher's term "chest voice" as opposed to "head voice", etc. reflects a rather ancient belief that certain modes of phonation, especially as used in singing, emanated from the chest, the forehead, etc. This, of course, is physiological nonsense (as explicitly acknowledged by Rush 1827:104) although it may have pedagogical, i.e., heuristic, value.

4.3. Loudness

Research by Ladefoged and his colleagues and others have established that, other things being equal (e.g., the shape of the supraglottal configuration), loudness is correlated with Ps (Ladefoged & McKinney 1963, Allen 1971). This by itself does not necessarily mean the pulmonic system is the exclusive cause of voice loudness due to the above-discussed fact that increased Ps can also come about due to increased resistance at the glottis. In fact, there is clear evidence that loud voice is accompanied by a change in the glottal waveform, specifically, one in which the vocal cords are closed for a longer time during each glottal cycle. However, there is also evidence of increased activity by the pulmonic system itself during loud voice (Hixon 1973, Hixon, Goldman, & Mead 1973, Hixon, Mead, & Goldman 1976).

4.4. Syllables

Stetson (1929, 1951 [retrospective ed. 1988]) claimed that individual syllables were produced with ballistic pulmonic pulses which resulted in 'breath pulses' (momentary increases in the pressure and flow rate of expiratory air) which were further shaped (released, checked) by the oral articulators, thus giving rise to the segments or phonemes of speech.[4] He presented a variety of physiological records in support of his claim, including chest wall movements, aerodynamic, and electromyographic data. However, Draper, Ladefoged & Whittridge (1959) and Ladefoged (1962), using more refined EMG methods and a wide variety of speech sample types, showed that Stetson's claims could not be supported. Stetson had evidently been misled by the pressure variations that develop in the lungs due to variations in the resistance encountered by the expiratory airflow due to "downstream" segmental closures and openings.

---

4 A related but inverse claim regarding syllable production was proposed by James Rush (1827)--inverse in the sense that he seemed to suggest that the respiratory muscles released air in syllable-sized chunks. In his own words: "By our voluntary power over the muscles of respiration, the breath in speech is frugally dealt out to successive syllables, in such small portions as may be requisite for the time and force of each." [107]. In some of his works Henry Sweet also proposed an account of syllables similar to that of Stetson's, e.g., "A syllable is a vowel, either alone or in combination with consonants, uttered with a single impulse of stress" (1891:229). As noted below, he also attributes 'stress' to an action of the respiratory system. Neither of these authors elaborate on their views to the same extent as Stetson nor do they present any supporting instrumental evidence.

4.5. Stress and Other Aspects of Intonation

4.5.1. Introduction. The belief that stress (or the prominence given to individual syllables) is implemented by greater respiratory effort is quite old. I don't know exactly how old, but I have found references to emphasis as 'a mode of sound requiring a greater profusion of breath' in Foster (1763) (and carefully differentiating this from accent which he relates to the pitch of voice and from quantity or the duration of speech sounds). Foster refers to Scaliger (1610) as expressing similar views in his Diatriba de Europaeorum linguis. Sweet (1911) declared "Stress is, organically, the result of the force with which the breath is expelled from the lungs; while acoustically it produces the effect of loudness..." Jones (1922) essentially follows Sweet in this view although he acknowledges (p. iv) the work of Coleman (1914) which, in the case of English, emphasizes instead the importance of pitch (F0) changes for the production of stress. Others who thought pitch primary for stress include Mitford (1804), Abas (1925), and Morris (1925), Muyskens (1931), Scott (1939). The importance of pitch variations for the perception of stress (and secondarily, greater duration) has been overwhelmingly supported by modern investigations (Bolinger 1958, Fry 1958) involving a wide variety of languages (including English, Polish, Czech, German, and Dutch; see references in Ohala 1977a). Perhaps the ultimate verification of these findings has been the fact that they have led to highly successful speech synthesis-by-rule (Mattingly 1966, Umeda 1976).

Stetson, while allowing that there might be pitch changes on accented syllables, thought that it would be a secondary effect of the increased expiratory force since

> ... the heavy stroke of the accent involves the chest pressure and is apt to change the pitch because the laryngeal musculature is often affected by tensions in the other musculatures of speech. [141]

However, he cites no evidence supporting the claim of special sensitivity of the laryngeal muscles to other muscles' contraction and I know of no such evidence that has been presented in the meantime. Nevertheless, it has been known since Müller's research in 1837 (see Mueller 1851:vol. 2, p. 212ff) that increases in Ps can cause slight increases in F0, so the views that greater expiratory force and pitch changes are essential for stress are not logically incompatible.

Stetson presented evidence (records of oral and subglottal air pressure and chest movements) that these allegedly separate 'breath pulses' for each spoken syllable had greater magnitude on stressed syllables.[5]

---

5 It is interesting to compare Stetson's views with those of Roudet (1910:182ff), published earlier, who was one of the first to study a wide range of aerodynamic parameters in speech (Roudet 1900a,b). Like

Fonagy (1958), studying Hungarian, reported that increased respiratory muscular activity was a more reliable correlate of stress than variations in intensity, F0, or duration.

Ladefoged (1962b, 1967) and his colleagues found a momentary increase in the activity of the expiratory muscles (sampled via EMG needle electrodes) and in Ps during and sometimes immediately before a stressed syllable, whether emphatically stressed or not. He calibrated the effect that variations in Ps would have on both intensity (see above) and F0 (Ladefoged 1963). Although Ps was found to be primarily responsible for the major variations in intensity of voice, the major part of the F0 variations including those on stressed syllables could be attributed to action of the larynx via changes in the tension of the vocal cords.

4.5.2. Recent Controversies. Lieberman (1967) attributed a greater role for the pulmonic system and the variations in Ps which it created than had previously been considered--not only in causing the F0 variations associated with stress but also the utterance-final F0 fall found in declarative statements. The basis for his claim was recordings from three speakers of American English of the acoustic speech signal and Ps during a variety of utterances. He noted that there was generally a close temporal coincidence between a momentary F0 rise on stressed syllables and a momentary increase in Ps and likewise the fall in F0 at the end of declarative sentences and the fall in Ps. (Similar observations had been made previously by others who sampled Ps during speech, e.g., Smith (1944).)

There were numerous problems with Lieberman's claim (see Isshiki 1969, Vanderslice 1970, Ohala 1970, 1977a, 1978, 1982, Ohala & Ladefoged 1970):

1. The inference that Ps was responsible for certain F0 variations in speech rested on the assumption that the laryngeal muscles (which were not sampled in his study) were inactive. There was, however, no basis for this assumption.

2. Controlled calibrations of the effect which independent Ps variations could have on F0 in the modal register (i.e., where care was taken to insure that laryngeal tension was not varying) gave values in

---

Stetson, he also found variations in either Ps or in oral airflow related to the production of separate syllables. However he more reasonably (from our present viewpoint) attributed these to variations in either glottal or supraglottal resistance. One finds Ps peaks when some closure or constriction is made downstream, i.e., during the production of a consonant. Oddly, Stetson made no reference to Roudet's work in either the first or second edition of Motor Phonetics even though both Rousselot (1897-1908) and Scripture (1902), to whom Stetson makes extensive reference, give reviews of Roudet's work.

the range of 2 to 5 Hz/cm H2O (Mueller 1851:vol 2, p. 212ff, Ladefoged 1963, Öhman & Lindqvist 1966)--too low to account for most of the observed F0 variations in speech, given the magnitude of accompanying Ps variations. Öhman and Lindqvist commented, in addition,

> ...the fo changes during the stressed syllables do not correlate well with the stress-induced ΔP changes either in phase or in amplitude.

Lieberman reported a 16 to 22 Hz/cm H2O influence of Ps on Fo based on the unsupported assumption that vocal cord tension was constant during the samples he obtained. In fact, if his measurements had been done according to the criteria he specified (and any reader can check this since, commendably, Lieberman published all the data from his study), one would obtain values of up to 28 to 33 Hz/cm H2O. These are an order of magnitude higher than values obtained under controlled conditions (then or now--see Titze 1989).

3. Most importantly, (as mentioned earlier) Ps varies not only due to changes in the pulmonic force but also due to changes in the resistance to airflow downstream. There is abundant evidence that changes in F0 and intensity do involve changes in glottal pulse shape, including mean glottal aperture (Roudet 1910: 183, Sonesson 1960, Isshiki 1969, Ishizaka & Flanagan 1972). Indirect evidence of increased glottal resistance during stress syllables (in comparison to unstressed syllables) comes from the occasional (but not consistent) finding that oral airflow is reduced on stressed syllables (Klatt, Stevens, & Mead 1968--see their figure 5--, Broad 1968; this same relation is evident in aerodynamic data obtained by Celia Scully from some of her subjects, personal communication). Atkinson (1973, 1978), whose work Lieberman cites as support for his earlier claims, also fails to consider how Ps is partially dependent on laryngeal state.

Subsequently, there have been several electromyographic studies which have uniformly indicated that F0 variations in speech are implemented by activity of the laryngeal muscles (Ohala & Hirano 1967, Hirano, Ohala, & Vennard 1969, Hirano & Ohala 1969, Fromkin & Ohala 1968, Ohala 1970, 1972, Lieberman, Sawashima, Harris, & Gay 1970, Atkinson 1973, Netsell 1969, 1973, Collier 1975, Maeda 1975, Shipp 1975, Erickson & Atkinson 1975, Kakita & Hiki 1975, Erickson, Baer, & Harris 1983). The muscles involved in raising F0 are the cricothyroid, thyrohyoid, lateral crico-arytenoid, and vocalis; those involved in lowering F0 are the sternohyoid and sternothyroid.

In spite of the array of evidence which fail to support Lieberman's claim that Ps can contribute in any significant way towards the F0 variations in stress and utterance-final fall, Lieberman and Blumstein (1988) still maintain that

> [when] stressing a particular word..., the primary determinant of fo variation is the subglottal air pressure. [107]

> ... the "easiest" or "most natural" way of producing a breath-group appears to be the state of minimal [laryngeal] control that results in a terminal falling fo contour...[i.e., due to falling Ps]. [201]

Even though it is clear now that any pulmonic contribution to the F0 changes observed on stressed syllables must be small, this does not rule out the possibility of a pulmonic contribution to other aspects of stress, in particular increases in intensity. Ladefoged (1963, 1967, 1968) citing his own extensive electromyographic investigations of the activity of the expiratory muscles during speech, maintains that there is a momentary increase in expiratory activity during or slightly preceding each stressed syllable, whether emphatically stressed or not. However, Munro & Adams (1971), using needle electrodes to record respiratory activity during speech in four speakers of English, failed to find any 'obvious localised bursts of internal intercostal activity with each stressed syllable.' Van Katwijk (1974) obtained surface electromyographic recordings of the activity of some of the expiratory muscles during speech (of two native speakers of Dutch) and found appreciable increases in their activity only on emphatically stressed syllables, not on conversationally stressed syllables.

It is difficult to evaluate such EMG records by themselves for two reasons. First, there is no simple way of estimating the magnitude of lung volume decrement caused by any bursts of expiratory muscle activity (if found). After all, it is only by decreasing the lung volume at a faster-than-normal rate that these muscles can have any influence on Ps. It is possible that short-term fluctuations in the level of expiratory muscle activity produce only negligible variations in Ps. Second, it is difficult to know <u>why</u> the expiratory muscles might show increased activity near stressed syllables. Is its long-term function to maintain a constant pressure in the lungs or to maintain a constant rate of lung volume decrement, or some combination of the two? If its task is to maintain a steady rate of lung volume decrement, then the momentary increase in expiratory activity on stressed syllables could simply be a compensatory reaction (even in advance) to the increased glottal resistance to airflow expected, which increased resistance would otherwise momentarily slow the rate of lung volume decrease. If this were so, the increased activity of the pulmonic muscles would not, strictly speaking, be an independent feature of stress; it would be dependent upon downstream changes in resistance to airflow.

In order to clarify some of these issues, I obtained records of lung volume during speech.

4.5.3. Plethysmographic Studies. A whole-body pressure plethysmograph was used. It consisted of 1 and 2/3 steel drums welded together with a housing that fitted on top to accommodate the standing subject's head and in one wall of which was face mask through which the subject breathed atmospheric air. The inside volume was approximately 325 l.

Speech samples were obtained from four adult speakers: two male speakers of American English (LB and JJO), one male Swedish speaker, and one female speaker of Hindi. In all cases recordings of the acoustic speech signal were also obtained although these were not of very high quality given the distortions caused by the face mask. Approximate calibrations of the lung volume signal was attempted only in the case of the recording with the Hindi speaker. In the case of one of the English speakers (JJO) and the Hindi speaker, combined oral and nasal airflow was sampled simultaneously and averages of some of the parameters sampled were computed (based on 9 or 10 tokens each) along with the first derivative of the averaged lung volume signal. (Further details may be found in Ohala, Riordan, & Kawasaki 1980 and M. Ohala 1980.)

Figs. 2a,b shows microphone and lung volume curves during single tokens of the sentence 'deem unilluminable real', first in conversational style (a), and then with emphatic stress on the syllable [lum] (b). Figs. 3a-d shows the same parameters during single tokens of 'deem __oon real' [dĩm ____ũn ɹiɫ],[6] where the blank was filled in with [m, s, h, t^h], respectively. Fig. 4 shows averaged lung volume and its first derivative (slope) for the Hindi nonsense word [ʔib^hi] (here and with other averaged curves, the vertical dotted line indicates the synchronization point used to form the averages). Figure 5 shows averages of F0, lung volume and the oral plus nasal airflow for the English sentence 'Lure Bill near'. Fig. 6. gives averaged airflow and differentiated lung volume for 5 Hindi nonsense words with [p, b, m, p^h, b^h] in intervocalic position. Fig. 7. exhibits averaged lung volume and airflow for the sentences 'You drew Will near' (on the left) and 'you drew pill near' (on the right).

Based on such data, the following observations were made:

1. There are relatively large rapid decreases in lung volume during moments of high oral airflow, e.g., during aspiration, [h], and fricatives. See Figs. 3, 4, 6, 7. These presumably represent a passive collapse of the lungs due to the rapid flow of air out of the lungs and the consequent decrease in lung pressure. This contrasts with the case of sonorant consonants (nasals, laterals, etc.) where the rate of lung volume decrement is the same as that for surrounding vowels. See Figs. 3, 5, 6, 7.

2. There are also moments of less than normal lung volume decrement during periods of reduced airflow, e.g., as occurs during the closure phase of stops or affricates. See Figs. 2, 4, 5, 6, 7.

3. There are momentary greater-than-normal decreases in lung volume during emphatically stressed syllables. See portion delimited by broken

---

6 The use primarily of high vowels in the speech samples was necessitated by the face mask.

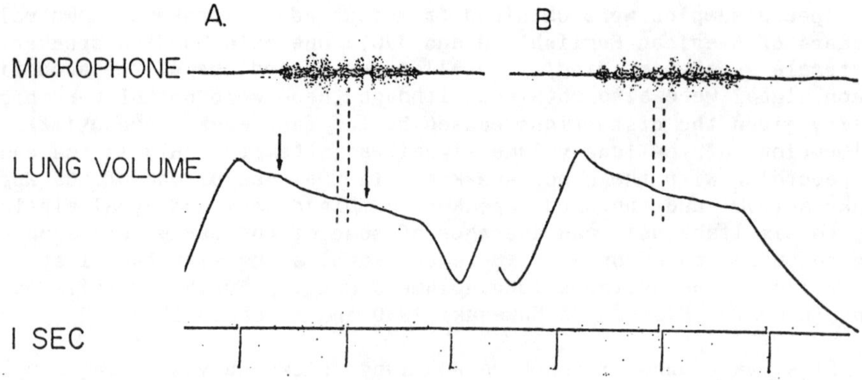

Figure 2. Single tokens of 'deem unilluminable real', in normal conversational style (left) and with heavy stress on the syllable 'lum' (right). Top: microphone signal; bottom: lung volume. Dotted lines delimit the stressed syllable 'lum' in both cases. Arrows mark the momentary plateaus in the lung volume parameter due to the stop closures of the [d] and [b].

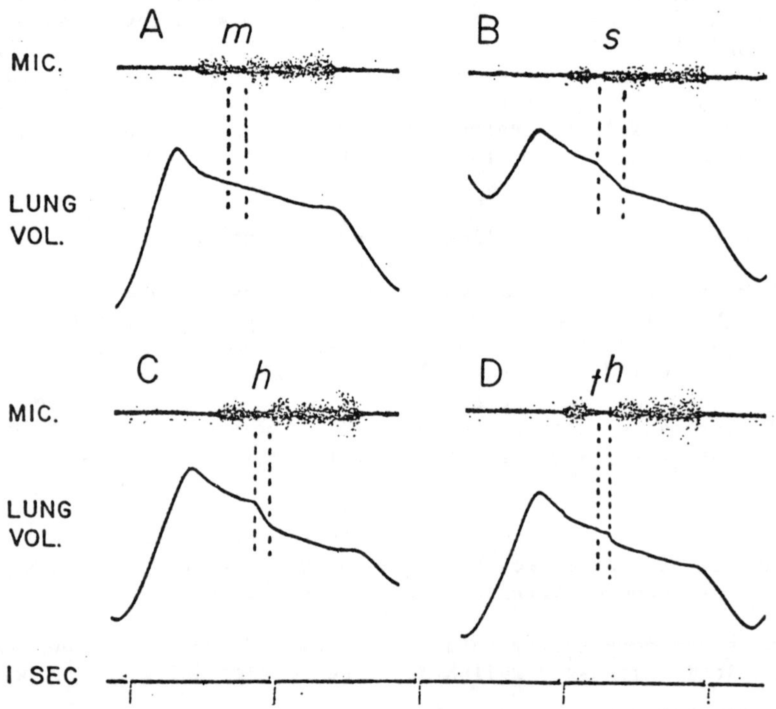

Figure 3. Individual tokens of the frame sentence 'deem __oon real' where the blank was filled by the consonants [m s h t^h]; dotted lines delimit the consonants' onsets and offsets. Top: microphone; bottom: lung volume.

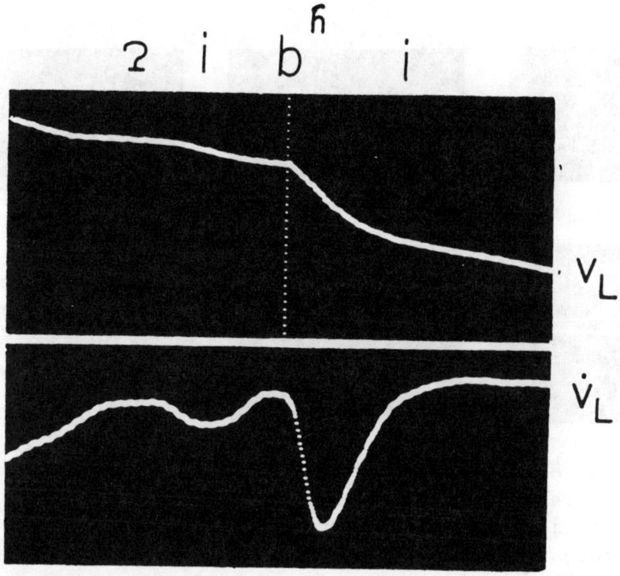

Figure 4. Averaged traces of lung volume (top) and differentiated lung volume (bottom) during the Hindi nonsense utterance [ʔibʱi]. The dotted line marks the consonant release and the point of synchronization for the averages.

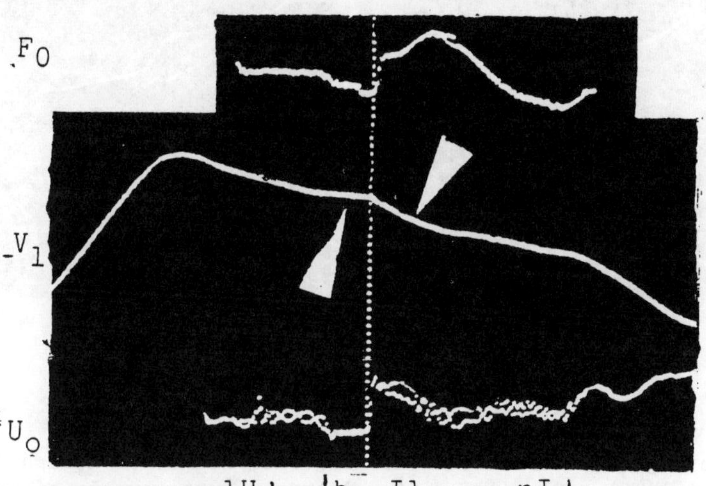

Figure 5. Averaged traces of fundamental frequency (top), lung volume (middle), and oral plus nasal airflow (bottom) during the utterance 'lure Bill near'. Dotted line marks the consonant release and the point of synchronization for the averages.

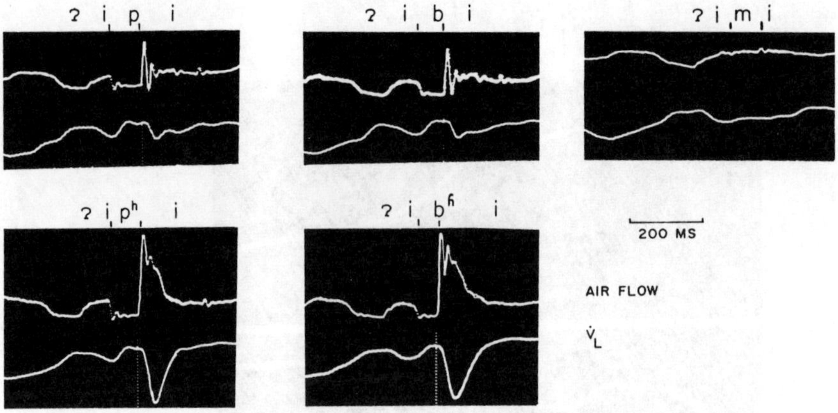

Figure 6. Records of airflow (top) and differentiated lung volume (bottom) during five Hindi utterances: [ʔipi], [ʔibi], [ʔimi], [ʔip^{h}i], [ʔib^{h}i]. Oral airflow represents a single token; the differentiated lung volume is the average of 9 or 10 tokens of each utterance. The dotted vertical line marks the consonant release and the synchronization point for the averages.

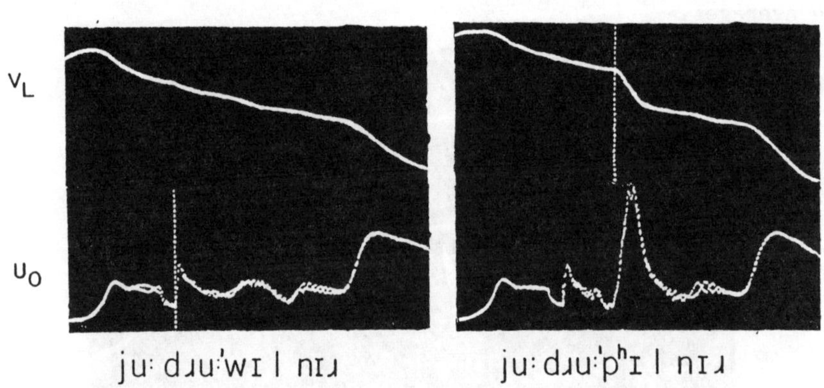

Figure 7. Averages of lung volume (top) and oral plus nasal airflow (bottom) during the utterances 'you drew Will near' (left) and 'you drew pill near' (right). Vertical dotted line marks the synchronization point used to form the averages (left: the release of the [d] in 'drew'; right: the release of the [p] in 'pill').

vertical lines in Fig. 2b. ('Emphatically stressed' means the kind of very forceful emphasis given in heated arguments, dynamic oratory, and the like.)

4. There need not be any obvious change in the rate of lung volume decrement on non-emphatically stressed syllables, i.e., during normal conversational speech. This is easiest to see in sentences consisting entirely of sonorants since the segments themselves induce no perturbation in the rate of airflow out of the lungs. See portion delimited by broken vertical lines in Fig. 2a which shows no change in rate of decrement on the stressed vowel.

5. In general, there is a close short-term (and inverse) relationship between lung volume decrement and the volume of airflow exiting through the mouth and nose. See Figs. 5, 6 and 7.

I interpret these results as showing that the primary function of the pulmonic system during speech is simply to produce Ps that is reasonably constant and above some minimal level. Short-term passive variations in lung volume decrement occur in reaction to variations in lung pressure which in turn are reactions to changing downstream resistance to airflow created by oral articulations. Active short-term variations in lung volume decrement are probably limited to the production of variations in the loudness of speech: lesser decrement for less intense speech, greater decrement for very loud speech.

4.6. F0 Declination

It has been widely observed that the F0 of voice gets progressively lower from beginning to end of a sentence (Pike 1945: 77, Maeda 1974, Pierrehumbert 1979). In African tone languages this results in tones at the end of a sentence being lower in F0 than those at the beginning. Sometimes a high tone may be even lower at the end than a low tone at the beginning (Silverstein 1976). Among Africanists this is called 'downdrift'; the phenomenon as it applies (potentially) to any language is called 'declination' (Cohen & 't Hart 1965). Declination may be absent in questions where instead the failure of F0 to fall helps to differentiate them from declarative utterances.

It should be mentioned that doubt has been expressed about the generality of declination, especially in spontaneous speech--as opposed to planned "laboratory" speech (Lieberman, Katz, Jongman, Zimmerman, Miller 1985a; but see the ensuing exchange between these authors and their critics: Repp 1985, Lieberman, Katz, Jongman, Zimmerman, Miller 1985b, 't Hart 1986, Lieberman 1986). Nevertheless, there is good evidence that listeners expect declination in sentences; for two accented syllables in the same sentence to sound equally prominent the second should have a lower F0 than the first (Pierrehumbert 1979; however, cf. Cohen, Collier & 't Hart 1982).

Assuming that declination does exist in at least some utterances and is wide-spread among languages (Weitzman 1970, Silverstein 1976), there has been considerable speculation on the physiological mechanisms which cause it. In particular, given its apparent pervasiveness, many have sought to identify automatic, i.e., mechanical and non-purposive, physiological causes: something which changes subtly and progressively during an utterance and which could influence F0. Naturally, attention was focused on the respiratory system. Maeda (1974) suggested that the larynx progressively lowers during a single speech expiration ("breath group") due to its muscular linkage to the sternum which should lower as the lung volume decreases. Given this and the fact that there is often a correlation between larynx height and F0 [7], this movement could cause a gradual lowering of F0. However, the strength of this posited mechanical linkage between the sternum and larynx is called into question by the observation of Ewan (1979 p:36) that the larynx normally moves upward, not downward, during expiration.

Collier (1975), while acknowledging that the major linguistically-significant F0 variations in speech (e.g., those signalling stress, question-vs-statement) are implemented by the laryngeal muscles, suggested that the F0 variation in declination was caused by the gradually falling Ps which in turn was undoubtedly connected to the progressive increase in expiratory effort required to maintain a sufficiently high Ps as the elastic recoil force decreases when the lung volume approaches the FRC and then opposes expiration as it passes below the FRC. Consistent with this view, he pointed out, were two other facts: (a) there seemed to be no indication in EMG records from the laryngeal muscles during speech that they show any change that can be correlated with the F0 decline and (b) that the rate of F0 change with respect to Ps change during declination was in the range cited above (2-5 Hz/cm H2O) which is compatible with independent Ps-caused F0 variation.

This issue cannot be regarded as settled. It is true that more systematic EMG studies addressing the mechanisms underlying declination are needed but there already exist some suggestive data on the issue. In fact, some of the figures published in Collier (1974; an expanded version of Collier 1975) seem to show graded muscle activity during a period of declination; see Fig. 8 ('Contour 15'in the original). In this figure (as well as in 'Contour 11' of the original) the F0 in the latter portion of the utterance is high but gradually falling at about the same rate as in other instances of declination. The Ps contour does not match the F0 trend but the cricothyroid activity does. Maeda (1975) has also presented EMG data which could be interpreted as showing progressive lessening of cricothyroid activity and increase in sternohyoid activity during speech showing the typical F0 declination.

Collier's conclusion that the covariation between Ps and F0 during declination implies that it is Ps causing the F0 change is subject to the

---

7 The reasons for this correlation are not clear but see Ohala (1972, 1977a, 1977b, 1982), Ohala & Eukel (1987).

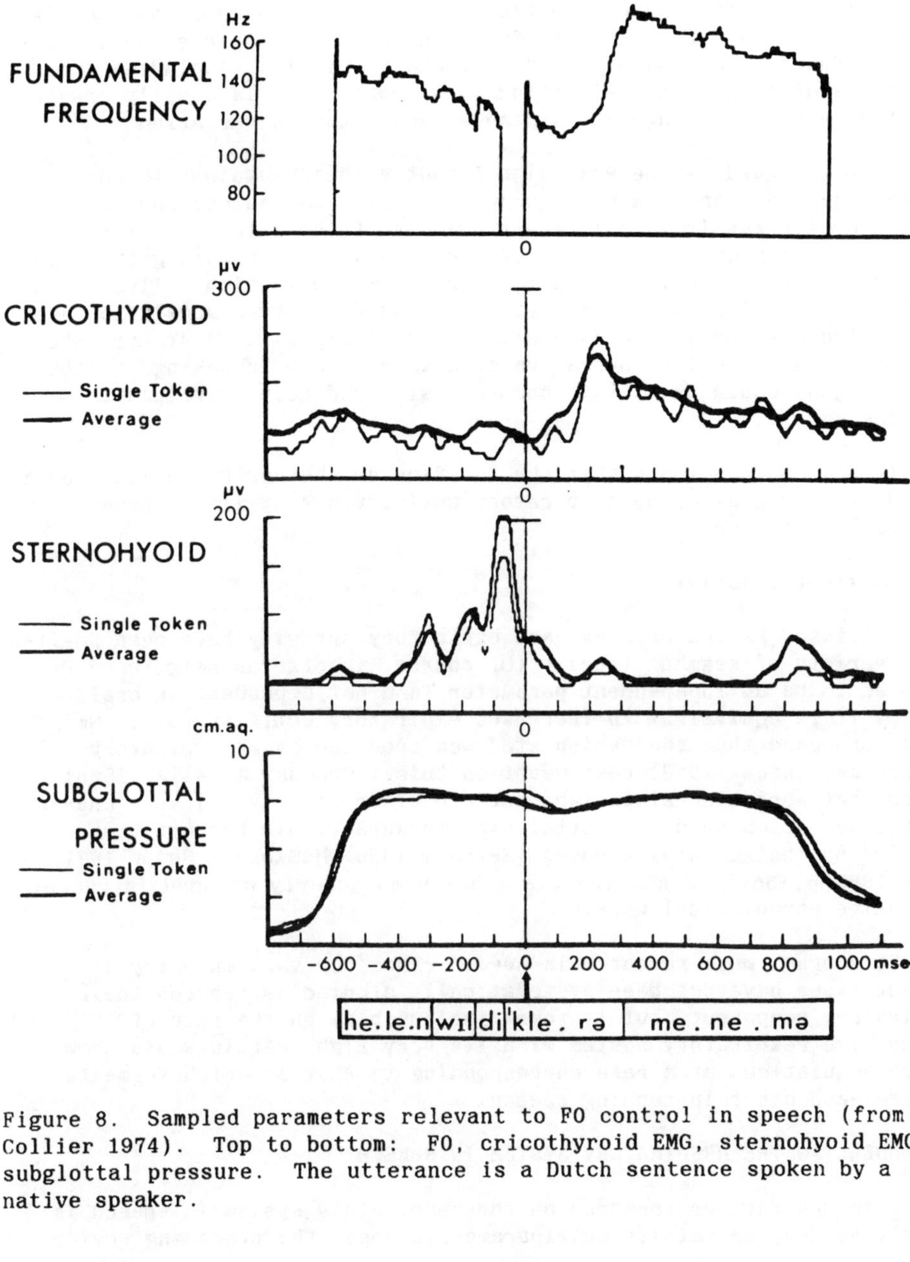

Figure 8. Sampled parameters relevant to F0 control in speech (from Collier 1974). Top to bottom: F0, cricothyroid EMG, sternohyoid EMG, subglottal pressure. The utterance is a Dutch sentence spoken by a native speaker.

same criticism offered above to Lieberman's claims (except that Collier is careful to make sure that the F0/Ps ratio is a plausible one): Ps is partly a function of downstream resistance to airflow, including the resistance offered by the vibrating vocal cords. It is equally possible that the changing laryngeal resistance causes the Ps variation.

What I regard as the most significant evidence against the hypothesis that declination is a consequence of purely mechanical factors connected with respiration is the finding that the rate of declination varies inversely with the length of the utterance or breath group: the shorter the utterance the steeper the rate of declination (Silverstein 1976). It is quite plausible that level and rate of F0 modulation are implemented purposely by speakers in order to signal to listeners where they are in a logical or narrative structure. Such "chunking" of the speech signal could facilitate comprehension and help to regulate turn-taking.

It must be concluded that the question of whether F0 declination is caused by laryngeal or by respiratory activity has still not been answered definitively.

4.7. Segmental Activity

Increased Ps and/or increased expiratory activity have been posited for a variety of segment types. (Of course, insofar as heightened Ps is supposed to be an independent parameter (and not dependent on oral events) it is equivalent to increased expiratory activity [8].) Smith (1944) proposed that the Danish stød was produced by an expiratory pulse; Fischer-Jørgensen (1988) cast doubt on this. Chomsky & Halle (1968) stated that aspirated stops (and certain other stop types) were characterized by 'heightened subglottal air pressure'; this has not been verified by instrumental studies (Netsell 1969, McGlone & Shipp 1971, M. Ohala 1983 pp. 155-160) and the claim has been quietly dropped in recent generative phonological work.

Of course, more research is needed since the vast majority of segment types have not been systematically studied as regards their respiratory component, but it seems implausible, on the face of it, that the massive respiratory system with its very high inertia would show active modulations at a rate corresponding to that at which segments succeed each other in running speech.

5. MODELLING THE RESPIRATORY SYSTEM IN SPEECH

Although further research on the respiratory system in speech is clearly needed, especially multiparametric ones, the preceding review

---

8 Heightened Ps as a dependent, and therefore non-distinctive, feature of voiceless (as opposed to voiced) stops was demonstrated by Ohala (1975); see below.

suggests that from a functional aerodynamic point of view the respiratory system can be modelled in a very simple way. That is, considering the respiratory system simply as a mechanism for supplying the oral articulators with air under pressure, the system can be modelled as a piston and piston-chamber system driven by a constant force (for a given level of voice intensity) with the piston having a certain inertia. Due to a combination of variations in downstream resistance and the inertia of the respiratory 'piston', variations in Ps will occur, although generally these would not exceed +/- 2 cm H2O.

Speech aerodynamic models which incorporate this scheme of the respiratory system have been implemented (Rothenberg 1968, Ohala 1975, 1976, 1983, Müller & Brown 1980) and are capable of generating aerodynamic parameters (airflow and air pressure at various sites in the vocal tract) which agree with those obtained from real speech. Fig. 9 shows representative output from the model presented by Ohala (1975). In the figure, the bottom three parameters, oral aperture and glottal aperture (expressed as resistances) and the oral volume (which increases by 2 cm during the voiced stop closure) are the input to the model; the remaining parameters, oral and glottal airflow, subglottal and oral pressure, are the output of the model. The figures shows on the left the simulation of a sequence like /ɑpʰɑ/ and,on the right, /ɑbɑ /. A comparison with the Ps curve in Fig. 1 reveals the same increase during the stop closures due to the pulmonic system momentarily encountering a greater resistance to the exiting airflow. The Ps peak is also slightly higher for the voiceless than for the voiced stop since the resistance is higher in the former (effectively infinite) whereas in the latter it is less and must be so as long as some air may still flow through the glottis, i.e., as long as voicing can be maintained.[9] Also modelled correctly is the precipitous drop in Ps upon the release of the aspirated stop and its gradual return to a higher level for the following vowel. In contrast to that, the Ps is initially high after the release of the voiced stop and then gradually returns to a lower level for the following vowel. From both Figs. 1 and 9 and the relation between Ps and voice intensity, reviewed above, one would predict that, other things being equal, the amplitudes of vowels should be higher after a voiced stop than after a voiceless aspirated stop, due to the different Ps levels. This has, in fact, been found (House & Fairbanks 1953, Lehiste & Peterson 1959).

Further evidence of the primacy of glottal and oral articulations in implementing the linguistic units of speech is given by the success of articulatory synthesis which concentrates on articulatory movements from the glottis and above (Scully 1987).

---

9 This does not constitute confirmation of the Chomsky & Halle claim that aspirated stops differ from unaspirated by the feature [+ Heightened Subglottal Pressure]. The higher Ps here would be true of any voiceless obstruent (vis-a-vis a cognate voiced one), aspirated or not, and, moreover, is a feature <u>dependent</u> on the specified glottal and supraglottal events.

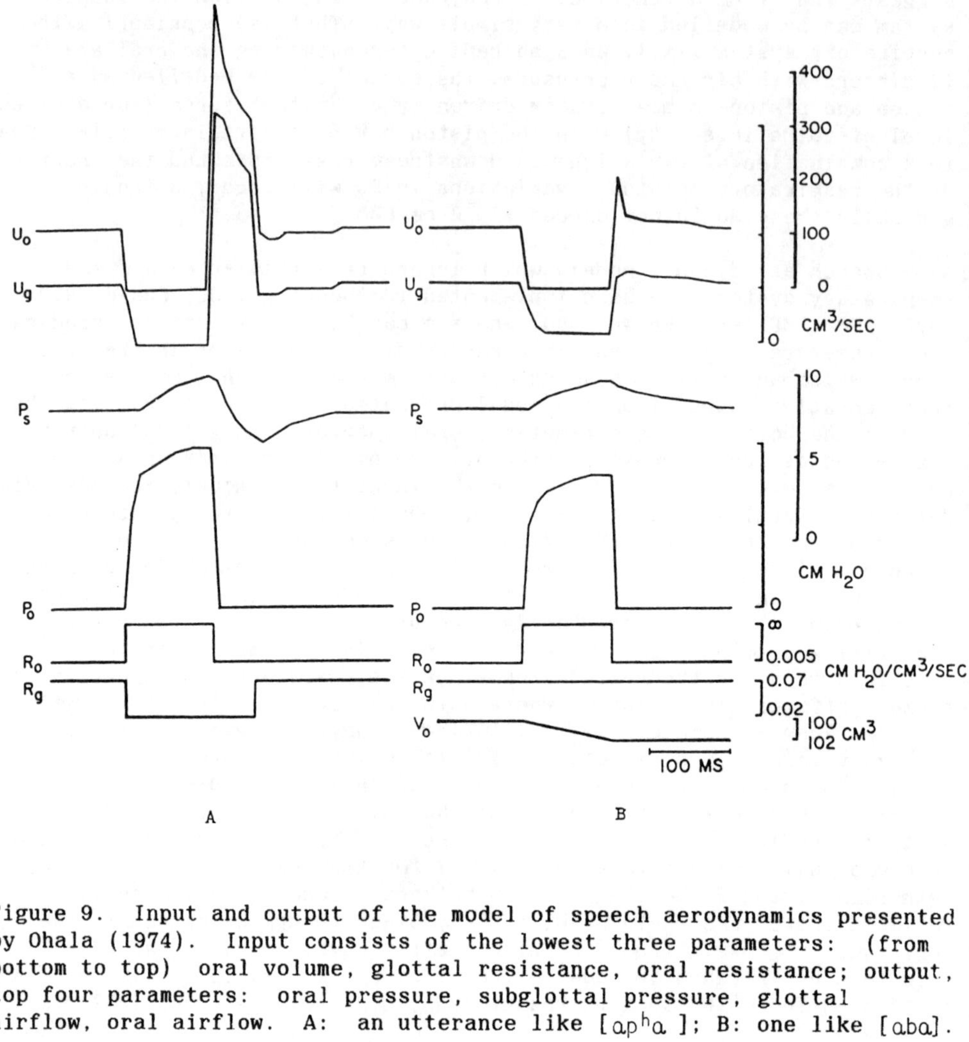

Figure 9. Input and output of the model of speech aerodynamics presented by Ohala (1974). Input consists of the lowest three parameters: (from bottom to top) oral volume, glottal resistance, oral resistance; output, top four parameters: oral pressure, subglottal pressure, glottal airflow, oral airflow. A: an utterance like [ɑpʰɑ]; B: one like [ɑbɑ].

No doubt we can expect future models of speech respiratory mechanics to go beyond simple functional models and to parcel out the expiratory force to elastic recoil, chest and abdominal muscles and to incorporate sufficient physiological and anatomical detail to be able to calculate the biomechanical efficiencies of different modes of expiration, as discussed by Weismer (1985).

6. CONCLUSION

The picture that emerges from the evidence reviewed is that the respiratory system (the pulmonic cavity and the forces which change its volume and the air pressure inside) is functionally quite simple. For a given intensity of voice, the force applied to the lungs is essentially constant; observed short-term variations in the rate of lung volume decrement and in subglottal air pressure are thus caused by two factors, (a) variations in the resistance to airflow offered downstream, i.e., by the oral articulations and (b) the inertia of the respiratory system which prevents it from recovering immediately from the momentary variations in internal lung pressure. Claims that independent action of the respiratory system underlies the production of syllables, stress, certain sentence-final F0 contours, F0 declination, and certain segment types are called into question.

7. ACKNOWLEDGEMENTS

I thank the participants of the NATO ASI, especially Celia Scully and Ray Kent, for helpful comments on an earlier draft of this paper. Some of the research reported here was conducted in collaboration with Minoru Hirano, Haruko Kawasaki, Manjari Ohala, and Carol Riordan and was supported by the National Science Foundation.

8. REFERENCES

Abas, A. (1925) 'L'accentuation syllabique en néerlandais', Arch. Néerl. de Phys. 10.82-154.

Allen, G. D. (1971) 'Acoustic level and vocal effort as cues for the loudness of speech', J. Acoust. Soc. Am. 49.1831-1841.

Arutjunjan, E. A., Granstrem, M. P., & Kozhevnikov, V. A. (1967) 'Some characteristics of the respiratory system acting in speech', Digest of the 7th International Conference on Medical and Biological Engineering. Stockholm. p. 346.

Atkinson, J. (1973) Aspects of intonation in speech: implications from an experimental study of fundamental frequency. Ph.D. diss., Univ. of Connecticut.

Atkinson, J. (1978) 'Correlation analysis of the physiological factors controlling fundamental voice frequency', J. Acoust. Soc. Am. 63.211-222.

Baglioni, G. (1922) 'Movimenti respiratori e fonazione', in Livre Jubilaire offert à H.Zwaardemaker Czn. [Archives Néerlandaises de Physiologie de l'Homme et des Animaux, vol. 7], pp. 484-487.

Bolinger, D. L. (1958) 'A theory of pitch accent in English', Word 14.109-149.

Broad, D. J. (1968) Some physiological parameters for prosodic description. Speech Communication Research Laboratories, Santa Barbara, Monogr. No. 3.

Catford, J. C. (1939) 'On the classification of stop consonants', Le Maitre Phonetique 3rd Ser., No. 65.

Catford, J. C. (1977) Fundamental problems in phonetics. Bloomington: Indiana University Press.

Chomsky, N. & Halle, M. (1968) The sound pattern of English. New York: Harper & Row.

Cohen, A., Collier, R., & 't Hart, J. (1982) 'Declination: Construct for intrinsic feature of speech pitch?', Phonetica 39.254-273.

Cohen, A., & 't Hart, J. (1965) 'Perceptual analysis of intonation patterns', in D. E.Commins (ed.), Proceedings of the Fifth International Congr. on Acoustics, Liège, 1964. Paper A16.

Coleman, H. O. (1914) 'Intonation and emphasis', Miscellanea Phonetica 1.6-26.

Collier, R. (1974) 'Laryngeal muscle activity, subglottal air pressure, and the control of pitch in speech', Status Rep. on Speech Res. [Haskins Labs.] SR-39/40.137-170.

Collier, R. (1975) 'Physiological correlates of intonation patterns', J. Acoust. Soc. Am. 58.249-255.

Darwin, E. (1803) The temple of nature; or the origin of society: a poem with philosophical notes. London: Printed for J. Johnson by T. Bensley.

Di Cristo, A. (1975) Soixante-et-dix ans de recherches en prosodie. Travaux de l'Institut de Phonétique d'Aix-en-Provence. Études Phonétiques, I.

Draper, M., Ladefoged, P., & Whitteridge, D. (1959) 'Respiratory muscles in speech', J. Speech & Hearing Res. 2.16-27.

Erickson, D. & Atkinson, J. (1976) 'The function of the strap muscles in speech', Status Rep. on Speech Res. [Haskins Labs.] SR-45/46.205-210.

Erickson, D. , Baer, T., & Harris, K. S. (1983) 'The role of the strap muscles in pitch lowering', in D. M. Bless & J. H. Abbs (eds.), Vocal fold physiology. Contemporary research & clinical issues. San Diego: College-Hill Press. 279-285.

Ewan, W. G. (1979) Laryngeal behavior in speech. Rep. Phonology Laboratory [Univ. of California, Berkeley] No. 3.

Fischer-Jørgensen, E. (1988) 'A phonetic study of the stød in Standard Danish', Phonetica 45.

Fonagy, I. (1958) 'Elektrophysiologische Beiträge zur Akzentfrage', Phonetica 2.12-58.

Foster, J. (1763) An essay on the different nature of accent and quantity. [2nd ed.] Eton: J. Pote.

Fromkin, V. A. & Ohala, J. J. (1968) 'Laryngeal control and a model of speech production', Preprints of the Speech Symposium, Kyoto, 1968. C3:1-5.

Fry, D. (1958) 'Experiments in the perception of stress', Language & Speech 1.126-152.

Gray, G. W. (1936) Studies in experimental phonetics. [Louisiana State University Studies No. 27.] Baton Rouge.

't Hart, J. (1986) 'Declination has not been defeated--A reply to Lieberman et al', J. Acoust. Soc. Am. 80.1838-1840.

Hirano, M. & Ohala, J. J. (1969) 'Use of hooked-wire electrodes for electromyography of the intrinsic laryngeal muscles', J. Speech & Hearing Res. 12.362-373.

Hirano, M., Ohala, J. J., & Vennard, W. (1969) 'The function of laryngeal muscles in regulating fundamental frequency and intensity of phonation', J. Speech & Hearing Res. 12.616-628.

Hixon, T. J. (1971) 'An electromagnetic method for transducing jaw movements during speech', J. Acous. Soc. Am. 49.603-606.

Hixon, T. J. (1972) 'Some new techniques for measuring the biomechanical events of speech production: One laboratory's experiences', Am. Speech & Hearing Assoc. Rep. 7.68-103.

Hixon, T. J. (1973) 'Respiratory function in speech', in F. D. Minifie, T. J. Hixon, & F. Williams (eds.), Normal aspects of speech, hearing, and language. Englewood Cliffs, NJ: Prentice-Hall. 73-125.

Hixon, T. J ., Goldman, M. D., & Mead, J. (1973) 'Kinematics of the chest wall during speech production: volume displacements of the rib cage, abdomen, and lung', J. of Speech & Hearing Res. 16.78-115.

Hixon, T. J., Mead, J., & Goldman, M. D. (1976) 'Dynamics of the chest wall during speech production: function of the thorax, rib cage, diaphragm, and abdomen', J. of Speech & Hearing Res. 19.297-356.

House, A. S. & Fairbanks, G. (1953) 'The influence of consonant environment upon the secondary acoustical characteristics of vowels', J. Acoust. Soc. Am. 25.105-113.

Ishizaka, K. & Flanagan, J. L. (1972) 'Synthesis of voiced sounds from a two-mass model of the vocal cords', Bell System Technical Journal 51.1233-1268.

Isshiki, N. (1969) 'Remarks on mechanisms of voice intensity variations', J. Speech & Hearing Res. 12.669-672.

Ishizaka, K. & Flanagan, J. L. (1972) 'Synthesis of voiced sounds from a two-mass model of the vocal cords', Bell System Technical J. 51.1233-1268.

Jaeger, M. J. & Matthys, H. (1970) 'The pressure flow characteristics of the human airways', in A. Bouhuys (ed.), Airway dynamics, physiology and pharmacology. Springfield, IL: Charles C. Thomas. 21-32.

Jones, D. (1922) An outline of English phonetics. 2nd ed. Leipzig: B. G. Teubner.

Kakita, Y. & Hiki, S. (1976) 'Investigation of laryngeal control in speech by use of thyrometer', J. Acoust. Soc. Am. 59.669-674.

van Katwijk, A. (1974) Accentuation in Dutch. An experimental linguistic study. Utrecht.

Keller, K. C. (1971) Instrumental articulatory phonetics. An introduction to techniques and results. Norman, OK: Summer Institute of Linguistics.

von Kempelen, W. (1791) Mechanismus der menschlichen Sprache. Wien. J. B Degen.

Klatt, D. H., Stevens, K. N., & Mead, J. (1968) 'Studies of articulatory activity and airflow during speech', in A. Bouhuys (ed.), Sound production in man. Annals of the New York Academy of Sciences 155.42-54.

Ladefoged, P. (1962a) A phonetic study of West African languages. Cambridge: Cambridge University Press.

Ladefoged, P. (1962b) 'Subglottal activity during speech', Proc. 4th Int. Congr. of Phonetic Sciences. The Hague: Mouton. 73-91.

Ladefoged, P. (1963) 'Some physiological parameters in speech', Lang. & Speech 6.109-119.

Ladefoged, P. (1967) Three areas of experimental phonetics. Oxford University Press.

Ladefoged, P. (1968) 'Linguistic aspects of respiratory phenomena', in A. Bouhuys (ed.), Sound production in man. Annals of the New York Academy of Sciences 155.141-151.

Ladefoged, P. & McKinney, N. P. (1963) 'Loudness, sound pressure and subglottal pressure in speech', J. Acous. Soc. Am. 35.454-460.

Lehiste, I. & Peterson, G. E. (1959) 'Vowel amplitude and phonemic stress in American English', J. Acoust. Soc. Am. 31.428-435.

Lieberman, P. (1967) Intonation, perception, and language. Cambridge: M.I.T. Press.

Lieberman, P. (1986) 'Alice in declinationland--A reply to Johan 't Hart', J. Acoust. Soc. Am. 80.1840-1842.

Lieberman, P. & Blumstein, S. E. (1988) Speech physiology, speech perception, and acoustic phonetics. Cambridge: Cambridge University Press.

Lieberman, P., Katz, W., Jongman, A., Zimmerman, R., & Miller, M. (1985a) 'Measures of the sentence intonation of read and spontaneous speech in American English', J. Acoust. Soc. Am. 77.649-657.

Lieberman, P., Katz, W., Jongman, A., Zimmerman, R., & Miller, M. (1985b) 'Reply to Bruno H. Repp', J. Acoust. Soc. Am. 78.1116-1117.

Lieberman, P., Sawashima, M., Harris, K. S., & Gay, T. (1970) 'The articulatory implementation of breath group and prominence: Cricothyroid muscular activity in intonation', Language 46.312-327.

Luchsinger, R. & Arnold, G. E. (1965) Voice--speech--language. Belmont: Wadsworth.

Maeda, S. (1974) 'A characterization of fundamental frequency contours of speech', Quart. Prog. Rep., Res. Lab. of Electronics, M.I.T. 114.193-211.

Maeda, S. (1975) 'Electromyographic study of intonational attributes', Quart. Prog. Rep., Res. Lab. of Electronics, M.I.T. 115.261-269.

Mattingly, I. G. (1966) 'Synthesis by rule of prosodic features', Lang. & Speech 9.1-13.

McGlone, R. E. & Shipp, T. (1971) 'Comparison of subglottal air pressures associated with /p/ and /b/', J. Acoust. Soc. Am. 51.664-665.

Mead, J. (1960) 'Volume displacement body plethysmograph for respiratory measurements in human subjects', J. Appl. Physiol. 15.736-740.

Mead, J., Bouhuys, A., & Proctor, D. F. (1968) 'Mechanisms generating subglottal pressure', in A. Bouhuys (ed.), Sound production in man. Annals of the New York Academy of Sciences 155.177-181.

Mitford. William. (1804) Inquiry into the principle of harmony in language, and of the mechanism of verse, modern and antient. 2d ed. London : T. Cadell & W. Davies.

Morris, A. R. (1925) The orchestration of the metrical line. Boston: Badger.

Mueller, J. (1851) Manuel de physiologie. [Transl. from German by A.-J.-L. Jourdan] 2nd ed. Paris: Chez J.-B. Baillière.

Müller, E. M. & Brown, W. S., Jr. (1980) 'Variations in the supraglottal air pressure waveform and their articulatory interpretations', in N. J. Lass (ed.), Speech and language: Advances in basic research and practice. Vol. 4. New York: Academic Press. 317-389.

Munro, R. R. & Adams, C. (1971) 'Electromyography of the intercostal muscles in connected speech', Electromyography 11.365-378.

Muyskens, J. H. (1931) 'An analysis of accent in English from kymograph records', Vox 17.2.55-65.

Netsell, R. (1969) 'Subglottal and intraoral air pressures during the intervocalic contrast of /t/ and /d/', Phonetica 20.68-73.

Netsell, R. (1973) 'Speech physiology', in F. D. Minifie, T. J. Hixon, & F. Williams, (eds.), Normal aspects of speech, hearing, and language. Englewood Cliffs, NJ: Prentice-Hall. 211-234.

Ohala, J. J. (1970) Aspects of the control and production of speech. Working Papers in Phonetics, UCLA, No. 15.

Ohala, J. J. (1972) 'How is pitch lowered?', J. Acous. Soc. Am. 52.124.

Ohala, J. J. (1975) 'A mathematical model of speech aerodynamics', in G. Fant (ed.), Speech Communication. Vol. 2: Speech production and synthesis by rule. Stockholm: Almqvist & Wiksell. 65-72.

Ohala, J. J. (1976) 'A model of speech aerodynamics', Report of the Phonology Laboratory (Berkeley) 1.93-107.

Ohala, J. J. (1977a) 'The physiology of stress', in L. M. Hyman (ed.), Studies in stress and accent. [Southern California Occasional Papers in Linguistics, No. 4] Los Angeles: University of Southern California. 145-168.

Ohala, J. J. (1977b) 'Speculations on pitch regulation', Phonetica 34.310-312.

Ohala, J. J. (1978) 'Production of tone', in V. A. Fromkin (ed.), Tone: A linguistic survey. New York: Academic Press. 5-39.

Ohala, J. J. (1982) 'Physiological mechanisms underlying tone and intonation', in H. Fujisaki & E. Garding (eds.), Preprints, Working Group on Intonation, 13th Int. Cong. of Linguists, Tokyo, 29 Aug.-4 Sept. 1982. 1-12.

Ohala, J. J. (1983) 'The origin of sound patterns in vocal tract constraints', in P. F. MacNeilage (ed.), The production of speech. New York: Springer Verlag. 189-216.

Ohala, J. J. & Eukel, B. (1987) 'Explaining the intrinsic pitch of vowels', in R. Channon & L. Shockey (eds.), in honor of Ilse Lehiste. Ilse Lehiste Pühendusteos. Dordrecht: Foris. 207-215.

Ohala, J. J. & Hirano, M. (1967) 'Studies of pitch change in speech', Working papers in phonetics (UCLA) 7.80-84.

Ohala, J. J. & Hirose, H. (1969) 'The function of the sternohyoid muscle in speech', Reports of the Autumn Meeting of the Acous. Soc. of Japan, 359-360. [Reprinted in: Ann. Rep., Inst. of Logopedics & Phoniatrics, Faculty of Medicine, Univ. of Tokyo 4.41-44 (1970).]

Ohala, J. J. & Ladefoged, P. (1970) 'Further investigation of pitch regulation in speech', Working Papers in Phonetics, UCLA, 14.12-24.

Ohala, J. J., Riordan, C. J., and Kawasaki, H. (1980) 'Investigation of pulmonic activity in speech', Rep. of the Phonology Lab. [Univ. of Calif., Berkeley] 5.89-95.

Ohala, M. (1980) 'Phonological features of Hindi stops', Rep. of the Phonology Lab. [Univ. of Calif., Berkeley] 5.96-105.

Ohala, M. (1983) Aspects of Hindi phonology. Delhi: Motilal Banarsidass.

Öhman, S. & Lindqvist, J. (1966) 'Analysis-by-synthesis of prosodic pitch contours', Quart. Prog. & Status Reports, Speech Transmission Lab., Stockholm 4/1965.1-6.

Panconcelli-Calzia, G. (1924) Die experimentelle Phonetik in ihrer Anwendung auf die Sprachwissenschaft. [2nd ed.] Berlin: Walter de Gruyter.

Pierrehumbert, J. (1979) 'The perception of fundamental frequency declination', J. Acoust. Soc. Am. 66.363-369.

Pike, K. L. (1945) The intonation of American English. Ann Arbor, MI: University of Michigan Press.

Repp, B. H. (1985) 'Critique of 'Measures of the sentence intonation of read and spontaneous speech in American English' [J. Acoust. Soc. Am. 77, 649-657 (1985)]', J. Acoust. Soc. Am. 78.1114-1116.

Rothenberg, M. (1968) The breath-stream dynamics of simple-released-plosive production. Bibliotheca Phonetica 6.

Roudet, L. (1900) 'Recherches sur le rôle de la pression sousglottique dans la parole', La Parole 2.599ff.

Roudet, L. (1900) 'De la dépense d'air dan la parole et de ses conséquences phonétiques', La Parole 2.209ff.

Roudet, L. (1910) Éléments de phonétique générale. Paris: H. Welter.

Rousselot, P. J. (1897-1908) Principes de phonétique expérimentale. 2 Vols. Paris: H. Welter.

Rush, James. (1827) The philosophy of the human voice. Philadelphia: J. Maxwell.

Scott, N. C. (1939) 'An experiment on stress perception', Le Maitre Phonetique. July 1939, p. 44.

Scully, C. (1987) 'Linguistic units and units of speech production', Speech Communication 6.77-142.

Scripture, E. W. (1902) The elements of experimental phonetics. New York: Charles Scribner's Sons. [Reissued 1973, New York: AMS Press.]

Shipp, T. E. (1975) 'Vertical laryngeal position during continuous and discrete vocal frequency change', J. Speech & Hearing Res. 18.707-718.

Silverstein, R. O. (1976) 'A strategy for utterance production in Hausa', Studies in African Linguistics, Suppl. 6.233-241.

Smith, S. (1944) The Danish stød. Copenhagen: Kaifer.

Sonesson, B. (1960) On the anatomy and vibratory pattern of the human vocal folds. Acta Otolaryngologica, Suppl. 156.

Stetson, R. H. (1928) Motor phonetics. La Haye: Martinus Nijhoff. [2nd ed. 1951. Amsterdam: North Holland Publishing Co.; retrospective ed. 1988, ed. by J. A. S. Kelso & K. G. Munhall. Boston: College-Hill.]

Sweet, H. (1891) A New English grammar. Oxford: Oxford Univ. Press.

Sweet, H. (1911) 'Phonetics', Encyclopedia Britannica. 11th ed. 458-467.

Titze, I. R. (1989) 'On the relation between subglottal pressure and fundamental frequency in phonation', J. Acous. Soc. Am. 85.901-906.

Umeda, N. (1976) 'Linguistic rules for text-to-speech synthesis', Proc. IEEE 64.443-451.

Vanderslice, R. (1970) 'Review of Intonation, perception, and language, by Philip Lieberman', J. Linguistics 6.138-144.

Weismer, G. (1985) 'Speech breathing: Contemporary views and findings', in R. G. Daniloff (ed.), Speech Science, College Hill Press, San Diego, pp. 47-72.

Weitzman, R. S. (1970) Word accent in Japanese. Studies in the Phonology of Asian Language [Univ. of So. Calif.] 11.

Willis, R. (1830) 'On the vowel sounds, and on reed organ-pipes', Transactions of the Cambridge Philosophical Soc. 3, 229-268.

Willis, R. (1833) 'On the mechanism of the larynx', Transactions of the Cambridge Philosophical Soc. 4.323-352.

Zemlin, W. R. (1968) Speech and hearing science. Englewood Cliffs: Prentice Hall.

ACQUISITION OF SPEECH PRODUCTION: THE ACHIEVEMENT OF SEGMENTAL INDEPENDENCE

PETER F. MACNEILAGE & BARBARA L. DAVIS
The University of Texas at Austin
Austin Tx. 78712
U.S.A.

ABSTRACT. As a primate communicative event, the repetitive, rhythmic, open-close alternation of the mandible, accompanied by phonation, is observable in three forms; some variants of the lipsmack, which is widespread in other primates, the initial babbling of human infants, and the production of the syllables of adult speech. In the first two of these, successive cycles tend to be uniform but in the third there is a highly variegated pattern in successive cycles. Adult segmental serial ordering errors, (e.g. spoonerisms) the effects of which are strongly constrained in terms of syllable structure, suggest that variegation is achieved by placement of independently controlled "Content" elements in syllable "Frames". This paper considers implications of the view that development of infant communicative vocalizations from initial reduplicated babbling to variegated babbling and then to speech, primarily involves a gradual functional differentiation of segmental and subsegmental content elements from a phylogenetically prior basis consisting of 'pure' syllable frames; that is, mandibular oscillations without internal articulatory modulation.

1.Background

Serial ordering errors in normal adult speech that involve single speech segments moving around in an otherwise correct utterance tell us beyond doubt that the individual segment is an independent unit in the control of adult speech. However, there is no evidence that when infants first sound as if they are producing speech segments with speech-like durations, they already have segmental units under independent control. The question addressed in this paper is simple; how do infants achieve this segmental independence ?

We wish to address this question in an evolutionary perspective. The independent control of the ordering of segments in a vocal utterance is a new development in primate evolution, a development specific to hominids. However vocalization itself is far from new. Consequently there has been a phylogenetically recent development of superimposition of segmental permutations on vocal output. A clue to the way that this may have been achieved in evolution comes from a well known set of constraints on segmental speech errors known as syllable-position constraints. The constraints are best observed in spoonerisms, but also apply in similar fashion to other serial ordering errors. In spoonerisms, it is almost always true that initial consonants exchange with other initial consonants, vowels exchange with vowels, and final consonants exchange with final consonants. The existence of these syllable position constraints on speech errors gives rise to the "Frame/Content" metaphor for speech motor control; speech production includes a

W. J. Hardcastle and A. Marchal (eds.), Speech Production and Speech Modelling, 55–68.

stage of insertion of segmental "Content" units into syllable structure "Frames" (MacNeilage, Studdert-Kennedy and Lindblom, 1985; MacNeilage, 1987).

The syllable can be described as the interface between the segmental and the suprasegmental levels of speech production. At the suprasegmental or prosodic level, the syllable serves as the unit of rhythmic organization, and the point of reference for the intonation contour and semantic markings. At the segmental level, as we have seen, it serves as a structured repository controlling the serial position of individual segments. If the capacity for vocalization and thus, basically, the capacity for suprasegmental control is not new in hominids, but the segmental capacity is new, how did this change come about ? How did the syllable evolve as the interface between the two levels ?

Intensive study of the neural mechanisms of monkey vocalization (summarized by Jurgens, 1987) has shown that vocal output is controlled by a system that is medially placed in the neuraxis involving the cerebral cortex and subcortical centres in the midbrain and lower brain stem. At the 'head' of this system are two regions of medial cortex; 1. the anterior cingulate gyrus (ACG), and 2. a neocortical outgrowth of the ACG, the Supplementary Motor Area (SMA). Lesion studies have shown that in monkeys there is no major role of lateral frontal cortex - either lateral premotor cortex, analogous to Broca's Area, or primary motor cortex of the precentral gyrus - in vocalization control. (Jurgens, Kirzinger and von Cramon, 1982).

Important evidence regarding the evolutionary progression from monkey to modern hominids comes from the study of the SMA in humans. A number of different investigators, most notably Penfield and his collaborators (e.g. Penfield and Welch, 1951) have reported evoking rhythmic syllable-like vocalization from electrical stimulation of the SMA. This has not been reported from any other cortical site, to our knowledge. The possibility that these responses reflect the natural function of this area is strengthened by 8 different reports of a total of 17 patients with irritative lesions of the SMA who also involuntarily produced rhythmic syllable-like vocalization (for a summary see Jonas, 1981). These findings lead us to conclude that the SMA plays a leading role in the generation of syllable frames for speech.

A possible precursor to this function in other primates is the well-known lipsmack. This communicative gesture is extremely widespread in higher primates (Redican, 1975). It is a rhythmic series of open-close alternations of the mandible, with the noise that gives it its name being produced either by the release of lip closure or tongue contact with the roof of the mouth. When accompanied by phonation, it is known as the girney (Green, 1975). It seems appropriate that the SMA would serve in an interfacing role between older and newer vocal capacities as it is at once part of the older vocalization control system but unique in that system in being neocortex just as the perisylvian cortex most responsible for language is neocortex. To complete this picture from the motor control standpoint, we would propose that a major event in the evolution of the capacity required for the context-dependent production of segmental content elements in speech is the development of lateral frontal cortex, particularly Broca's Area. (For an expanded discussion of these and other points regarding the neural control of speech production see Lindblom, MacNeilage and Studdert-Kennedy, forthcoming).

As to the relevance of this background for speech acquisition, we suggest, in brief, that an infant recapitulates the phylogenetic steps, beginning without frames or context, developing frames at the onset of babbling, and, much later, following a developmental progression, achieving independent segmental content. The intent of this paper is to see how much of speech acquisition can be understood in terms of a distinction between the role of frames and content elements in the developmental process.

2.Babbling

The most salient event in speech acquisition from a motor control standpoint is the relatively sudden onset of babbling, typically around the end of the 7th month. (For a description of earlier stages of infant vocalization see Oller, 1980, and Stark, 1980). At this point, children suddenly begin to give the impression that they are alternating between consonant-like sounds (which we will call closants after Martin, 1981) and vowel-like sounds, (called vocants by Martin) in a rhythmic manner. In motor control terms, the close-open cycle of the mandible that underlies most of these closant-vocant sequences, accompanied by phonation, gives rise to the impression of a monosyllabic word. Reduplications of the cycle make the entire event sound like a multisyllabic word. For the first time, the infant vocalizations sound truly speech like.

This is a remarkable change, not only because it happens so suddenly (Van Der Stelt and Koopmans-van Beinum, 1986) but because it has such radical consequences. As Holmgren et al. (1986) have stated: "It is remarkable that - although the child does not yet use words that carry meaning and are motorically controlled, as in adult speech, by strings of discrete phonemic segments, vocalizations appear to present precursors of attributes universal in adult speech in the form of "segmental" and "syllabic" organization, "polysyllabicity" and "timing regularities" (p 61). We wish to argue that all this is achieved primarily by the advent of the motor correlate of the syllable frame. As a step towards the full recognition of the importance of this event, we suggest specifically that it is a result of the ontogenesis of the device that has evolved for interfacing the phylogenetically older suprasegmental level of primate vocalization and the newer segmental level, specific to hominids. It has been suggested by Ploog (1979) that prebabbling vocalizations are primarily produced by early maturing subcortical components of the medial vocalization subsystem common to most mammals. We suggest that the onset of babbling is associated with the maturation of the neocortical head of this medial subsystem in humans, the SMA.

This sequence is a case of ontogeny recapitulating phylogeny, at least to the degree that later evolving neural structures mature later in development. But we would emphasize that this is not simply a result of maturation. There is now ample evidence that deaf children are typically delayed by several months in the onset of babbling (e.g. Oller and Eilers, 1988) and that blind children are delayed in the acquisition of consonants with an important visible component (Mills, 1983). It seems to us to be a profound rather than a trivial question as to what the learning process is that enables normal infants to babble 'on schedule' when they are still so far away (typically about 6 months) from the ultimate use to which this elaborate structuring will be put.

The initial two or three months of babbling (8-10 months) has been termed the reduplicative babbling stage. Reduplication refers to repetition of the same sequence of closant and vocant, most typically beginning with a closant (C) and ending with a vocant (V); that is, the typical sequence is "CVCV...V". This is considered to be followed (10-13 months) by a stage of variegated babbling (Oller, 1980). As we mentioned earlier, the alternation between closants and vocants, in reduplicated babbling, may be produced almost completely by mandibular oscillation alone. This has led us to suggest that babbling begins with a stage of production of syllable precursors regarded as "pure frames" (MacNeilage and Davis, 1989). The frames are considered to be "pure" in the sense that there may be no neuromuscular activity specific to the production of the individual vocants and closants during the utterance, beyond that controlling the mandible. This activity is analogous in its reduplicative character and may be homologous in its neurobiological attributes to monkey lipsmacks, and the involuntary utterances of patients with artificial activation of the SMA.

A variety of vocants and closants can be observed during babbling. However, contrary to the earlier view, fostered by Jakobson (1968), that the closants and vocants are a random sample of the sounds in the world's languages, it is now clear that there is a definite

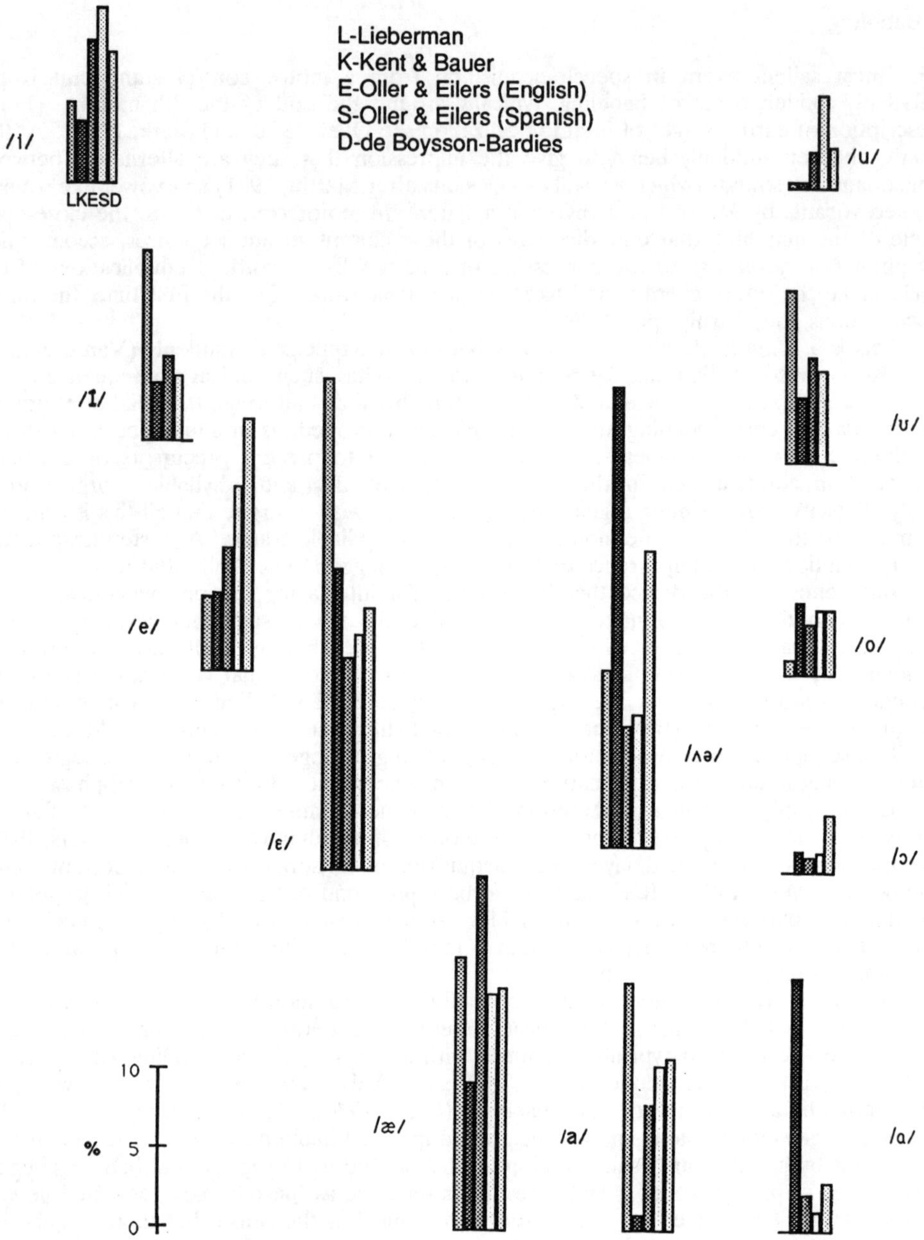

Figure 1. Percentages of various vowels transcribed in 4 studies of babbling.

hierarchy of sound preferences in babbling (see Locke, 1983). Favoured supraglottal closants are labial and apical stops, (voiceless unaspirated) glides and nasals (see Locke, 1983). A note on the description of stop consonants is appropriate at this point. The first sounds of the English words "bill" and "dill" are labial and apical voiceless unaspirated stops respectively and are phonetically transcribed in square brackets as [p] and [t]. However it is a confusing fact about English that such sounds are called voiced from a phonemic standpoint, and transcribed phonemically in angled brackets as /b/ and /d/. They are called voiced at the phonemic level because they contrast with a pair of other stops that have less voicing than they do, and are consequently called voiceless - namely the first sounds of the words "pill" and "till". These latter sounds are voiceless aspirated stops and are transcribed phonetically and [pʰ] and [tʰ], but phonemically they are called voiceless and transcribed as /p/ and /t/.

The pattern of vocant preferences during the babbling period has received little attention, as has the acquisition of vowel production in general. As a step towards increasing the amount of attention given to vowels, we will consider vocant preferences in babbling in some detail. Figure 1 is a summary of aspects of 4 studies of vocant preferences during babbling which classified vocants by means of phonetic transcription. The studies, and the respective linguistic environments of the infants, were as follows (Number of subjects is given in parentheses); Lieberman, 1980, English (1); Kent and Bauer, 1985, English (5); Oller and Eilers, 1982, English and Spanish subgroups (8 and 8); and de de Boysson-Bardies, Sagart and Bacri, 1981, French (1). (Lieberman also reported acoustical data). The vowel symbols in Figure 1 were used in all studies except that de Boysson-Bardies et al did not use [ɪ] and [ʊ], Lieberman did not use [ɑ], [ə] and [ɔ], and Kent and Bauer did not use [ə] in analyzing vowels in CV syllables (approximately one-half of their corpus). The symbols [ʌ] and [ə] are grouped together because they were grouped together by Kent and Bauer and by Oller and Eilers. Some phonetic categories used by some investigators were omitted from the figure as they were used in a negligible proportion of occasions. In addition the following steps were taken with the results of de Boysson-Bardies et al. Vocants transcribed as high front rounded vowels [y] and [ø] were included with their unrounded counterparts [i] and [e], and the high back unrounded vowels [ɤ] and [ɯ] were included with their rounded counterparts [u] and [o]. The low back rounded vowel [ɒ] was included in the category [ɑ]. The symbol [a] designated as a low front vowel, was placed in the category [æ], and the symbol [A] designated as a low central vowel was placed in the category [a].

Results are relatively consistent across studies and language environments. The only major discrepancy, the high frequency of [ɑ] coexisting with a very low frequency of [a] in the Kent and Bauer study presumably arises from a different transcription convention. The main trend is a concentration of preferences in the lower left quadrant of the vowel space. There is a secondary trend towards preference for high front vowels. High and low back vowels are strongly disfavoured. Additional evidence suggests that the preference for high front vowels develops in later stages of babbling, suggesting that the only early babbling preference is for the lower left quadrant of the vowel space. The mean of the preferences for the high front vocants [i] and [ɪ] in Lieberman's subject increased from approximately 5 to 10% from the first to the second half of the longitudinal study, while the mean of the preferences for [ɛ] and [æ] reduced from about 30% to about 20%. Lieberman (1980) states that a similar trend was noted by George (1978) in a study of 29 children (see also Labov and Labov, 1978) .

It follows from this description of babbling preferences that the individual episodes of reduplicated babbling can contain one type of vocant and closant on one occasion and other types of vocant and closant on other occasions. How can this variety of perceived sound qualities, from occasion to occasion, be reconciled with the claim that reduplicative babbling consists of pure frames simply produced by mandibular oscillation? Does this variety mean that there are a number of pure frames ? We suggest instead the following possibilities. 1. vocant sounds that differ across utterances in height, and closant sounds that differ in amount of constriction are produced primarily by differences in the amplitude of mandibular oscillation from one utterance to the next; 2. closants that differ across utterances in place of constriction, and vocants that differ on the front-back dimension are a joint result of either particular resting positions of articulators other than the mandible, or non-rest articulatory presettings that are basically maintained through the entire utterance. This maintained position results, for example, in the illusory impression that there are two gestures of the tongue apex itself in [tata] just as there would be in adults while in fact the tongue was preset in the apical position and simply maintained in that position throughout the utterance. The impression that the [t] was produced twice may thus be due to the fact that neither the position of the tongue nor the amplitude of mandibular oscillation changes during an utterance.

The consensus in the literature that babbling is first reduplicative, then variegated, led us to suggest a scenario for this period which was "frames first, then content" (MacNeilage and Davis, 1989). Elbers (1982) has reported a longitudinal study of a single child who exhibited the sequence, reduplicative babbling, then variegated babbling. However there is some question as to how typical this sequence is. In a study of 10 children, Smith et al (1989) found practically as many variegated utterances as reduplicative utterances during the 6-9 month period, and no significant change in these proportions in the 10-13 month period.

Whether this study shows that there is no initial reduplicative stage at all in some or all of these children is not clear. In the study each child was sampled only once in each period, and some children may have already made a transition from an initial reduplicative stage to a mixed reduplicative and variegated stage during the age period of 6-9 months before their output was sampled. But it does show that simply regarding 8-10 months as a reduplication stage and 10-13 months as a variegation stage is an oversimplification. However this is resolved, we believe that the conception of pure frames may not only remain appropriate for reduplicated babbling, but also provide a principled approach to the understanding of the nature of variegation. Let us consider further the implications of the conception, first for reduplicative babbling, and then for variegated babbling.

One prediction from the characterization of reduplicated babbling as pure frames is that there should be extensive coarticulation between vowels and lingual consonants from the beginning of babbling, because no tongue movement independent of the mandible is predicted during an utterance. In a case study of a single child, during the period 14-20 months (Davis and MacNeilage, 1989) we observed a tendency to favour co-occurrence of front vowels with apical stops, and back vowels with velar stops. We have also noted a tendency for preferential co-occurrence of [t] with [æ] which contrasts with a co-occurrence of [p] with [a] in the initial babbling of one subject (unpublished observations). In addition, Stoel-Gammon (1983) has noted an affinity between apical consonants and front vowels in the form of a number of cases in the literature in which apical stops were substituted for labial stops in front vowel environments.

The possibility that there are some inherent coarticulatory constraints between vowels and consonants is of interest from the standpoint of adult speech motor control because the question of coarticulation has been perhaps the main single question in the area. Kent (1983) has suggested that early speakers lack coarticulation after failing to observe adult-like formant transitions into the post vocalic consonant in the word "box" in 4 year old children. However a lack of a formant transition may be evidence for coarticulation insofar as it means that different movements are not being made for the two segments. Nittrouer et al (1989) have noted that younger children have more coarticulation between fricatives and following vowels.

As to variegated babbling, our position suggests that we should apportion the variance due to frame and content within variegated utterances in the same way as we did across different reduplicated utterances. That is, 'vertical' changes of vocant height and degree of constriction of closants within a variegated utterance would be due to the frame variable while changes in place of articulation along the 'horizontal' (front-back) axis would be due to content factors. At early stages of the development of syllabic vocalizations we would expect vertical adjustments to predominate over horizontal adjustments. Evidence consistent with this possibility has been obtained by Hodge, Kent and Forrest (1989). They studied formant frequency displacements during the production of series of [dɑ] syllables in adults and a number of groups of children the youngest of which ranged from 14 to 22 months of age. They concluded that "When these patterns of displacement for the first and second formant are combined, one might infer that the younger children are accomplishing the opening and closing gestures with a relatively greater contribution of mandibular rather than lingual movement while, conversely, the adults and the older children are using a greater relative contribution of the tongue versus the jaw."

To the extent that one would wish to embroider the basic hypothesis one could develop a number of predictions:
Two examples; 1. As the frame component is most closely linked with suprasegmental control, one might expect changes in amplitude of the mandibular cycle to covary with acoustic amplitude (beyond that expected due to intrinsic factors) as Labov and Labov (1978) reported in one subject; 2. A 'limited resource' assumption would predict that simultaneous variation in both the horizontal and the vertical axis (e.g. [dɪd] or [dɪdu]) would be rare, at least initially. In any event, the framework could perhaps assist in the interpretation of various results when the requisite level of detail of analysis is reached.

3. Early Speech

Children produce their first words around one year of age, and for the next 2 1/2 years or so they move slowly towards sound patterns of adult words that seem correct to a listener. A good deal of work has been done on this stage of acquisition (for reviews see Macken and Ferguson, 1983; Menn, 1983). Even though the importance of <u>motor</u> constraints in determining the infants' versions of adult words is widely conceded, the dominant perspective in the area has been one of linguistic theory; "Generally the formalisms of linguistic theory were adopted for the analysis of child data and in large part dictated not only the questions of interest but also the methods of data collection and analysis. In short, the focus of these analyses was on purely linguistic regularities in the child corpus". Macken and Ferguson, 1983, p 255). A great deal of attention has been devoted to the formulation of rules or the characterization of "processes" (Stampe, 1969) applying to an input representation consisting of the correct adult word, to produce an output which is the child's word.

In Table 1, some of the most common generalizations arising from this approach are given on the left hand side, accompanied by an example of the child's production. Note that in all of these examples except number 6, the unit in terms of which the rules are expressed is either the segment or a subsegmental attribute. Although this is not explicitly stated, such descriptions (sometimes even regarded as explanations) give the impression that segments and features are independently manipulable by infants. If true, this would be important information, bearing on the main question of this paper. However it is notable that all of the generalizations in Table 1 can be derived from the very brief summary of babbling propensities given earlier. On the right hand side of the table we have listed an alternative babbling-oriented set of descriptions for the observed trends. These descriptions make it seem unlikely that these children are applying rules to segmental or subsegmental inventory items in order to derive their own words from adult forms. They are simply approximating adult forms with the best fit to their babbling-derived motor repertoires, and

Table 1. Typical ways in which sound patterns of infant words differ from adult patterns.

Traditional Description	Example	Babbling-derived Description
1.Assimilation	"duck" [dʌk]→[gʌk]	Reduplication preference (place)
2.Deaspiration	"top" [tʰap]→[tap]	Unaspirated preference
3.Substitutions:		
a)stop for fricative	"sock" [sak]→[tak]	Closant preference
b)glide for liquid	"lip" [lɪp]→[yɪp]	
4.Final consonant deletion	"bat" [baet]→[bae]	CV preference
5.Cluster simplification	"stick" [stɪk]→[tɪk]	Single consonant preference
6.Weak syllable deletion	"banana" [bənaenə]→ [naenə]	Rhythmic uniformity preference

these performances are, on the face of it, uninformative as to whether they include independent control over segments or subsegmental units.

Another reason why much of the work done during this period serves implicitly to provide possibly spurious evidence for segmental independence is that virtually all the work has been done on consonants and not on vowels. Because the most typical speech at these ages involves an alternation between consonants and vowels, the lack of attention to vowels means that the question of the context sensitivity of consonants to vowels and of vowels to consonants (and the complementary question of the context-independence of each from the other) has not been systematically addressed.

One way of looking at developments during this period from a frame/content perspective is in terms of the way in which the child responds to the high frequencies of variegation required in adult words. If one takes adult English disyllabic words, for example, an analysis of the Daniel Jones Pronouncing Dictionary shows that 18% of consonant pairs separated by a vowel reduplicate place of articulation. Only 3% of vowel sequences are reduplications. Presumably a child only reaches these levels of variegation when virtually all the words in the lexicon are correctly pronounced, at the age of about 3 1/2 years. Kent and Bauer (1985) found that five children at the age of the first words (13 months) were producing an average of 30% of disyllables which were completely reduplicated, another 30% in which only consonants were reduplicated and a further 14% in which the vowels were reduplicated. Thus about 3/4s of the utterances had reduplication of at least one of the major class content elements. As sound patterns of babbling are closely related to those of the first words, these children still had a long way to go to reach adult levels of variegation.

We suggest that the relative frequency of reduplication in multisyllabic words can be taken as an index of the degree to which infants are using pure frames to approximate adult words. Reduplication of both vowels and consonants in an utterance is the clearest case. Reduplication of one major class element without the other could be a modification of frame structure if it involves variegation in the vertical dimension, and a modification at the content level if it involves variegation in the horizontal dimension. An example of frame modification would be [mamə] in contrast to [mama]. As an example of content modification, the child we studied made heavy use of [dɪ du] for "Big Bird" for a couple of months. (Davis and MacNeilage, 1989).

The ratio of variegation to reduplication may not only be a useful index of the course of development of individual children, but a useful basis for comparison of different children at comparable stages of lexical growth. A number of writers have distinguished between two modes of infant speech designated Gestalt and Analytic by Peters (1967). The Gestalt child seems to successfully approximate prosodic contours including syllable structure, with little attention to the correctness of segments. These children readily simulate multisyllabic utterances. The Analytic child tends to be more successful in producing a range of different segmented types, but with fewer multisyllabic utterances.

The hypothesis that frame control is responsible for variation in the vertical dimension is obviously an oversimplification. If taken to its extreme, frame control would be accorded a major role in the precise adjustments in the vertical dimension required for fricatives, and this is probably not appropriate. Frame control is probably not normally responsible for fine details of movement. A useful analogy (though probably not a homology) can be made between the control of the phonological level of speech production and certain kinds of bimanual coordination in which the nonpreferred (frame) hand serves in a supporting role for the dominant (content) hand (MacNeilage, 1986). As Guiard (1988) has described it, the two hands often cooperate as two motors assembled in series, with the variable position of the left hand serving as a reference position for the right hand. In these situations (e.g. the golf swing) the movements of the frame have a longer time constant than content movements. This is the image that we wish to project for phonological production, and it fits the evolutionary sequence of more gross control followed by more fine control that we have attempted to develop. Content components 'ride' on a carrier composed of the frame, and provide phasic local modulations or modifications.

We have suggested a number of ways in which these local modifications might develop during speech acquisition (MacNeilage and Davis, 1989). These include changes in relative timing of two events (as in development of new voice onset time categories - see Macken and Barton, 1980) increase in amount of muscle contraction (more inferior genioglossus contraction gives a higher, more front vowel) temporospatial modulation of contraction (more tongue elevation following a high front vowel may give an alveolar fricative) and new movements (unrounded vowels become rounded). Most importantly, these modifications must be introduced at the right moment in relation to the basic close-open alternation of the mandible. In addition, at the end of the segment to which they are relevant, these effects may often need to be actively eliminated by compensatory muscle contraction (if we are to judge by adult coarticulation results) in order for the following segment to be produced correctly.

One problem in understanding the organization of speech which does not exist for bimanual coordination is that speech does not have anatomically separate frame and content components which can therefore be observed separately. In normal adult speech, these two components are smoothly combined. Consequently, in order to understand them they must be picked apart by statistical analysis, study of errors, consideration of phylogeny and pathology, and we hope, a search for their relative roles in ontogeny, before they are optimally combined for adult-like speech.

4. Later speech

In children who have begun to increase their vocabulary beyond about 50 words, there are numerous instances of steps in phonological development that would not be expected simply on the basis of choice of approximations to adult words from the babbling repertoire. One subclass of these instances are those in which the child seems to develop a nonreduplicative multisyllabic whole-word template which is used to approximate a number of different adult words, often doing considerable violence to the adult phonotactic pattern. For example, Macken (1978) reports a Spanish child who favoured the sequence of word-initial labial consonant and then alveolar consonants in words, so strongly that a number of very differently formed adult words were subject to this constraint. Examples;

"sopa" [pota]

"Telefono" [fantino]

Priestly (1977) reports a child who favoured a CVyVC sequence.
Examples;

"sucker" [fayak]

"hanger" [hayaŋ]

"elephant" [eyat]

"lizard" [ziyan]

We would like to describe the basic form of these utterances as "special purpose nonreduplicative frames" and suggest that the basic form is being generated by the medial control subsystem. But however they are described, they constitute evidence for the existence of a dichotomy between a general-purpose suprasegmental support structure and special-purpose segmental level units at stages well beyond the onset of babbling. Analysis of the latter components can lead to specific conclusions regarding segmental independence in these subjects.

Another phenomenon sometimes observed beyond the 50 word stage has been called "overgeneralization" (Macken and Ferguson, 1983). In these cases "the child extends a role for a particular set of sounds or words to a set of sounds or words where the extension is inappropriate from the adult's view." (p 267) Macken and Ferguson report the following sequence;

1. [r] is "deleted" in initial stop plus [r] clusters.

e.g. "pretty" [pʰɪ tʰi]

2. Initial [tr] and [dr] are produced as [f].

The authors suggest that " the strong aspiration (we would say affrication) combined with lability (in the [r]) in adult production of /tr/ and /dr/ led to the child's decision to switch from [t] to [f] productions for /tr/ and /dr/."

3. [f] is used for all initial /pr/, /tr/, /dr/ and /kr/ words.

Phenomena such as those noted above from Priestly, Macken, and Macken and Ferguson, have led to an increasing emphasis on cognitive factors in speech acquisition. We fear that just as the formalisms of linguistic theory have overshadowed motor factors in past approaches to speech production acquisition, so will a cognitive emphasis overshadow them in current approaches. This is not to say that motor factors are the only relevant ones in theories of speech production acquisition but only to say that an adequate theory in this domain must explicitly take them into account. Some elaboration of this claim is necessary. When a child who possess the word "big" [bɪ] says [m ɪ mu] for "big moose" (Matthei, 1987) it is clear that the [m] in [mɪ] is appearing as a result of a motor constraint favouring reduplication. It is equally clear that the child has an internal linguistic representation for the lexical item "big" which is closer to [b ɪ] than to [m ɪ]. Consequently it is necessary to postulate some independence between motor and lexical levels in this child's speech production, just as it is in the other examples given in this section, and numerous others in the literature (for a discussion of this issue see Menn, 1983).

The term "cognitive" applies most readily to the lexical level of this the motor-lexical dichotomy. However we wish to draw attention to the motor level here not only because such examples establish its independent status, but because the particular form of the translations it demands from lexical representations must have important implications for the nature of lexical structure itself. For example, in the case discussed here, motor factors would seem to force a [b] - [m] relationship (or a [b ɪ]-[m ɪ] relationship on the lexicon. And in Macken and Ferguson's case of overgeneralization by cluster simplification, motor factors were obviously influential in determining the functional unit of lexicomotor interfacing. Even the choice of a labial-alveolar sequence in Macken's subject probably had a motor motivation. In our case study (Davis and MacNeilage, 1989) the labial-alveolar sequence was the favourite nonreduplicative consonant sequence, being produced about 3 times as often as the labial-alveolar sequence in both speech and concurrent babbling. With regard to Priestly's child, it has been noted that continuants are relatively favoured in intervocalic position in infant speech, presumably for motor reasons. We suggest that motor factors exert crucial influences throughout the entire span of the speech acquisition process. In fact, given the availability of motor evidence, relative to perceptual evidence, regarding later stages of speech acquisition, such evidence may be the best means of determining how sound-related aspects of a child's mental lexicon are formed.

References

Davis, B.L. and MacNeilage, P.F. (1989) Acquisition of correct vowel production: a quantitative case study, Journal of Speech and Hearing Research, (In press).

de Boysson-Bardies, B., Sagart, L., and Bacri, N. (1981) Phonetic studies of late babbling: a case study of a French child, Journal of Child Language, 8, 511-524.

Elbers, L. (1982) Operating principles in repetitive babbling: a cognitive continuity approach, Cognition, 12, 45-63.

George, S.L. (1978) The relationship between cranial base angle morphology and infant vocalizations, Doctoral Dissertation, University of Connecticut, Storrs.

Green, S.L. (1975) Variations of vocal pattern with social situation in the Japanese Monkey (Macaca fuscata): A field study, In L.A. Rosenblum (Ed.) Primate Behaviour: Developments in Field and Laboratory Research, Vol. 4, New York, Academic Press.

Guiard, Y. (1988) The kinematic chain as a model for human asymmetrical bimanual coordination, in A. Colley and J. Beech, (Eds.) Cognition and Action in Skilled Behaviour, Amsterdam, North Holland.

Hodge, M., Kent, R. and Forrest K. (1989) Phonetic constants in phonological development, Paper presented at Child Phonology Conference, Evanston, Illinois, May 1989.

Holmgren, K., Lindblom, B., Aurelius, G., Jalling, B. and Zetterstrom, R. (1986) On the phonetics of infant vocalization, in B. Lindblom and R. Zetterstrom (Eds.) Precursors of Early Speech, New York, Stockton Press.

Jakobson, R. (1968) Child language, aphasia and phonological universals, The Hague, Mouton, 1968.

Jonas, S. (1981) The supplementary motor area and speech emission, Journal of Communication Disorders, 14, 349-373.

Jurgens, U. (1987) Primate communication: Signalling, Vocalization, In G. Adelman (Ed.) Encyclopedia of Neuroscience, Vol.2 Boston, Birkhauser.

Jurgens, U., Kirzinger, A. and Cramon, D. von. (1982) The effects of deep reaching lesions in the cortical face area on phonation: A combined case report and experimental study, Cortex, 18, 125-140.

Kent, R.D. (1983) The segmental organization of speech, In P.F. MacNeilage, (Ed.) The Production of Speech, New York, Springer Verlag.

Kent, R.D. and Bauer, H.R. (1985) Vocalizations of one year olds, Journal of Child Language, 1985, 12, 491-526.

Labov, W. and Labov, T. (1978) The phonetics of cat and mama, Language, 54, 816-852.

Lieberman, P. (1980) On the development of vowel production in young children, In Yeni-Komshian, G., Kavanagh, J.F. and Ferguson, C.A. (Eds.) Child Phonology: Volume 1; Production, New York, Academic Press, 1980.

Lindblom, B., MacNeilage, P.F. and Studdert-Kennedy, M.G. Evolution of Spoken Language (In preparation).

Locke, J. (1983) Phonological Acquisition and Change, New York, Academic Press.

Macken, M.A. (1978) Permitted complexity in phonological development: one child's acquisition of Spanish consonants, Lingua, 44, 219-253.

Macken, M.A. and Barton, D. (1980) A longitudinal study of the acquisition of-the voicing contrast in American-English word-initial stops, as measured by voice onset time, Journal of Child Language, 7, 41-74.

Macken, M.A. and Ferguson, C.A. (1980) Cognitive aspects of phonological development: Model, evidence and issues, In K.E. Nelson, (Ed.) Children's Language, Volume 4. Hillsdale, N.J., Erlbaum.

MacNeilage, P.F. (1986) Bimanual coordination and the beginnings of speech, in B. Lindblom and R. Zetterstrom (Eds.) precursors to Early Speech. New York, Stockton Press.

MacNeilage, P.F. (1987) Speech, motor control, In G. Adelman, (Ed.) Encyclopedia of Neuroscience, Vol. 2. Boston, Birkhauser.

MacNeilage, P.F., Studdert-Kennedy, M.G. and Lindblom, B. (1985) Planning and Production of speech: an overview, in Planning and Production of Speech in Normally Hearing and Deaf People, J. Lauter (Ed.) ASHA Reports, 15-21.

MacNeilage, P.F., and Davis, B.L. (1989) Acquisition of speech production: Frames, then content. In M. Jeannerod (Ed.) Attention and Performance, XIII; Motor representation and control.

Martin, J.A.M. (1981) Voice, Speech and Language in the Child: Development and Disorder, New York, Springer-Verlag.

Matthei, E.H. (1987) Crossing boundaries: More evidence for phonological constraints on early multi-word utterances. Paper presented at the annual meeting of the Linguistic Society of America, December 1987.

Menn, L. (1983) Development of articulatory, phonetic, and phonological capacities, In Butterworth, B. (Ed) Language Production 2: Development, writing, and other processes, New York, Academic Press.

Mills, A.E. (1983) Acquisition of speech sounds in the visually handicapped child, in A.E. Mills (Ed.) Language acquisition in the blind child: normal and deficient, London, Croom Helm.

Nittrouer, S. Studdert-Kennedy, M.G. and McGowan, R.S. (1989) The emergence of phonetic segments: Evidence from the spectral structure of fricative-vowel sequences spoken by children and adults. Journal of Speech and Hearing Research, (In Press).

Oller, D.K. (1980) The emergence of the sounds of speech in infancy, In G.H. Yeni-Komshian, J.F. Kavanagh, and C.A. Ferguson (Eds.) Child Phonology, Vol 1. New York, Academic Press, 1980.

Oller, D.K. and Eilers, R.E. (1982) Similarity of babbling in Spanish and English-learning babies. Journal of Child Language, 9, 565-577.

Oller, D.K. and Eilers, R.E. (1988) The role of audition in infant babbling, Child Development, 59, 441-449.

Penfield, W. and Welch, K. (1951) The supplementary motor area of the cerebral cortex, A.M.A. Archives of Neurology and Psychiatry, 66, 289-317.

Peters, A. (1977) Language Learning Strategies: Does the Whole Equal the Sum of its Parts, Language, 53, 560-73.

Ploog, D. (1979) Phonation, emotion, cognition, with reference to the brain mechanisms involved, Brain and Mind: Ciba Foundation Series 69, Excerpta Medica 90, 79-98.

Priestly, T.M.S. (1977) One idiosyncratic strategy in the acquisition of phonology, Journal of Child Language, 4, 45- 66.

Redican, W.K. (1975) Facial expressions in nonhuman primates, In Rosenblum, L.A. (Ed.) Primate Behaviour: Developments in Field and Laboratory Research, Vol.4 New York, Academic Press.

Smith, B.L., Brown-Sweeney, S. and Stoel-Gammon, C. (1989) A quantitative analysis of reduplicated and variegated babbling, Journal of Child Language, (In Press).

Stampe, D. (1969) The acquisition of phonetic representation, Paper presented at the 5th regional meeting of the Chicago Linguistic Society, Chicago, Ill.

Stark, R. (1980) Stages of speech development in the first year of life, In Yeni-Komshian, G., Kavanagh, J.F. and Ferguson, C.A. (Eds.) Child Phonology: Volume 1; Production, New York, Academic Press.

Stoel-Gammon, C. (1983) Constraints on consonant-vowel sequences in early words, Journal of Child Language, 10, 455-457.

van der Stelt, J.M. and Koopmans van Beinum, F.J. (1986) The onset of babbling related to gross motor development, In B. Lindblom and R. Zetterstrom (Eds.) Precursors to Early Speech, New York, Stockton Press, 1986.

SEGMENTAL REDUCTION IN CONNECTED SPEECH IN GERMAN:
PHONOLOGICAL FACTS AND PHONETIC EXPLANATIONS

K. J. KOHLER
Institute of Phonetics and
Digital Speech Processing
University of Kiel
Olshausenstr. 40
D-2300 Kiel
West Germany

ABSTRACT. This paper deals with /r/ vocalization, weak forms, elisions and assimilations in connected speech in German and provides phonological rules within the generative framework for the derivation of any observable reduction forms, given certain phonetic environments. Some syntactic constraints are also discussed. This phonological description is then supplemented by an enquiry into the production and perception constraints of real speakers and real listeners to find answers to the questions as to why certain articulatory modifications occur and others do not, and why they take particular directions. This phonetic explanation is sought in a minimization of energy expenditure in relation to the perceptual demands of the communicative situation.

1. INTRODUCTION

The Sanskrit grammarians provided a very elaborate corpus of rules of combination, rules of *putting together*, i.e. *Sandhi* rules [1], distinguishing those that apply to the internal make-up of a word, by the addition of derivative and inflectional endings to roots and stems, from the ones that apply to the more external putting together of stems to make compound stems, and the yet looser and more accidental collocation of words in the sentence. In both classes, the general principles of combination are the same, namely hiatus, deaspiration, assimilation, which provides by far the greatest body of combinatorial changes in Sanskrit, extension and abbreviation of consonant groups, permitted finals in words, increment and decrement, i.e. strengthening or weakening processes. These Sanskrit rules were described as synchronic sound structure rules.

The diachronically oriented linguists of the 19th and the early 20th century worked with the same rule types, especially assimilation, strengthening and weakening, but now in the diachronic perspective (of historical sound change), which, by subsuming individual instances of

W. J. Hardcastle and A. Marchal (eds.), Speech Production and Speech Modelling, 69–92.

phonetic changes in words under these general principles, purported to explain them. The more fundamental and far more interesting explanatory question as to why particular sound changes occur whereas others do not was, of course, not solvable by this attribution of tokens to types of change. The static surface phonetic descriptions attached to these categories of change, in terms of places and modes of articulation as well as types of phonation, as for instance in Grammont's *Traité de Phonétique* [2], did not capture the dynamic processes of speech production and its variability, as constrained by the human vocal apparatus and its degrees of freedom, but simply constituted identification tags for changes in the graphic representation of speech phenomena.

Besides this dominant stream of historical linguistics we find the development of practical phonetics in connection with foreign language teaching, which led to the foundation of the International Phonetic Association in 1886. Learning modern foreign languages beside or instead of Latin and Greek also made a concern for pronunciation mandatory, at the level of isolated words as well as their concatenation in utterances. So assimilations and reduction phenomena were studied in the European languages, especially in English because the English School of Phonetics from Henry Sweet and Daniel Jones to their later successors assumed a leading role in this development of practical phonetics. But its association with foreign language teaching set the following severe limitations to the scientific goals.

1. The correct empirical observation and description of speech phenomena were at stake, not their explanation. Practical phonetics did not, and did not want to, explain why the observed reductions occurred and no others.
2. Empirical observation and description were impressionistic, based on the techniques of ear and articulation training. Instrumental and experimental techniques played a very minor role.
3. Although it is extremely important for learners of a foreign language to learn the rules that govern the use of phonetic reductions - namely sociolinguistic, stylistic, pragmatic situational, semantic, syntactic and phonetic contextual constraints - practical phonetics of the English School largely limited itself to the ad hoc enumeration of instances of assimilations, elisions and weak forms. It was left to the learner to internalise the rules in the practical performance by induction. This was partly due to educational principles, but largely to the anchoring of practical phonetics in the tradition of English linguistic scholarship, which, until quite recently, excelled in the provision of large data bases, of which Murray's *Oxford English Dictionary* and all its followers and derivatives are outstanding examples, and which treated the elaboration of rules for linguistic behaviour as something marginal, rather than central, and often with great suspicion.
4. The orientation towards the foreign language learner also necessitated the restriction to standard social and stylistic forms of pronunciation, imparted under classroom conditions, which lack the more extreme types of phonetic reductions. Therefore the lists of observed phenomena of assimilations, elisions and weak forms in the

standard works by Jones [3] and Gimson [4] are highly incomplete. So, although practical phonetics dealt with the synchronic aspects of sound reduction it did not do so adequately and it provided no more illuminating accounts than did diachronic linguistics.

With the advent of phonology and the polarization of the study of spoken language into an arts discipline called phonology, and a science discipline called phonetics, the latter was relegated to an ancillary function of filling the primary phonological categories with phonetic substance. The explanation of language phenomena became the main goal, but it rested on phonological structures, established according to linguistic principles such as otherness, i.e. differentiation rather than identification, not on measurable phonetic phenomena, which were denied any explanatory power in their own right in linguistic analyses. This standpoint received its most extreme and at the same time formalised expression in generative phonology. The dichotomization of language into performance and competence, and the restriction of linguistics proper to the latter excluded, or at least confined to a minor place, all enquiries into phonetic variation determined by sociolinguistic, physiological and perceptual factors, lying outside the phonological system. Thus reduction phenomena were not a primary concern.

The sociolinguistic and stylistic studies by Labov [5] and Trudgill [6] stayed outside the mainstream of generative phonology and its various descendants, and questions into the biological foundations of phonological structures and processes did not enter into phonological explanations within generative phonology. In the phonological literature common sound patterns were and still are "explained" by reference to phonology-internal concepts such as "markedness", leading to statements of the sort "obstruents tend to devoice because voicelessness is the unmarked case, i.e. it is normal for obstruents to devoice." As Ohala [7] has rightly pointed out, such statements are worthless as explanations since they reduce to the tautology "X because X". Instead, the <u>explicandum</u> should be referred to <u>explanans</u> categories that are independently motivated, and among these the constraints the human vocal apparatus and hearing exert on observable phonological facts are of prime inportance. Just as there have been departures from the all-embracing dominating paradigm of generative phonology with regard to the sociolinguistic aspects of sound systems, there have been similarly new explorations into their biological foundations in the writings of Lindblom [8-11] and Ohala [7,12-15].

What we need is the application of this new way of thinking to sound reduction phenomena at the sentence level in connected speech in order to provide a principled explanation for their occurrence and manifestation in synchronic sound systems. Questions to be answered are whether segmental reduction is rooted solely in the variability of speech production, governed by an economy of effort principle, and/or whether perceptual constraints intervene to filter out those articulatory processes from among the possible ones that then become established in a speech community. In this connection, we must also give an answer as to whether diachronic sound change follows from the same principles that determine segmental reduction in the synchronic perspective. Sound change may be viewed as a changed sociolinguistic grouping of the phonetic variability caused by biological factors, and as the result of a deviant articulatory reinterpretation of auditory impressions received from speech productions of others, involving successive generations of language users. Synchronic articulatory reductions, on the other hand, occur within the same speakers under various contextual and situational conditions. So the acoustic and perceptual aspects of speech may play a much more prominent role in the diachronic shifts than in the synchronic processes.

I am now going to present reduction rules for German in a series of steps that retrace the historical stages of dealing with reduction phenomena, which I have just presented. I shall begin with ad hoc utterances in careful and in casual pronunciation, list the differences, refer them to general phonological rules covering any German utterance and finally explain their occurrence with reference to production and perception constraints. I shall also include some remarks about syntactic/semantic and sociolinguistic/pragmatic/stylistic conditions, the main stress, however, being on biological constraints.

2. REDUCTION RULES

The reduction processes in connected speech in German have been described in great detail in a number of publications [16-20]. Some of the rules have been incorporated by this author into the INFOVOX [21-25] German text-to-speech system to improve the intelligibility and naturalness of its output. The synthesis sounds very satisfactory, providing additional support to the model derived from the analysis of production data. Let me start the presentation and discussion of this model with a few examples comparing careful and reduced pronunciations in IPA phonetic transcription in relation to standard orthography. The sentences in Table 1 illustrate the four linguistic categorizations of reduction phenomena: */r/ vocalization, weak form, elision* and *assimilation* to be dealt with in 2.1 - 2.4.

2.1. /r/ vocalization

When /r/ is not followed by a vowel it is vocalized to [ɐ]; /ər/ in this context also develops to [ɐ]. This rule still applies when /ə/ is deleted after /r/, as in 'gefahren'.

2.2. Weak form

Function words (pronouns: 'ich', 'du', 'er', 'sie', 'mir'; articles: 'dem' (definite, dative), 'einen' (indef., accusative); auxiliary verbs: 'bin', 'hast', 'hat'; prepositions: 'mit', 'nach'; conjunctions: 'und') are reduced when unstressed. This reduction involves duration changes, opening, centralization and monophthongization of vowels, vowel and consonant elision, and often results in several phonetic realisations of the same lexical item forming a reduction hierarchy, where the degree of reduction correlates with the lowering of the stylistic level and with the degree of familiarity in the communicative situation, with extreme reductions also being less common at utterance beginning or after pause.

TABLE 1. Orthography and IPA transcription of German sentences in careful and reduced pronunciations (stress marks, ' = primary, " = secondary, in front of stressed vowel); English translations

Orthography

(1) Ich bin mit dem Wagen nach Bonn gefahren.
"I went to Bonn by car."
(2) Er hat mir geholfen.
"He has helped me."
(3) Sie hat jedem zehn Mark gegeben.
"She has given each one ten marks."
(4) achtundvierzig
"48"
(5) Hast du einen Moment Zeit?
"Have you got a moment to spare?"

Careful

(1) ʔɩç bɩn mɩt^{h} de:m v'a:gən na:χ b'ɔn gəf"a:ʁən.
(2) ʔe:ɐ hat^{h} mi:ɐ gəh'ɔlfən.
(3) zi: hat^{h} j'e:dəm ts'e:n m'aʁk^{h} gəg"e:bən.
(4) ʔ'aχt^{h} ʔunt f'ɩʁtsɩç.
(5) hast^{h} du: ʔaɩnən mo:m'ɛnt^{h} ts'aɩt^{h}?

Reduced

(1) ç b̥ɩ mɩ m v'a:ŋ naχ b̥'ɔŋ gəf"a:ɐn.
(2) ɛɐb mɐ gəh'ɔlfm̩.
(3) zɩ at ç'e:m ts'e: m'aɐ g̊əg"e:m.
(4) ʔ'aχtnf'ɩɐtsɩç.
(5) hasp m mom'ɛn ts'aɩt^{h}?

For example, the inflected definite and indefinite articles 'dem' and 'einen' undergo the following progressively stronger reductions:

(1) de:m → dem → dęm → dəm → dm → bm → m

→ an → ən
(2) ʔaɩnən → anən → n
→ ənən → nən

The articles 'einem' and 'dem' (dat.) or 'einen' and 'den' (acc.) can be reduced to the same extreme forms [m] or [n] so that there may be coalescence of definite and indefinite articles. In the case of 'Ich bin mit _____ Wagen nach Bonn gefahren.' the probability for the indefinite article is very low, whereas in 'Sie ist mit ______ Freund in Urlaub gegangen.' ("She has gone on holiday with a boy friend.") it is very high because otherwise one would expect the possessive pronoun 'ihrem' instead of the definite article 'dem'. In 'Hast du ______ Moment Zeit.' the definite article is ruled out, whereas in 'Er macht ______ besten Eindruck ("He gives the best impression.") it is the only possible form. So, in all these cases the extreme reductions are tolerated in spite of formal coalescence. When syntactic and semantic redundancies do not weight one word class more highly than the other and when a differentiation is essential for effective speech communication, the reductions stop at points where the phonetic forms are still distinct.

The reduction of 'dem' to [m] is of very old standing because here the orthography has partly followed suit: 'im' ← 'in dem', 'am' ← 'an dem', 'zum' ← 'zu dem', 'beim' ← 'bei dem', 'überm' ← 'über dem', 'unterm' ← 'unter dem'; even 'auf'm' and 'aus'm' are possible spellings in the representation of more colloquial speech. [mɩm] for 'mit dem' is, however, different from these because one would expect [mɩtm] or [mɩpm], i.e. the plosive should still be represented. Its disappearance constitutes the same further development as in 'guten Tag' ("good day"): [gu:tn] → [gɷdn] → [gɷn], which will be dicussed in 3.3.

Since there is an orthographically represented reduction 'zum' beside 'zu dem', the two forms can be used for different syntactic and semantic purposes, the latter in an anaphoric function, as in

(3) 'Er kam zu dem Schluß, daß alles verloren war.'
("He reached the conclusion that everything was lost.")
[tsɷdę̧m → tsɷdəm]

(4) 'Er kam zum Schluß, die andern waren schon da.'
("He arrived last, the others were already there.")
[tsɷm]

(5) 'Nach vielen Abschweifungen kam er endlich zum Schluß.'
("After many digressions, he finally came to the end of his speech.")
[tsɷm]

If there is no anaphoric function, 'zu dem' [tsɷd(ę)m] is impossible.

As these inflected article forms illustrate, the centralization of vowels in function words goes as far as [ə] for /e:/ and /aɩ/, and like any other [ə], it can be elided before /n/. In the case of /e:r/ in 'er' and 'der', the central half-open vowel [ɐ] may result as an extreme reduction, which is more common in the article than in the pronoun. In the pronoun its probability of occurrence is higher in postverbal than in preverbal position (commonly [ɛɐ] in 'er hat' vs. [ɐ] in 'hat er'). That means that preverbal [ɛɐ] occupies the same stylistic position as postverbal [ɐ].

/i:ɐ/ in 'ihr' (poss. pron. or pers. pron. 2nd pers. pl. nom. or pers. pron. 3rd pers. sing. fem. dat.), 'mir' (pers. pron. 1st pers. sing. dat.), 'dir' (pers. pron. 2nd pers. sing. dat.), 'wir' (pers. pron. 1st pers. pl. nom.), as well as /y:r/ in 'für' (prep.) can be weakened to [ɐ], but this does not apply to the possessive pronoun 'ihr' (e.g. 'ihr [ɩɐ] Kleid' "her dress") and is far less likely in the dative than in the nominative personal pronouns (e.g. 'Er hat ihr [ɩɐ] geholfen.' "He has helped her."). With regard to the pre- and postverbal positions of the nominative personal pronouns 'wir', 'ihr' the same holds as for 'er'.

'du' (pers. pron. 2nd pers. sing. nom.) can be reduced to [də] with the same restriction in proclitic position. In 'ihm' /i:m/ and 'ihn' /i:n/ (pers. pron. 3rd pers. sing. masc. dat./acc.) the vowel may develop to [i] → [ɩ] → [ə] and even be lost, but in the latter case a syllabic nasal is retained; 'ihnen /i:nən/ (pers. pron. 3rd pers. pl. dat.) can, however, only be weakened as far as [ɩnn]. The conjunction 'und' /ɷnt/ is generally reduced to [ɷn] and further to [ən], particularly in compound numerals, where [ə] may also be deleted.

2.3. Elision

Elision covers the cases of /ə/ deletion in endings and weak forms, the reduction of geminates to single consonants and the elimination of aspiration between stops, as well as before nasals after /ə/ deletion.

In /ə(nə)n,m,l/ the first /ə/ is deleted, except after a sequence of obstruent + nasal. The deletion is illustrated by the words of Table 2.

TABLE 2. Examples of /ə/ deletion

(a) from the sentences of Table 2
'Wagen' [v'a:gn], 'gefahren' [gəf'a:ɐn]
'geholfen' [gəh'ɔlfn], 'jedem' [j'e:dm]
'gegeben' [gəg'e:bn]

(b) further instances illustrating additional contexts
'ebenen' "even" (inflected) ['e:bənən] → ['e:bnən]
'lernen' "to learn" [l'ɛɐnən] → [l'ɛɐnn]
'zeichnen' "to draw" [ts'aɩçnən]
'Adel' "nobility" ['a:dəl] → ['a:dl]

Both /ə/ of /ənən,m/ can be deleted if the sequence occurs in function words, cf. 'hast du einen' [hastd̥ənən] → [hastnn] in sentence 5 of Table 1.

The reduction of geminates is a general process in German and even applies to those cases that arise through /ə/ deletion if syntactic conditions do not interfere (cf. 2.2.). Geminate reduction is blocked if it causes a semantic confusion by eliminating a function word altogether, e.g. in compound numerals. For example, in the numeral "98" ['aχtnn'ɔɩntsɩç] a geminate is necessary to differentiate it from the short form for the money amount "8 Mark 90 Pfennig", i.e. "8.90" ['aχtn'ɔɩntsɩç] with single [n]. In sentence 5 of Table 1 the [m] representing the function word 'einen' is likely to stay, except in very fast and relaxed speech. The first stop in a sequence of two stops is not aspirated, even across word boundaries. Table 3 lists some instances of geminate reduction and elimination of aspiration.

TABLE 3. Examples of geminate reduction and elimination of aspiration

(a) from the sentences of Table 1
'mit dem' [mιtd̥əm] → [mιd̥əm], hat mir [hatmιɐ]
'Mark gegeben' ['maɐkg̊əg"e:bn] → ['maɐg̊əg"e:bn]
'hast du einen' [hastn]
'Moment Zeit' [mom'εnt ts'aιt] → [mom'εn ts'aιt]

(b) further instances illustrating additional contexts
'lernen' [l'εɐn]
'98' ['aχtnn'oιntsιç]
'Er hat ihnen geholfen.' "He has helped them."
[εɐt ιnən gəh'ɔlfən] → [εɐt ιnn gəh'ɔlfn]
'Er will ihn nehmen.' "He wants to appoint him."
[εɐ vil ιn n'e:mən] → [εɐ vιlṇ n'e:m]

TABLE 4. Examples of assimilation

(a) progressive, place
'Wagen' [v'a:gn] → [v'a:gŋ]
'geholfen' [gəh'ɔlfn] → [gəh'ɔlfɱ]
'gegeben' [gəg'e:bn] → [gəg'e:bm]

(b) regressive, place
'bin mit' [bιn mιt] → [bιmmιt] (→ [bιmιt], see 2.3.)
'mit dem' [mιtm] → [mιpm]
'hat mir' [hat mɐ] → [hap mɐ]
'jedem' [j'e:dm] → [j'e:bm]
'zehn Mark' [ts'e:n m'aɐk] → ['ts'e:m m'aɐk]
(→ [ts'e: maɐk], see 2.3.)
'hast du einen Moment' [hastn mom'εnt] → [haspm mom'εnt]

(c) regressive, manner
'mit dem' [mιpm] (→ [mιbm], see (e)) → [mιmm]
(→ [mιm], see 2.3.)
'Wagen' [v'a:gŋ] → [v'a:ŋŋ] (→ [v'a:ŋ], see 2.3.)
'jedem' [j'e:bm] → [j'e:mm] (→ [j'e:m], see 2.3.)
'gegeben' [gəg'e:bm] → [gəg'e:mm]
(→ [gəg'e:m], see 2.3.)

(d) progressive, devoicing
'ich bin' [ιç bιn] → [ιç b̥ιn]
'hat jedem' [hat j'e:m] → [hat ç'e:m]
'Mark gegeben' [maɐk gəg'e:m] → [maɐk g̊əg'e:m]

(e) regressive, voicing, in unstressed syllables and function words
'mit dem' [mιpm] → [mιbm]

2.4. Assimilation

Assimilation may be progressive or regressive and may affect place and manner of articulation as well as phonation. Table 4 lists tokens of the different types occurring in German.

2.5. Generative rules

2.5.1. Formalism. The reduction phenomena presented as observational data in 2.1. - 2.4. can now be accounted for by a set of ordered rules that generate correctly any weak forms, elisions and assimilations that may occur in Standard German, and thus add descriptive to observational adequacy. Only the most general rules will be given. They need further specifications for syntactic, semantic and pragmatic conditions, and they are, of course, not necessarily applicable to all dialectal varieties of German.

The formalism follows that of generative phonology [26] and has the following characteristics.

1. Rules are of the form

(1) $$X \rightarrow Y \;/\; W \_\_\_\_ Z,$$

where X, Y, W, Z are (strings of) symbols corresponding to sequential sound units and boundaries, and where X is rewritten as Y after the left-hand context W and before the right-hand context Z. If W and Z are missing the rule is a context-free rule. If Y is missing, the rule is a deletion rule, in the absence of X, it is an insertion rule.

2. Each symbol is composed of a phonetic segment or boundary name (e.g. i:, n, ʔ; # for word boundary; V or C for the whole class of vowels or consonants) and binary phonetic features marked as + or - (e.g. + STOP, + NASAL, - VOICE, +/- CORONAL, +/- ANTERIOR, +/- HIGH, +/- LOWSTRESS to separate [ə] from other vowels, +/- OBSTRUENT to distinguish obstruents from sonorants; +/- FUNC, function/ non-function word marker). Features are arranged in vertical order, with the symbol name at the top (when given), and enclosed in square brackets.

(2) $$\begin{bmatrix} \text{-CONT} \\ \text{-NAS} \\ \text{+VOICE} \end{bmatrix} \rightarrow [\text{+NAS}] \;/\; \_\_\_\_ \begin{bmatrix} \text{C} \\ \text{+NAS} \end{bmatrix}$$

3. Symbols in the rule context may be optional and are then enclosed in parentheses. If a symbol is to occur a variable number of times it is followed by (m,n) with m, n=0, 1, 2... The first figure refers to the minimal, the second to the maximal number of repetitions. (,) means "any number of occurrences", (m,) "m or more occurrences".

(3) $$\text{V} \rightarrow \text{ə} \;/\; \begin{bmatrix} \# \\ \text{+FUNC} \end{bmatrix} \text{C(,)} \;\_\_\_\_$$

(4) $$\text{ə} \rightarrow \;/\; \text{V C} \;\_\_\_\; \text{(n (d) ə)} \begin{bmatrix} \text{C} \\ \text{+NAS} \end{bmatrix}$$

4. Several symbols may constitute context alternatives for rule application. The alternatives are enclosed in braces.

(5) $$\text{ə} \rightarrow \begin{bmatrix} \text{V} \\ \\ \text{-LSTR} \end{bmatrix} \;/\; \left\{ \begin{matrix} \begin{bmatrix} \text{C} \\ \text{+VOC} \end{bmatrix} \text{(0,1) C} \\ \text{(C) C} \begin{bmatrix} \text{C} \\ \text{-NAS} \end{bmatrix} \end{matrix} \right\} \;\_\_\_\_\; \text{n}$$

5. To indicate agreement or disagreement between the signs of features in different symbols, variables α, ß, ... are used. This symbolization is particularly useful in assimilation rules.

(6) $$\begin{bmatrix} \text{C} \\ \text{+COR} \\ \text{+ANT} \\ \text{+NAS} \end{bmatrix} \rightarrow \begin{bmatrix} \text{-COR} \\ \alpha\text{ANT} \end{bmatrix} \;/\; \begin{bmatrix} \text{C} \\ \text{-COR} \\ \alpha\text{ANT} \\ \text{+STOP} \end{bmatrix} \;\_\_\_\_$$

6. Rules are applied in ordered sequence, the output of one rule being the input to the following one. In the case of optional contexts, the longest string is tried first, i.e. (4) expands into the ordered sequence

(7a) $$\text{ə} \rightarrow \;/\; \text{V C} \;\_\_\_\_\; \text{n d ə} \begin{bmatrix} \text{C} \\ \text{+NAS} \end{bmatrix}$$

(7b) $$\text{ə} \rightarrow \;/\; \text{V C} \;\_\_\_\_\; \text{n ə} \begin{bmatrix} \text{C} \\ \text{+NAS} \end{bmatrix}$$

(7c) $$\text{ə} \rightarrow \;/\; \text{V C} \;\_\_\_\; \begin{bmatrix} \text{C} \\ \text{+NAS} \end{bmatrix}$$

7. Any one rule is successively tried for application on each symbol of an input string, starting with the left-most one.

2.5.2. System of reduction rules in German

(1) $$\begin{bmatrix} \text{C} \\ \text{+OBST} \\ \text{+VOICE} \end{bmatrix} \rightarrow \begin{bmatrix} \text{-VOICE} \\ \text{+DEVOICED} \end{bmatrix} \;/\; \begin{bmatrix} \text{C} \\ \text{-VOICE} \end{bmatrix} \#\text{(0,1)} \;\_\_\_\_$$

Devoicing of voiced stops and fricatives after
voiceless consonants

(2a) ʝ̊ → ç

(2b) z̥ → s

Coalescence of the devoiced non-labial fricatives with their
fortis counterparts

(3)

$$\text{t}^{\text{h}} \rightarrow \; / \; \begin{bmatrix} \# \\ +\text{FUNC} \end{bmatrix} \; \text{ʊ n} \; \underline{\qquad}$$

[t] elision in the function word 'und'

(4)

$$\text{ʔ} \rightarrow \; / \; \begin{bmatrix} \# \\ +\text{FUNC} \end{bmatrix} \; \underline{\qquad}$$

Elision of glottal stop in function words

(5)

$$\text{ɩ} \rightarrow \; / \; \begin{bmatrix} \# \\ +\text{FUNC} \\ +\text{INITIAL} \end{bmatrix} \; \underline{\qquad} \; \text{ç}$$

Elision of utterance-initial [ɩ] in the function word 'ich'

(6)

$$\text{V} \rightarrow \text{ə} \; / \; \begin{bmatrix} \# \\ +\text{FUNC} \end{bmatrix} \; \text{C(,)} \; \underline{\qquad}$$

Vowel reduction in function words

(7a)

$$\text{ə} \rightarrow \; / \; \begin{bmatrix} \text{V} \\ \\ -\text{LSTR} \end{bmatrix} \left\{ \begin{matrix} \text{C} \\ \begin{bmatrix} \text{C} \\ +\text{VOC} \end{bmatrix} \; \text{C} \\ \text{(C) C} \; \begin{bmatrix} \text{C} \\ -\text{NAS} \end{bmatrix} \end{matrix} \right\} \; \underline{\qquad} \; \begin{bmatrix} \text{C} \\ +\text{NAS} \end{bmatrix}$$

(7b)

$$\text{ə} \rightarrow \; / \; \begin{bmatrix} \# \\ +\text{FUNC} \end{bmatrix} \; \text{(C)} \; \underline{\qquad} \; \begin{bmatrix} \text{C} \\ +\text{NAS} \end{bmatrix}$$

[ə] elision before nasal consonant after a vowel that is not [ə];
in function words the latter restriction does not hold

(8)

$$\text{ə} \rightarrow \; / \; \text{C}\underline{\qquad}\text{l}$$

[ə] elision before [l]

(9)

$$\text{əʁ} \rightarrow \text{ɐ} \;/\; \underline{\quad\quad} \left\{ \begin{matrix} C \\ \# \end{matrix} \right\}$$

(10)

$$\text{ʁ} \rightarrow \text{ɐ} \;/\; \underline{\quad\quad} \left\{ \begin{matrix} C \\ \# \end{matrix} \right\}$$

[ʁ] vocalization before consonant or word-final

(11)

$$\begin{bmatrix} C \\ +ASP \end{bmatrix} \rightarrow \begin{bmatrix} -ASP \end{bmatrix} \;/\; \underline{\quad\quad} \;\#(0,1)\; \begin{bmatrix} C \\ -CONT \end{bmatrix}$$

Elimination of aspiration before stops and nasals (which are both categorised as -CONTINUANT), also across word boundaries

(12)

$$h \rightarrow \;/\; \begin{bmatrix} \# \\ +FUNC \\ -INITIAL \end{bmatrix} \underline{\quad\quad}$$

[h] elision in the various forms of the function word 'haben' when it is not utterance-initial

(13)

$$\text{ə} \rightarrow \;/\; \text{ɐ} \begin{bmatrix} \# \\ +FUNC \end{bmatrix} \underline{\quad\quad}$$

Vowel elision initially in function words after /r/ vocalization in the preceding word, e.g. in 'hat' of 'er hat' and in 'es' of 'hat er es'

(14)

$$\begin{bmatrix} C \\ +COR \\ +ANT \\ -CONT \\ -ASP \end{bmatrix} \rightarrow \begin{bmatrix} -COR \\ \alpha ANT \end{bmatrix} \;/\; \underline{\quad\quad} \begin{bmatrix} C \\ +COR \\ +ANT \\ -CONT \end{bmatrix}(,) \;\#(,)\; \begin{bmatrix} C \\ -COR \\ \alpha ANT \\ -CONT \end{bmatrix}$$

Assimilation of apical nasals and unaspirated stops to following labial or dorsal nasals or stops; cf. the extreme example 'mit bunten Papierschlangen' ("with coloured paper streamers") [mɪt b'ɷntn pap'i:ɐʃl"aŋn] → [mɪpb'ɷmpm pap'i:ɐʃl"aŋŋ].

(15)
$$\begin{bmatrix} \text{C} \\ \text{+COR} \\ \text{+ANT} \\ \text{-CONT} \\ \text{-ASP} \end{bmatrix} \rightarrow \begin{bmatrix} \text{-COR} \\ \alpha\text{ANT} \end{bmatrix} \; / \; \begin{bmatrix} \text{C} \\ \text{-COR} \\ \alpha\text{ANT} \\ \text{-CONT} \end{bmatrix} \; \_\_\_\_ \; \begin{Bmatrix} \# \\ \text{C} \end{Bmatrix}$$

Assimilation of apical nasals and unaspirated stops to preceding labial or velar nasals or stops; e.g. 'Hemden' ("shirts") [h'ɛmdn] → [h'ɛmbm], 'Beamten' ("civil servants") [bəʔ'amtn] → [bəʔ'ampm], 'behaupten' ("assert") [bəh'aɷptn] → [bəh'aɷpm]. This assimilation also occurs after fricatives, but less frequently, e.g. 'geholfen' ("helped") [gəh'ɔlfɱ], 'rauchen' ("to smoke") [ʁ'aɷxŋ], 'achten' ("to pay attention") [ʔ'axkŋ]. To cover these cases as well, -CONT has to be deleted from the rule context.

(16)
$$\begin{bmatrix} \text{C} \\ \alpha\text{COR} \\ \beta\text{ANT} \\ \text{-NAS} \\ \text{+OBST} \\ \varepsilon\text{CONT} \\ \text{-VOICE} \end{bmatrix} \rightarrow \; / \; \_\_\_\_ \; \# \begin{bmatrix} \text{C} \\ \alpha\text{COR} \\ \beta\text{ANT} \\ \text{-NAS} \\ \text{+OBST} \\ \varepsilon\text{CONT} \\ \text{-VOICE} \end{bmatrix}$$

Reduction of double consonants that are voiceless or devoiced

(17)
$$\begin{bmatrix} \text{C} \\ \text{+STOP} \\ \text{-VOICE} \end{bmatrix} \rightarrow \begin{bmatrix} \text{+VOICE} \end{bmatrix} \; / \; \begin{array}{l} \begin{bmatrix} \# \\ \text{+FUNC} \end{bmatrix} \text{C(,)} \\ \begin{bmatrix} \text{V} \\ \text{-STRESS} \end{bmatrix} \begin{bmatrix} \# \\ \text{+FUNC} \end{bmatrix}^{(0,1)} \; \_\_\_\_ \\ \begin{bmatrix} \# \\ \text{+FUNC} \end{bmatrix}^{(0,1)} \begin{bmatrix} \text{-OBST} \\ \text{-STRESS} \end{bmatrix} \end{array}$$

Voicing of plosives in unstressed function words

(18)
$$\begin{bmatrix} \text{C} \\ \text{-CONT} \\ \text{-NAS} \\ \text{-COR} \\ \text{+VOICE} \end{bmatrix} \rightarrow \begin{bmatrix} \text{+NAS} \end{bmatrix} \; / \; \_\_\_\_ \begin{bmatrix} \text{+NAS} \end{bmatrix}$$

Nasal assimilation of voiced non-coronal stops before nasals; coronal stops assimilate far less readily, except in function words, e.g. 'werden' [vɛɐn] vs. 'reden' [ʁ'e:dn].

(19)
$$\begin{bmatrix} C \\ \alpha COR \\ \beta ANT \\ \gamma NAS \\ \partial OBST \\ \varepsilon CONT \\ \zeta VOICE \end{bmatrix} \rightarrow \emptyset \; / \; \_\_\_\_ \; \# \begin{bmatrix} C \\ \alpha COR \\ \beta ANT \\ \gamma NAS \\ \partial OBST \\ \varepsilon CONT \\ \zeta VOICE \end{bmatrix}$$

Reduction of double consonants

2.5.3. Application of the rules. The rules listed in 2.5.2. are certainly more adequate than an ad hoc list of collected reduction forms because they provide a model for the generation of any reduced segmental pronunciations in German, and since the algorithm may be stopped at different points, we get a great variety of alternative forms, all actually occurring in various styles of German. As has been pointed out in 2.2., further refinements and restrictions have to be introduced in accordance with semantic, syntactic and stylistic factors, e.g. different degrees of approximation to [ə] in the function word vowel reduction of rule (7), whereby the competing forms [e·ɐ, eɐ, e̜ɐ, ɛɐ] besides [ɐ] are generated from /e:r/. In certain styles and under particular semantic and syntactic conditions, some of these purely phonetic rules are blocked. These restrictions will have to be investigated and incorporated in the rule system. Ultimately, this will result in a number of modules for segmental reduction which represent different speaking styles and are internally coherent with regard to the stylistic level, combining all the rules that concur to generate a particular style.

Such a generative model is of great value in the development of text-to-speech systems, although it has very little explanatory power beyond its observational and descriptive adequacies, because it does not explain why these rules rather than others occur. But its segmental orientation and linear rule application constitute a powerful device for the successful transformation of strings of alphabetic symbols, through various levels of phonetic transcription, into acoustic patterns. It is irrelevant for this task that human speech production does not proceed along the same lines. What counts is the success of such systems in their practical application, and the English and German versions of INFOVOX, among others, testify to the feasibility of the approach. The concept of stylistic modules has also been introduced into the INFOVOX framework, first for English [27], and it is now being pursued for German.

The working of the rules presented in 2.5.2. will now be illustrated by applying them to some data of Table 1.

1. mɪtʰ de:m v'a:gən
   mɪtʰ d̥e:m v'a:gən (1)
   mɪtʰ d̥əm v'a:gən (6)
   rule (6) is blocked in [mɪt]
   mɪtʰ d̥m v'a:gn (7a,b)

mɩt	d̥m	v'a:gn	(11)
mɩp	b̥m	v'a:gn	(14)
mɩp	b̥m	v'a:gŋ	(15)
mɩ	b̥m	v'a:gŋ	(16)
mɩ	bm	v'a:gŋ	(17)
mɩ	mm	v'a:ŋŋ	(18)
mɩ	m	v'a:ŋ	(19)

2.	ʔe:ʁ	hat^{h}	mi:ʁ	gəh'ɔlfən	
	e:ʁ	hat^{h}	mi:ʁ	gəh'ɔlfən	(4)
	ɛʁ	hət^{h}	məʁ	gəh'ɔlfən	(6) modified
	ɛʁ	hət^{h}	məʁ	gəh'ɔlfn	(7a)
	ɛʁ	hət^{h}	mɐ	gəh'ɔlfn	(9)
	ɛɐ	hət^{h}	mɐ	gəh'ɔlfn	(10)
	ɛɐ	hət	mɐ	gəh'ɔlfn	(11)
	ɛɐ	ət	mɐ	gəh'ɔlfn	(12)
	ɛɐ	t	mɐ	gəh'ɔlfn	(13)
	ɛɐ	p	mɐ	gəh'ɔlfn	(14)
	ɛɐ	b	mɐ	gəh'ɔlfn	(17)

Rules (18), (19) are blocked in [ɛɐbmɐ] (although they are applied in [mɩbm] of 1.) because the syntactic cohesion between the auxiliary verb 'hat' and the dative pronoun 'mir' is weaker than between the preposition 'mit' and the article 'dem'.

3. PHONETIC EXPLANATIONS

In spite of the usefulness phonological observations and generative rules can be put to in, e.g., text-to-speech systems, they do not explain the occurrence of reduction phenomena because they do not refer them to general principles of speech behaviour that determine certain processes and rule out others. To gain these insights into the ways speech functions phonetically, and thus to achieve explanatory adequacy we have to turn to the physiological, articulatory and perceptual constraints of speech and to the social conditions that impose selection and codification on biological potentials. I shall be concerned solely with the former, and shall restrict myself to a discussion of only a few rules under these general phonetic aspects, namely (1) vowel reduction in function words and [ə] shortening, (2) deletion of [ə] and aspiration, and assimilation of place, (3) shortening of consonants: nasalisation, lenition of stops and reduction of geminates.

In dealing with phonetic processes rather than changes of strings of phonetic symbols in a sequence of rules, we can also avoid a serious defect of generative phonology, namely the decomposition of a unitary process into a series of segmental steps. For example, the change from [k^{h}ən] to [kŋ] is achieved by the reorganization of articulatory gestures which eliminates an opening-closing movement involving two articulators - the tongue dorsum and the tongue apex -, replaces it by a long oral closure of the dorsum and transfers the stop release to the velum. This new coordination of articulatory structures and processes results in the simultaneous deletion of aspiration, schwa and apical closure, which is misrepresented by the segmental sequence [kʰən] → [kʰn] → [kn] → [kŋ] provided by the rules.

3.1. Vowel reduction in function words and [ə] shortening

In unstressed function words the distances the articulators travel are reduced to spaces closer to their neutral positions. This results in schwa-like vowel productions, which are at the same time shorter than the unreduced forms they are related to because of the shorter trajectories into and out of the more central vowel targets and because of the general shortening in unstressed position. This follows from the stress-timing principle of a language such as German, i.e. from the tendency to make stressed syllables follow each other at similar intervals [28]. Although this isochrony cannot be complete, it is a strong factor in the language, achieved by the compressibility of function words, and of unstressed syllables in general, especially those containing [ə].

It has been shown [28,29] that [ə] adjusts its duration reciprocally to the duration of a preceding stressed syllable: it is shorter after long than after short syllables. In the phonetic structure C + ə + NASAL, the durational variability of [ə] may be even more extreme and result in the total elimination of the vowel when velic opening and oral occlusion for the nasal are so much advanced in time that they are established at the offset of the preceding C. As regards the timing of nasal offset, it may be equally advanced, or it may stay the same. In the former case, the total duration of the sequence is decreased by curtailing the movement in and out of [ə]; in the latter case, the deletion of [ə] is compensated for by a lengthening of the nasal. Thus the structure C + ə + NASAL + ə may either be reduced to one syllable or remain disyllabic with a syllabic nasal (e.g. 'die verbotene Partei' "the forbidden party" [tʰənə] → [tnə], [tn̩nə]). This temporal plasticity of unstressed syllables is a strong indication of a tendency towards isochronous timing.

3.2. Deletion of [ə] and aspiration, and assimilation of place

In addition to the durational variability of [ə] discussed in 3.1., a reorganization of articulatory gestures may take place in the phonetic structure STOP or NASAL + ə + NASAL if there is not a word boundary before the second nasal [n] or [m]. In such a non-initial position in the word, where articulations tend to be temporally less well defined and less precise (as is also evidenced by sound changes affecting final rather than initial consonants, cf. [15]), the closing movement for the nasal may be completely integrated into the preceding consonantal gesture. This is achieved in two ways:

(a) The oral closure for the final nasal is advanced to such an extent that coarticulation results, i.e. partial overlap of two closures. So the labial occlusion precedes the dorsal release in the type D examples of Table 5, whereby the schwa vowel and also the aspiration are deleted. In the type E examples of Table 5, an opening-closing movement between two identical closure places is simply cut out, without any further adjustments of the articulators.

TABLE 5. STOP (+STOP) + ə + NASAL, NASAL (+ STOP) + ə + NASAL combinations in German

A

labial + apical

[bən] → [bm] 'eben' ("even"), but: 'ebene' (inflected) [bnə], [bm̩nə]

[pʰən] → [pm] 'Lappen' ("cloth")

[ptʰən] → [pm] 'behaupten' ("to assert")

[mən] → [mm] 'kommen' ("to come")

[mdən] → [mbm] 'Hemden' ("shirts")

[mtʰən] → [mpm] 'die Beamten' ("the civil servants")

B

apical + labial

[dəm] → [bm] 'mit jedem' ("with everybody")

[tʰəm] → [pm] 'mit fettem Speck' ("with greasy bacon")

[nəm] → [mm] 'mit schönem Geschwätz' ("with nice blether")

C

dorsal + apical

[gən] → [gŋ] 'Regen' ("rain"), but: 'gelegene' ("situated", inflected) [gnə], [gŋ̩nə]

[kʰən] → [kŋ] 'trocken' ("dry"), but: 'trockene' (inflected) [knə], [kŋ̩nə]

[ŋən] → [ŋŋ] 'die langen Hosen' ("the long trousers")

D

dorsal + labial

[gəm] → [gm] 'mit ewigem Geschwätz' ("with continual blether")

[kʰəm] → [km] 'mit welkem Gras' ("with dried-up grass")

[ŋəm] → [ŋm] 'mit langem Gesicht' ("with a long face")

E

apical + apical, labial + labial

[dən] → [dn] 'leiden' ("to suffer"), but: 'geladene' ("invited", inflected) [dnə], [dn̩nə]

[tʰən] → [tn] 'leiten', ("to lead"), but: 'verbotene' ("forbidden", inflected) [tnə], [tn̩nə]

[nən] → [nn] 'nennen' ("to name")

[bəm] → [bm] 'mit gelbem Papier' ("with yellow paper")

[məm] → [mm] 'mit dummem Geschwätz' ("with stupid blether")

(b) In the type A and type B examples of Table 5 the apical occlusion is superseded by the dominant labial or dorsal one, which is established at the beginning of the sequence and kept right to the end, provided the structure C + ə + NASAL is clearly final, i.e. followed by a consonant or a word boundary, not by a vowel. In the latter case, the articulatory restructuring either follows (a), if [ə] is deleted and the nasal stays non-syllabic (e.g. 'ebene' ['e:bənə] → ['e:bnə]), or - if a syllabic nasal develops (see 3.1) - it follows (b) as far as the nasal syllable nucleus, but leaves an

apical nasal off-glide to the following vowel (e.g. 'ebene' ['e:bənə] → ['e:bm̩nə].
The release of a stop is taken over by velic action. If its timing, as well as that of voice onset, stay the same as in the unadjusted pronunciation, a fortis stop will not have nasal aspiration, because the lowering of the velum and voice onset only occur after the aspiration phase. This in turn means that the fortis stop closure duration is longer before nasal than before oral plosion, which is supported by empirical data.

TABLE 6. STOP, NASAL # STOP, NASAL combinations in German

	A labial + dorsal, dorsal + labial	
[p#g]	'abgeben'	("to hand over")
[p#k]	'abkaufen'	("to buy from")
[k#b]	'Deckblatt'	("cover")
[k#p]	'Packpapier'	("wrapping paper")
[m#g]	'umgehen'	("to go round")
[m#k]	'rumkriegen'	("to win over")
[ŋ#b]	'Sprungbrett'	("diving board")
[ŋ#p]	'Rangplatz'	("rank")
[k#m]	'zurückmelden'	("to report back")
[ŋ#m]	'rangmäßig'	("by rank")
	B labial + apical, dorsal + apical	
[p#d]	'abdrehen'	("to turn off")
[p#t]	'abtreten'	("to retire")
[k#d]	'zurückdrehen'	("to turn back")
[k#t]	'zurücktreten'	("to resign")
[m#d]	'umdrehen'	("to turn over")
[m#t]	'rumtreiben'	("to flit around")
[ŋ#d]	'Gangdecke'	("corridor ceiling")
[ŋ#t]	'Jungtier'	("young animal")
[p#n]	'abnehmen'	("to lose weight")
[k#n]	'zurücknehmen'	("to take back")
[m#n]	'kaum noch'	("hardly")
[ŋ#n]	'das ging nicht'	("it did not work")
	C apical + labial, apical + dorsal	
[t#b] → [p#b]	'mitbringen'	("to bring along")
[t#p] → [p#p]	'mitpfeifen'	("to join in the whistling")
[t#g] → [k#g]	'mitgehen'	("to go with")
[t#k] → [k#k]	'mitkommen'	("to come with")
[n#b] → [m#b]	'anbringen'	("to attach")
[n#p] → [m#p]	'Bahnpreise'	("train fares")
[n#g] → [ŋ#g]	'angeben'	("to declare")
[n#k] → [ŋ#k]	'ankommen'	("to arrive")
[t#m] → [p#m]	'mitmachen'	("to join")
[n#m] → [m#m]	'anmelden'	("to register")

TABLE 6, continued

	D
	labial + labial, apical + apical
[p#b]	'abbitten' ("to apologise")
[p#p]	'abputzen' ("to clean")
[t#d]	'Raddampfer' ("paddle-streamer")
[t#t]	'mittanzen' ("to join in the dancing")
[k#g]	'zurückgeben' ("to give back")
[k#k]	'zurückkehren' ("to return")
[p#m]	'abmachen' ("to take down")
[t#n]	'mitnehmen' ("to take along")
[m#m]	'Strommenge' ("amount of electricity")
[n#n]	'annehmen' ("to accept")

The two types of articulatory restructuring also occur in sequences of stops and nasals across word boundaries. (a) is represented by the type A, B and D examples, (b) by the type C examples of Table 6. The structures of Table 6 differ from those of Table 5 by having a word boundary before the second consonant. Since initial consonants are articulated more precisely than final ones the articulatory reorganization only affects the first, not the second consonant. Otherwise the principle is the same: the apical occlusion is cancelled and superseded by the dominant labial or dorsal one, which is kept right through the sequence thus creating a new integrated consonant gesture. This implies that the release and aspiration of a stop in first position are suppressed, or contrariwise, the articulatory adjustment does not take place if aspiration is present.

The following restrictions apply to the articulatory reorganization in (b):

(1) Only stops and nasals are affected; fricatives do not assimilate (e.g. 'Ausfahrt' ("exit") and 'Auffahrt' ("entry") are always distinguished as [sf] and [(f)f]).

(2) Only apicals are changed; i.e. the sequences LABIAL # DORSAL and DORSAL (#) LABIAL stay unaltered.

(3) Apicals in APICAL # LABIAL, DORSAL are assimilated, whereas the sequences LABIAL, DORSAL # APICAL are not affected.

(4) Apicals in APICAL + ə + LABIAL NASAL and in LABIAL, DORSAL (+ APICAL) + ə + APICAL NASAL are assimilated.

The questions that have to be answered with regard to these restrictions are:

- Why are apical, rather than labial or dorsal gestures eliminated?
- Why is this process restricted to stops and nasals?
- Why does it only occur <u>before</u> labials and dorsals across word boundaries and not to stops <u>after</u> labials and dorsals?

Some of the answers are found in speech production, others in acoustics and perception.

Speech articulation makes continuous use of dorsal tongue and of lip movements in temporally varying ways. Superimposed on these continuously ongoing basic articulatory processes are additional apical tongue movements in certain limited cases, mainly in consonants. Tongue dorsum and lip gestures, in close connection with jaw movements, constitute large, global processes, associated with, e.g. the large extrinsic tongue muscles. Tongue apex articulations, on the other hand, require much more finely controlled and more precisely tuned, and therefore also more costly, adjustments, linked with the smaller intrinsic tongue muscles. When, for perceptual reasons, the fine control is not absolutely necessary and can thus be neglected without harm to the communicative goal, it may be dispensed with under certain situational conditions of, e.g., higher speech rate or relaxed speech, when there is less time for finer control or when the energy expenditure per unit time [9] is reduced to the minimum necessary for a successful speech act. Thus, it is the additional, occasional and special apical gestures, requiring greater precision of timing and muscular coordination than the basic, continuously present labial and dorsal movements, that are weakened and eliminated. This reference to an economy of speech effort within the physiological conditions of the vocal apparatus can explain why only sequences containing apicals are affected by these assimilation processes, and why they do not result in the preservation of the apicals. Thus type D of Table 5 and type A of Table 6 are only coarticulated, not assimilated in either direction, and types A, B, C of Table 5 and type C of Table 6 show the assimilation of apicals to labials and dorsals, not vice versa.

Furthermore, these assimilations do not affect word and syllable initial apicals because especially the initial position in a word has a high signalling value for a listener and thus demands greater articulatory precision from a speaker than the final position. There are thus stronger and more numerous cues as to the identity of place of articulation, particularly for stops, which are differentiated by spectral characteristics of the burst and (in the case of voiceless ones) of the aspiration, over and above the differences in formant transitions. Thus, whereas type C of Table 6 shows assimilation, type B does not, and although there is progressive assimilation in types A and C of Table 5 (besides regressive assimilation in type B), it only applies to nasals and unreleased stops, and an apical gesture stays if a vowel follows. So, we get 'eben' ['e:bm], but 'ebene' ['e:bnə] or ['e:bm̩nə], and always 'Akt' ("act") ['aktʰ], 'Abt' ("abbot") ['aptʰ], 'Beamter' ("civil servant") [bəʔ'amtʰɐ], although [mpm] without oral release and aspiration of [p] is possible in 'Amtmann' ("head of administrative office") and 'Beamten' ("civil servants").

Finally, fricatives are not assimilated under any conditions, because they are acoustically and auditorily far more distinct than nasals and unreleased plosives with regard to place cues so that their articulatory reduction would be too salient and is, therefore, not tolerated.

It has been argued that the synchronic place assimilation processes observable in German can be connected with an articulatory restructuring due to a minimization of energy expenditure which limits the extent of movements and reduces the number of moving organs involved in speech production. So in 'mit dem' becoming [mɪpm] the apical gesture is eliminated altogether and replaced by a labial occlusion, which has to be carried out anyway, and the lowering of the velum, which is necessary for the nasal consonant, now at the same time takes over the plosive release and thus dispenses with an oral opening and closing movement. But all these articulatory reductions are checked by the perceptual demands of the communicative situation. They are only accepted (1) if they bear an auditory similarity to their points of departure and (2) if the situational context does not force the speaker to rate the cost of a misunderstanding or a break-down of communication very high.

In view of this acoustic and auditory corrective, it may be objected that these reduction phenomena are not primarily caused by physiological and articulatory conditions, but are entirely constrained in the acoustic and auditory domains. Ohala has voiced such an opinion with regard to historical sound change involving places of plosives in, e.g., Lat. octo - Ital. otto, Lat. aptu - Ital. atto [13,15], where he assumes coarticulatory adjustments which, because of their acoustic and auditory similarity, are mistaken for apical geminates by the listener and then introduced into production. But I think these cases are not comparable at all to the synchronic processes found in German and in many of other languages, including English. Coarticulated [dm] in 'jedem' is certainly very similar to [bm] and may be mistaken for the latter by a listener, but the same applies to coarticulated [gm] in 'regem' ("alert", inflected) and assimilated [bm]. However, the former process occurs in synchronic assimilations, the latter does not. Similarly, coarticulated [tk] in 'mitkommen' or coarticulated [kt] in 'wegtun' ("to remove") are very similar to [kk] and [tt], respectively, but again only the former constitutes a synchronic reduction process. So there must be something beyond acoustic and auditory similarity that constrains the observable productions. And it is at least very plausible that the driving force lies in speech production itself and in a general economy of effort principle that simplifies complex articulations, e.g. apical gestures, whenever the demands of communication do not impose extra precision on speech production. This means that synchronic reduction processes and diachronic sound changes are not always reducible to the same causes, the former being primarily determined by physiological and articulatory constraints on variability in the speaker, the latter more by transmission from speaker to hearer and by the tranformation of auditory impressions into articulations.

3.3. Shortening of consonants

The nasal assimilations of rule (18) are tied to lenis stops, which have short closure durations. If in fast or relaxed speech the raising of the velum for a stop before an immediately following nasal consonant is weakened because the trajectories are shortened, the short stop closure may be decreased still further and eliminated altogether. This does not happen in fortis stops outside function words, because here the closure, and therefore the temporal safety zone, is much longer. What may happen, however, is a lenition and voicing through a shortening of the closure duration.

In unstressed function words, articulatory movements are weakened even more, resulting in reduced closure durations and less firm contacts for voiceless stops so that passive voicing may continue through the occlusion. Thus 'mit dem' [mɩpm] may change to [mɩbm], and this lenis stop may then be subject to nasal assimilation. The same applies to content words that have become devoid of meaning in set phrases, such as adjectives in greetings, e.g. 'guten [gɷn] Tag'. The fact that [mɩpm] may develop to [mɩm] whereas 'Lappen' always retains a stop closure can therefore find its explanation in a greater reduction of articulatory energy expenditure under lack of stress, and in the higher probability of lenis stops, rather than the much longer fortis ones, losing their closure before nasals.

Geminate consonants that occur across word boundaries or as the result of [ə] deletion are shortened to the durations of single consonants. This is a language-specific process in German of very high frequency and is the more complete the faster and/or the more relaxed speech production is. This process can also be related to a minimization of energy expenditure, but whereas the reductions discussed in 3.1. and 3.2. are very widespread and, for instance, also apply to English, geminate shortening seems to be more restricted.

4. CONCLUSION

It will have become clear after this presentation and interpretation of reduction rules in German that the phenomena of articulatory weakening in fluent colloquial speech in German can, at least partly, be explained with reference to a criterion of motor economy in relation to the demands of the communicative situation. The vocal apparatus is not put to more effort than is necessary in order to maintain the linguistic contrasts in the particular listener-oriented goal. Its tendency to reduce articulatory movements to a minimum is checked by perceptual and social constraints, of which only the former have been the subject of this paper. The listener decides whether reductions are permissible in certain speech situations or whether they interfere with the transmission of information, and the linguistic group judges the stylistic and social acceptability. To explore further the biological foundations of speech production and perception as well as their integration with sociolinguistic theories of sound variability is a challenging task for phoneticians, and reduction phenomena are a particularly rewarding area for trying out new, unorthodox, and even "un-linguistic" ways of dealing with speech and language.

5. REFERENCES

[1] Whitney, D. W. (1896) A Sanskrit Grammar, 3rd ed., Breitkopf & Härtel, Leipzig.

[2] Grammont, M. (1971) Traité de Phonétique, Librairie Delgrave, Paris.

[3] Jones, D. (1969) An Outline of English Phonetics, 9th ed., Heffer, Cambridge.

[4] Gimson, A. C. (1980) An Introduction to the Pronunciation of English, 3rd ed., Edward Arnold, London.

[5] Labov, W. (1972) Sociolinguistic Patterns, University of Pennsylvania Press, Philadelphia.
[6] Trudgill, P. (1974) The Social Differentiation of English in Norwich, University Press, Cambridge.
[7] Ohala, J. J. (1979) 'The contribution of acoustic phonetics to phonology', in B. Lindblom and S. Öhman (eds.), Frontiers of Speech Communication Research, Academic Press, London/New York/ San Francisco, pp. 355-363.
[8] Lindblom, B. (1980) 'The goal of phonetics, its unification and application', Phonetica 37, 7-26.
[9] Lindblom, B. (1983) 'Economy of speech gestures', in P. F. MacNeilage (ed.), The Production of Speech, Springer, New York/ Heidelberg/Berlin, pp.217-245.
[10] Lindblom, B. (1984) 'Can the models of evolutionary biology be applied to phonetic problems', in M. P. R. van den Broecke and A. Cohen (eds.) Proc. of the Tenth Inter. Congress of Phonetic Sciences, Foris Publications, Dordrecht, pp. 67-81.
[11] Lindblom, B. (1986) 'On the origin and purpose of discreteness and invariance in sound patterns', in J. S. Perkell and D. H. Klatt (eds.), Invariance and Variability in Speech Processes, Lawrence Erlbaum, Hillsdale, pp. 493-510.
[12] Ohala, J. J. (1974) 'Phonetic explanation in phonology', in A. Bruck, R. A. Fox and M. W. Lagaly (eds.), Papers from the Parasession on Natural Phonology, Chicago Linguistic Society, pp. 251-274.
[13] Ohala, J. J. (1981) 'Articulatory constraints on the cognitive representation of speech', in T. Myers, J. Laver and J. Anderson (eds.), The Cognitive Representation of Speech, North Holland Publishing Company, Amsterdam/New York/Oxford, pp. 111-122.
[14] Ohala, J. J. (1983) 'The origin of sound patterns in vocal tract constraints', in P. F. MacNeilage (ed.), The Production of Speech, Springer, New York/Heidelberg/Berlin, pp. 189-216.
[15] Ohala, J. J. (1984) 'Prosodic phonology and phonetics', Phonology Yearbook 1, 113-127.
[16] Kohler, K. J. (1974) 'Koartikulation und Steuerung im Deutschen', in Sprachsystem und Sprachgebrauch. Festschrift für H. Moser, Teil 1, Schwann, Düsseldorf, pp. 172-192.
[17] Kohler, K. J. (1976) 'Die Instabilität wortfinaler Alveolarplosive im Deutschen - eine elektropalatographische Untersuchung', Phonetica 33, 1-30.
[18] Kohler, K. J. (1977a) Einführung in die Phonetik des Deutschen, Erich Schmidt, Berlin.
[19] Kohler, K. J. (1977b) 'Investigating coarticulation', in Phonologica 1976, Innsbrucker Beiträge zur Sprachwissenschaft 19, pp. 243-247.
[20] Kohler, K. J. (1979) 'Kommunikative Aspekte satzphonetischer Prozesse im Deutschen, in H. Vater (ed.), Phonologische Probleme des Deutschen, Studien zur deutschen Grammatik 10, Gunter Narr, Tübingen, pp. 13-39.

[21] Carlson, R. and Granström, B. (1975) 'A phonetically oriented programming language for rule description of speech', in G. Fant (ed.), Speech Communication, Almquist & Wiksell, Stockholm, Vol. 2, pp. 245-253.
[22] Carlson, R., Granström, B. and Hunnicutt, S. (1982) 'A multi-language text-to-speech module', in Proc. ICASSP82, Vol. 3, Paris, pp. 1604-1607.
[23] Carlson, R., Granström, B. and Hunnicutt, S. (1989) 'Multilingual text-to-speech development and applications', in W. A. Ainsworth (ed.), Advances in Speech, Hearing, and Language Processing, to be published by JAI Press, London.
[24] Kohler, K. J. (1988a) 'An intonation model for a German text-to-speech system', in Proceedings Speech'88, 7th FASE Symposium, Edinburgh, pp. 1241-1247.
[25] Kohler, K. J. (1988b) 'Zeitstrukturierung in der Sprachsynthese, in Digitale Sprachverarbeitung - Prinzipien und Anwendunge, ITG-Fachbericht 105, VDE-Verlag, Berlin/Offenbach, pp. 165-170.
[26] Chomsky, N. and Halle, M. (1968) The Sound Pattern of English, Harper & Row, New York/Evanston/London.
[27] Bladon, A., Carlson, R. Granström, B., Hunnicutt, S. and Karlsson, I. (1987) 'A text-to-speech system for British English and issues of dialect and style, in J. Laver and M. A. Jack (eds.), European Conference on Speech Technology, Vol. 1, pp. 55-58.
[28] Kohler, K. J. (1986) 'Invariance and variability in speech timing: from utterance to segment in German', in J. S. Perkell and D. H. Klatt (eds.), Invariance and Variability in Speech Processes, Lawrence Erlbaum, Hillsdale, pp. 268-289.
[29] Kohler, K. J. (1983) 'Prosodic boundary signals', Phonetica 4, 89-134.

V-C-V LINGUAL COARTICULATION AND ITS SPATIOTEMPORAL DOMAIN

E. FARNETANI
Centro di Studio per le Ricerche di Fonetica del CNR
35122 Padova
Italy

ABSTRACT. Recent literature on coarticulation shows that VCV coarticulatory effects are the result of the cooccurrence of context-independent overlapping gestures and of context-dependent gestures. Interlanguage and inter-category differences in degree of coarticulation seem to be due to differences in the competing demands on common articulators, which in turn depend on the articulatory/acoustic requirements for both the influencing and the influenced sounds. As for VCV lingual coarticulation, this paper proposes a coarticulation index defined palatographically, which quantifies the spatial range of coarticulation at different points in time within a VCV sequence. For some selected Italian consonants it is shown that, in agreement with Recasens' model (1987, 1989), coarticulation of tongue body varies inversely with the degree of tongue dorsum elevation. This holds for the voiced/voiceless distinction in single and in geminated dentoalveolar stops, for the place-of-articulation distinction in laterals, and for variations in stress and speaking rate. It is argued that the coarticulatory behaviour in the spatial domain can largely explain the temporal extent of coarticulation in both the anticipatory and the carryover directions.

1. INTRODUCTION

Coarticulation is one of the major sources of acoustic variability in speech and is manifested as a complex pattern of reciprocal influences among speech segments. Understanding why and how it occurs has been the goal of a great number of studies in the last twenty years. The most important theoretical issue concerns the cognitive (linguistic) vs the physiological nature of coarticulation. Two issues related to the first

W. J. Hardcastle and A. Marchal (eds.), Speech Production and Speech Modelling, 93–130.

concern the universal vs the language specific nature of coarticulation and whether coarticulation is learned or is present in early speech and decreases with age. Questions on how coarticulation is manifested, on its temporal extent, on the interindividual and interlanguage differences are thus of primary importance for throwing light on the fundamental issues.

At present there are two main lines of thought, with divergent views on the nature of the "abstract" units of speech and consequently on the nature of coarticulation; they are referred to by Fowler as translation theories and intrinsic timing theories (Fowler, 1977, 1980). Within the former theoretical framework, where a clear cut distinction is made between the plan of an utterance and its actualization in speech, a number of models of coarticulation have been proposed and the most relevant differences among them concern the extent to which coarticulation has to be ascribed to the plan of an utterance or to its execution. What is common to all these models is the idea that coarticulation is a modification of the "units" or canonical segments. The most popular of these models is the "feature spreading" model proposed by Daniloff and Hammarberg (1973); Hammarberg (1976). Within the intrinsic timing framework there is no dichotomy between plan and execution and coarticulation is not an alteration of canonical forms but just the way the segments are serially ordered in speech. This is at the basis of the "temporal" or "coproduction" models proposed by Bell-Berti and Harris (1981), Fowler (1980, 1981). The classical formulations of the two groups of models are as follows:

In the feature spreading models the encoding units are segments defined as bundles of binary features. In running speech, segments are linearly ordered and the features vary discretely over time: coarticulation is a spreading of features from elements to neighbour ones, and occurs in order to accomodate sounds to each other. Daniloff and Hammarberg (1973) incorporate Henke's look-ahead scanning mechanism in their model (Henke, 1966). This operator allows anticipation of a feature to all the segments that are not specified (i.e. "neutral") for that feature; so anticipatory coarticulation is always an active process. In Henke's model carryover effects are passive, a consequence of the inherent sluggishness of the system. For Daniloff and Hammarberg (1973) carryover effects are partly passive and partly deliberate.

In the coproduction models the segments are defined dynamically, i.e. in space and in time, "The movements toward an articulatory goal are considered to be as much a part of the specification of the segment as is the goal itself" (Bell-Berti and Harris, 1981). A characteristic of these units is that in speech they are not linearly ordered but overlap in time. Thus coarticulation is not an alteration of canonical forms, but the way they are organized in their actualization. The coarticulatory rules are implicit in the above definition. The onset of a movement of an articulator, when there is no articulatory antagonism, is time-locked to the phoneme to be produced and is independent of the length or the number of the preceding phones. Thus, the anticipatory effects are predicted to start at a constant time before the target sound, but will have different durations according to the articulators involved in the gesture. Fowler's account of consonant-vowel coproduction is largely drawn from Ohman's articulatory model (Ohman, 1966) which posits that consonants are superimposed on an underlying vowel-to-vowel gesture. According to Fowler (1980, 1981, 1985) the constant durations of the vowel-to- vowel gestures (or of stressed vowel to stressed vowel gestures) are due to the coordinative structures underlying their production, which operate in a cyclical way and this implies that the system will keep the vowel-to-vowel gestures as synchronous as possible. Thus, for instance, an increased length of the consonantal period should extend the superposition and hide more of the vocalic gesture, but the vowel gesture onset time, relative to the vowel acoustic onset will be constant, i.e. independent of the length of the consonant.

As seen above, the two groups of models make quite different predictions on the temporal extent of anticipatory coarticulation: it is constant in the coproduction models, and variable, i.e. dependent on the number of neutral preceding segments in the feature spreading models.

However, the data from the experimental studies which have focused on the temporal domain of coarticulation, are disparate, contradictory with one another, and support and disconfirm in turn the predictions of either model. Reviews of coarticulation models and of experimental data are in Kent and Minifie (1977); Sharf and Ohde (1981); Fowler (1980, 1985, 1989); Kent (1983); Harris (1983).

Altogether these studies reveal the inadequacy of feature specification as an index of articulatory conflict (Bell-Berti and Harris, 1981): segments specified as neutral to a feature were found to condition the spreading of that feature (see Perkell [1986]) for lip protrusion; onset of coarticulatory lip rounding gestures were observed to occur earlier in antagonistic than in neutral segments (Sussman and Westbury, 1981). Needless to say, inadequate specification of segments along a featural or articulatory dimension makes the interpretation of temporal data problematic: short extents of anticipatory coarticulation could be ascribed to the "time-locking" mode of production as well as to the blocking effect of the context. On the other hand, little is still known about the temporal extent of articulatory periods of different articulators, or class of sounds to be able to decide whether long coarticulatory extents are to be ascribed to long articulatory periods or to feature spreading.

A better understanding of coarticulation has come from investigations where the spatial and the dynamic aspects of coarticulation (besides its temporal extent) have been considered: they have shed more light on the effects of articulatory constraints on the coarticulatory movements i.e. on the complex interplay between influencing and influenced sounds, and on the relationship between coarticulation and articulatory antagonism. Another important aspect of coarticulation, - which, however, has received much less attention - is the relationship between coarticulation and suprasegmental changes, such as stress and speaking rate. This might throw some light on the relationship between coarticulation and reduction, the other factor most responsible for acoustic variations.

The paper will address these issues, mainly in relation to the VCV lingual coarticulation (sections 2, 3 and 4). Electropalatographic data (section 5) will illustrate the effects of articulatory antagonism and those of suprasegmental variations on the spatiotemporal coarticulatory patterns of some lingual consonants in Italian. As a conclusion, a tentative model of the distribution of stop and lateral consonants in the articulatory space as a function of their coarticulatory behaviour will be presented.

2. VCV LINGUAL COARTICULATION

2.1. The Vowel-to-vowel Gesture

Lingual coarticulation differs from both velar and labial coarticulation in many respects: It is ubiquitous - it occurs whenever a syllable is produced; mutual - vowels affect consonants and viceversa; it has a large bidirectional temporal extent - it spans from the one to the other transconsonantal vowel; it is complex - coarticulatory gestures occur simultaneously along both the front/back and the high/low dimension. Mutual V to C and C to V effects were observed in Kiritani, Itoh, Hirose and Sawashima (1977), and Recasens (1989). These studies also indicate that C to V effects are smaller and have a more restricted temporal range than V to C effects. Transconsonantal vowel-to-vowel effects in both the anticipatory and the carryover directions have been observed in several articulatory studies (Carney and Moll, 1971; Sussman, MacNeilage and Hanson, 1973; Butcher and Weiher, 1976;

Kiritani et al., 1977; Recasens, 1984a, 1984b, 1989; but see Gay, 1974, 1977 for contrary evidence), and have acoustic consequences (Ohman, 1966; Fowler, 1981; Vayra, Fowler and Avesani, 1987; Recasens, 1987; 1989). Ohman's acoustic data on Swedish subjects indicate that context- dependent changes affect the frequencies of the first, the second and the third formant transitions of transconsonantal vowels, and that changes in F2 are larger and more systematic than changes in the other two formants. This indicates that V to V coarticulation occurs mainly along the front/back dimension, but that the high/low dimension is also involved. His X-ray data on a Swedish subject confirm that during the intervocalic consonant the tongue body is displaced by the flanking vowels along both dimensions. Transconsonantal effects on the frequencies of F1 and F2 were also observed by Manuel and Krakow (1984) in speakers of Swahili, Shona and English. Figure 1 illustrates very clearly the trajectories of the tongue dorsum along the front/back and the high/low dimensions during the production of /V1pV2/ where V1 is schwa. It can be seen that the coarticulatory gestures in the horizontal direction are much larger and have a greater temporal extent than those in the vertical direction (Alfonso and Baer, 1982).

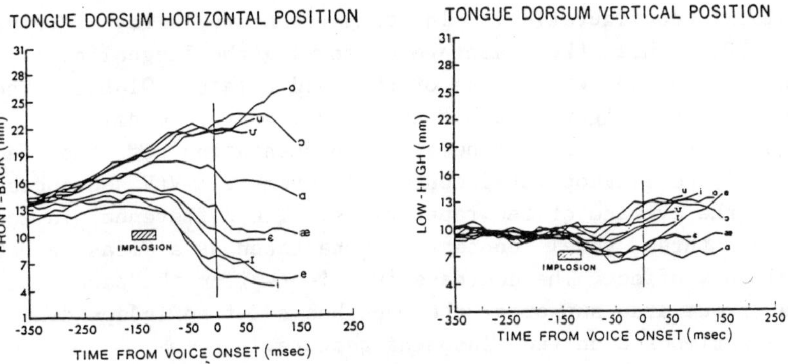

Figure 1. Tongue dorsum trajectories in the production of /V1pV2/ in English (V1=schwa), recorded with cinefluorography (from Alfonso and Baer, 1982).

Ohman's V-to-V gesture model appears to account for the occurrence of the V to C, the C to V and the V to V effects reported in the literature. His model is based on acoustic and X-ray data on Swedish,

American English and Russian stop consonants. According to his model, the overlap of vocalic and consonantal gestures is possible because the neural commands for vowels and consonants are transmitted over different channels to different, though partially overlapping tongue regions, the apical and the dorsal for the production of consonants and the tongue-body region for the production of vowels. The independence of commands is the condition "sine qua non" for the coarticulatory gesture to occur and explains why velar consonants do not block coarticulation: under two independent commands the tongue body is able to perform a "distorted" vowel gesture while executing the consonant closure. The gestural overlap explains the mutual V to C and C to V effects, while the greater temporal extent of the vocalic gestures explains the stronger effects of vowels on consonants. Coarticulation is blocked when a consonant command is transmitted over the vocalic channel, in addition to the consonantal channel; this is the case, for instance, for consonants with distinctive secondary articulation, like the Russian palatalized stops, which block coarticulation.

Figure 2 illustrates some basic VCV coarticulatory patterns for lingual consonants, recorded by means of electropalatography.

EPG records the pattern of tongue to palate contact over time. Unlike X-ray techniques, the articulatory movements are not directly detected, but must be inferred from the changing pattern of the linguopalatal configuration. In comparison to X-ray techniques, however, EPG has the advantage of showing the lingual activity in great detail over the whole area of the hard palate. Globally the data illustrated in the figure support the V-to-V gesture model.

Figure 2 (top) is a schematic representation of the temporal evolution of the linguopalatal contact in symmetric VtV sequences. The data are the average of ten repetitions. The difference in contact across the three vowel contexts can be taken as a measure of the coarticulatory effect. The decrease in effects from the back to the mid to the front area supports the hypothesis of an independent vowel gesture superimposed on the consonant gesture.

The traces in the mid of the figure show the evolution of the contact for the utterances "ania" and "ancia". Here the palate surface is divided in two areas, the front (solid lines) and the mid-back (dashed lines). The utterance on the left is a clear example of overlapping independent gestures (notice that the decrease of front contact for the opening of the /n/ closure occurs simultaneously with the increase of back contact for the production of the semivowel). The utterance on the right is instead an example of assimilation of /n/ to the following prepalatal affricate (the cluster is produced with one tongue gesture). In this case the involvement of the tongue body in the consonant production blocks the V-to-V gesture.

The EPG frames at the bottom show the areas of contact for /k/ in asymmetric VCV (five repetitions). Measurements were taken at two points in time, C and R, corresponding to the first and the last frames of the consonant closure. The dotted areas correspond to the electrodes activated 4 or 5 times over five, the striped areas to those

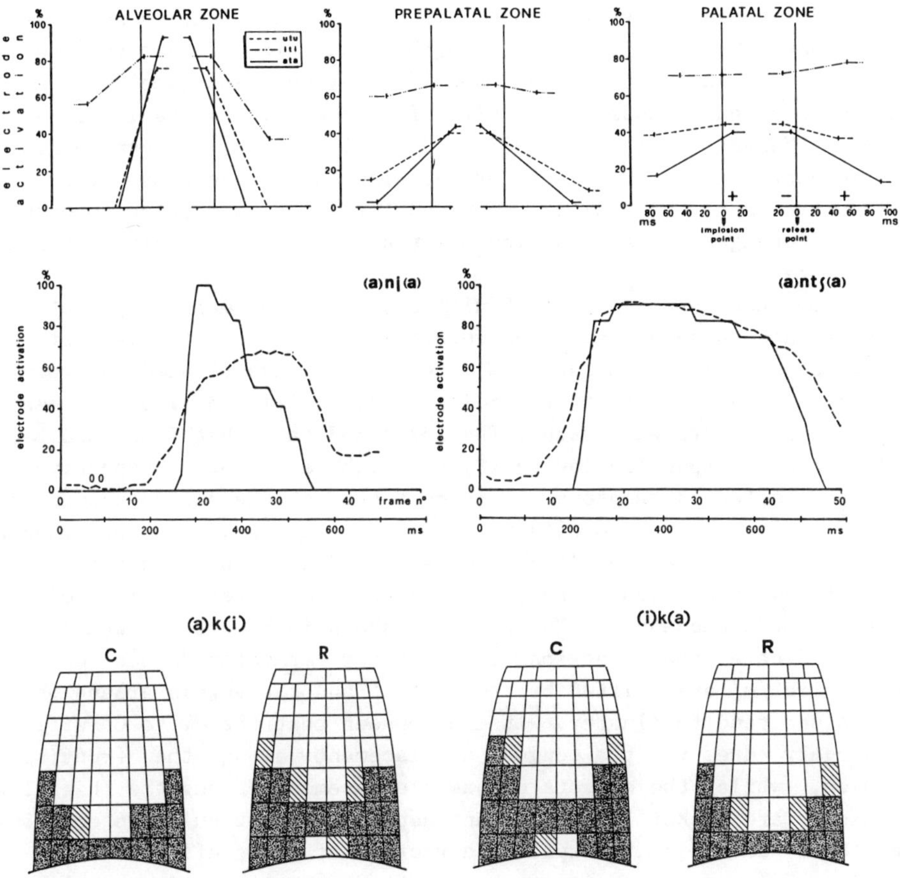

Figure 2. Coarticulatory patterns of some lingual consonants in Italian and French, recorded with EPG.

activated 2 or 3 times. For both items the contact varies during the closure along the front/back dimension, according to the vowel context, and reflects quite clearly the gesture from V1 to V2, in agreement with Ohman's model (but see counter-examples in Butcher and Weiher, 1976).

2.2. Tongue and Jaw Movements

From the effects observed acoustically, or by means of EPG or X-ray, it is not possible to infer that coarticulation is the result of the tongue-body vocalic gesture alone, if the jaw behaviour is not taken into account. The jaw is involved in the production of both consonants and vowels; both seem to have a specific jaw position (see Kiritani, Itoh, Hirose and Sawashima [1977] for vowels; Kiritani, Tanaka, Hashimoto, Masaki and Shirai [1983] and Keating [1983] for consonants). The jaw performs the VC gesture by carrying the tongue body from the vowel position to the position required for the constriction to occur. Contextual changes in jaw position may thus affect the resultant tongue body gestures.

An important study on coarticulation, where the main components of the vowel and consonant gestures were taken into account, is that of Kiritani et al. (1977) on Japanese. The authors investigated the vowel-consonant mutual effects by recording tongue, jaw and lip movements with the X-ray microbeam system. The data indicate that front and back vowels have a specific jaw position, high and low, respectively. Consonants affected the tongue component but not the jaw component of vowels. As for the V to C effects, the authors considered the tongue component along the front/back dimension and the jaw component along the high/low dimension. The test consonant was intervocalic /m/ in asymmetric CVC sequences. The data are shown in Figure 3, where the arrows indicate the directions in which the parameter values of the intervocalic /m/ are shifted by the context and the vowels around which these values tend to cluster. It can be seen that the V to C effects are largely due to the tongue displacement along the front/back dimension, while the effects of jaw displacement (along the high/low dimension) are weaker. An important aspect of coarticulation emerging from these data is the spatial direction of the effects along a dimension: /u/ and /i/ appear to be the strongest attracting vowels and a great part of the data points shift towards /i/ or /u/. But /u/ is stronger than /i/, as can be seen from the position of the /imu/ /umi/ items, clustering around /u/.

A parallel study by Kiritani, Sekimoto and Imagawa (1977) reveals an important feature of the Japanese vowels in relation to the American English vowels. The study indicates that the distinction along the

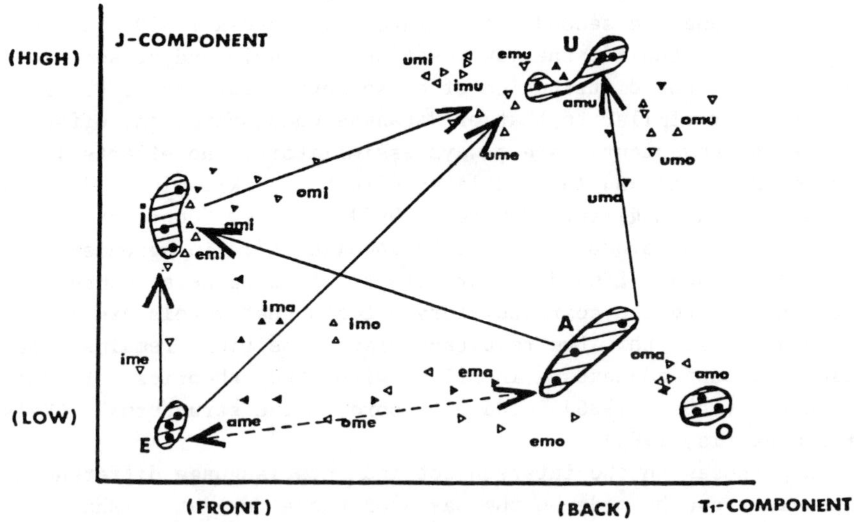

Figure 3. Vowel coarticulatory effects on tongue and jaw position of intervocalic /m/ in Japanese, recorded with X-ray microbeam (from Kiritani et al., 1977).

high/low dimension is brought about almost exclusively by the jaw movements in Japanese, and by both the tongue-body and the jaw movements in American English. This can explain why in the previous study by Kiritani et al. no vowel effects on the tongue high/low movements were found. From the results of the two studies it is reasonable to hypothize that in languages where the tongue and the jaw components are both active in distinguishing the vowels along the high/low dimension, the V to C effects along this dimension may be larger than in Japanese.

In a subsequent study (Kiritani et al., 1983) the authors focused on the vowel effects on the jaw position of different consonants. They examined consonants /s/, /t/, /k/ and /p/ in symmetric and asymmetric CVC utterances. Their data indicate that the overall jaw position is higher in /s/ and /t/ than in /k/ and /p/. In consonants with high jaw position, high vowels induce a jaw lowering and low vowels induce an increase in height, whilst for consonants with low jaw position, vowels induce the opposite effects, i.e. high vowels increase and low vowels decrease the jaw height. The authors defined the first effect "contrastive" and the second "assimilatory". Condax [1980, cited by Keating, 1983] studied the jaw position of vowels and consonants in Fijian; whilst the distribution of consonants according to their jaw position is very similar to that of Japanese consonants, the effects of vowels, when they occur, are always assimilatory: no effects in /s/, small effects in /t/ and /p/ and large effects in /k/. Instead, in a study on American English, Edwards (1985) observed "contrastive" jaw adjustments in the production of intervocalic /t/, in agreement with the Japanese study. Edwards interprets the finding as a compensation made by the jaw to the coarticulatory effects that vowels exert on the tongue blade, so that the resultant blade position remains nearly invariant across contexts, in line with the theories of "motor equivalence" (Abbs, 1986), and of "coordinative structures" (Kelso, Tuller and Harris, 1983).

No hypotheses on the intersubject and interlanguage differences in jaw behaviour can be made on the basis of these few data. What these studies reveal is that the jaw position of consonants is affected by the context. The data agree that in consonants with an overall low jaw position the effects tend to be assimilatory, but are contradictory on the effects on consonants with a high jaw position: for these consonants the effects can be small, absent or compensatory.

In any case these studies show that the jaw displacements can increase or decrease the resultant effects of the tongue body and indicate that the VCV lingual coarticulatory effects are the result of the cooccurrence of context independent gestures at the lingual level (at least for alveolar and labial stops) and of context dependent gestures at the jaw level.

3. ANTAGONISM AND DEGREE OF COARTICULATION

Articulatory antagonism occurs whenever an articulator is simultaneously under competing, conflicting demands. How is the conflict resolved? The early models of coarticulation all agreed that the coarticulatory gestures are blocked in cases of gestural conflict. The feature spreading models (Daniloff and Hammarberg, 1973) posit that when the requirements for the production of a phone conflict with the gestural overlap for the following segments, there is always a priority for the goals of the more immediate unit, and coarticulation does not take place. The temporal model (Bell-Berti and Harris, 1981) states that in the case of gestural antagonism the contextual gesture is delayed and altered, for example, the velum rising movement for the production of

/k/ is delayed and becomes faster if the oral consonant is immediately preceded by a nasal consonant.As seen above, in Ohman's model the occurrence of the V-to-V gesture is governed by the presence in the consonant of vowel features with distinctive function, that is, coarticulation is blocked when the intervocalic consonant is specified for distinctive secondary articulation. Keating, in an investigation of the relationship between coarticulation and feature specification (1985), tested in an acoustic study the blocking effect of English dark [l], which is specified for secondary articulation, like the Russian palatalized stops, but, unlike these, the specification is non-constrastive: she found that dark [l] blocked the V-to-V gesture along the front/back dimension (i.e. F2). She also tested the blocking effect of /s/ - Ohman's investigation had shown intervocalic fricatives to be resistant to coarticulation - and found that the consonant blocks the V-to-V gesture along the high/low but not along the front/back dimension. Thus she shows that consonants specified for secondary non-contrastive features (i.e. backness in dark [l]), and consonants unspecified for vowel features but requiring a high tongue blade position for their production (such as /s/), interfere with coarticulatory overlapping, and that in these cases coarticulation is blocked only in one direction, the direction from which antagonism arises.

A great number of experimental studies, (see also the investigations reported in the preceding section) show, however, that the response to articulatory conflict is rarely a discrete response, and that coarticulation can vary in degree also along one dimension. In other words they show that competing demands on the same articulator do not necessarily prevent coarticulation occurring. Ohman (1966) had made a distinction between factors that block coarticulation and those that modify the coarticulatory pattern. His explanation of the distorted vowel gesture during velar stop closures is that "the dynamic response of the tongue to a compound instruction is a complex summation of the responses to each of the components of the instruction". This means that the trajectory of such gestures will be the result of the relative strength of the two or more competing forces, and that coarticulatory effects can vary in degree.

The literature shows that this is true both for the same sound categories across languages, and for different sound categories in the same subjects.

3.1 Interlanguage Differences in Coarticulation

Clumeck (1976), in a cross-language investigation of coarticulatory velar movements, observed that the onset of velar lowering tends to occur earlier in speakers of American English and Brazilian Portuguese than in speakers of French, Hindi, Amoy Chinese and Swedish. Lubker and Gay (1982) compared the patterns of upper lip movements in Swedish and American subjects and observed interlanguage differences, both in the onset time and in the amplitude of the coarticulatory lip rounding gesture. Interlanguage differences were explained in terms of language specific articulatory/phonetic requirements both for the influencing and the influenced segments. For instance, lip rounding can be the only distinctive feature between a pair of vowels in Swedish, but not in American English. Hence in Swedish the rounding gesture needs to be longer and more precise than in American English. A language with distinctive vowel nasalization will tend to delay the lowering of velum associated with a nasal for as long as possible during oral vowels in order to avoid confusions between distinctive and coarticulated nasality; in a language like Swedish with a very crowded vowel space, long periods of coarticulated nasality might result in perceptual confusion among vowels (Clumeck, 1976). The hypothesis that the magnitude of coarticulatory effects in different languages is governed by the principle of perceptual distinctiveness was tested in an acoustic study by Manuel and Krakow (1984) . The authors compared V-to-V coarticulation in languages with a different number of vowel phonemes in order to test whether languages with smaller vowel inventories allow stronger V-to-V effects than languages with larger vowel inventories. Their data support the hypothesis: Swahili and Shona (five-vowel systems) have a greater range of V-to-V coarticulatory effects than English. The authors point out, however, that the number of vowels is not the sole factor affecting coarticulation: it is the distribution of vowels in the vowel space that determines the extent to which any individual vowel is allowed to vary. Thus English allows larger coarticulatory variations along the front/back than along the high/low dimension because more vowels are distributed along the vertical than the horizontal plane.

There are, however, some exceptions to the distinctiveness principle: coarticulatory nasalization, for instance, has a very restricted range in standard Italian (Farnetani, 1986), a language with no apparent necessity for distinctiveness since it has only seven vowel phonemes and no nasal vowels. Evidently the cross-language data are too limited to allow any conclusion on the principles that govern coarticulatory variations in different languages.

From these studies we can infer that language specific control of the temporal and the spatial range of coarticulation aims at the safeguard not only of perceptual distinctiveness but also of the exact phonetic quality expected and accepted in a given language or dialect.

3.2. Inter-segment Differences in Coarticulation

Some investigators have been directly concerned with the issue of articulatory antagonism and have compared and quantified the coarticulatory effects on different sound categories within the same language.

Bladon and Al-Bamerni (1976) were the first to emphasize that the nature of coarticulation is continuous, rather than discrete. Their data on the RP allophones of /l/, show that coarticulatory effects decrease from clear to dark to dark syllabic [l]. The authors postulate that coarticulation is governed by an articulatory control principle - coarticulation resistance -, which is associated with speech segments in the form of different values. Unlike Keating's data, dark [l] shows contextual variations also in the second formant, i.e. along the front/back dimension; the data on Catalan dark [l] (Recasens, 1987) are very similar to those of Bladon and Al-Bamerni, the discrepancies between British English and Catalan on one hand, and American English on the other, are probably due to a stronger degree of velarization of dark [l] in the American English speakers. A subsequent cinefluorographic study on English alveolar consonants by Bladon and Nolan (1977) indicated once again that there are different degrees of articulatory constraints on coarticulation, which depend on the articulatory characteristics of the sounds. Their data on tongue blade position show that /n/ and /l/ are more affected than /t/ by an adjacent /s/, and that no changes occur in /s/ in the context of the other alveolars. Perkell (1986) approached the problem of coarticulation resistance in a study of anticipatory lip protrusion, which aimed to test the look-ahead vs the time-lock coarticulatory strategy. Subjects were speakers of English, Spanish and French. Previous data by Gay (1977) had shown that lip rounding across an /utu/ utterance is discontinuous and decreases during /t/. If troughs occur because alveolar consonants, presumed to be neutral to lip rounding, are instead non-neutral, the two strategies cannot be tested because coarticulation could not extend beyond the intervening consonant. If, alternatively, the consonants are indeed neutral but produce troughs, this would support time locking. Perkell did find troughs of various depths for most of the consonants as well as relevant interlanguage differences (the consonants of American English were produced with a more retracted lip position than the Spanish or the French ones). This indicated that alveolars are not neutral to lip rounding and that their resistance to coarticulation varies across subjects and across languages. Comparisons between a rounded and a non-rounded context indicated that /n/ is the most neutral consonant, hence the most susceptible to contextual effects, while the consonant most resistant to vowel effects was the /s/ produced by the Spanish subject. Two points are of particular interest here. Firstly, the finding that a lip protrusion target specification for non-labial consonants is much more common than expected. Secondly, the author's attempt to relate the target specification value to the degree of coarticulatory resistance, i.e. the resistance to variations along the protrusion/retraction dimension: a very high value for protrusion, such as that of Spanish /s/, makes the sound very resistant to coarticulatory effects of unrounded vowels. A rather low value for protrusion, such as that of English /s/, has an analogous though weaker effect on the other side, i.e. makes the sound rather resistant to the rounding effect of /u/; while French /s/ which has a target value between the other two, is highly affected by the context. This suggests that the degree of coarticulation resistance may be associated with the distance of the target specification value from either ends of the protrusion/retraction continuum.

3.3. Recasens' Articulatory Model of Lingual Coarticulation

As for lingual coarticulation the most extensive investigation on articulatory antagonism has been carried out by Recasens in a number of studies (1983, 1984a, 1984b, 1987, 1989), where Spanish and Catalan lingual consonants of different place and manner of articulation were analyzed by means of electropalatography. He showed that the V-to-V coarticulatory effects vary inversely and monotonically with the degree of tongue dorsum elevation required for the consonant, so that a progressive decrease in the degree of tongue dorsum contact with the palate induces a progressive increase of coarticulatory effects. Recasens compares vowel-related coarticulatory effects with consonant-related coarticulatory effects and finds that (in comparable neutral environments) vowel effects are always more extensive than consonantal effects, in agreement with Ohman's model. Recasens then relates the consonant-specific articulatory antagonism to the vowel-specific articulatory antagonism. For both vowels and consonants the degree of tongue dorsum activity can be taken as a measure of articulatory resistance to coarticulatory effects. Recasens' findings and propositions indicate that the sounds that block or reduce the coarticulatory effects of the neighbour sounds are also those exerting the strongest effects, in other words they exhibit the least contextual variation and induce the greatest. Examples are vowel /i/ and consonant /ʃ/. EPG data on other languages seem to confirm that the degree of tongue dorsum elevation can be a measure of coarticulatory resistance:

/i/ was found to be the strongest influencing vowel and the most resistant to the coarticulatory effects of other vowels (see Butcher and Weiher [1976] for German; Farnetani, Vagges and Magno Caldognetto [1985] for Italian). However, in the above mentioned study on Italian, we found that /a/ is more resistant than /u/ to the transconsonantal effects of /i/. Engstrand (1983) has shown that during the production of intervocalic /p/ in Swedish, the tongue has to keep rather low, and exhibits some kind of coarticulatory resistance to the raising effects of /i/. Similar troughs during intervocalic /p/ were observed in the vowel-to-vowel gesture along the high/low dimension in the X-ray study by Alfonso and Baer (1982) (see Fig. 1). This might suggest the existence of a second (though weaker) pole of articulatory resistance for coarticulatory tongue movements and could explain the resistance of non-lingual consonants specified for a low tongue position to the coarticulatory effects of /i/, like the /i/ pole explains the resistance of the palatoalveolar fricatives of Catalan to the lowering effects of /a/. In analogy with the labial coarticulatory effects along the protrusion/retraction dimension described by Perkell, it is possible that also for lingual coarticulation more extreme target specifications along the high-front/low-back dimension induce greater resistance to the coarticulatory effects from the opposite pole, while less extreme specifications allow larger coarticulatory variations. An extension of Recasens' articulatorv model could thus account for the coarticulatory patterns of non-lingual consonants.

3.4. Compound Gestures

These investigations indicate that complete neutrality is almost absent in speech and that competing demands over an articulator are the norm, rather than the exception. They also show that in the majority of cases conflict is not resolved with the blocking of coarticulation but with a reduction of its effects. The response of an articulator to competing demands can be viewed as a compound gesture, or a blend (Kent, 1986), i.e. a movement adjusted to accomplish simultaneously two or more phonetic objectives, as suggested by Ohman. Coarticulatory differences among languages appear to be due to differences in the articulatory/acoustic goal to be achieved. So, different languages may not use exactly the same set of articulators for the production of the same class of sounds, or they may use the same set of articulators but the competing demands over the same articulator may differ in their relative strength among languages.

In the coarticulation literature objections to the coproduction models have been repeatedly raised concerning the fundamental proposition that coarticulation is the temporal overlapping of independent gestures. If this is so, spatial overlapping, that is, displacements as a function of the context are left unexplained by this approach (Kent, 1983). Diehl (1986) argues that, if it is incompatibility of gestures that prevents coproduction, the spatio-temporal overlapping may still be interpreted as coproduction of vowel-consonant effects, but not as coproduction of effects of stressed vowels on transconsonantal unstressed vowels (as was proposed by Fowler, 1981). This is because their gestures are essentially incompatible, and therefore cannot be simultaneously and independently realized. This effect requires an explanation based on assimilation, not on coproduction. In more recent formulations of the coproduction model (1985, 1989), Fowler also views spatial overlapping as coproduction. She hypothizes that multiple influences on the same articulator are wholly independent and they simply sum up at the periphery. This is in agreement with the task dynamic model of speech production where "coarticulatory effects result from the temporally overlapping demands (conflicting or synergistic) made by the same or different tract variables on a common articulator subset" (Kelso, Saltzman and Tuller, 1986). In this view, however, a complete block on coarticulatory effects could not occur without invoking a mechanism that completely suppresses the coarticulatory demands.

Within the featural theory, the "coarticulation resistance" model proposed by Bladon and Al-Bamerni (1976), accounts for the graduality of coarticulatory effects in space, but, like the binary feature models, it cannot account, in VCV sequences, for the continuously changing tongue configurations between the steady state configuration of V1 and that of V2, unless it applies different weightings of a vocalic feature throughout the consonant, with the weightings of V1 decreasing and that of V2 increasing across the consonant (Fowler, 1985).

A very promising alternative to the coarticulation resistance model is the "window" model, recently proposed by Keating (1988). I shall briefly summarize the main points: In this model segments are specified for a range of values that each feature is allowed to take in a given articulatory dimension. This range of possible spatial variations is a "window". The width of the window associated with a feature can vary among segments: narrow windows reflect little contextual variation and wide windows large contextual variation. It will be the context, with its associated window that will determine the value that the feature of a segment will take and hence the trajectory, the "path" through the windows. It is assumed that such trajectories are constrained by requirements of continuity and of minimal articulatory effort, in agreement with the general principle of articulatory economy (Lindblom, Pauli and Sundberg, 1975; Lindblom, 1983). So, whilst the traditional models propose that the canonical forms are altered by contextual effects, in this model, the context constrains the variability associated with the segment. A "window" analysis provided by Keating of articulatory data (velum and jaw movements) accounts both for the various degrees of contextual variations in the space domain, which are made possible by the "window", and for the changes in the time domain, described as "paths".

4. COARTICULATION, STRESS AND SPEAKING RATE

4.1. Effects of Stress

As for the effects of stress on coarticulation the data in the literature concerning the V-to-V coarticulatory effects indicate that unstressed vowels are much more affected by the flanking consonants and/or by the transconsonantal vowels than stressed vowels (Nord, 1974, Fowler, 1981, Recasens, 1989). The decrease in degree of coarticulatory resistance can be viewed as a side effect of vowel "reduction" in unstressed position, which may indicate target undershooting or less extreme target specification. So far it seems that very few studies have been concerned with effects of stress on the coarticulatory behaviour of consonants. In the EPG study of Farnetani et al. (1985) on intervocalic /t/ in isolated VCV sequences we found that the position of stress has some effects on coarticulation: carryover effects were found to be stronger when V1 was stressed, and the anticipatory effects inconsistent when V2 was stressed. Coarticulatory effects on consonant in absolute unstressed position (between two unstressed vowels) do not seem to have been investigated. It may be expected that, like unstressed vowels, consonants under this condition are reduced and consequently less resistant to coarticulatory effects.

4.2. Effects of Speaking Rate

There are a number of studies concerning the effects of speaking rate on articulatory movements and articulatory configurations, but very few on the effects of speaking rate on coarticulation. Speakers may use different strategies for increasing speech rate. As reported in reviews on the effects of speaking rate on the dynamics of speech movements (Ostry and Munhall, 1985; Wieneke, Janssen and Belderbos, 1987), the following articulatory changes can occur: a) a reduction in the amplitude of the movements with no changes in velocity (Kent and Moll, 1972; Wieneke et al, 1987; b) a reduction in the amplitude and an increase in velocity (Gay, 1981); c) no changes in the amplitude and an increase in velocity (Gay, Ushijma, Hirose and Cooper, 1974); d) reductions in both amplitude and velocity. The studies reporting reductions in the amplitude of the articulatory movements are in agreement with the undershoot hypothesis put forth by Lindblom (1963).

It is difficult to predict what the coarticulatory changes might be when a higher speaking rate is achieved by reorganizing the system as in b) and c), probably none. But when fast speech is achieved by reducing the amplitude of the articulatory gestures it is reasonable to expect stronger coarticulatory effects as a result of a decrease in the articulatory resistance (Recasens, 1989). The few available data suggest that this is the case. Engstrand (1988) tested the effects of a number of variables on the articulatory movements in VpV sequence by means of X-ray films and found stronger coarticulatory effects for a fast than for a normal speech rate. Stronger coarticulatory effects in spontaneous speech than in isolated words (Krull. 1989) also point to a close relationship between coarticulation and reduction, which usually characterizes spontaneous speech (Engstrand and Krull, 1988).

5. THE VCV LINGUAL COARTICULATION IN ITALIAN : VOWEL TO CONSONANT EFFECTS ANALYZED BY MEANS OF ELECTROPALATOGRAPHY

5.1. Aim of the Experiment

One aim of the experiment was to test Recasens' articulatory model with two sets of Italian consonants, laterals and dentoalveolar stops. For laterals the effects of place of articulation (alveolar vs palatal) were tested, for stops, the effects of voicing.

A second aim of the experiment was to assess in a very preliminary way the relationship between coarticulation and changes in the temporal domain. This was done by comparing the coarticulatory effects 1) in single vs geminated apical stops and laterals (/t, d, l/ vs /tt, dd, ll/); 2) in consonants flanked by one stressed and one unstressed vowel vs consonants flanked by two unstressed vowels (test consonants were /l, t, d/); 3) in normal vs fast speaking rate (test consonants were the single and geminated dentoalveolar stops).

EPG data from a study on voicing (Farnetani, 1989), indicated that voiceless and voiced stops exhibit relevant spatiotemporal differences, with larger linguopalatal contact areas and longer closure durations for voiceless than for voiced stops, but that the spatial differences between the two categories are much greater in the context of /a/ than in the context of /i/. This can be seen in Figure 4, which represents the temporal evolution of the tongue to palate contact on the global surface of the artificial palate. The number of activated electrodes (ordinate) have been taken in five points in time (abscissa).The lining up points correspond to the EPG frames where changes in activation indicate the onset of the V to C movement. The other four points correspond to the first frame of complete closure, the frame of maximum activation within the closure, the last frame of complete closure, and the frame indicating the conclusion of the C to V movement. Each point is the mean of five repetitions.

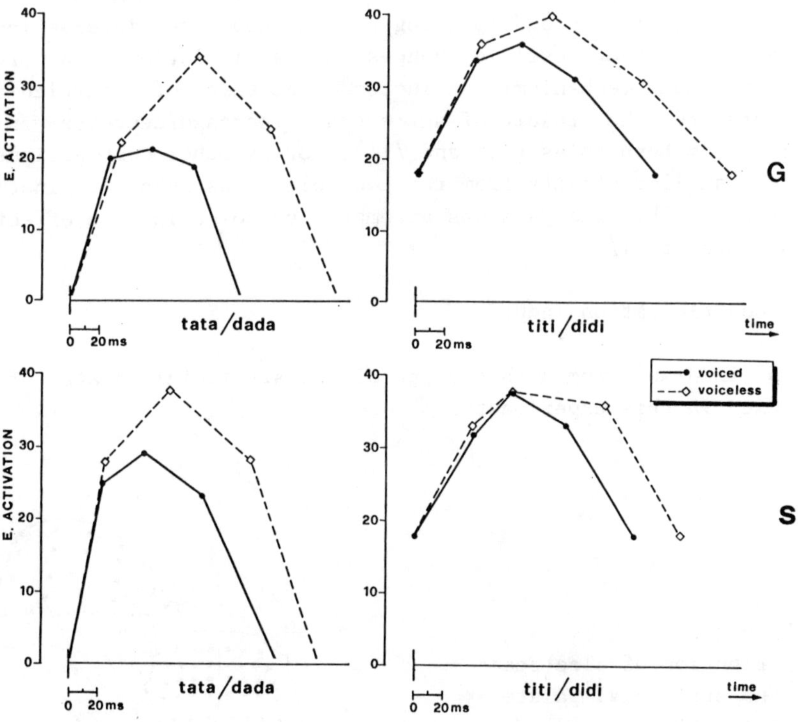

Figure 4. Temporal evolution of the tongue to palate contact in the production of intervocalic /t/ and /d/ by two Italian subjects.

The EPG patterns of the two consonants indicate that the vowel effects are stronger on /d/ than on /t/: this can be seen by comparing /dada-didi/ (solid lines) with /tata-titi/ (broken lines). These data did not indicate whether the differences in degree of coarticulation related to voicing were a side effect of the different durations of the consonants or were due to their inherent articulatory characteristics: in this case the difference in coarticulatory effects should be maintained also when the duration of consonants is varied. In the present experiment, as seen above, the consonant durations have been systematically manipulated. With variations in stress and speaking rate the aim was to investigate not only whether (and to what extent), coarticulation varies as a function of these factors, but also how such changes are brought about. Figure 4 suggests that the resultant greater range of coarticulatory variation of /d/ is due to the stronger coarticulatory effects from one side only, the side of the open vowel. With changes in stress and speaking rate it would be interesting to know whether or not possible changes in coarticulation are brought about by the same mechanism: an increase/decrease of coarticulatory effects may be the result of an actual increase/decrease in the influence from both poles (/a/ and /i/), or by other combinations of greater and smaller effects from the two poles, as seen, for instance, in /d/ vs /t/. This analysis was attempted by comparing the effects of /a/ with those of /i/.

5.2. The Coarticulation Index

Figure 5 is drawn from a photograph of the artificial palate of the subject used in this experiment.

Figure 5.
Distribution of electrodes on the artificial palate (Reading EPG System)

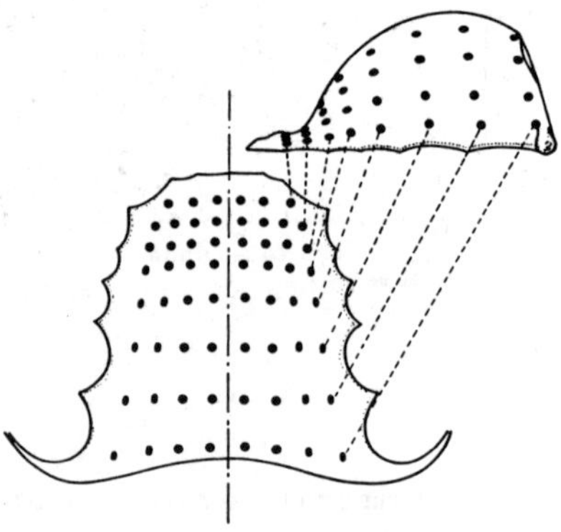

The EPG system used in this experiment is the Reading system, developed by Hardcastle (1984). It can be seen that the electrodes are ordered in eight rows, with four of them concentrated over the alveolar and the postalveolar region of the hard palate and the other four covering the remaining surface of the palate.

The contextual effects are expressed by a coarticulation index (CI). The index has to capture what is "non-common" between two contexts, that is, variations in position of contact as well as variations in amount of contact. In order to do so the CI was calculated on the basis of the difference between contexts (/i/ vs /a/), within each row, as illustrated in the profiles of intervocalic /l/ shown in Figure 6.

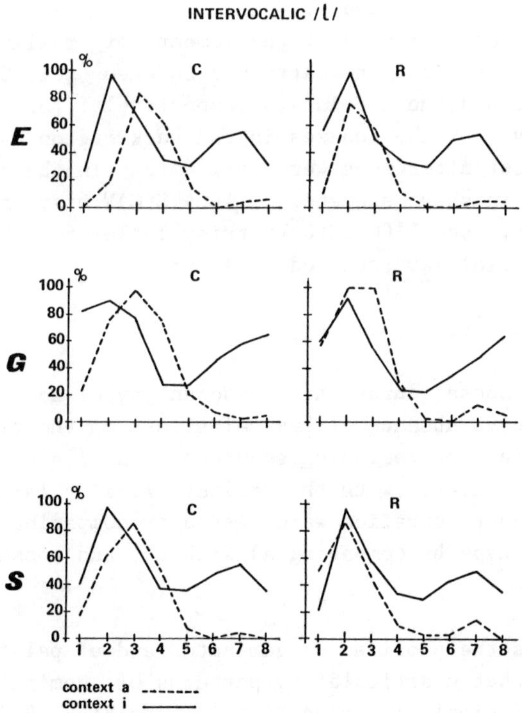

Figure 6. Palatographic profiles of intervocalic /l/ produced by three Italian subjects (see text for expansion).

The profiles are based on the percent frequency of electrode activation within each row in five repetitions. "C" represents the first frame of complete closure and "R" the frame immediately before the release.

It can be seen that for all three subjects in the context of /a/ intervocalic /l/ is produced with a more retracted tongue tip/blade position and with a lower tongue body position than in the context of /i/. Both these context effects are captured in the index by summing up the differences between /a/ and /i, be they positive or negative. The differences are averaged over the eight rows, thus obtaining the global CI, or over the four front rows (front CI) or the four back rows (back CI). Thus the CI represents the percent amount of contextual variation in electrode activation at a given point in time and for a given palatal region. For each utterance five EPG frames have been selected for measurements, the same as in the study of voicing (see above), and referred to as V1, C, Max, R and V2.

The CI has been computed at the moment of maximum activation (Max), the anticipatory and the carryover indexes (ACI, CCI) have been calculated on the C and the R points, respectively: for the purpose of the present study only the indexes in the back region have been used for comparisons. Test utterances were real words (in the great majority of cases) embedded in short sentences, '(C)VC(C)V bisyllables for the stressed condition, and '(C)VC(C)VV trisyllables for the unstressed condition. The material was repeated five times.

5.3. Lateral Consonants

The effects of consonant duration, consonant type and word prosodic structure (presence vs absence of lexical stress on the flanking vowel) were investigated in the following sequences: a) /'ala/, b) /'alla/, c) /'aʎa/, and d) /ala/, with the variants /i-i/, /a-i/, /i-a/. The effects of consonant duration were tested by comparing a) with b), those of consonant type by comparing a) with c), and those of stress by comparing a) with d).

Figure 7 shows the profiles of geminated and of palatal laterals. It can be seen that coarticulatory patterns of geminated are very similar to those of single intervocalic /l/ (see Fig. 5 Subj. E), with CI=35 for /l/ and CI=33 for /ll/. The profiles of the palatals look quite different, with a linguopalatal contact extending over a very large area and very small effects from the flanking vowels (CI=12). The anticipatory and carryover effects can be detected by comparing the profiles of symmetric with those of asymmetric sequences at the moment

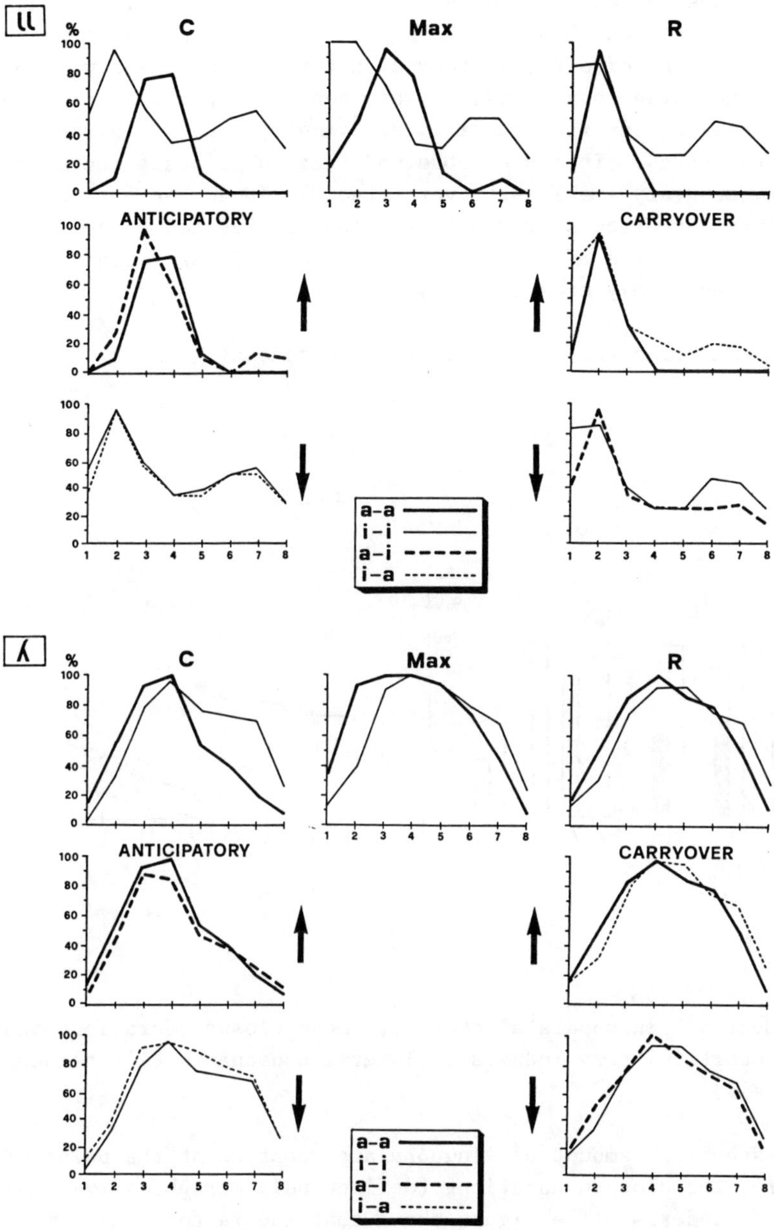

Figure 7. EPG profiles of geminated and palatal laterals.

of closure for the anticipatory effects and at the moment of release for carryover effects, as shown in the figure. If the profiles tend to coincide, there are no effects, if they do not coincide, there will be effects from the transconsonantal vowel when the profile of the asymmetric sequence differ from that of the symmetric in the direction indicated by the arrow. It can be observed that in palatals there are no anticipatory and very small carryover effects. In geminates instead there are carryover effects from both /a/ and /i/ (as can be seen at point R from the profiles of /alli/- illi/ and of /illa/-/alla/) and anticipatory effects only from the following /i/ (i.e. at point C in /alli/- /alla/).

Figure 8 summarizes the data on lateral consonants.

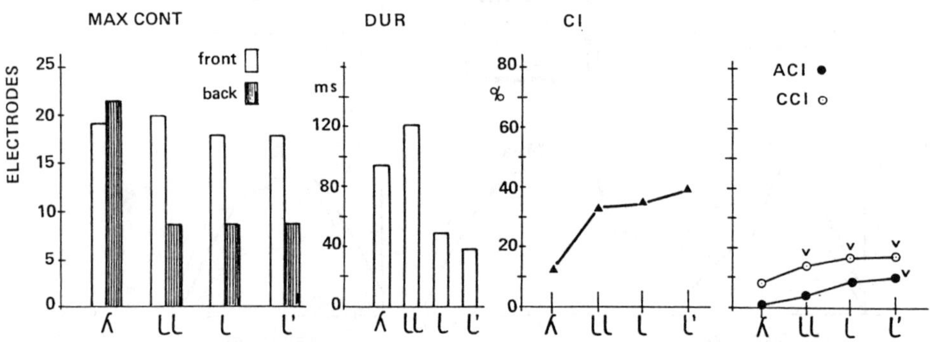

Figure 8. Amount of linguopalatal contact, mean closure durations and values of the coarticulatory indexes in lateral consonants (l'= between unstressed vowels)

Figure 8 shows the amount of linguopalatal contact at the point of maximum closure, the closure durations (defined palatographically), and the values of indexes. The figure shows that the factor that most affects coarticulation is, as expected, the degree of tongue dorsum activity (compare the pattern of Max. contact in the back region with that of CI). The palatal consonant exhibits the greatest degree of coarticulatory resistance and the unstressed intervocalic /l/ the least. The carryover effects are always greater than the anticipatory and extend till the following unstressed vowels for both geminate and single laterals. The anticipatory effects extend to the preceding vowel only when it is unstresssed (see the "v" marks).

5.4. Voiced and Voiceless Stops

These consonants were analyzed under three speech conditions: in words spoken in isolation, in the same words embedded in short sentences, produced at normal and at fast speaking rate. For each condition the three factors, consonant type (/t/ vs /d/), consonant duration (single vs geminate) and stress were tested.

The global effects of speech condition on the duration of the closures and on the amount of linguopalatal contact is illustrated in Figure 9, where the closure durations (abscissa) are plotted against the amount of electrode activation (ordinate).

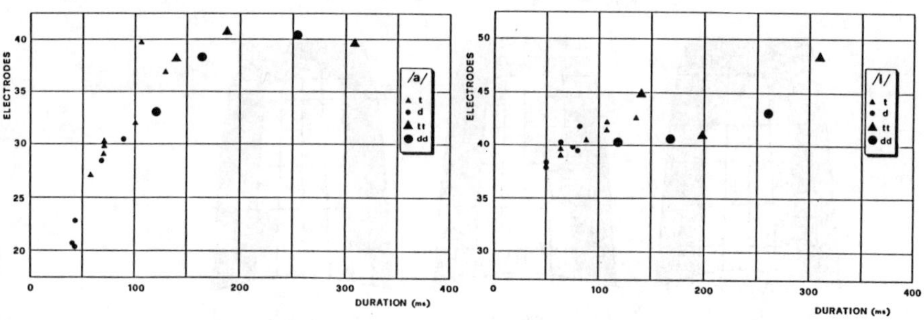

Figure 9. Relationship between closure duration and amount of electrode activation in the production of voiced and voiceless stops.

It can be seen in Figure 9 that both geminated and single consonants are affected in their duration and in the degree of linguopalatal contact by the speech conditions, and that the amount of contact is strongly related with duration of the closure, especially in the context of /a/. More interesting is the way in which the speech condition affects the tongue to palate contact, i.e. on which regions, on which consonants, and in which vowel environment the speech condition exerts the greatest effects, and this is illustrated in Figure 10.

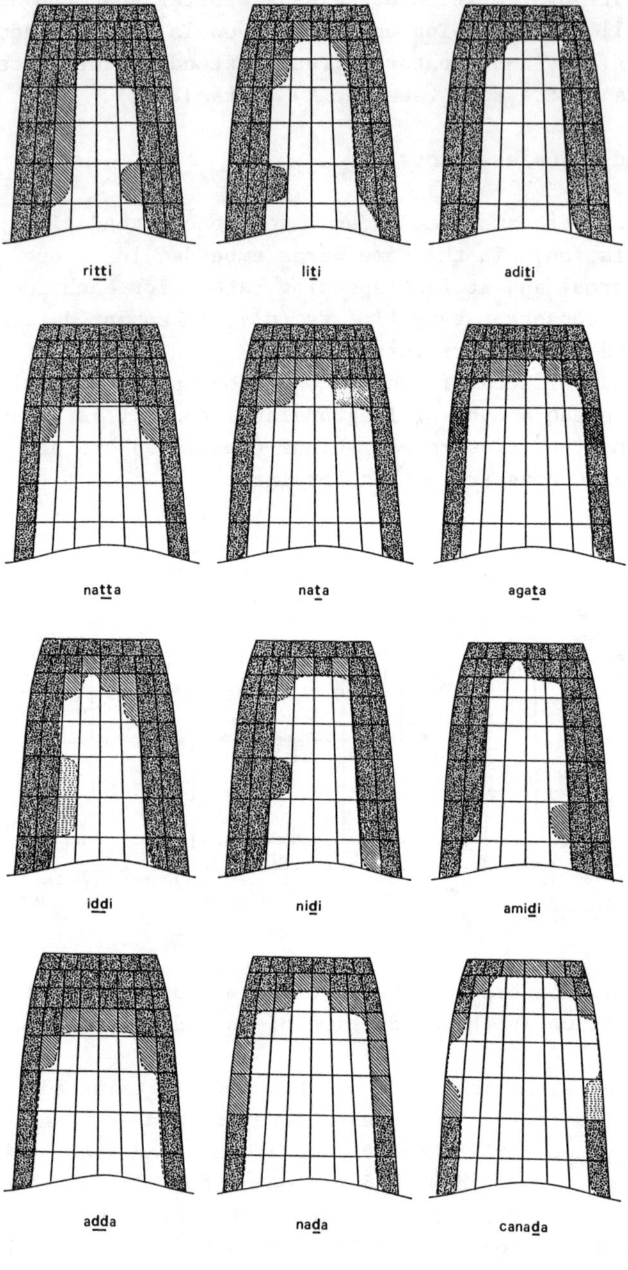

Figure 10. Linguopalatal configurations of voiced and voiceless stops at normal and fast speaking rate.

Figure 10 compares the two speech conditions that most differentiate the linguopalatal configuration, isolated words vs fast speaking rate. The linguopalatal configurations in the EPG frames represent the electrodes that have been activated with a frequency no less than 80% , i.e. that have been activated at least four times over the five repetitions. The sum of the dotted and the striped areas represents the contact in isolated words while the dotted areas represent the contact at fast speaking rate. It can be seen that the contact decreases from normal to fast in both the /i/ and the /a/ contexts but that /a/ allows greater reduction of the contact. The single consonants are reduced more than the geminated, and single /d/ is affected more than single /t/. The region where the contacts are most reduced is the front region and in single /d/ the contact can be reduced to such a point that complete closure may not be achieved; this occurs when /d/ is preceded and followed by an unstressed vowel. In the back region the contact is reduced relatively less than in the front region, but once again the greatest effects are observed in /d/. In unstressed /d/ (see item "Canada") the very light back contact associated with incomplete front closure strongly suggests that such a pattern results from a reduction of the rising movement of the jaw.

The data relating all the factors are illustrated in Figure 11. On the right of the figure it can be seen that the CI increases gradually across speech conditions for voiced but not for the voiceless stops: this can be explained by the fact that the reduction in contact for voiceless stops mostly occurs in the front region (the CI refers to the back region). For voiced stops fast speaking rate reduces the contact both in /a/ and in /i/ contexts, but the reduction is greater in the context of /a/: so the larger CI is simply the outcome of the reduction of the consonant. The figure shows that under all conditions /d/ is more affected than /t/ by the context (see the CI values), and this is mostly due to the stronger coarticulatory effects of /a/ on /d/ than on /t/, as clearly shown in Figure 9. Looking at the left side of Figure 10 it can be observed that the average tongue dorsum contact (back electrodes) is more extensive in the production of /t/ than in the production of /d/, so the data are in agreement with Recasens' model. Consonant duration has greater effects on stops than on laterals: in single consonants the average CI value increases by 14% for voiceless stops and by 38% for voiced, so the duration of the closure tends to decrease the contextual effects but by a very small amount when compared to the much more noticeable differences in duration (in nearly all instances the durations of geminates are more than double the durations of single consonants).

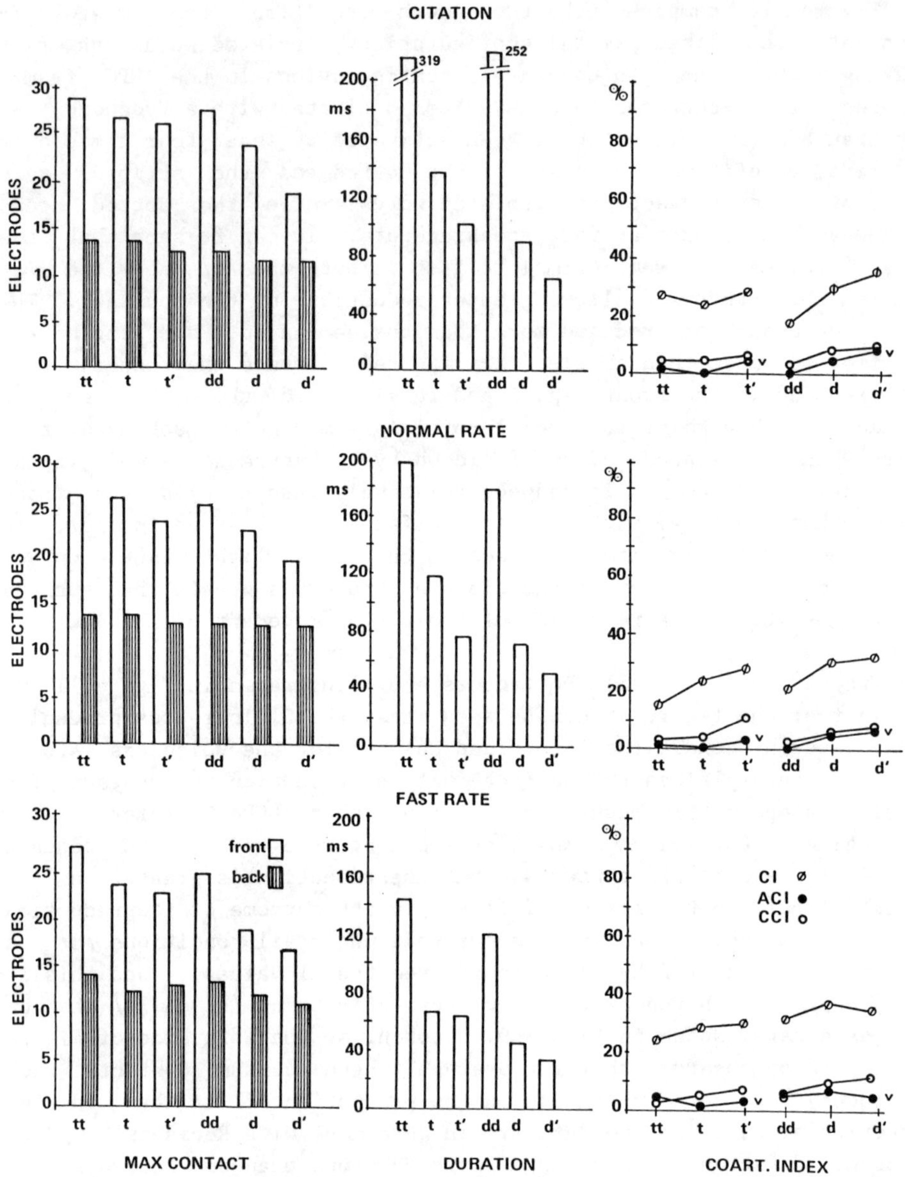

Figure 11. Amount of linguopalatal contact, mean closure durations and values of the coarticulatory indexes in voiced and voiceless stops under three speech conditions (t', d'= between unstressed vowels).

Finally, the effect of stress is strong and systematic: for both /t/ and /d/ and all the speech conditions CI increases when the flanking vowels are both unstressed, indicating that the unstressed condition makes not only vowels, but also consonants less resistant to the contextual effects. The only exception to the general trend is /d/ spoken at a fast speech rate. Under this condition the CI has a very high value (37) when V1 is stressed and does not increase any further. The comparison between the effects from /a/ and those from /i/ in stressed vs unstressed condition for /l/, /d/ and /t/, indicates that in /l/ the CI has increased in unstressed condition because the effects from /i/ have increased, while in /d/ the increase in CI is due to larger effects from /a/. So the higher coarticulation indexes are the results of greater coarticulatory effects from one or the other vowel. Instead, the larger coarticulatory effects in unstressed /t/ are brought about by a decrease of the effects from vowel /i/, accompanied by a larger increase in the effects from /a/. In this case the larger CI is a side effect of reduction, exactly like the effects of speaking rate. This aspect of coarticulation in relation to reduction, and these unexpected different strategies for different consonants need further and more systematic investigations with larger sets of consonants.

As for the temporal extent of coarticulation, it can be seen in the figure that once again the carryover effects tend to be larger than the anticipatory: in geminated consonants and in /t/ spoken at normal speech rate anticipatory effects are practically absent, while for consonants in unstressed condition they occur systematically and reach the preceding vowel (see the mark "v").

6. CONCLUSION: THE SPATIOTEMPORAL RANGE OF V-C-V LINGUAL COARTICULATION

6.1. Temporal Range

With the EPG technique alone it is not possible to determine the temporal range of coarticulation. However, the presence and the amount of anticipatory and carryover effects at the C and R points in time (and also the presence vs absence of the effects at the V1 or V2 points) can give some indications of the temporal extent of coarticulation. The two flanking vowels used in this investigation are very well differentiated palatographically: /a/ is characterized by an extremely small number of on-electrodes concentrated at the sides of the back edge of the artificial palate, whilst /i/ is characterized by a relatively great amount of activation all over the back region, so that their effects can be detected quite easily. Absence of anticipatory and carryover effects at the C and R points respectively, suggests that the V2 directed gesture does not start before implosion and is concluded before the release: in this case the consonant would reflect only V1 at the implosion and only V2 at the release. This never occurs. What usually occurs, instead is the presence of both anticipatory and carryover effects, which suggests that the V2 directed gesture starts before implosion and is concluded after the release: this V-to-V gesture is more continuous than the hypothetical gesture described above, and in Keating's terms (1988) can be considered a more economical path than the former. The greater amount of carryover than of anticipatory effects that characterizes this subject suggests a temporal asymmetry of this movement in relation to C and R: there is a delay in the V2 directed movement, a delay of its onset with respect to implosion and a delay of its conclusion with respect to release, and this automatically reduces the anticipatory effects from V2 respect to the carryover effects from V1. This occurs in all utterances with stress on V1 and to a much higher degree when the intervocalic consonant is resistant to coarticulatory effects (i.e. geminates spoken at a normal speech rate and palatals): in this extreme case, there is no sign of anticipatory coarticulation, whereas carryover effects are always present.

Altogether these data indicate that both strong vowels (stressed vowels) and strong consonants cause a delay in the V2 directed gesture, and when they combine, the V2 effects are completely absent. When both vowels are unstressed, the anticipatory effects tend to increase in magnitude with respect to carryover effects. This indicates that when the two flanking vowels have comparable articulatory strength, the anticipatory effects tend to increase in magnitude respect to carryover effects, suggesting that the V-to-V gesture has become more symmetric, that is, it tends to start rather early during V1 and is concluded rather late in V2.

Also the duration of the consonant affects the V2 directed gesture: in geminated stops spoken at fast speaking rate and in geminated laterals the occurrence of both anticipatory and of carryover effects suggests that the long duration of the consonantal interval has reduced the speed of the V-to-V gesture (the average duration of these consonants is 125 ms): if it was not so, if the gesture had started during the closure and had the same speed as gestures across shorter consonants, both effects would be absent and at the release the consonant would reflect only the V2 effects.

The overall data indicate that the anticipatory and carryover effects reflect the timing and the dynamics of one gesture, i.e. of the

V2 directed gesture, whose characteristics are determined by the result of the complex interplay of the articulatory strength and resistance associated with the consonant and the two flanking vowels.

6.2. Spatial Range

The overall data indicate that the major factor that determines the amount of the coarticulatory effects of vowels on lingual consonants is the degree of tongue body elevation for the consonant, in agreement with Recasens' articulatory model. Tongue body elevation as measured with palatography can be the composite result of degree of tongue body activity and degree of jaw elevation. Stronger and weaker coarticulatory effects represent the range of variation allowed to each consonant according to the articulatory/acoustic requirements for its production. The coarticulatory index decreases from alveolars to palatals and, within alveolars, it decreases from laterals to voiced stops to voiceless stops. Another factor that controls coarticulation, though at a much lesser extent than tongue body elevation, is the duration of the consonant closure. Within each category the coarticulation index can increase or decrease as a function of stress and of speaking rate. The increase of coarticulatory effects in unstressed sequences seems to result from an actual increase in the coarticulatory effects from /i/ or from /a/, at least in laterals and voiced stops; the higher values of the coarticulation index in fast speaking rate results from a reduction of the consonant, i.e. a decrease in the area of contact in both contexts. It seems that the more a consonant is allowed to vary as a function of the context, the more it will vary as a function of other factors: this suggests that coarticulation and articulatory reduction are strongly related factors.

The diagram in Figure 12 shows how consonants could be distributed in the articulatory space as a function of their coarticulatory behaviour with respect to vowels /a/ and /i/.

The two vowels represent two poles of coarticulatory resistance (the position of vowels is based on the number of electrodes activated in the back region). The data points relative to each consonant are the intersection between the CI values (abscissa) and the number of electrodes activated in the back region, averaged over the contexts /a/ and /i/ (ordinate). It can be seen that there is an inverse relationship between the two variables, in agreement with Recasens' predictions. The consonants may be distributed along an ideal trajectory connecting the two opposite poles. The range of context dependent variation associated with each consonant (expressed by the CI

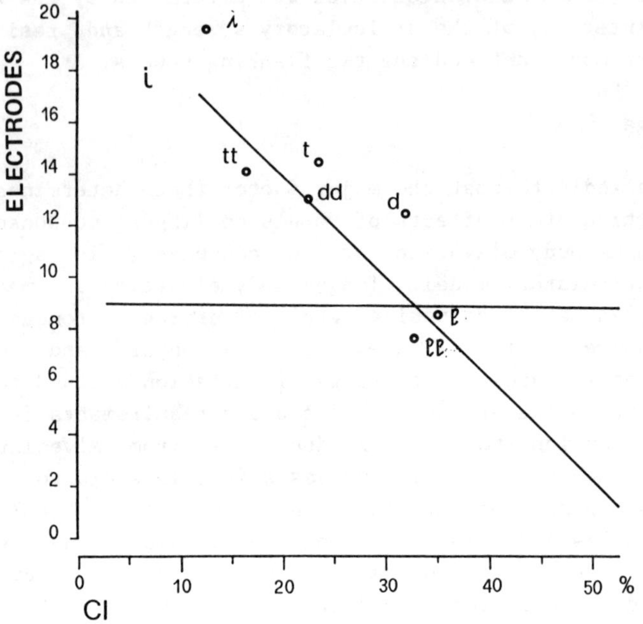

Figure 12. Relationship between degree of tongue elevation (ordinate) and CI (abscissa) for laterals and dentoalveolar stops.

value) could itself be visualized on the line connecting the two poles, by a segment whose extremes are the values of the number of activated electrodes in the two contexts and whose length expresses the range of variation. Fricatives can be a good test for this model. As seen in the literature, they are more resistant to coarticulation than stops, but if the principle that relates tongue body elevation and coarticulatory resistance has a general validity, fricatives could also be distributed in the articulatory space like stops and laterals, possibly on a different regression line. I believe that further elaboration of this sort of representations might yield global profiles of consonant coarticulatory behaviour that could easily be compared among individuals and among languages and give a more exact measure than traditional methods of the individual- , language- and age-specific aspects of coarticulation.

ACKNOWLEDGMENTS

My thanks to Christian Abry, Olle Engstrand, Alain Marchal, Peter Roach, Daniel Recasens and Mario Vayra for their very helpful comments. A special thank-you to Bill Hardcastle and his assistants for welcoming me in the Reading laboratory and for generously helping me in collecting the EPG data.

This research was supported in part by a grant from NATO Scientific Affairs Division (RG 0235/88).

REFERENCES

ABBS, J.H. (1986) "Invariance and variability in speech production: a distinction between linguistic intent and its neuromotor implementation", in J. Perkell and D.H. Klatt (eds.), Invariance and variability in speech processes, Lawrence Erlbaum Ass., New Jersey, pp. 202-219.

ALFONSO, P.J. and BAER, T. (1982) "Dynamics of Vowel Articulation", Language and Speech, 25, 151-173.

BELL-BERTI. F. and HARRIS, K.S. (1981) "A Temporal Model of Speech Production", Phonetica, 38, 9-20.

BLADON, R.A.W. and Al-BAMERNI, A. (1976) "Coarticulation resistance in English /l/", Journal of Phonetics, 4, 137-150.

BLADON, R.A.W. and NOLAN, F.J. (1977) "A video-fluorographic investigation of tip and blade alveolars in English" Journal of Phonetics, 5, 185-193.

BUTCHER, A. and WEIHER, E. (1976) "An electropalatographic investigation of coarticulation in VCV sequences", Journal of Phonetics, 15, 111-126.

CARNEY, P. and MOLL, K.L. (1971) "A cinefluorographic investigation of fricative consonant-vowel coarticulation", Phonetica, 23, 193-202.

CLUMECK, H. (1976) "Patterns of soft palate movements in six languages" Journal of Phonetics, 4, 337-351.

CONDAX, I.D. (1980). Correlation of mandible position and vowel durations", Journ. Acoust. Soc. Am., 67, Suppl.1, S93.

DANILOFF, R.G. and HAMMARBERG, R.E. (1973) "On defining coarticulation" Journal of Phonetics, 1, 239-248.

DIEHL, R.L. (1986) " Coproduction and direct perception of phonetic segments: a critique", Journal of Phonetics, 14, 61-66.

EDWARDS, J. (1985) "Contextual effects on lingual-mandibular coordination", Haskins Lab. Status Rep. Speech Res., 79/80, 39-48.

ENGSTRAND, O. (1983) "Articulatory coordination in selected VCV utterances: A means-end view", Rep. Uppsala Univ. Dep. Linguistic, 10, 1-145.

ENGSTRAND, O. (1988) "Articulatory correlates of stress and speaking rate in Swedish VCV utterances", Journ. Acoust. Soc. Am., 83, 1863-1875.

ENGSTRAND, O. and KRULL, D. (1988) " On the systematicity of phonetic variation in spontaneous speech", PERILUS V (Inst. Linguistics, Stockholm University), 34-47.

FARNETANI, E., VAGGES, K. and MAGNO-CALDOGNETTO, E. (1985) "Coarticulation in Italian /VtV/ Sequences: A Palatographic Study", Phonetica, 42, 78-99.

FARNETANI, E. (1986) " A pilot study of the articulation of /n/ in Italian using electropalatography and airflow measurements", 15e Journées d'Etudes sur la Parole, GALF, Aix en Provence, 23-26.

FARNETANI, E. (1989) "An articulatory study of "voicing" in Italian by means of dynamic palatography", Speech Research International Conference. Linguistic Institute of Hungarian Academy of Sciences, T. Szende, Budapest, 395-398.

FOWLER, C.A. (1977) Timing control in speech production, Indiana University Linguistics Club, Bloomington.

FOWLER, C.A. (1980) "Coarticulation and theories of extrinsic timing", Journal of Phonetics, 8, 113-133.

FOWLER, C.A. (1981) "Production and perception of coarticulation among stressed and unstressed vowels", Journal of Speech and Hearing Research, 24, 127-139.

FOWLER, C.A. (1985) "Current Perspectives on Language and Speech Production: A Critical Overview", in R.G. Daniloff (ed.), Speech Science, Taylor & Francis, London, pp. 193-278.

FOWLER, C.A. (1989). Speaking. (in press).

GAY, T.J. (1974) "A cinefluorographic study of vowel production", Journal of Phonetics, 2, 255-266.

GAY, T.J. (1977) "Articulatory movements in CVC sequences", Haskins Lab. Status Rep. Speech Res., 49, 121-147.

GAY, T.J. (1981) "Mechanisms of the control of speech rate", Phonetica, 38, 148-158.

GAY, T., USHIJIMA, T., HIROSE, H., and COOPER, F. (1974) "Effect of speaking rate on labial consonant-vowel articulation", Journal of Phonetics, 2, 46-63.

HAMMARBERG, R.E. (1976) "The methaphysics of coarticulation", Journal of Phonetics, 4, 353-363.

HARDCASTLE, W.J. (1984) " New methods of profiling lingual palatal contact patterns with electropalatography", Work in Progress, Phon. Lab. Univ. Reading, 4, 1-40.

HARRIS, K.S. (1983) "Coarticulation as a component in articulatory description", in R.G. Daniloff (ed.), Articulation Assessment and Treatment Issues, College-Hill Press, San Diego, pp. 147-167.

HENKE, W.L. (1966) Dynamic articulatory model of speech production using computer simulation, Doct. Diss., MIT.

KEATING, P.A. (1983) "Comments on the jaw and syllable structure", Journal of Phonetics, 11, 401-406.

KEATING, P.A. (1985) "CV phonology, experimental phonetics, and coarticulation", paper presented at the Colloque Phonologie Pluri-Linaire, Lyon.

KEATING, P.A. (1988) "The window model of coarticulation: articulatory evidence", UCLA Working Papers in Phonetics, 69, 3-29.

KELSO, J.A.S., TULLER, B. and HARRIS, K.S. (1983) "A 'dynamic pattern' perspective on the control and coordination of movement", in P. MacNeilage (ed.), The Production of Speech, Springer Verlag, New York, pp.137-173.

KELSO, J.A.S., SALTZMAN, E.L. and TULLER, B.(1986) " The dynamical perspective on speech production: data and theory", Journal of Phonetics, 14, 29-59.

KENT, R.D. (1983) "The Segmental Organization of Speech", in P.F. MacNeilage (ed.), The Production of Speech, Springer Verlag, New York, pp. 57-89.

KENT, R.D. (1986) "Is a paradigm change needed?", Journal of Phonetics, 14, 111-115.

KENT, R.D. and MOLL, K.L. (1972) "Cinefluorographic analysises of selected lingual consonants". Journal of Speech and Hearing Research, 15, 453-473.

KENT, R.D. and MINIFIE, F.D. (1977) " Coarticulation in recent speech production models". Journal of Phonetics, 5, 115-133.

KIRITANI, S., ITOH, K., HIROSE, H. and SAWASHIMA, M. (1977) "Coordination of the consonant and vowel articulations - X-ray microbeam study on Japanese and English", Res. Inst. Logopedics and Phoniatrics, Tokio Univ. Ann. Bull., 11, 11-21.

KIRITANI, S., SEKIMOTO, S. and IMAGAWA, H. (1977) "Parameter description of the tongue movements for vowels", Res. Inst. Logopedics and Phoniatrics, Tokio Univ. Ann. Bull., 11, 31-37.

KIRITANI, S., TANAKA, T., HASHIMOTO, K., MASAKI, S. and SHIRAI, K. (1983) " Contextual variation of the jaw movement for the intervocalic consonant in VCV utterances", Res. Inst. Logopedics and Phoniatrics, Tokio Univ. Ann. Bull., 17, 45-53.

KRULL, D. (1989) "Consonant-vowel coarticulation in spontaneous speech and in reference words" STL- QPSR, KTH, Stockholm, 1, 101-105.

LINDBLOM, B. (1963) "Spectrographic study of vowel reduction", Journ. Acoust. Soc. Am., 35, 1773-1781.

LINDBLOM, B. (1983) "Economy of Speech Gestures", in P.F. MacNeilage (ed.) The Production of Speech, Springer-Verlag, New York, pp. 217-245.

LINDBLOM, B., PAULI, S. and SUNDBERG, J. (1975) "Modeling coarticulation in apical stops", in G. Fant (ed.) Speech Communication, Almqvist & Wiksell, Stockholm, Vol. 2, pp. 87-94.

LUBKER, J. and GAY, T. (1982) "Anticipatory labial coarticulation: experimental, biological and linguistic variables", Journ. Acoust. Soc. Am., 71, 437-448.

MANUEL S.Y. and KRAKOW, R.A. (1984) "Universal and language particular aspects of vowel-to-vowel coarticulation", Haskins Laboratories Status Rep. Speech Res., 77/78, 69-78.

NORD, L. (1974) "Vowel reduction-centralization or contextual assimilation?", in G. Fant (ed.) Speech Communication, Almqvist & Wiksell, Stockholm, Vol. 2, pp. 149-154.

OHMAN, S. (1966) "Coarticulation in VCV utterances: Spectrographic measurements", Journ. Acoust. Soc. Am., 30, 151-168.

OSTRY, D.J. and MUNHALL, K.G. (1985) "Control of rate and duration of speech movements", Journ. Acoust. Soc. Am., 77, 640-648.

PERKELL, J.S.(1986) "Coarticulation strategies: preliminary implications of a detailed analysis of lower lip protrusion movements", Speech Communication, 5, 47-68.

RECASENS, D. (1983) "Timing and coarticulation for alveolo-palatals and sequences of alveolar [j] in Catalan", Haskins Laboratories Status Rep. Speech Res. 74/75, 97-112.

RECASENS, D. (1984a) "V-to-C coarticulation in Catalan VCV sequences: an articulatory and acoustical study", Journal of Phonetics, 12, 61-73.

RECASENS, D. (1984b) " Vowel-to-Vowel coarticulation in Catalan VCV sequences", Journ. Acoust. Soc. Am., 76, 1624-1635.

RECASENS, D. (1987) " An acoustic analysis of V-to-C and V-to-V coarticulatory effects in Catalan and Spanish VCV sequences", Journal of Phonetics, 15, 299-312.

RECASENS, D. (1989) "Long range coarticulatory effects for tongue dorsum contact in VCVCV sequences", Speech Communication (in press).

SHARF, D.J. and OHDE, R.N. (1981) " Physiologic, acoustic and perceptual aspects of coarticulation: implications for the remediation of articulatory disorders", in N.J. Lass (ed.) Speech and Language, Advances in Basic Research and Practice, Academic Press, New York, Vol. 5, pp. 153-247.

SUSSMAN, H.M., MACNEILAGE, P.F. and HANSON, R.J.(1973) "Labial and mandibular dynamics during the production of bilabial consonants: Preliminary observations", Journal of Speech and Hearing Research, 16, 397-420.

SUSSMAN, H.M. and WESTBURY, J.R. (1981) "The effects of antagonistic gestures on temporal and amplitude parameters of anticipatory labial coarticulation", Journal of Speech and Hearing Research, 24, 16-24.

VAYRA, M., FOWLER, C.A. and AVESANI, C.(1987) "Word-level Coarticulation and Shortening in Italian and English Speech" Haskins Laboratories, Status Rep. Speech Res., 91, 75-89.

WIENEKE, G., JANSSEN, P. and BELDERBOS. H. (1987) "The influence of speaking rate on the duration of jaw movements", Journal of Phonetics, 15, 111-126.

COMPENSATORY ARTICULATION DURING SPEECH: EVIDENCE FROM THE ANALYSIS AND SYNTHESIS OF VOCAL-TRACT SHAPES USING AN ARTICULATORY MODEL

Shinji Maeda
Département de Recherche en Communication par la Parole
Centre National d'Etudes des Télécommunications (CNET)
22301 Lannion, France

ABSTRACT. Temporal variations observed in the vocal-tract profiles during continuous speech were studied using a factor analysis. Roughly 1000 frames of cineradiographic and labiofilm data corresponding to 10 French sentences uttered by two speakers have been analyzed. The factor analysis permits us to describe the observed profiles as the sum of a small number of linear components. Since these components can be interpreted by articulatory terms, the linear model is considered as an articulatory one. For example, the tongue profiles can be specified by the following four parameters; the mandibular and tongue-dorsal positions, the dorsal shape, and the apex position. The entire vocal-tract configurations including the frontal lip-opening shapes can be described with reasonable accuracy by as few as seven parameters in total.

With this model the temporal variations are described in terms of the frame-by-frame samples of these articulatory parameters. We observe that "target" parameter values for the same vowel vary significantly presumably due to different phonetic contexts. An acoustic calculation with the model predicts that a particular pair of articulators can compensate acoustically each other. For example, by an appropriate adjustment of the tongue-dorsal position, the model is capable of producing the same F1-F2 pattern for different jaw position, or *vice-versa*. The compensation between the jaw and the dorsal positions, however, is possible only for unrounded vowels. In the case of rounded vowels, the jaw position can be compensated by the lip aperture.

The measured "target" values of the paired parameters indicated a linear relationship, suggesting that the speakers actually exploit the inter-articulator compensation in the speech production. This explains the observed large "target" value variability. The comparison of parameter trajectories for the same sentences uttered by the two speakers indicates similitude rather than difference, suggesting that the manner of the production involving the compensatory articulation could be relatively invariant.

1. INTRODUCTION

In continuous speech, we observe that the acoustic manifestation of each phoneme deviates considerably from its intrinsic characteristics, and depends on the phonetic context. Öhman (1966) has shown that the formant patterns in VCV (vowel-consonant-vowel) syllables are determined by a vowel-to-vowel gesture plus the consonant articulation. The acoustics of the consonant is, therefore, influenced by the surrounding vowels. Martin *et al.* (1982) have shown that the decoding process takes into account these coarticulatory modifications: the reaction time is longer for stimuli without coarticulatory transitions than stimuli with them. Reciprocally, the acoustics of a vowel can be modified as a function of the articulatory place

W. J. Hardcastle and A. Marchal (eds.), Speech Production and Speech Modelling, 131–149.

of the following consonant, *e.g.*, the palatalization or velarization of the vowel. When the intervocalic consonant is labial, the preceding vowel can be coarticulated heavily with the second vowel to such extent that its formant trajectories during the entire portion are modified (Vaissière, 1987). The degree of the coarticulation tends to increase with speaking rate. The acoustic observation is that phonemes in a sentence are not produced in a piece-by-piece fashion, rather they are "coarticulated" with the neighboring phonemes. In addition to the phonetic context, prosodic factors can affect on the individual articulatory positions. For example, lexical or emphatic stress can influence on the jaw position as demonstrated by Condax (1979) for Fijian and by Stone (1981) for English.

The term "coarticulation" is employed to explain specifically the contextual variations of the acoustic manifestation of consonants in the intervocalic position. A fixed target position of each articulator for a given vowel is assumed. In this way, the vowel-to-vowel transition is specified as the displacement of each articulator from one target to another. Mermelstein (1973) has adopted this scheme to drive his articulatory model for the synthesis of VCV syllables. The movements of the articulators which participate directly in the consonant production and which correspond to the vowel-to-vowel transitions are orchestrated on the basis of a look-ahead operator, proposed by Henke (1967). In effect, the acoustically complex coarticulatory effects are described by merely the specification of which articulator, for example the lips for the labial consonants, should participate in the realization of the consonantal closure in the model. What is crucial here is the validity of a set of "fixed" articulatory targets for each vowel, which provides anchor points in the specification of the movements and which makes the description of the coarticulation simple and explicit.

There are, however, numerous counter indications against fixed articulatory targets for vowels or in general for phonemes. The vocal-tract configuration necessary for the production of a consonant or of a vowel can be realized by many different combinations of individual articulatory positions. Hughes *et al.* (1976) have shown, in the case of a closed vowel, that the magnitude of the variation of the individual articulators, specifically the mandibular position and the degree of the labialization, is much greater than that of the lip aperture, which is the "true" target, in the sense that a right aperture must be achieved to produce a correct vowel. Presumably, a deviation in the position of one articulator is corrected by a repositioning of another articulator to achieve an appropriate lip aperture. They demonstrated, in effect, that the "motor equivalence" is operating in the articulatory process, and that the speech articulation is essentially a "goal oriented" process (Gay *et al.*, 1981). The implication is that fixed targets cannot be assumed.

If this is the case, we must postulate an articulatory strategy in which the target positions of the articulators are calculated as a function of their current states and the goal. In the case of the stop consonants for example, the calculation of a precise target may not be necessary, since the sensory mechanism can detect whether an articulator, such as the lips or the tongue tip, has reached the target or not, *i.e.*, whether a closure is accomplished or not. In fact such a feedback mechanism is implemented in the model of Henke (1967) and of Mermelstein (1973), which contributed to the simplification of their articulation strategy. But how to determine the variable targets in the vowel cases remains to be solved. The problem is that the phonemically same vowel can be produced with many, perhaps infinitely, different positions of the individual articulators. It is hard to accept, however, that a speaker arbitrarily and randomly determins a set of target values for a given vowel during speech. Rather it is reasonable to assume the influence of certain constraints, such as a principle of economy in articulatory effort, imposed on the articulatory strategy in the determination of the targets.

Lindblom *et al.* (1971) have demonstrated the existence of jaw-tongue compensation from their bite-block experiments. On the one hand, they showed that a speaker can produce almost identical acoustic patterns of vowels regardless of the fact that the jaw position is fixed by the bite-block or not(also in Lindblom *et al.*, 1979; Gay *et al.*, 1981). On the other hand, a model analysis of x-ray data on various sustained vowel articulations indicated an ordered relation between the jaw position and vowel height in the phonological sense. Without some constraints such ordered relation should not exist, since a vowel can be produced with almost any arbitrary jaw position. In order to account for the observed ordered jaw position, they have postulated a principle of "minimum antagonism". In the articulation of a vowel with a given height, there is an optimum jaw position so that the antagonism between the jaw and the tongue-body gesture becomes minimum. The corollary is that the optimum jaw position is proportional to vowel height.

Our previous study on the analysis of x-ray data for VCV logatoms has indicated such ordered relation between jaw position and vowel height (Maeda, 1979). In this particular study, it has been shown that not only the jaw position, but also tongue-dorsal position is proportionally related to vowel front-back position. In a simulation study with an articulatory model, Majid *et al.* (1986) were able to select a set of articulatory parameter values for French vowels having a coherent structure in terms of articulator positions as well as of formant patterns. These results can be explained nicely by the minimum antagonism principle. However their validity could be limited to the case of vowels in isolation or in VCV syllables. If such a principle were generalized literally to arbitrary sentences, then we would expect every vowel to be produced with their locally optimum and thus fixed targets, which is unlikely the case. Nevertheless, it is interesting to assess the presence or absence of fixed articulatory targets for vowels in the utterance of sentences.

In this paper, we shall document data in favor of the premise of the compensatory articulation operating in the production of vowels in "natural" speech (without a bite-block), and against the fixed target hypothesis. A salient feature of this study is that the simultaneous cineradiographic and labiofilm data are analyzed by using an articulatory model. The observed time-varying vocal-tract shapes are described in terms of frame-by-frame variations of model parameters, such as the jaw and tongue-dorsal positions *etc.*. Since the model specifies the entire tract shapes, the acoustic characteristics are calculated for a given set of parameter values. This is important in that it allows us to describe and evaluate the consequences of the articulatory compensation in terms of the acoustics, for example, of formant patterns, which are the "final" goal. Recall that what's available to listeners is the acoustic signal and the perceptual process must interpret such patterns to identify a vowel.

2. ARTICULATORY DATA

The data consist of more than 1000 digitized tracings of vocal tract shapes corresponding to 10 French sentences uttered by two female speakers, PB and DF. Each of the data frames representing the vocal tract profiles, from the glottis to the lip opening, and the frontal lip shapes was obtained by manually tracing the simultaneous radiofilm and labiofilm with 50 frames per second. An example of the data frames is shown in Fig. 1a and 1b. The digitized version of those data has been kindly provided by the Phonetic Institute of Strasbourg. Detailed information, such as the list of the sentences, selected x-ray tracings and the corresponding acoustic data, is found in Bothorel *et al.* (1986).

3. MEASUREMENT OF VOCAL-TRACT SHAPES

The first step of the data analysis is to measure the vocal-tract shapes with appropriate references. This procedure is necessary because there is no frame-to-frame correspondence in a series of points which constitutes a particular contour representing, for example, the midsagittal tongue shapes. Recall that the articulatory motions are represented cinematographically in our data. The measurements are done such that they contain sufficient information to calculate the acoustic characteristics. The vocal tract is divided into three sections, the lip-opening tube, the principal tract corresponding to the buccal and pharynx cavities, and the laryngeal extreme. The mandibular position is also measured for each frame.

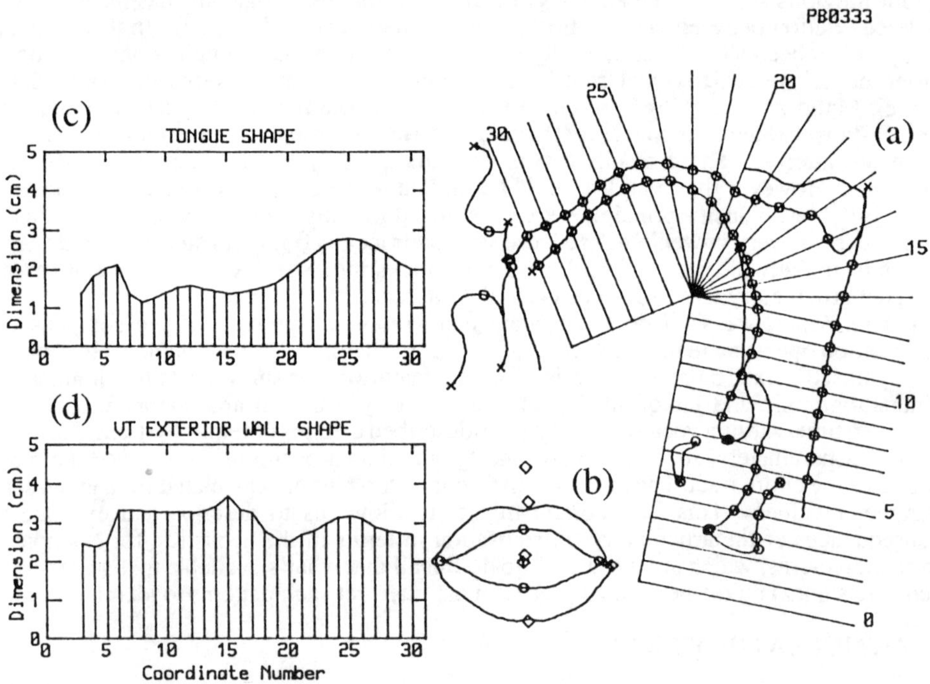

Figure 1. (a) Vocal-tract profile superimposed with a semipolar coordinate system, where the circles indicate the measured points. (b) Frontal view of the lips. (c) Vector representation of the measured tongue shape including the upper laryngeal region. (d) Vector representation of the exterior walls.

3.1. THE PRINCIPAL TRACT

The form of the interior and exterior tract walls is measured using a semipolar coordinate system as shown in Fig. 1a. The interior walls consist of the tongue tip, body, root, and the upper part of the larynx and the exterior contour, the upper incisors, hard and soft palates, rear pharynx and laryngeal walls. The coordinate is fixed to the rigid maxillary structure. All movements, or deformations of the tongue, of the velum, of the rear pharynx walls, and of the larynx are therefore measured with relation to the maxilla. The coordinate grids are spaced by 0.5 cm in the two linear regions and by 11 degrees in the polar region.

The intersections between the contours and the grids, indicated by the circles in Fig. 1a, are detected by an automatic procedure based on a binary search algorithm. A set of the coordinate values of the intersections in the order of the grid number then constitutes a vector representing the measured interior contour as shown in Fig. 1c and the exterior contour in Fig. 1d. The vectors are considered as the tract shape sampled at the coordinate grids.

3.2. LIP AND LARYNX SECTIONS

In the case of these two sections, the lip and the larynx tubes, the semipolar coordinate is useless, since these articulators tend to move in the direction perpendicular to the grid lines. For example, the larynx edge can move up and down. This leads to two undesirable consequences: first, the accuracy in the vertical position of the vocal-tract extreme is limited by the grid spacing, *i.e.*, 0.5 cm; second, since the edge can be above or below certain grids depending on the frame, the total number of the intersections, and therefore, the size of vector varies as a function of frames. Because we intend to analyse the observed vectors statistically, the vector size must be kept constant through all data frames for a given speaker.

For this reason, each of the two points corresponding to the anterior and the posterior extremes of the larynx edge is specified by two values (x, y) on the linear part of the coordinate system instead of sampling the shape at the grid lines. The form of the larynx tube is described by combining these two points and the intersections detected in the previous procedure. Admittedly this specification of the larynx shape is not a very satisfactory one. The larynx tube should have been modelled and the associated variables could be identified in this stage, as in the case of the lip-opening tube as described in the followings.

The lip-opening tube is modelled by a uniform elliptic tube with three variables, height, frontal width, and length (protrusion). The height and width are defined as, respectively, the distance between the highest and lowest points, and the distance between the most left and right points on the front inner lip contours. The four detected points are indicated by the circles in Fig. 1b. The protrusion is measured on the lip profile as the distance between the upper incisors and the point where the vertical separation of the upper and lower lips becomes minimal. These definitions of the three variables enable us to use a simple algorithm in the automated measurement.

3.3. REGENERATION OF VOCAL-TRACT SHAPES FROM THE MEASUREMENTS AND ACOUSTIC CALCULATION

Figure 2a illustrates the regenerated vocal-tract profile by projecting the measured vectors, shown in Fig. 1b and 1c, on the reference coordinate system. The frontal view of the lip-opening is shown in Fig. 2b. The area function (in Fig. 2c) and then the transfer function and the formant frequency values (in Fig. 2d) are calculated from the regenerated vocal-tract shape (Maeda, 1982). There is a relatively good agreement between the calculated formants

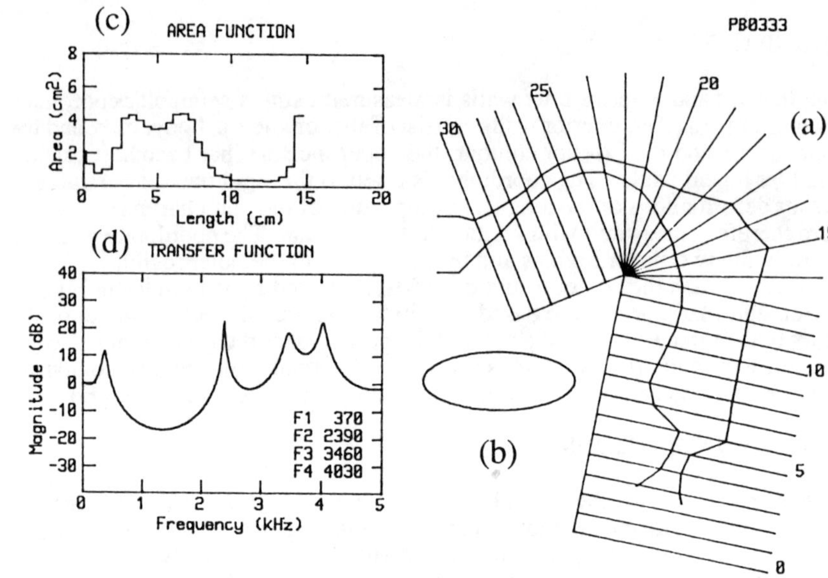

Figure 2. (a) Regenerated vocal-tract profile by projections of the tongue and exterior wall vectors (shown respectively at Figs. 1c and 1d) on the semipolar coordinate system. (b) An elliptic approximation of the frontal lip-opening shape. (c) The corresponding area function. (d) The calculated transfer function and the first four formant frequencies in Hz.

and their estimation from the corresponding spectrogram (Bothorel *et al.*, 1986). The estimated values are approximately, F1 = 300 Hz, F2 = 2800 Hz, F3 = 3200 Hz, and F4 = 4000 Hz.

It must be noted that the conversion from sagittal dimension to area has not been really established (Heinz *et al.*, 1964; Gauffin *et al.*, 1978; Perrier *et al.*, 1989). The formula, $A = \alpha x^{\beta}$, where A and x are respectively the cross-sectional area in cm^2 and the corresponding midsagittal dimension in cm, was used. The values of the coefficients α and β vary along the length of the tract, and are determined by an *ad hoc.* manner. The relative correspondence between the calculation and the estimation merely indicates that there is no gross error in the specification of the tract profile, and in the sagittal-to-area conversion. However the calculated vocal-tract transfer is important in that we actually model the measured vectors, and not directly the original vocal-tract contours; therefore the transfer function can serve as a reference in the evaluation of the model in terms of its acoustic performance.

4. LINEAR COMPONENT MODEL OF THE TONGUE

As described below, we have employed a factor analysis technique to construct an articulatory model on the basis of the cineradiographic and labiofilm data. This does not necessarily mean that an articulatory model can be derived automatically by such an analysis. We must have a basic idea about how to represent the vocal tract. In this regard, we follow a jaw based model proposed by Lindblom *et al.* (1971), in which vocal tract shapes are determined as a function of parameters such as jaw, tongue-body, tongue-tip, lip height and width, and larynx height. The underlying hypothesis is that, during speech production, the complex activities of the articulatory organs are organized into a limited number of independently controllable functional blocks and that the tract shapes are determined by the state of these blocks. Let us call these blocks "elementary articulators" and their action "elementary gestures". A descriptive model proposed by Coker *et al.* (1966), by Coker (1976), or by Mermelstein (1973) consists of elementary articulators which are defined by geometrical elements. For example, the tongue body is represented by a cylinder of which position and diameter can be varied to emulate the gesture. In our approach, we attempt to extract these elementary articulators from the data by means of a statistical analysis.

The factor analysis being founded on the correlation structure of the data, the lips, tongue, larynx, and velum which can act independently, must be analyzed separately. Since the same analysis technique is employed for all articulators, we shall describe here only the analysis of the tongue contour, considering that the tongue body is undoubtedly the most important articulator.

The tongue shape of each frame can be described by the measured vector with about 30 variables, such as shown in Fig. 1c. Their values depend on the states of the individual elementary articulators. The basic assumption is that the influence of each elementary gesture on the deviation of the tongue from its neutral shape is proportional to a parameter representing a strength of that gesture, and that the influences from the different articulators can be added up to produce the final tongue shapes. In short, the measured vectors, thus the vocal-tract shapes, are represented by a linear combination of the effects of elementary gestures.

The model, therefore, can be regarded as a linear transformation of the measured vectors. An advantage of the linear formulation is that once the transformation is determined, the parameter values can be calculated from the measured data. The linear transformation is not unique however. There exist, in fact, an infinite number of possible transformations. We qualify a linear model as an articulatory one only when each of the determined components represents, or can be interpreted as, an elementary articulator.

The first linear model has been proposed by Liljencrants (1971). In his study, the vector representing the tongue-body shape is expanded into a cosine series. Only the first three components suffice to specify the observed tongue shapes with good accuracy. On the vector representing the tongue shapes, the effect of each elementary gesture appears as a cosine pattern with a specific spatial frequency, whose magnitude is proportional to the parameter value. Although the resultant components are similar to those from a factor analysis, specifically from the principal component analysis, the decomposition of the mid-sagittal profile in terms of a cosine series can not have conceptual significance.

The elementary articulators can be extracted without a prescribed pattern using a factor analysis technique, for example, by the classical principal component analysis (Maeda, 1978; Zerling, 1979) or by the PARAFAC analysis (Harshman *et al.*, 1977; Jackson, 1988). The principal component analysis imposes the following two constraints to derive a unique decomposition: components are orthogonal in a statistical sense, *i.e.*, they are uncorrelated to each other; for each component, the variance is maximally extracted. PARAFAC does not impose, in principle, the orthogonality condition. Although there are no constraints on the patterns, both statistical methods do not guarantee that every component can be interpreted by articulatory terms. Since they provide a unique solution for a given set of data, one can only hope that the analysis results in an articulatory model. Unfortunately this is often not the case. For example in our principal component analysis (Maeda, 1978), it was impossible to extract the component representing the influence of the jaw position, even if the measured jaw position was included in the analysis data.

The situation was considerably improved by using the so-called arbitrary factor analysis (Overall, 1962; Maeda, 1979). The idea of using this particular technique was inspired by Kiritani *et al.* (1977). In their study, the microbeam x-ray data, representing the trajectories of the three or four lead pellets placed on the tongue surface, were subjected to the principal component analysis. Before subjecting the data to analysis, the effects of the jaw gesture, detected by tracking an additional pellet placed behind the lower incisors, upon the tongue pellet positions are extracted by means of the linear regression and then subtracted from the data. In this way, the jaw component was sure to be extracted. In addition the principal component analysis of the residual after the subtraction resulted in components which could be easily interpreted as the front-back tongue body position and the flat-arched tongue shape.

The arbitrary factor analysis can be considered as a generalized and somewhat more formal version of the extraction-subtraction procedure of known parameters. By loading a specific variable, in our case the measured jaw position, the jaw component is extracted. Then the effect of the jaw component is subtracted from the correlation matrix. After the subtraction, for example, a tongue variable in the pharynx region, the coordinate number from 8 to 14 in Fig. 1a, can be assumed to represent the tongue-dorsal position, if the effects of other components are minimal. If this is the case, the dorsal component can be extracted by similar fashion and then subtracted. This procedure is repeated until all relevant components are extracted one-by-one. Or at some stage, the residual can be subjected to the principal component analysis to extract the variance maximally with a minimal number of components. An important point in this approach is that it provides us with a certain degree of control over the analysis to obtain a satisfactory model.

The results of our analysis of the cineradiographic data indicated similar patterns representing the effects of the jaw, tongue-body position, and the tongue shape, except our model includes the fourth component that accounts for the raising and lowering of the tongue tip.

In Figs. 3a - 3d, the influence of each component extracted for the speaker PB is illustrated by the projection on the coordinate system varying systematically the standardized parameter value in three steps, -2, 0, and 2. In this particular example, the residual after the extraction of the jaw component was subjected to the principal component analysis. The parameter values are normalized by the corresponding standard deviations. The central trace in each figure represents a "neutral" tongue position, *i.e.*, the averaged tongue contour over the data. The two other traces indicate the deformation of the tongue due to the activation of a component in question. The exterior walls plotted in each figure correspond to their neutral shape.

4.1. INFLUENCE OF THE JAW POSITION

The tongue movement due to the displacement of jaw position is shown at Fig. 3a. The open and close mandibular gesture corresponds to the concomitant vertical displacement of the tongue in the buccal region and the horizontal in the pharynx region. It should be noted that the variance extracted (denoted by "v.e." in the figure) is 15% for this speaker PB. This means that only 15% of the variance observed in the tongue data is explained as the consequence of the jaw articulation. For the second speaker DF, the variance is 24%. It can be interpreted that the second speaker uses more mandibular gestures than the first speaker in the production of speech.

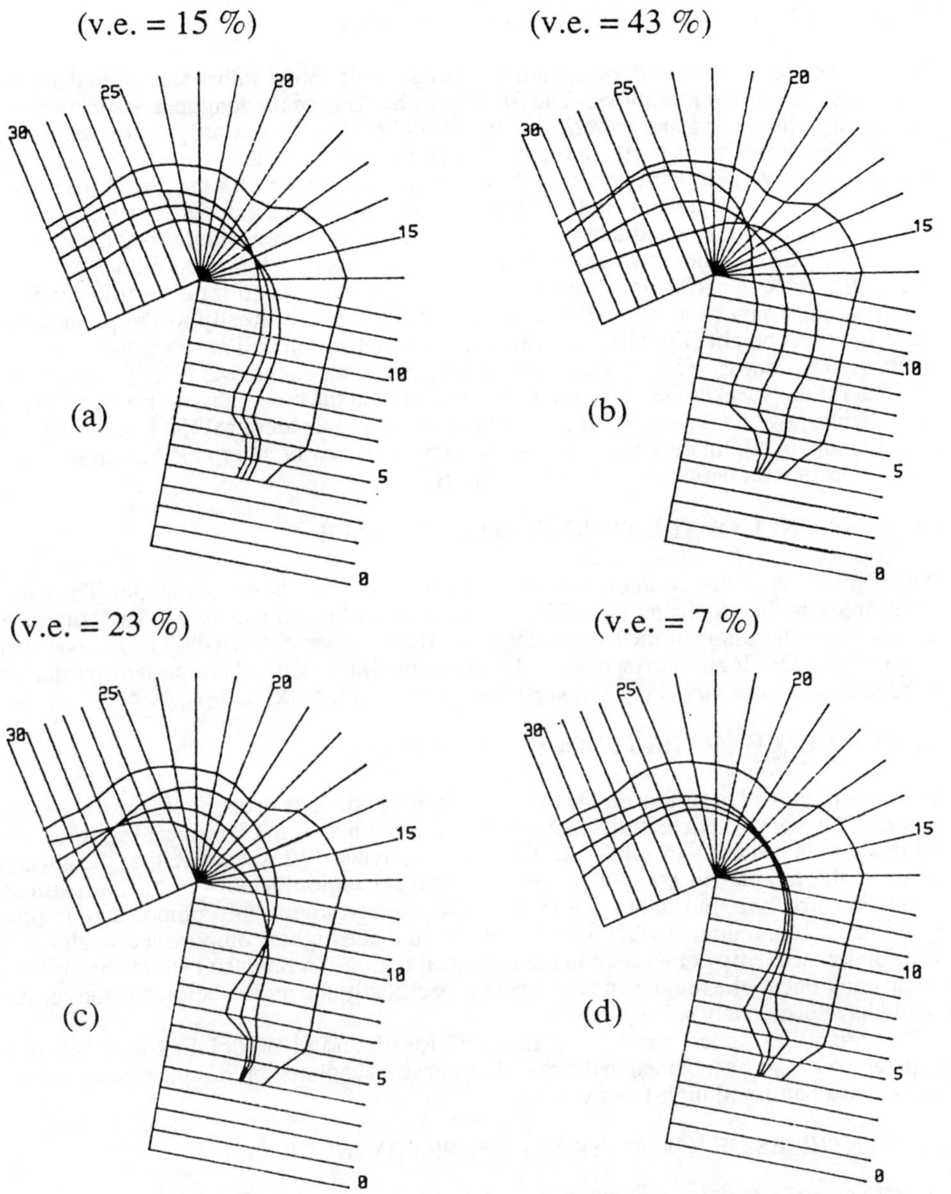

Figure 3. Influence of the four articulatory parameters upon the tongue shape for the speaker PB: (a) Jaw position (open/close). (b) Tongue-dorsal position (front/back). (c) Tongue-dorsal shape (arched/flat). (d) Tongue-tip position (raised/lowered). In each figure, "v.e." indicates the variance explained by the component.

4.2. INFLUENCE OF TONGUE-DORSAL POSITION

The influence of the tongue dorsal position, shown in Fig. 3b, is rather similar to that of the jaw position described just above. The fronting or backing of the tongue is associated with, respectively, the raising and lowering in the buccal region. Contrarily to the jaw gesture however, the emphasis is more on the horizontal displacement than the vertical one in the sense that the invariant point occurs at the coordinate number 19 for the dorsal position (in Fig.3b) and 16 for the jaw position (in Fig. 3a). Considering the anatomical structure of the tongue, it cannot be expected that the tongue can move only in the horizontal directions.

As described later, these two components, jaw and dorsal positions, coordinate with each other during speech and their acoustic effects can be compensated for a specific condition. The open-close jaw gesture, therefore, does not correspond directly to the phonological notion of vowel height. Equally, the front-back dorsal position will not correspond to vowel frontness or backness. The variance extracted by this component is 43% for the speaker PB and 32% for the speaker DF. It is interesting to note that the cumulative variance of the two components, jaw and dorsal, of PB and that of DF are almost identical (58% *vs.* 56%). This almost identical cumulative value for the two speakers can be interpreted as an indication of the coordination between the two articulators.

4.3. INFLUENCE OF THE TONGUE-DORSAL SHAPE

The contribution of this component on the tongue contour is shown in Fig. 3c. The pattern of the three contours indicates an arching or a flattening of the tongue body. This component thus modifies the shape of the tongue dorsum. The variance extracted by this dorsal shape component is 23% for the two speakers. The contribution is relatively important, particularly in the articulation of vowel [u] and velar consonants such as [k] and [r].

4.4. INFLUENCE OF THE TONGUE-TIP POSITION

As shown in Fig. 3d, its effect on the tongue contours manifests primarily in the tongue apex region. This component exhibits a considerable influence in the laryngeal region as well, but this is rather due to an artifact of the factor analysis. We speculate that the apparent effect of the tongue tip component in the laryngeal region is not due to the particular correlation structure, but rather due to the orthogonality among the components, imposed by the statistical method. In fact we often observed that a higher component cancels out the contribution of the tip component in the laryngeal region. In order to extract appropriately the tip component, the tongue and the larynx, specifically the region below the tongue root, should have been analyzed separately.

The variance extracted is relatively small, 7% for PB and 11 % for DF. Nevertheless this tip parameter must be included in the model, because it accounts for the apical gesture during dental consonants and high-front vowels.

4.5. SYNTHESIS OF VOCAL-TRACT SHAPES BY MODEL

The first four components described above together explain 88% of the variance for PB and 90 % for DF. As mentioned already, the fifth component, which accounts for about 4% of the variance, contributes to improve the description in the laryngeal region. The variance extracted is a convenient statistical measure indicating only an "average" performance of the model. Even when its value is high, it is still possible that some individual frames exhibit a catastrophic error. In order to judge the adequacy of the model, the tract shapes must be synthesized, and its acoustic characteristics calculated.

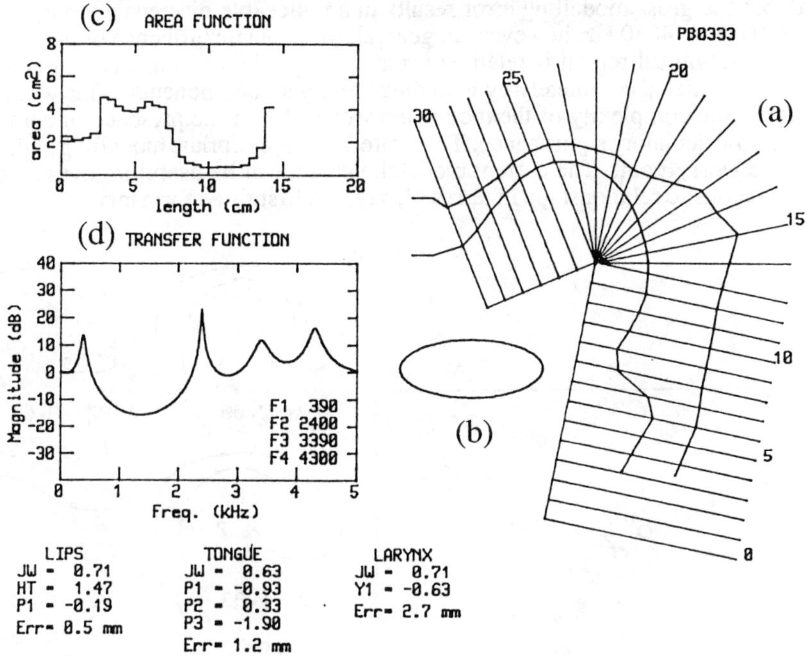

Figure 4. The vocal-tract profile at (a) and the frontal lip-opening shape at (b), synthesized by the model with seven parameters. The parameter values are calculated from the data frame shown in Fig. 2a and 2b. The exterior walls are fixed to the "neutral" shape. (c) The corresponding area function. (d) The transfer function and the estimated first four formant frequencies in Hz. The normalized parameter values and averaged absolute errors in mm are indicated at the bottom.

An example of the synthesized vocal-tract with seven most significant components is illustrated in Fig. 4a and 4b, where the parameter values are calculated from the frame shown in Fig. 2a and 2b. Any discrepancy of the model generated vocal-tract shapes in relation to the original is solely due to the truncation of the higher components. The lip opening tube is specified by the jaw position plus two parameters, the height and protrusion. The tongue profile is specified by the four parameters already described in detail above. Each of the two laryngeal extreme points depends on the two parameters; the jaw and its own height. An averaged contour is used for the vocal-tract exterior walls. The calculated values of the parameters and absolute average errors (in mm) are indicated at the bottom in Fig. 4. Due to the truncation of the higher components, the calculated values of the jaw parameter differ somewhat depending on the sections, *i.e.*, 0.63 for the tongue and 0.71 for the lip and larynx sections.

Note that a relatively large error occurring in the larynx (2.7 mm) is due to a poor prediction of the horizontal position, which is further aggravated by the poor modelling of the larynx tube. This error in shape is manifested naturally on the calculated area function in Fig. 4c, and then on the transfer function to a certain extent. Comparing the transfer function calculated from the model in Fig. 4d with that from the measured shape in Fig. 2d, it is observed that the gross modelling error results in a noticeable discrepancy only in the F4 values, 4380 Hz *vs*. 4030 Hz, however. In general, the acoustic influence of the vocal-tract shapes in the laryngeal region is relatively small except higher formants. Of course, the error can be eliminated completely by including the higher components. This will result in an increase in the complexity of the model and worse still in the presence of components which have no articulatory significance. Therefore, more appropriate modelling of the larynx tube must be worked out. The current model, however, can be useful in investigating the articulatory-acoustic relationships in terms of, say, the first three formants.

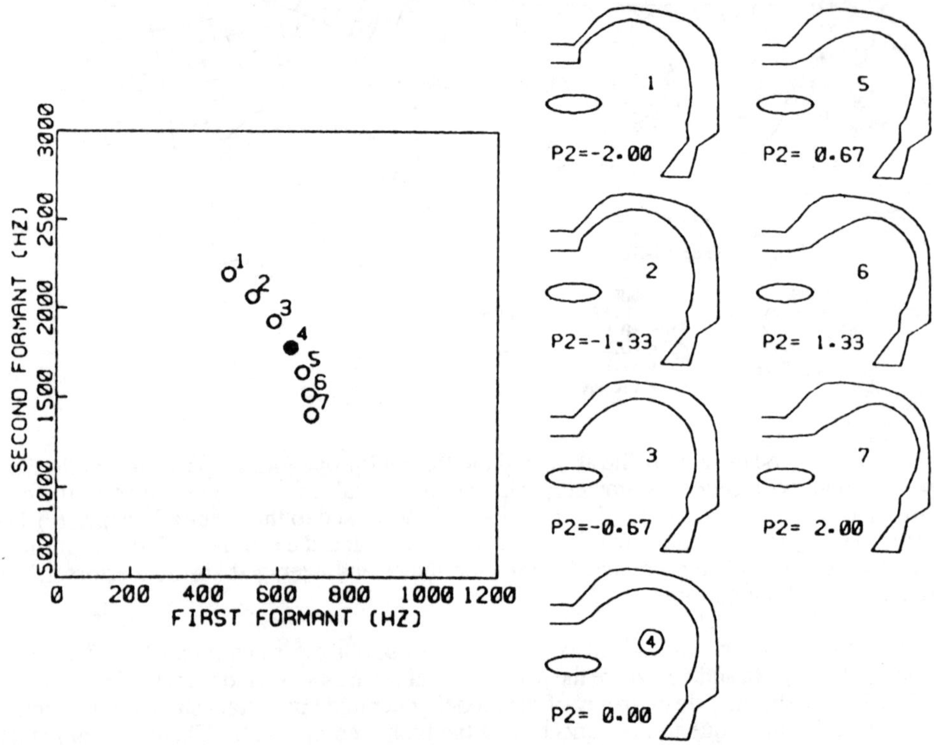

Figure 5. Acoustic effects, in terms of F1-F2 plots, of a variation in the tongue-dorsal position around the neutral vowel marked by number 4.

5. ARTICULATORY-ACOUSTIC RELATIONSHIPS

An important advantage of an articulatory model specifying the entire vocal-tract is that the relationships between the articulation and its acoustic consequences can be studied independently of the data. In fact, such predicted relations help to understand better the data representing the frame-by-frame samples of the individual articulatory gestures.

In Fig. 5, the acoustic effects of a change in the tongue-dorsal position are illustrated in terms of the F1-F2 diagram. All parameter values, except that of the dorsal, are set equal to zero. The dorsal parameter value is varied from -2.0 to 2.0 in seven steps. The magnitude of variation is well within that observed during speech, since it rarely exceeds the range between -3.0 and 3.0 for any parameter. The correspondence between the model profiles and the F1-F2 plots is marked by the numbers from 1 (most front) to 7 (most back). The profile marked by 4 with the circle corresponds to the neutral shape. The trajectory of the F1-F2 plots is highly vowel dependent, because the acoustic effects of a component depend upon the states of the other components. This particular trajectory in Fig. 5 is, therefore, regarded as an acoustic manifestation of the dorsal component, which is valid only around the neutral position.

In Fig. 6a, the acoustic effects of a variation in the two articulatory parameters, jaw position marked by the open circles and dorsal position by the closed diamonds, are plotted on the F1-F2 plane. The effects are calculated around the five cardinal vowels. The acoustic consequences of the two articulator movements are distinctively different for the two vowel groups, rounded vowels [u and o] and unrounded vowels [i, e, and a].

For the rounded vowels [u and o], on the one hand, the trajectories indicating the effects of the jaw parameter variation are almost vertical. Observe the dispersion of the open circles. The dispersion of the diamonds indicating the effects of the dorsal parameter variation, on the other hand, are almost horizontal. This means that for the rounded vowels the effect of the jaw appears on the F1 value alone, and those of the dorsal position primarily on F2.

For the unrounded vowels [i, e, and a], the F1-F2 dispersions of the two articulators tend to overlap each other. This means that a deviation in the position of one of the two articulators can be compensated by an adjustment of the remaining articulator to produce a "target" F1-F2 pattern of these unrounded vowels.

Interestingly the situation is exactly opposite for the jaw and the lip aperture (as the results of the lip rounding or spreading gesture). In Fig. 6b, the F1-F2 dispersions around the same five vowels as acoustic affects of the lip parameter, indicated by the closed triangles, are superimposed on these of the jaw. In contrast to the jaw-dorsal case, the overlap occurs for the rounded vowels [u and o], and the F1-F2 dispersions of the two gestures are roughly orthogonal for the unrounded vowel series, [i, e, and a]. A deviation of the jaw position therefore can be acoustically compensated for any vowel either by the adjustment of the dorsal position or by that of the lip rounding, depending on the two vowel categories. Then it was not so surprising after all to find that the mandibular gesture of the speaker PB was much weaker than that of the second speaker DF, as mentioned before. The weak mandibular gesture of PB can be compensated by an appropriate adjustment of the tongue-dorsal or of the lip rounding gesture.

It is noted that the F1 and F2 frequencies are quite stable against the variation of the lip aperture, in particular for the vowel [i] where the triangles are closely clustered together. This doesn't mean however that the lip rounding gesture has no effect on the phonetic value of that vowel. In fact, the effects of the rounding can appear as the lowering of F3 frequency. The rounding puts the F3 closer to F2 and apart from F4, resulting in the sound change from [i] to [y].

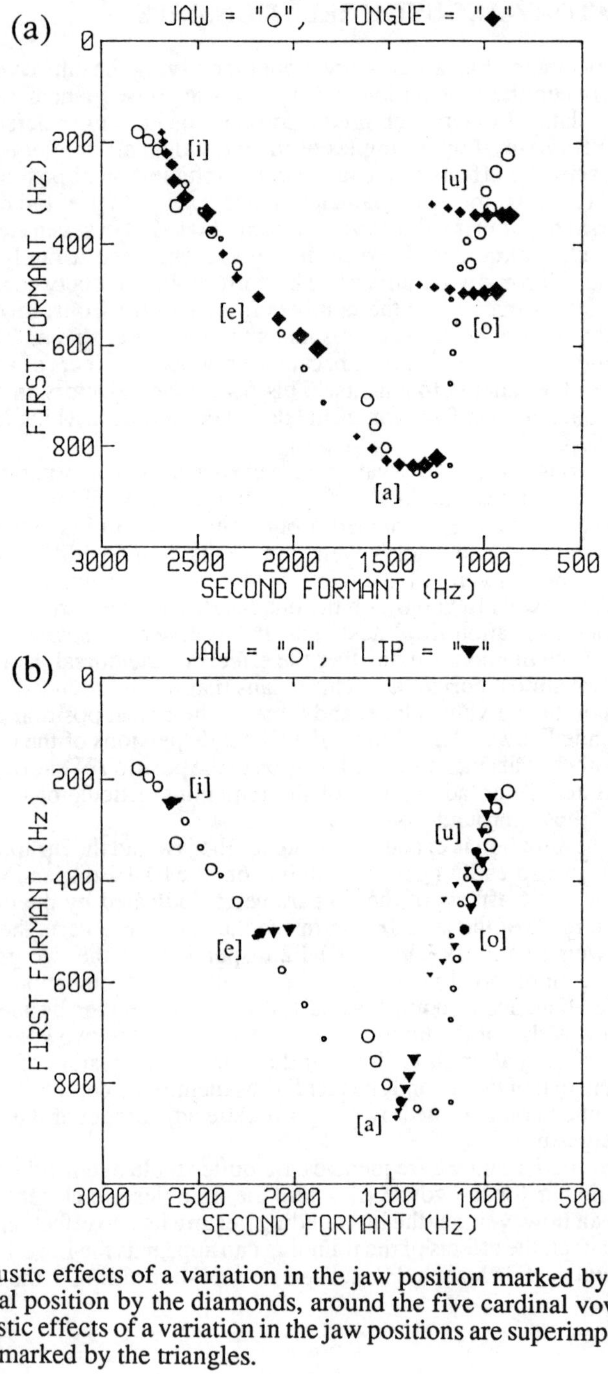

Figure 6. (a) Acoustic effects of a variation in the jaw position marked by the circles and in the tongue-dorsal position by the diamonds, around the five cardinal vowels. (b) Same as (a), except acoustic effects of a variation in the jaw positions are superimposed with those in the lip aperture marked by the triangles.

6. COMPENSATORY ARTICULATION

The articulatory-acoustic relationships have indicated that the compensation (or one might prefer the term, inter-articulatory coordination) between the jaw and tongue-dorsal positions is acoustically effective for the unrounded vowels, and between the jaw and lip positions for the rounded vowels. One might raise the question whether or not speakers actually exploit the compensatory possibility during the production of speech. In order to provide an answer to this question, we have calculated the frame-by-frame variation of the articulatory parameters for the 10 sentences pronounced by the two speakers. Temporal patterns of parameters during vowels are then investigated. Unfortunately, the number of occurrences of the same vowel is small, ranging from only once to at most seven times. The data for only two vowels, [i and a] therefore, were used for the investigation. As a consequence of this, only the jaw-dorsal coordination is tested.

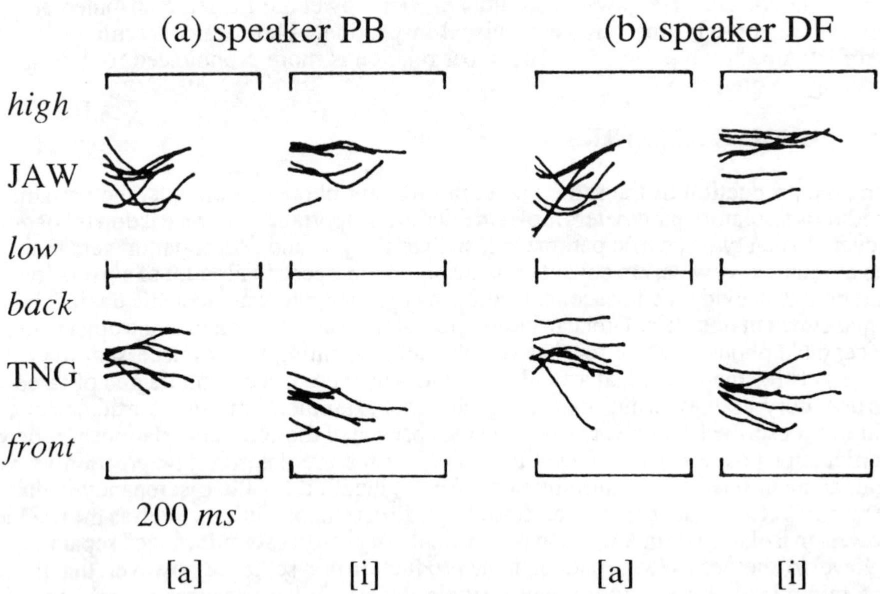

Figure 7. Superimposed temporal patterns of the jaw and tongue-dorsal parameters during the vowels [i and a] in different phonetic contexts, for the speaker PB on the left and DF on the right.

The calculated temporal patterns of the jaw and dorsal parameters concerning the two vowels are superimposed by aligning them at each vowel onset, as shown in Fig. 7. They exhibit a large variability for the same vowel, which are in different phonetic contexts. It is noted that the vertical space between the two abscissas for each parameter equals to six times the observed standard deviation, which corresponds roughly to the maximum range of the parameter variation. What's more striking however is the relatively close range of the two bundles of the contours corresponding to these two extreme vowels. As seen in the upper part of Fig. 7, the vertical ranges of the jaw parameter patterns of these vowels overlap considerably in particular for speaker PB. At the lower part of the figure, the dorsal position tends to be more fronted for [i] than [a], but the ranges of the variations still overlap to some extent.

Even though we do not expect an invariant target for individual articulators, it is still surprising to find such a poor contrast between high and low vowels. Indeed, the figures nicely fit with the prediction from the articulatory-acoustic relationships described above; the poor contrast can be interpreted as a manifestation of the jaw-dorsal coordination. Such a magnitude of the variations and thus the poor contrast is possible, because the acoustic consequences can be compensated by the coordination of the other articulator. If this is the case, the two parameters, jaw and dorsal positions, should exhibit a linear or at least a monotonous relation for each vowel.

In order to verify this hypothesis, the tongue-dorsal positions are plotted as a function of the corresponding jaw positions, as shown in Fig.8a for the speaker PB. The dorsal and jaw positions are sampled at minimum or maximum point of the contours within each vowel token. In the case where there is no such point, the parameter values are sampled at the center of the vowels. The jaw-dorsal plots of each vowel are indeed distributed along the corresponding straight lines drawn by visual inspection. The data concerning the second speaker DF are shown in Fig. 8b. The linear relation is more pronounced for this speaker, supporting the prediction.

7. CONCLUDING REMARKS

During the production of the same vowel in different phonetic contexts, the variability of individual articulatory parameters is of considerable importance. The variation is not random however. When two specific parameters, such as the jaw and dorsal parameters in the case of the unrounded vowels, are considered, they indicate a certain amount of correlation. This is interpreted as evidence for acoustic compensation such that an acoustic deviation due to a displacement of one articulator can be compensated by an adjustment of the other to produce an acceptable phonetic value for that vowel. Consequently, it is not necessary to postulate fixed targets for the individual articulators, and thus there is a certain degree of freedom in the articulatory strategy, which can be exploited, for example, to minimize articulatory effort.

The data described here clearly indicate the absence of the coherent relationships between the articulator position and a vowel phonological feature, *i.e.*, the jaw position *vs.* vowel height, as mentioned in the introduction. An explanation for the discrepancy is that such coherent vowel structure can be manifested only in certain conditions, such as the production of vowels in isolation or in VCV's, where articulatory efforts are minimized separately from one vowel to another. We speculate, in the production of a sentence however, that the range of the minimization spans more than a single vowel. What speakers intend to minimize could be the articulatory effort over a sequence of phonemes rather than effort within single phonemes, resulting in the enormous variation of the temporal patterns during individual vowels and probably consonants. Since the compensation cannot be perfect, the variations in the articulatory process must manifest as a certain amount of the acoustic deviations, for example, on formant transitions. Such deviations are recognized as the context dependent coarticulatory effects.

The speaker's strategy in the articulation of sentences turned out to be more complex than previously proposed strategies. With the possibility of the compensatory articulation, we cannot assume a set of fixed targets for each vowel and probably not for each consonant either. In other words, the articulatory strategy can have multiple solutions. It might be possible that every speaker has different solutions even under the constraint such as the minimal effort. If this is the case, every speaker might articulate the same sentence differently.

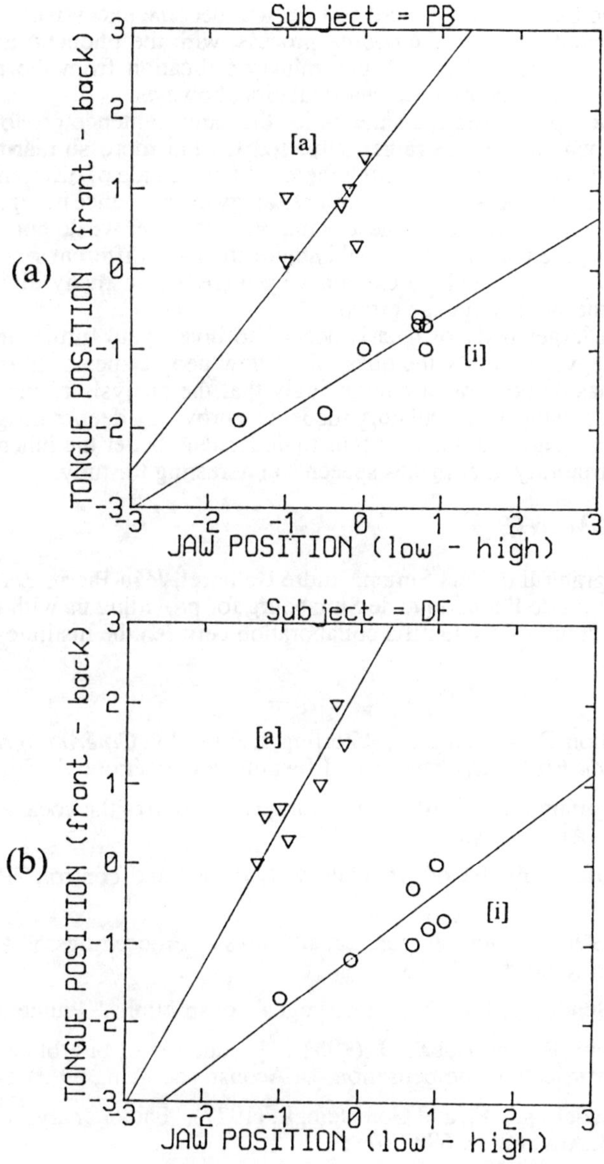

Figure 8. A linear relationship, indicated by the straight lines, between "target" positions of the jaw on abscissa and the tongue-dorsal on the ordinate, for the two unrounded vowels [i and a].

Consequently, the line of study described here will become excessively complex in trying to relate the articulations, *i.e.*, the coding process, with the phonetic universal, which is essentially speaker independent. A preliminary indication from the present data is in opposition to such a chaotic manner in articulation, however.

When calculated parameter movements for the same sentence pronounced by the two speakers are compared, we are rather surprised to find more similarity than difference between them. This means that although there are a multitude of different ways to produce the same sentence, both speakers use a similar strategy to pronounce it, suggesting a relatively invariant rule issuing commands to the articulators. However we do not expect a unique set of such rules for a given language. Probably there are many different sets of rules. Without such multiplicity, it is difficult to explain why there are so many different dialects and idiolects within the same language group.

We feel that a larger body of data is needed to investigate further the question of the articulatory strategy, or simply the question of how people speak. In our hope, the work presented here has demonstrated convincingly that the analysis of the conventional cineradiographic data using an articulatory model can provide a deeper insight into the speech production process, and that it can help us to understand better the inherent characteristics of speech, the variability, that makes speech so interesting to study.

ACKNOWLEDGMENTS

The author is grateful to Péla Simon, André Bothorel, Jean-Pierre Zerling, and François Wioland at l'Institut de Phonétique de Strasbourg for providing us with the digitized x-ray data base, as the result of a scientific collaboration between the institute and CNET.

REFERENCES

Bothorel, A., Simon, P., Wioland, F., et Zerling, J. P. (1986). *Cinéradiographie des Voyelles et Consonnes du Français,* Travaux de l'Institut de Phonétique de Strasbourg.

Coker, C. and Fujimura, O. (1966). "Model for specification of the vocal tract area function," J. Acoust. Soc. Am., 40, 1271.

Coker, C. (1976). "A model of articulatory dynamics and control," Proc. IEEE, 64(4), 452-460.

Condax, I. D. (1979). "Mandible position and stress," Proceedings of ASA50, J. J. Wolfs and D. H. Klatt (editors), 131-134.

Gauffin, J. and Sundberg, J. (1978). "Pharyngeal constrictions," Phonetica, 35, 157-168.

Gay, T., Lindblom, B., and Lubker J. (1981). "Production of bite-block vowels: Acoustic equivalence by selective compensation," J. Acoust. soc. Am., 69(3), 802-810.

Harshman, R., Ladefoged, P., and Goldstein, L. (1977). "Factor analysis of tongue shapes," J. Acoust. Soc. Am., 62(3), 693-707.

Heinz, J. M. and Stevens, K. N. (1964). "On the derivation of area functions and acoustic spectra from cineradiographic films of speech," J. Acoust. Soc. Am., 36, 1037.

Henke, W. L. (1967). "Preliminaries to speech synthesis based upon an articulatory model," Proc. IEEE Conf. Speech Commun. Process., 170-182.

Hughes, O. M. and Abbs, J. H. (1976). "Labial-mandibular coordination in the production of speech: Implication for the operation of motor equivalence," Phonetica, 33, 199-221.

Jackson, M. T. T. (1988). "Analysis of tongue positions: Language-specific and cross-linguistic models," J. Acoust. Soc. Am., 84(1), 124-143.

Kiritani, S., Sekimoto, S., Imagawa, H., and Fujisaki, H. (1977). "Parameter description of the tongue movements for vowels," Contribution papers of 9-th ICA, Madrid, Vol. 1, I13, 419.

Liljencrants, J. (1971). "Fourier series description of the tongue profile," Speech Transmission Laboratory, Royal Institute of Technology, Stockholm, Sweden, QPSR-4, 9-18.

Lindblom, B. E. F. and Sundberg, J. E. F. (1971). "Acoustical consequences of lip, tongue, jaw, and larynx movement," J. Acoust. Soc. Am., 4 (Part 2), 1166-1179.

Lindblom, B., Lubker, J., and Gay, T. (1979). "Formant frequencies of some fixed-mandible vowels and a model of speech motor programming by predictive simulation," J. Phonet. 7, 147-161.

Maeda, S. (1978). "Une analyses statistique sur les positions de la tongue: Etude préliminaire sur les voyelles françaises," 9èmes JEP, GALF, 191-199.

Maeda, S. (1979). "Une modèle articulatoire de la tongue avec des composantes linéaires," 10émes JEP, GALF, 152-164.

Maeda, S. (1982). "A digital simulation method of vocal-tract system," Speech Communication, 1, 199-229.

Majid, R., Boë, L.J., and Perrier, P. (1986). "Fonctions de sensibilité, modèle articulatoire et voyelles du français," 15émes JEP, GALF, 59-63.

Martin, J. G. and Bunnell, H. T. (1982). "Perception of anticipatory coarticulation effects in vowel-stop consonant-vowel sequences," J. Exp. Psychol.: Human Percept. Perform. 8, 473-488.

Mermelstein, P. (1973). "Articulatory model for the study of speech production," J. Acoust. Soc. Am., 53, 1070-1082.

Öhman, S. E. F. (1966). "Coarticulation in VCV utterances: Spectrographic measurements," J. Acoust. Soc. Am., 39, 151-168.

Overall, J. E. (1962). "Orthogonal factors and uncorrelated factor scores," Psychological Reports, 10, 651-662.

Perrier, P. et Boë, L. J. (1989). "Passage de la coupe sagittale à la fonction d'aire: les zones de faibles dimensions," J. Acoustique, 2, 59-67.

Stone, M. (1981). "Evidence for a rhythm pattern in speech production: Observations of jaw movement," J. Phonetics, 9, 109-120.

Vaissière, J. (1987). "The use of allophonic variations of /a/ in automatic continuous speech recognition of French," Speech Communication Group Working Papers, Research Laboratory of Electronics, Massachusetts Institute of Technology, Volume V, 15-25.

Zerling, J. P. (1979). *Articulation et Coarticulation dans les Groupes Occlusive-Voyelle en Français,* Doctorat de Troisième Cycle (phonétique), Université de Nancy II.

ARTICULATORY SYNTHESIS

CELIA SCULLY
Department of Linguistics & Phonetics
University of Leeds
Leeds LS2 9JT, UK

ABSTRACT

Ways in which speech synthesis might go beyond acoustic source-filter theory are considered. The physical processes of speech production to be represented and the linguistic units to be used in articulatory synthesis are considered. As an example of the kinds of analyses and simulations required, the characterisation of two contrasting speakers and simulation of some of the processes in their productions of a plosive-affricate contrast in R.P. English are attempted.

1. Some aspects of Articulatory Synthesis

1.1. INTRODUCTION

Any attempt to produce synthetic speech is based on some view or model of how natural speech is made. Diphone synthesis, for example, exploits the fact that some portions of the acoustic signal change less rapidly and are less influenced by phonetic context than others. Acoustic based synthesis has generally assumed that there are sources of sound which are separable from filter shapes and that the resonance processes linking them are linear.

One path towards better synthesis requires more complex context-dependent rules for acoustic domain parameters such as formant frequencies. Another approach is to try to go beyond linear acoustic theory in order to construct a signal model more closely matched to the physics of speech production. This means that stages which come conceptually before acoustic processes must be modelled, to include some or all of the following: neuromuscular, articulatory, aerodynamic and acoustic source generation processes. This second approach is likely to yield a better understanding of both normal and disordered speech, as well as, eventually, more natural sounding synthetic voices.

If we include in this review speech synthesis in which the articulatory shapes and movements are explicitly represented, then we are bound to include aerodynamic and acoustic models as well; a simulation of the geometry will not, in itself, generate sounds. Articulation will be defined as the positions and movement paths for all the moving structures used in speech production, not just of those in the vocal tract, above the glottis. It is quite difficult to define the terms and to draw clear distinctions between those synthesis methods which are entirely in the acoustic domain and those others which go further back in the stages of speech production as a signal model for speech. A distinction can be made between terminal-analogue and line-analogue synthesis. In the former the resonances of the vocal tract are controlled independently and realism is achieved by matching formant patterns in the synthesiser to those seen on spectrograms of natural speech. In the latter, the acoustic patterns are derived from a simulation of a resonating acoustic tube, the vocal tract. When this changes its shape all the formants change together, so they are constrained to covary in a natural kind of way. Control parameters for both these kinds of synthesis are obtained from the acoustic signal of speech, the sound pressure waveform. For the purposes of this paper, therefore, they will be considered to be acoustic-domain methods for synthesis and will not be included under the name of articulatory synthesis. Articulatory synthesis will be defined here as quantitative simulations or models of the processes of speech production which start from articulation, as defined above, and generate speech-like soundwaves. Under the overall name of articulatory synthesis, we

W. J. Hardcastle and A. Marchal (eds.), Speech Production and Speech Modelling, 151–186.

shall need to consider articulatory models, aerodynamic models, acoustic source models and models for the acoustic filtering processes and the radiation of a soundwave, the output signal.

Probably the two most famous complete models of speech production are those of von Kempelen (1791) and of Flanagan and his co-workers (1975). Both of these articulatory synthesisers model as one system low frequency aerodynamics and audio frequency acoustics, though by very different means (see Linggard, 1985). Von Kempelen's model was a mechanical one, with bellows for the lungs, providing air pressure which made a reed vibrate as the voice source. It included narrow pipes for the generation of turbulence noise, and a mobile mouth made of india-rubber permitted changing filter shapes. By contrast, the model developed by Flanagan and his colleagues at Bell Laboratories is a digital simulation of the physical processes, implemented on a high-speed computer. It includes an applied subglottal air pressure, a pair of vocal folds set into vibration when conditions are suitable, and a mobile vocal tract which filters and interacts with the voice source, and into which turbulence noise sources are inserted at places and times where conditions are appropriate.

One disadvantage of numerical methods is the long computation time. For speech production modelling as a research tool this is not a crucial factor; numerical stability, file handling and graphics are far more important, but for a practical articulatory synthesiser near real-time operation would be essential. Computation times have been as high as five hours for about one second of synthetic speech on a Data General NOVA 4 (Bocchieri, 1983), less than an hour on a VAX 11/750, and only minutes expected on a SUN for the same utterance (Childers, pers. comm., 1989). Coker's model has operated at 360 times real time with a CSP-30 computer (Coker, 1976), at 200 times real time on an Eclipse S/230 (Flanagan et al., 1980), and at about twice real time with a Cray-1 (Sondhi and Schroeter, 1987). Comparisons may not be valid across synthesis systems but perhaps the downward trend justifies optimism that computation time alone will not be an insuperable obstacle.

Hardware mechanical models on which actual acoustic measurements can be made are of great value today. Mechanisms for fricatives are investigated in this way (Shadle, 1986); a human larynx was combined with a hardware model vocal tract to investigate source-filter interaction in voicing (Laine and Vilkman, 1987); the perceptual effects of cross-modes were investigated with a mechanical model by Holmes (1981).

Models of articulation and of neuromuscular control of the structures are immensely important in their own right, for gaining greater understanding of both normal and disordered speech production. Improved models of speech production are seen by Fant and Flanagan as important for speech technology also, adaptable to speakers, for synthesis, recognition, and coding of speech (van den Broecke and Cohen, 1984).

1.2. VOCAL TRACT ACOUSTICS AND ACOUSTIC MODELS

A brief consideration of vocal tract acoustics will show how complicated the real-life systems are and thus the magnitude of the synthesis task; there is a clear justification for severe simplification. It is not the case that there is one appropriate model of the filtering process; instead, there are several conflicting models, each applicable to a limited range of geometries and frequencies. During speech production conditions are constantly changing, both in distance along the vocal tract and with time, so that none of these models applies throughout and none is likely to give a perfect match with natural speech.

The most extreme lumped element model, the Helmholz resonator, applies if the vocal tract can be approximated by one or two cavities with narrow outlets and for frequencies such that the largest dimension of the cavity is much less than a wavelength. The exact relationship: velocity of sound equals frequency multiplied by wavelength always holds, but even here there is room for doubt in modelling: what is the velocity of sound inside the vocal tract? It depends on temperature and humidity, and its effective value within a pipe falls as the pipe diameter decreases (Regnault, 1868, cited by Richardson, 1940, p. 27). Yielding pipe walls, on the other hand, can be looked upon as increasing the effective velocity of sound in the pipe, or as tuning resonances to higher frequencies. Recent research supports a value of 350 m/s, little affected by flow profile or cross-section shape, but dependent upon gas composition, temperature and mean flow speed (Rice, 1980). Wavelengths are 350 cm at 100 Hz, 7 cm at

5 kHz and 3.5 cm at 10 kHz. Thus the dimensions of a vocal tract about 15 cm in length lie in the wavelength range of interest in speech. It must be divided up into portions, lumped elements, within each of which acoustic parameters of pressure and flow hardly vary. In practical terms, the maximum dimension of each lumped element along the vocal tract must be less than one eighth of a wavelength. For sections 0.5 cm long, the errors will be 5% at 6 kHz and less at lower frequencies. Since cross-modes of the vocal tract must be half-wave resonances with high pressure at each wall, the lowest frequency at which they occur, assuming a maximum diameter of 4 cm and a circular cross-section tube, is above 4 kHz (Stevens et al., 1953).

The Helmholtz model is appropriate only for the first formant F1 generally. Otherwise the vocal tract must be considered as a tube along which acoustic waves travel. As long as the largest cross-dimension is less than about half a wavelength, wave travel may be considered to be one-dimensional along the length of the tube. The Webster horn equation for loss-free conditions applies only if cross-section area changes are so gradual that internal reflections do not occur. Where there are noticeable changes of cross-section area within the vocal tract, reflections do occur and the acoustic tube behaves in the same sort of way as an electrical transmission line made up of a number of sections. Validity up to higher frequencies can be achieved by shortening each section, but this is at the cost of computation time and perhaps increased numerical instability (See Fant, 1960, especially A3, p. 281 ff. and Flanagan, 1972, especially sections 3.1, 3.2 and 6.23).

Chiba and Kajiyama (1941) solved the Webster equation numerically, but generally one of two basic approaches to an approximate solution is adopted (see Flanagan, 1972, p.85). The continuous tube may be represented by a multisection electrical transmission line in which each section has component values appropriate for an acoustic tube of particular cross-section area. The set of simultaneous differential equations is solved by numerical difference equation methods (Ishizaka and Flanagan, 1972). Alternatively, the continuous tube is approximated by a number of abutting cylinders of different cross-section areas. The magnitudes of incident, reflected and transmitted pressure waves are calculated at cylinder boundaries (Kelly and Lochbaum, 1962). In Mermelstein's (1972) method, reflections at tube section boundaries are expressed in z-transform terms and the computed transfer function is expressed as a recursive (linear predictor) filter.

But this is not all. The human vocal tract has at least the following acoustic complications: losses of various sorts - viscosity, heat conduction, at severe constrictions, through the glottis, and through radiation of sound from the mouth and nose outlets. The walls of the tube (which is not straight but curved) yield in response to slow changes of air pressure and vibrate, transmitting acoustic energy at higher, acoustic frequencies. There are aerodynamic and acoustic interactions between the voice source and the vocal tract filter and the subglottal airways may need to be included as part of the filter also. All these effects vary over time and with frequency, as discussed by, for example, Flanagan (1972, pp. 61-69).

What is included in a simplified representation of vocal tract acoustics, and how, is a matter of judgement and practicalities. Weighting would be given to the expected importance of the effects, the confidence in current knowledge of parameter values (although an empirical approach through the modelling itself may be justified), the goodness of available models to represent them, the additional computational load and the researcher's available time and expertise for implementation.

Fant (1980a) and Carré (1981) have reviewed the mapping from area functions to acoustic structure. The last few years have seen further research and development of acoustic models (see, for example, Maeda, 1982; Liljencrants, 1985; Wu et al., 1987; Laine, 1989). At least one implementation using a signal processing chip has been proposed (Cross et al., 1986).

Questions of stability and sensitivity of acoustic structures to particular changes of vocal tract shape (Stevens, 1972; Mrayati et al., 1988) must be of great importance in articulatory synthesis. The acquisition of speech by children and the acquisition of articulatory modelling skills by researchers must surely include tailoring articulation to vocal tract acoustics. The production of speech, natural or synthetic, is not a sequence descending from linguistic units at some high level in the central nervous system, but the result of some sort of near-optimal use of the properties and constraints of all stages of speech production , 'below' as well as 'above' that of articulation. All the considerations, including of course perceptual contrast required by listeners, come together, it may be supposed, in the actual movement paths and articulatory end point configurations chosen by a speaker.

1.3. ARTICULATORY MODELS

An important subset of articulatory models are those which describe vocal tract shapes. Whether the aim is articulatory synthesis directly or the derivation of articulatory information to match speech signals via area function to formant frequency mapping, some sort of articulatory model is nearly always used, since the number of degrees of freedom for a vocal tract acoustic tube quantised into sections 0.5 or 1.0 cm in length is inconveniently large. More importantly and more interestingly, the number of degrees of freedom in an articulatory model ideally should be equal to the number associated with real physiological constraints.

Perhaps the simplest representation of vocal tract shape was Dunn's (1950) transmission line model, where all the sections except one represented the same cross-section area. Stevens and House (1955) and Fant (1960) used three parameters to characterise vocal tract shape for vowels: the size of a constriction, its position, and the lip outlet area/length ratio. Parabolic or horn expressions specified the total area function. The vocal tract may be represented as four tubes (Fant, 1960; Badin and Boe, 1987).

Some modellers put the emphasis directly on control of the shape of the acoustic tube, motivated by the fact that the area function is the acoustically relevant aspect of the resonator, an approach advocated by Ladefoged (1979). Other models emphasise the interactions between the solid structures (for example, Lindblom and Sundberg, 1971; Mermelstein, 1972; Coker, 1976). Statistical models for tongue shape in vowels (for example, Maeda, 1975; Harshman et al., 1977; Kiritani, 1977) seem to be in broad agreement that two major tongue factors are needed, with a lip shape factor and perhaps a jaw associated factor in addition. About nine parameters seem to be appropriate to shape the vocal tract if consonants as well as vowels are modelled. For example, in Coker's (1976) model three parameters control the tongue body which is represented by a circle, two define lip actions, two are needed for raising and curling back the tongue tip, and another represents the velum. The ninth parameter controls the mapping between the side view of the vocal tract and its cross-section areas, still an important cause of uncertainty in articulatory models.

1.4. AERODYNAMIC MODELS

These are of extreme importance in the modelling of speech production processes, but will not be reviewed in any detail here. The whole respiratory tract needs to be represented here (see, for example, Rothenberg, 1968; Muller, 1980; Ohala, 1975; Scully and Allwood, 1984).

1.5. GENERATION OF ACOUSTIC SOURCES

If articulatory synthesis is to gain increased naturalness with more of the richness, complexity and redundancy of natural speech, the acoustic sources must not be specified in an ad hoc way; they must be determined by the prevailing physical conditions from which they arise.

1.5.1. *The voice source.* The larynx, with its several articulatory parameters and as the generator of the voice source, is a very large research area, as shown by the large number of conferences and symposia devoted to it in recent years. The voice source is the volume flow rate of air through the glottis, a quasi-repetitive waveform. In the complete articulatory synthesis system of Flanagan et al. (1975) the vocal folds are set into oscillation under suitable conditions. These include vocal tract configurations but the three controlling physiological parameters acting directly on the vocal folds and determining their mode of vibration in the model are subglottal air pressure P_s (called lung air pressure), tension of the vocal folds Q, and the rest area of the glottis A_g. The voice source glottal flow $U_g(t)$ is derived from the mathematical model. Alternatives to the two-mass model have been proposed and developed, for example a distributed mass model (Titze, 1981) and a three parameter continuous model of the vocal folds as a pair of hinged beams (Guérin, 1983).

There is clear evidence now that non-linear source and filter acoustic interactions occur. Inclusion of a suitable amount of acoustic interaction in terminal analog synthesis can improve the naturalness (Guérin et al., 1976; Childers et al., 1985). The voicing model may be made to interact with the filter; then true glottal flow is derived from a specified oscillatory glottal area (for example, Guérin et al., 1976). Reliable data for glottal area, both articulatory and oscillatory, are difficult to obtain. An alternative approach is a 'short circuit current' source. Here acoustic loading of the vocal folds by the supraglottal pressure of the first formant is not explicitly modelled, but the glottal flow waveshape can be skewed and can be made to vary in appropriate ways with phonetic context, thus including some of the effects of source-filter interaction (for example, Fant, 1980b; Fujisaki and Ljungqvist, 1986).

1.5.2. *Turbulence noise and transient sources.* Turbulence noise sources have been less fully investigated than the voice source in the past. In the articulatory synthesis of Flanagan and Ishizaka (1976) a noise pressure source is introduced into the transmission line at many points, at appropriate magnitudes. Sondhi and Schroeter (1987) have used a similar method for turbulence noise but inserted frication noise downstream of the most severely constricted portion of the vocal tract only, with aspiration noise inserted just above the glottis. Shirai and Masaki (1983) modelled the production of fricatives, varying the location of the frication noise source. The processes of noise generation have been analysed by Stevens (1971). The research of Shadle (for example, 1986) with mechanical and theoretical models is establishing a firmer basis for the realistic inclusion of turbulence noise sources in articulatory synthesis. Transient sources have been neglected also, but this coherent source, arising from the flow of air into a vocal tract constriction just after the release of a plosive for example, has been analysed by Maeda (1987).

1.6. SOME CONSIDERATIONS FOR ACOUSTIC SYNTHESIS TRANSFERRED TO THE ARTICULATORY DOMAIN

It would be unrealistic to expect articulatory synthesis by rule at this stage. Much copy synthesis must precede it. If articulatory synthesis by rule is nevertheless a research goal, automatic logging and analysis methods must be exploited to process large amounts of natural speech data, paralleling in some respects the current developments in automatic acoustic analysis. So the availability of good methods of observation and analysis is just as important as the availability of good models of speech production processes.

Perhaps even at this early stage it would be worth considering possible choices of linguistic units for articulatory synthesis. Would it be appropriate for example to consider the use of demisyllables or diphones as operational units of articulation? In the Holmes et al. (1964) acoustic based synthesis by rule system, the linguistic units are phonemes with associated prosodic markers. Each phoneme is represented by one or more phonetic elements. Nominal boundaries between these elements are established and formant transitions are created by interpolation between end point values. About eleven parameters, each with five properties, describe each phonetic element in tables to characterise a given speaker type (Holmes, 1988, pp. 84-90). How many parameters might be needed to represent articulatory end point states and how many such states might be needed in a complete description of a single speaker type for articulatory synthesis?

The synthesis of Holmes et al. (1964) has a hierarchy of dominance. Generally speaking consonants take precedence over vowels and more occlusive elements over less occlusive ones. Concepts of dominance according to manner and the criticality of particular articulatory transition end points were employed by Haggard (1979) in his articulatory synthesis. Important questions include whether transitions can be modified to enable an articulator to reach an essential state or region within a given time, or whether, on the contrary, articulatory transition durations are determined by the nature of the transitions themselves, whether these are in some sense inherent, or are matched to perceptual processes. The latter view seems more plausible, and emphasis has been placed recently on the properties of the actions

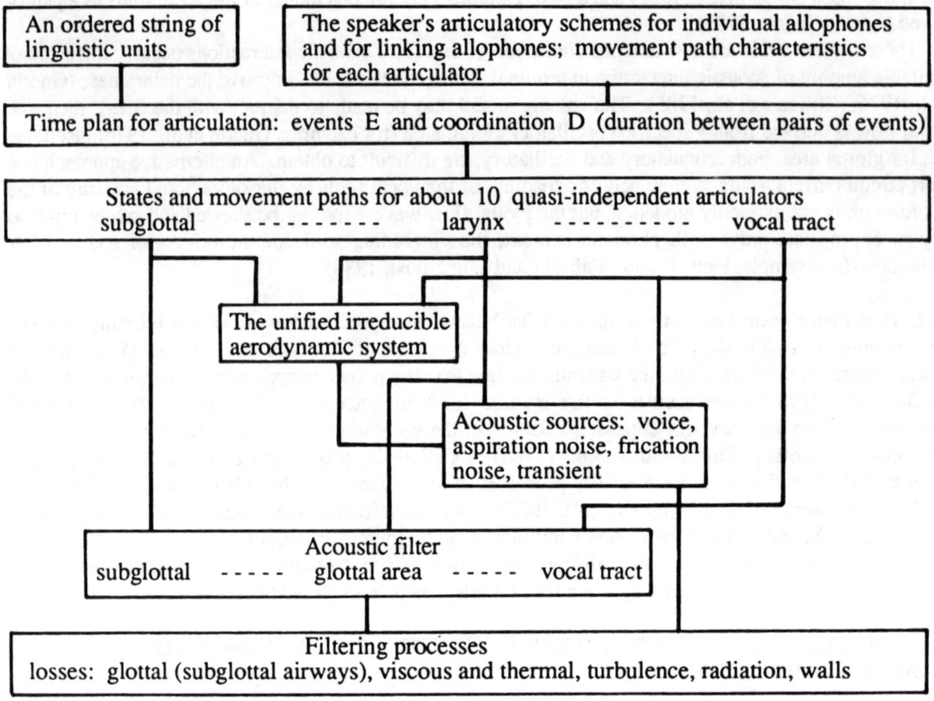

Figure 1 A view of the stages of speech production excluding neuromuscular processes and radiation of the output sound pressure wave. Information flows from the top downwards.

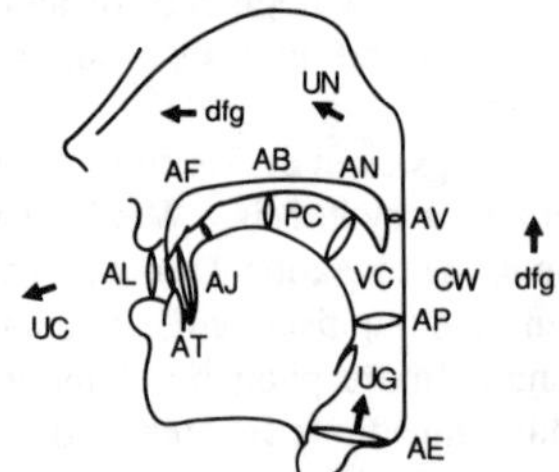

Figure 2 A composite model of speech production processes, described in the text. Articulatory parameters: PLU pressure of air in the lungs to represent subglottal control of the lung walls (alternatively DVLU rate of lung volume change); Q effective mass and stiffness of the vocal folds; AG articulatory component of glottal area; AE, EP, AN, AB, AC(not shown) and AF defining the tongue shape; AJ jaw region; AT teeth; AL lips; AV velum. Aerodynamic parameters: VLU lung volume; PSG subglottal air pressure, PC oral air pressure behind the vocal tract constriction of area AF; UG, UC and UN volume flow rate of air through the glottis, the oral constriction and the velopharyngeal port respectively; VC volume enclosed behind the vocal tract constriction; CW compliance of the walls of the vocal tract; GSG subglottal airflow conductance, increasing with lung volume. dfg is the distance from the glottis along the length of the vocal tract.

themselves (Fowler et al., 1980). Current research looks for invariance in transitions and in inter-articulator coordination across changes of context such as stress and across rate change (Nittrouer et al., 1988), and under externally applied perturbations (Gracco and Abbs, 1988). It seems likely that the sequence of auditory patterns generated is constrained, in its durational properties especially, by the actions required (Scully and Allwood, 1985a). Browman and Goldstein (1986) propose that patterns of overlapping articulatory gestures, gestural "constellations", can provide the basis on which to construct an articulatory phonology. Our own proposals for constructing articulatory time plans can, in principle, be applied to extended linguistic strings (Scully, 1987).

Articulatory synthesis should perhaps aim to be more ambitiously flexible than acoustic synthesis by rule, allowing for options available to a speaker within one style, changes of speech style, variations with phonetic context, and variability from one repetition to the next, in mapping from a linguistic string onto articulatory plans defined quantitatively along a time axis. It should be able to simulate a range of speaker types, characterising the speaker in ways which generate a multiplicity of internally coherent acoustic effects in the synthetic speech.

2. Preliminaries to Articulatory Synthesis: Characterisation of Two Different Speaker Types

In the acoustic domain individual speaker characteristics and phonetic structure are difficult to disentangle. Acoustic rules for synthesis do not explicitly differentiate between the physical laws governing all speech production, parameter values for the physical properties of each speaker type and options available to and choices made by an individual speaker. One aim of articulatory synthesis might be to help to make these distinctions explicit, if it is possible to do so.

Figure 1 suggests in outline the relationships between acoustic patterns of speech and an underlying linguistic symbolic description. Linguistic structures and speaker characteristics interact throughout the processes represented here. The physical laws governing the processes or stages of speech production as seen here are, by definition, independent of the linguistic elements; however, the application by the speaker of those laws and processes is likely to be determined by phonetic structure. For example, a speaker might make the walls of the vocal tract less or more compliant in order to assist the cessation or maintenance respectively of voicing during a plosive closure, if the phonological system of contrasts for plosives of the language requires it. We are not concerned so much with whether the speaker is able to choose certain options or is unable to vary certain physical characteristics; instead, we wish to explore quantitatively the multiplicity of ways in which different speaker characteristics, as deduced from any available data, operate under the same physical laws to generate the variation and variability of speech signals.

Bridle and Ralls (1984) consider these questions to be important in automatic speech recognition and believe that systematic estimation of a speaker's voice characteristics is needed. Components of their constructor model to generate detailed acoustic patterns are: a text of phonetic symbols, prosody, and speaker characteristics. Duration is particularly important and is handled separately. Parameters of their acoustic segmentally-based synthesis-by-rule system are adjusted to match different speakers.

We make a preliminary attempt at the same sort of adaptation of a model of speech to different speakers, operating with speech production instead of with speech acoustics. In our plan, two components feed into a model of speech production, as in Figure 1: a set of speaker characteristics and a sequence of phonetic symbols. The phonetic string includes, as well as segmental allophones, phonetic correlates of prosody and markers for speech onset and offset. Duration is not considered to be different in kind from the other parameters. Instead, time is part of the expression of several physical laws, for example equations of motion for the solid structures, simultaneous differential equations for the aerodynamics, and choices of timing and coordination of articulatory events. Since speech production is a motor skill mediated by sensory information, feedback dead times and other durational aspects of the control systems seem very likely to contribute also to the final durational structures of the output speech signal. In addition, duration is one of the phonetic correlates of prosody and of the phonological system of opposition, for example of vowel length in English. A model will not necessarily succeed in making all these aspects of duration separate and explicit; for example, when articulatory movement paths are described kinematically neuromuscular control properties and equations of motion may be compounded.

2.1. A COMPOSITE MODEL OF SPEECH PRODUCTION

In our modelling we attempt to simulate the processes of each of the blocks shown in Figure 1. The parameters of the model are shown in the sketch of Figure 2: some are articulatory, some are aerodynamic and some are mechanical properties. Speaker type information is supplied at several different points. There are three main computer programmes: for timing and articulation, for aerodynamics with acoustic sources and for the filtering of all the acoustic sources by the vocal tract with radiation of a sound pressure wave. The model has been described elsewhere (for example, Allwood and Scully, 1982; Scully and Clark, 1986; Scully, 1986, 1987). Here some features of particular relevance for this study will be mentioned.

2.1.1. *Articulation.* A number of solid structures, the articulators, are defined. Supraglottally, each parameter describes the cross-section area of a particular point along the vocal tract rather than the shape of the solid structure directly. The articulators are:

The lung walls, represented by changes of lung volume with time DVLU or, here, by air pressure in the lungs PLU; effective mass and stiffness of the vocal folds, Q, the laryngeal control factor for fundamental frequency F_0; velum position, as cross-section area of the velopharyngeal port, AV; lip outlet area, AL; the cross-section area of the vocal tract between the teeth, AT, and near the jaw just behind the teeth, AJ; the tongue body, made up of several parameter points, with cross-section areas defining vocal tract shape. These are chosen from AE, AP, AN, AB, AC and AF, but, as shown in Figure 13, only three points are needed to define some vowel-like shapes; here AE, AP and AC. AS is the cross-section area of the larynx tube. Other parameter points are brought in to define tongue configurations for consonants. Here AF represents the cross-section area of the vocal tract at the severe constriction for an alveolar plosive or affricate. Values given to AC, behind the F point and to AJ, in front of it, help to define the outline of the tip-blade-front cavity region of the vocal tract. The supraglottal parameter labels refer to different vocal tract regions: E is the entry to the larynx tube, P is the pharynx, N is the nasal region near the velopharyngeal port, B, C and F are for the back, centre and front respectively of the upper tongue surface, J is the jaw region, T the teeth, L the lips and V the velum. The main vocal tract constriction has the general label C with cross-section area AC, for constriction, when it is transferred from the articulatory to the aerodynamic block of the model. The last articulatory parameter is AG, the slowly changing articulatory component of glottal area. Rapid oscillations of the vocal folds with corresponding changes in glottal area at acoustic frequencies are excluded from AG. Vertical movements of the larynx, a third important aspect of larynx articulation, are not included in our model.

These articulators are treated as being quasi-independent; they interact with each other in a number of ways which contain some severe simplifications. For lung air pressure to be defined as an action under the speaker's control is clearly a simplistic view of respiratory control. Undoubtedly, the amount of airflow obstruction at the glottis and supraglottally in the vocal tract affects the build-up and decay of subglottal pressure. Further, an average value for air pressure is used and acoustic frequency changes are modelled separately. PLU is chosen here as a respiratory control parameter, rather than the more realistic DVLU option, for convenience. Q is considered to be independent of AG although there is evidence to suggest that when the vocal folds are parted and bowed out they are lengthened and stiffened in the process (see, for example, Titze, 1980, 1989; Honda et al., 1980). Inclusion of natural speech effects for this in our modelling is difficult, since it is not clear to what extent different speakers of English alter Q as a voluntary articulatory action for consonants such as voiceless fricatives and voiceless aspirated plosives, for which the vocal folds are abducted. The model does generate changes of F_0 under different aerodynamic

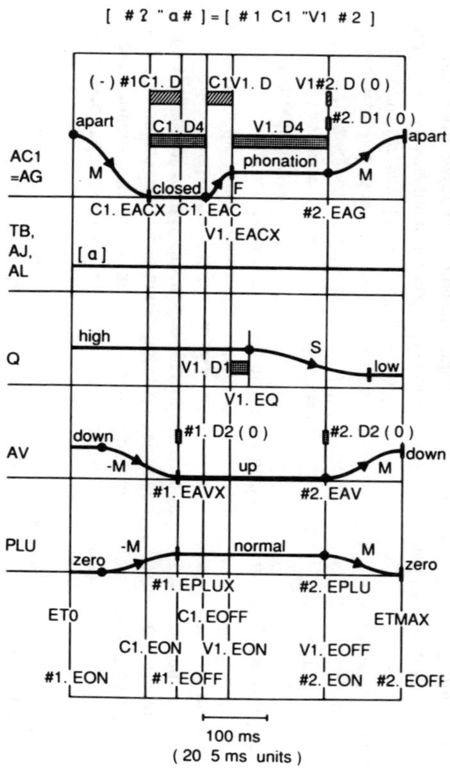

Figure 3 On the left, an articulatory time plan for "Ah" said in isolation with a falling pitch. Coordination durations internal to each allophone are shown stippled while linking durations are shown hatched. The labels for the articulators are: PLU air pressure in the lungs for respiratory articulation; AV area of the velo-pharyngeal port; Q effective mass and stiffness of the vocal folds; AL lip outlet area; AJ area near the jaw; TB areas defining the overall tongue shape; AC main vocal tract constriction, here AG for a glottal stop.

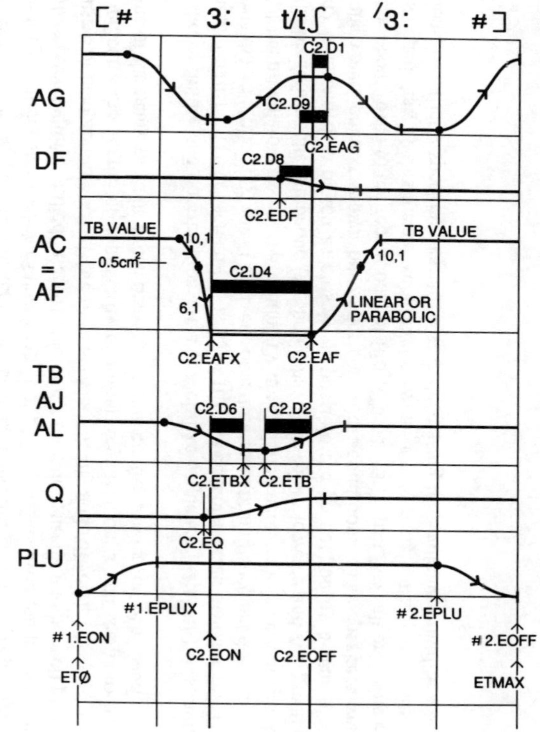

Figure 4 Part of an articulatory time plan for simulation of the sequence [#3:C'3:#] with C=[t] or [tʃ]. The stippled durations are for the coordinations to be varied here. The main vocal tract articulator AC is AF for an alveolar or post-alveolar place of articulation.

conditions at the glottis, as described below, but some AG-Q interaction seems to be called for in order to capture the high F_0 onset for voicing at high airflow, whether the effect is enhanced by additional Q control effects or not.

The various vocal tract parameter points are not always intended to be separate articulators; here E, P, C and J all move together to define a single tongue body articulator. They move through different distances; and constraints on tongue body shape, demonstrated by several researchers as discussed in Section 1.3., though not inherent in this under-constrained description, are respected. For example during diphthong-like tongue body movements, as AB and AC decrease AE and AP increase, while the cross-section area near the N point remains approximately constant. The complete vocal tract area function is obtained by linking together the parameter points. Intermediate points along the vocal tract are shaped by logarithmic interpolation.

In this block of the model some of the speaker geometry is defined. The vocal tract parameters are all given values for their distance from the glottis : DS (zero), DE, DP and so on. DL is the length of the vocal tract. Parameter points are allowed to move along the length of the vocal tract and this is done by setting up instructions for, say, DF as well as for AF.

An articulatory time scheme is defined. Pairs of events E are related by a duration D in a set of simple expressions. Inter-articulator coordination thus defined controls the relative timing of the various articulatory fragments needed for the production of each allophone. An offset event for a preceding allophone and an onset event for a following allophone are related in a similar way by a duration which determines the way in which adjacent allophones in the phonetic string are linked together. We should prefer to avoid the terms anticipatory (forward) and perseverative (backward) coarticulation for the influence of one allophone on another, except in the general sense of the ensemble of choices made by a speaker from the many different ways in which, it seems, the actions for individual allophones can be meshed together. In our framework, the linking duration for a pair of allophones is fixed if the link is homorganic, as in [s]-[t], but can be varied, including being made negative, if two different articulators are involved, as in [s]-[p]. This procedure may seem to suggest a single coarticulatory link, but in fact other articulators are involved also, so that allophone linking is multi-dimensional. It is, of course, important to note that the articulators do not all move simultaneously from one allophone state to the next. The particular allophone onset and offset events chosen by us are not the only ones that could have been chosen; it seems very likely that different speakers would have different organisational frameworks in the time domain. It seems to be just as complicated a task to start and end a speech sequence without omitting required sound patterns or allowing unwanted ones to intrude as it is to link allophone units within a string. Therefore [#], meaning the breathing state, is included in the allophone set for R.P. English. A few of the plausible ways of initiating and ending speech have been suggested (Scully, 1987). Here also there seems every reason to suppose that different speakers organise things in their own way, and there is some evidence for this in natural speech (Sawashima et al., 1975). Thus even the production of a static vowel on a falling pitch requires skilled inter-articulator coordination, needing to be well practised by the speaker if intrusive or inappropriate sounds are to be avoided (Scully and Allwood, 1984).

Figure 3 indicates the degree of complexity of timing organisation which seems to be needed for expressing some of the many ways in which a speaker might control timing for "Ah", said in isolation on a falling pitch.

Figure 4 shows timing and coordination for a speech-like sequence [#3:C'3:#] to simulate an R.P. speaker of British English. The consonant [C] is [t] or [tʃ], the consonants simulated here. The ' mark is meant to indicate that the following vowel is given some durational prominence but does not carry a pitch movement. All the articulators except the velum AV are included here. AV is kept at zero throughout, even though the breathing state allophone [#] should have the velum lowered and AV greater than zero. This modelling is intended to match as closely as possible the production by a real speaker of the consonant in the vowel context shown but under controlled conditions. The speakers were asked to produce a series, about seven to ten repetitions, of [#p3:C'3:p3:C'3:.....p3:C'3:#] all on a single expiratory breath. It was hoped that, by this means, subglottal pressure would remain nearly constant. Variability across repetitions, or, strictly speaking, variation with decreasing lung volume, could be assessed and a representative example chosen. For this study the second or third example in the sequence was chosen for analysis, well away from [#] so that the velum was expected to be raised. The middle portion of the time scheme of Figure 4 was adjusted in the modelling.

The ways in which parameter values were selected in an attempt to match the real speech are described below. Here, some features of the articulatory portion of the model will be noted. Each articulator is given a few lines of instructions from which its actual movement paths will be calculated.

The instructions for AG, AP and AF in [#3:tʃ3:#] are given as examples below:

Table 1 Examples of instructions for articulators' paths

First column instruction number; second column cross-section area for the articulator in cm^2; third column time point as an event or as time allowed for an action in 5 ms time units; fourth column transition type, with M for medium, PARAOUT for parabolic, or pairs of figures with the first specifying duration in 5 ms time units and the second a slope factor. A minus sign means that the transition is to end at the specified event; otherwise the transition starts at that event.

	AG		
0	0.25		
1	0.25	C1.EAG	M
2	0.06	24	24 150
3	0.06	V1.EAG	M
4	0.2	C2.EAGX	-24 150
5	0.2	C2.EAG	M
6	0.0376	V2.EAG	24 150
7	0.0376	V2.EAGX	M
8	0.25	C3.EAGX	-24 150
9	0.25	ETMAX	M

	AP (part)		
2	2.0	24	M
3	4.5	C2.ETBX	-M
4	4.5	C2.ETB	M
5	2.0	24	M

	AF		
0	AT 0		
1	AT 0	V1.EAF	M
2	TO 0.5	10	10 1
3	TO 0	C2.EAFX	6 1
4	0	C2.EAF	M
5	TO 0.5	16	PARAOUT
6	BY 0	10	10 1
7	AT 0	ETMAX	M

The value of AP defines the pharynx width and so the position of the tongue root. AP is fairly small for the back vowel [3:] and becomes larger as the tongue body moves forward to the configuration needed for an alveolar or post-alveolar consonant. At the same time AC, which was large for [3:], becomes smaller as the dorsum of the tongue is raised. AP is to reach its value of 4.5 cm^2 at the time of event C2.ETBX, not earlier. It moves with an S-shaped transition: starting from zero velocity, accelerating to reach maximum velocity midway in distance and time, then decelerating to reach zero velocity at C2.ETBX. It remains at 4.5 cm^2 until the time of event C2.ETB, then begins to move back towards its [3:] value, with the same movement path characteristics as before. This transition is of medium (M) duration, taking 120 ms (24 time units) to complete, whether the end point states are near together or far apart.

The justification for this simplified generalisation of movement paths has been made elsewhere (Scully, 1975, 1987). Undoubtedly this is an oversimplified view of articulatory transitions. It was adopted on the basis of cineradiographic data of the 1960s because, as a first approximation, constant transition time seemed to be nearer to the real system than, for example, constant velocity over most of the transition. Much recent evidence shows that in natural speech there is a rather strong positive correlation between peak velocity and distance traversed, for example for the jaw (Kiritani et al., 1982) and for the tongue dorsum (Ostry and Munhall, 1985), but that transition duration increases somewhat as distance increases. What seems to be most important is a good match to the middle portion of the transitions (Fujimura, 1981), and constant duration provides this reasonably well. Measurement of total transition duration is more difficult to do reliably; different studies use different criteria to define end points.

The second example of instructions, for AF, which represents the main vocal tract articulator for [t] and [tʃ], the tongue tip or tip-blade region, shows some other features of the articulatory transitions. AF is not explicitly defined in the overall shaping of the vocal tract. This is carried by AE, AP, AC, AJ, AT and AL. AF must move, from whatever value it happens to have as the tongue body is moving forward at the root and upwards at the dorsum, to a specific value for the consonant. Since both [t] and [tʃ] have complete closure of the vocal tract, AF must become zero. Indeed, it may be appropriate to move AF to an area value less than zero, to match the path characteristics for a firm contact, as shown in some mechanical pressure traces (McGlone et al., 1967). The negative area values are set to zero later.

Within the existing framework of transitions, it proved impossible to match the natural speech aerodynamically-derived data for any of the speakers analysed so far. It was decided that a good match must be made where the vocal tract constriction is severe, that is, where AF is small, but that the portion between AF being at its tongue body value (called AT 0 in the instructions for AF) and AF being 0.5 cm^2 was not crucial and could be approximated by a simple linear transition of duration 50 ms (10 time units). The upper boundary of 0.5 cm^2 for careful matching was chosen by experimenting with different values and selecting one which avoided noticeable discontinuities in the aerodynamic traces generated, yet permitted most of the natural speech aerodynamic data to be matched. It was necessary also to introduce a parabolic transition, because the S-shaped path with zero velocity at AF=0.5 cm^2 or some other value could not be made to match the natural speech paths. It is to be noted that cross-section areas for natural speech were obtained from volume flow rate of air through the constriction and air pressure drop across it using the same orifice equation as that used in the reverse direction when modelling. The model's aerodynamic equations were simplified (by setting viscous flow parameters to zero) in order to ensure this consistency. The philosophy is that it is more important that the speech analysis should be as nearly as possible consistent with a simple, manageable model than that the model should be as nearly as possible true to the physical system simulated.

The set of instructions for AF mean the following: start with area AF defined through the area -linking interpolation between the tongue body point C and the J point (AT 0) until the time point of event V1.EAF. Then start to move to a specific AF value of 0.5 cm^2 through a linear transition labelled 10 1 of duration 50 ms (10 time units); 10 time units are allowed for this. Next, move to value AF=zero by means of a linear transition of duration 30 ms (6 1), with 30 ms allowed for this. C2.EAFX is defined as the end of this transition. A rapid closure approximating the natural speech data is achieved. Maintain the complete closure until time point C2.EAF (with a notional medium M transition). Then move along a parabolic path as shown in Figure 4 , to reach AF=0.5 cm^2 in 80 ms (16 time units). Finally, return to the previously defined tongue body value, whatever it is at this later time (BY 0) by means of a linear transition of 50 ms duration with 50 ms allowed (10 10 1) and remain as defined by interpolation between AC and AJ until the end of the simulation ETMAX.

Instructions are provided, similarly, for all the articulators, including position D as well as area A if so desired, for example:

AE, AP, AN, AB, DB, AC, DC, AF, DF, AJ, DJ, AT, AL, AG, Q, AV, PLU.

Like AF, AN and AB are defined by interpolation as part of the whole vocal tract shape, then modified as they move to specific values for the consonant.

Once all the movement paths have been calculated and areas interpolated between parameter points to define vocal tract area functions at points 0.5 cm apart and at time intervals of 5 ms, the volume of the cavity VC enclosed behind the major constriction AF is calculated as a function of time. Paths for AG, AF (to be given the generalised label AC in the next, aerodynamics, block), Q, AG, AV, PLU and VC are saved as files needed for the aerodynamics, as well as the whole vocal tract area function, which is needed for the filtering stage. If more than one vocal tract parameter point is used for consonants in one run, then a composite constriction is calculated for use in the aerodynamics. For example, in [#p3:t'3:p#] AL is the constriction for [p] and AF that for [t]. A composite equivalent constriction is AC where

$$1/AC^2 = 1/AF^2 + 1/AL^2$$

With two points such as F and L close together, the volume enclosed behind the F point is a reasonable approximation for both portions of the utterance.

2.1.2. *Aerodynamics.* Since the aerodynamic stage of speech production links the actions to the sounds, it may be expected that two speakers who differ aerodynamically will have different basic parameter values and/or different operating conditions for the generation of voice, turbulence noise and transient sources. While articulators can be said to be quasi-independent of each other, being functionally linked for specific actions, the different portions of the respiratory tract cannot be said to be independent aerodynamically; this is a unified, irreducible system and the differential equations have to be solved simultaneously as a whole. The aerodynamics behaves very much like an electrical circuit, but with one important difference: instead of ohmic resistance for which voltage drop is directly proportional to current, flow (for electrical current) and pressure drop (for electrical voltage) are related by the orifice equation for the most important airflow obstructions of speech production. If this must be called a resistance, then it is a flow-dependent one; it is better to use the orifice equation which relates cross-section area of a severe constriction (A), volume flow rate of air through it (U) and air pressure drop across it (ΔP) by the expression

$$A = k \, . \, U / \Delta P^{0.5}$$

The same flow of air from the lungs provides subglottal air pressure to help drive the vocal folds for voicing and gives increased air pressure in the vocal tract P_o when its outlet is partially or completely blocked. For this reason the aerodynamic stage is strongly implicated in the multiplicity of contrasting acoustic pattern features arising from just one change of articulation.

The aerodynamic block of the model comes closest to being a true physical model based on physical laws, albeit represented in extremely simplified form (see Scully and Allwood, 1984). Inputs to this block are the files saved from the articulatory block. Another input file defines the speaker's respiratory system and its initial state, also a set of parameter values to establish a phenomenological model for generation of the voice source. The aerodynamic parameters calculated in this block are included in Figure 2. UG, UC and UN are the low frequency aerodynamic components of volume flow rate of air through the glottis, the oral constriction and the velopharyngeal port respectively. PC is the oral air pressure with atmospheric pressure called zero; PC is also the pressure drop across the oral constriction, with minimum cross-section area AC. Pressure drop across the glottis, area AG, equals PSG - PC , called PDIFF. Subglottal pressure PSG is not, in general, exactly the same as lung air pressure PLU. In this highly simplified respiratory model the resistance to airflow of the subglottal airways is included as an ohmic-type conductance GSG which is made to decrease linearly with lung volume from a maximum value (minimum resistance) of 2000 cm^3/s cm H_2O about half way up the vital capacity, down to zero conductance at residual volume.

Respiratory publications suggest strongly that there is strikingly clear separation of volume and flow resistance in the subglottal airways, with viscous flow resistance located almost entirely in generations 0, the trachea, to 12; and nearly all the volume contained in generations above 12 (Macklem and Mead, 1967; Pedley et al., 1970).

In the normal speech modelling described here, the effect of subglottal airway resistance is negligible, except at high flow rates after plosive and affricate releases, where PSG falls about 0.5 cmH_2O below PLU. But the high value of GSG seems to be crucial. Most respiratory research quotes total airways resistance, much of which is located in and above the glottis, including the nasal cavities. The expression for GSG used by us is based on a study by Vincent et al. (1970) who separated total from upper airway resistance. We found that, if the wrong, high value of subglottal resistance was used, subglottal pressure fell so catastrophically as the vocal folds were abducted that a good [s] could not be produced. In the respiratory control model used there the speaker was assumed to control rate of lung volume decrement; in the study reported here, less realistically, the speaker is assumed to control air pressure in the lungs.

The walls of the vocal tract yield as air pressure rises within the cavity of volume VC enclosed behind the vocal tract constriction, area AC. A value of 0.6 cm^3/cmH_2O is used for vocal tract wall compliance, based on Rothenberg (1968).

In a fuller form of the aerodynamic model, each of the two orifices, the glottis of area AG and the vocal tract constriction of area AC, has terms for viscous as well as turbulent flow losses, additively contributing to the pressure drop across the orifice. But here, for simplicity and for consistency with the analyses of the natural speech, specifically the area trace, no ohmic-type resistance was included. For both

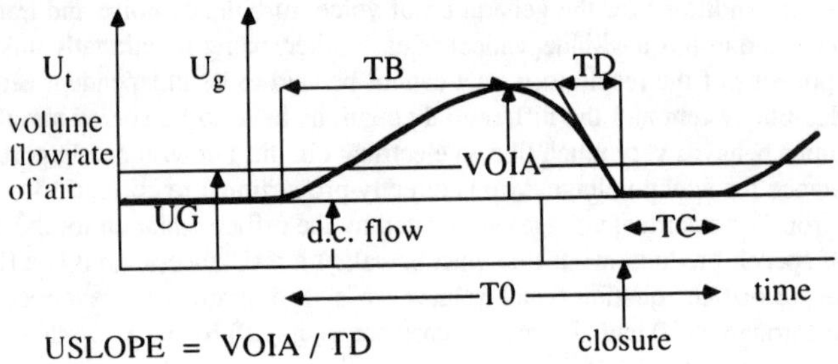

USLOPE = VOIA / TD

TCR = TC / TO = (1 - OQ)

TO = 1 / FO

$K = 0.5 + 0.125\ (TB/TD)^2$

Figure 5 The voice waveshape model. Total volume flowrate through the glottis U_t is the sum of the continuous d.c. flow of air and U_g the acoustic component of flow. UG is the mean volume flowrate of air through the glottis, an aerodynamic parameter.

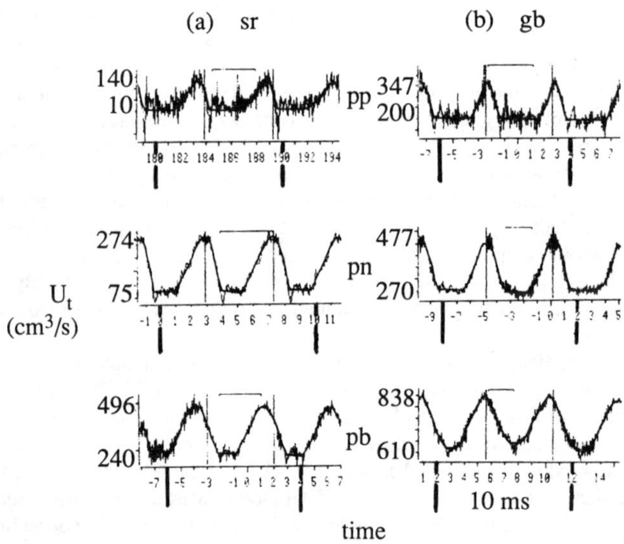

Figure 6 Examples of inverse filtering from total output volume flowrate of air, mouth and nose combined, for two speakers. Total volume flowrate of air through the glottis U_t is shown as a function of time in mid-vocoid portions of sequences [...p3:p3:...] for different phonation types. pp is pressed, pn is normal and pb is breathy phonation. (a) speaker sr; (b) speaker gb. A triangular approximation to the open phase of U_g ,the acoustic component of voicing, is used to derive the aerodynamic parameter UG from UG = dc flow + 0.5 (1 - TCR). VOIA

natural speech and modelling the orifice equation (Warren and DuBois, 1964) applies:

$$AC = 0.00076\ UC / PC^{0.5} \qquad \text{and} \qquad AG = 0.00076\ UG / (PSG - PC)^{0.5}$$

(PSG - PC) is called PDIFF in the modelling.
AC and AG are in cm^2, UC and UG are in cm^3/s, PC and PDIFF are in cmH_2O. The empirical constant is 0.00076 (Scully, 1986).

Given initial conditions, the aerodynamic variables UG, UC, UN, PSG, PC are obtained for each 5 ms time sample point by 'time marching' numerical methods for the solution of the aerodynamic equations. Problems of instability are likely to arise, especially where oral air pressure is very near zero (atmospheric). A combination of models expressing continuity of mass flow is used: excluding dPC/dt if PC is very low, but including this term if PC is above a threshold, usually 0.1 cmH_2O. Gear's method is used for numerical integration (NAG, 1987). This is suitable for 'stiff' equations in which the variables have very different magnitudes.

It is possible to estimate GSG and other airway properties for normal respiratory operating conditions. The methods are non-invasive, but respiratory parameter values were not measured for the speakers in this study.

2.1.3. *Analysis-by-synthesis of the aerodynamics.* In our modelling it is necessary to estimate larynx conditions from observations that can be made on the natural speech. Airflow through the glottis, for example, must be derived from the airflow measured at a mask covering the speaker's mouth and nose. In this study, total output flow from mouth and nose combined was measured, but it was confirmed, by use of an oral-nasal separated mask, that for the non-nasal consonant contexts modelled, nasal airflow UN was zero, so that total flow U_O measured for the real speaker was equated with oral airflow UC in the model.

In general UC, the flow through the vocal tract constriction, does not equal UG, the flow through the glottis. For expiratory flow, as during most of speech, UG is greater than UC if the vocal tract cavity volume VC is increasing, so that some air from the larynx is used in filling up a larger volume, or if oral pressure PC is rising, so that air from the larynx is used to increase the number of air molecules in a given volume VC. Thus, if VC and PC are both rising or one is constant and the other rising, UG is greater than UC; if VC and PC are both falling or one is constant and the other falling, UG is less than UC; if VC is rising and PC falling or if VC is falling and PC rising, it is not obvious except by modelling whether and, if so, how UG and UC differ. Changes of volume VC can be due to vocal tract articulatory movements or to movements of the vocal tract walls when oral air pressure departs from atmospheric, or both. Flow measured at a mask, even an oral-nasal separated one, is an approximation only to UC, the flow through a vocal tract constriction.

Oral air pressure was measured for the real speech by means of an orally inserted polyethylene tube, with its open end pointing across the airstream to minimise stagnation pressure effects. It is assumed that air pressure near the hard or soft palate equals air pressure just above the glottis. This is a reasonable approximation to make provided the vowels do not have extreme narrowing in the velar or pharyngeal portions of the vocal tract. The consonantal constrictions were chosen to be in front of the open end of the pressure tube. Subglottal pressure was estimated from oral pressure P_o measured at its peak value during closures for [p] consonants either side of the [VCV] sequence to be simulated, by linear interpolation between the two peak values, as shown in Figures 10 (a) and 11. A first estimate for a vocal fold path, AG as a function of time, is made from the estimated values of PDIFF and UG and the orifice equation above, at time points near the closing and release phases of the consonants.

2.1.4. *Voicing model.* Voicing is controlled through a phenomenological model based on the parametric representation shown in Figure 5 (Fant, 1980b). Three parameters besides F_0 are needed for the construction of each cycle of the acoustic component U_g of the total glottal volume flow rate U_t. The mean flow rate UG is the value computed in the aerodynamic block of the model. The (negative) gradient at closure, of magnitude VOIA / TD is called USLOPE in our modelling. Broadly speaking, VOIA indicates the acoustic intensity of low frequency components and USLOPE that of high frequency components of the voice excitation of vocal tract resonances. In recent developments, the LF model, a closer match to some natural speech waveforms is achieved by the inclusion of a gradual decay to zero flow after the steepest negative slope (Fant et al., 1985). Nevertheless, this simpler model is able to capture some important features of running speech. In previous applications of the whole composite model, parameter values were taken from the literature. Different kinds of phonation could be synthesised: breathy, normal (or modal) and pressed, with their associated spectral shapings and associated greater or lesser amounts of aspiration noise visible in the spectrograms of output vowel sounds (Scully and Allwood, 1985b). More generally, as controlling conditions at the larynx are changing continuously during a speech simulation the voice waveshape parameter values are changing in ways intended to be consistent with data from natural speech.

It is assumed in our modelling that three physiological parameters control the voice waveform. These are: PDIFF, the mean aerodynamic component of air pressure drop across the glottis, Q, the effective mass and stiffness of the vocal folds, and AG, the articulatory component of glottal area. This is a highly simplified approach. It is recognised, for example, that the effect on vibration mechanisms of raising oral pressure so as to reduce PDIFF is not the same as that of lowering PSG. The driving pressure is the air pressure within the glottis and at different portions of the voicing cycle this is nearer to either subglottal or supraglottal pressure (Titze, 1986; Rothenberg and Mahshie, 1986; Bickley and Stevens, 1986). But for simplification, all methods of changing PDIFF are taken as equivalent here.

It seems to be possible to observe some general trends in the way voice waveshape parameters VOIA, K, TCR and TD vary as the physiological conditions change. The evidence is drawn from inverse filtering of natural speech and singing for different efforts, pitches, phonation types and larynx states, (for example Lindqvist-Gauffin, 1970; Rothenberg, 1973; Sundberg and Gauffin, 1979; Gauffin and Sundberg, 1980; Sundberg, pers. comm., 1981; Karlsson, pers. comm., 1983; Karlsson, 1986), from measurements on excised larynges (for example van den Berg and Tan, 1959), from true physical modelling of the processes (for example Ishizaka and Flanagan, 1972; Monsen et al., 1978), and from perception experiments (for example Rosenberg, 1971).

In general, the literature up to about 1986 suggests the following trends: VOIA increases with PDIFF to a maximum value of about 700 or 800 cm^3/s, but to less than this for some women speakers; VOIA increases as AG increases but decreases as Q rises; TCR increases to a maximum value of 0.5 or 0.6 as PDIFF increases while TCR decreases when AG or Q increases; TD decreases to a minimum of 0.6 or 0.8 ms for speech and if TD is as low as 0.4 ms the synthetic speech lacks naturalness, but TD may go as low as about 0.3 ms for some loud professional singing; TD increases when AG or Q increases.

The general variation of USLOPE with the controlling variables can be predicted in some, but not all, cases. USLOPE = VOIA / TD and where VOIA and TD vary oppositely with a controlling variable the trend for USLOPE can be judged, but where VOIA and TD vary together the trend in USLOPE must depend on the magnitudes of the changes in VOIA and TD. As PDIFF increases VOIA increases and TD decreases so USLOPE should increase with PDIFF. By similar reasoning, USLOPE is expected to decrease as Q increases. It is not clear whether USLOPE should increase or decrease as AG increases.

Fundamental frequency F_0 is a dependent variable, controlled by PDIFF as well as by Q; this may explain some of the complexity of the findings in real speech, where F_0 is treated as an independent controlling variable for voice waveshapes. Increasing Q and PDIFF seem to have opposite effects on VOIA, TCR and TD, so that the effects of F_0 changes on these three parameters is not obvious.

Minimum values for transglottal pressure drop PDIFF and volume flow rate UG to permit voicing to start or stop are difficult to assess: the threshold values depend in complex ways on the mechanical state of the vocal folds, represented by Q in our modelling (van den Berg and Tan, 1959). As a first step, PDIFF has an onset threshold DPMON and a lower offset threshold DPMOFF. Values used previously were 4 and 2 cmH_2O. In addition, if the calculated VOIA is very small, a voice waveform is not generated. Previously the VOIA threshold for both onset and offset of voicing was VATON and VATOFF set to 0.1 cm^3/s.

From the published evidence, suitable values of voice wave parameters VOIA, K and TCR (or other combinations of three of the parameters shown in Figure 5) are inserted in a table showing selected sets of values for the controlling parameters, as follows: PDIFF 5 or 10 cmH_2O, Q 100 Hz or a higher value Q1, AG 0.05 or 0.15 cm^2 (see Figure 8). Linear mapping relationships are assumed, of the form:

c11 . PDIFF + c12 . Q + c13 . AG + c14 . VOIA + 1 = 0
c21 . PDIFF + c22 . Q + c23 . AG + c24 . VOIA + 1 = 0
c31 . PDIFF + c32 . Q + c33 . AG + c34 . VOIA + 1 = 0
c41 . PDIFF + c42 . Q + c43 . AG + c44 . VOIA + 1 = 0

Four different combinations of values for the three controlling and three controlled parameters are substituted here to give twelve simultaneous linear equations with twelve unknowns. This is solvable unless one equation is a multiple of another. F_0 is assumed to be determined by a laryngeal control parameter Q and an aerodynamic control parameter PDIFF through the expression

F_0 = Q + KF. PDIFF   where KF is an empirical constant, or
F_0 = Q (1 + KFX . PDIFF)   where KFX is an empirical constant.

This working formula will need to be revised. Titze (1989) has shown, through measurements on excised larynges and analysis of the effects of the non-linear stress-strain curve for vocal fold tissues, that subglottal pressure and the vocal fold tension factor interact in the opposite way to this: the largest changes in F_0 with subglottal pressure increase occur at low Q, when the vocal folds are very short and lax. The predicted value for KF is 0.5 to 6 Hz/cmH_2O.

It is assumed, based on a variety of published research, that there is a range of AG values within which full voicing can occur if other conditions are suitable. If AG becomes too small, less than a parameter AGMIN, then voice amplitude VOIA decreases and F_0 falls, to simulate glottalisation. If AG becomes too large, above AGMAX, then VOIA is gradually scaled down according to the factor

$$PDIFF^{1.5} / [\max \text{ of } (AG, AGMAX)]^{AEXP}$$

AEXP is a parameter determining how rapidly VOIA falls off as AG increases; it is intended to allow for different degrees of sensitivity to vocal fold abduction. VOIA is modified from its table values and then the voice waveshape is constructed from the revised set of parameter values.

As van den Berg and Tan (1959) observed: "...every larynx sang after its own fashion...". Bickley and Stevens (1986) found cross-speaker differences in the effects of vocal tract narrowing on voice source parameters; Laine and Vilkman (1987) found, among larynges of three men, that one was more sensitive to vocal tract shape than the others. So, our approach here is to try to characterise, for each of two speakers, the mode of vibration of the vocal folds under a range of conditions likely to occur during speech production. The aim is to make estimates consistent with their real speech data for as many of the parameters as possible in our model.

2.1.5. *Other acoustic sources.* The most important principle is that acoustic sources are not to be defined in an ad hoc way; in however severely simplified a manner, the model should capture in essence the fact that acoustic sources are present only when both geometry and aerodynamic conditions are in appropriate ranges; as they vary, the acoustic sources' properties should change.

The expressions for the amplitude envelopes of turbulence noise sources, based on Stevens (1971) are

aspiration noise, generated just above the glottis: $ASPA = AG^{0.5} . PDIFF^{1.5}$
frication noise, generated in front of a vocal tract constriction: $FRICA = AC^{0.5} . PC^{1.5}$
(at the teeth for alveolar and post-alveolar consonants)

ASPA or FRICA is multiplied by random numbers to give a white noise source. Some broad-bandwidth filtering has been applied sometimes, but not in this study, to give a slightly better match to spectral shaping, which should vary with the constriction geometry (Stevens, 1971; Shadle, 1986). For conditions alternating between suitable and unsuitable for turbulence, it is possible that the effective magnitude of turbulence noise generated will be reduced. For this reason, when voicing is present and VOIA is above an arbitrary fractional threshold VTOL, ASPA and FRICA are multiplied by a factor VFAC, less than one.

A transient source, earlier related to dPC/dt, has been modified recently in a first attempt to take account of the modelling of Maeda (1987). Here the transient is given an amplitude proportional to the rate of increase of constriction area dAC/dt at a consonant release. To reduce its effective duration, within the constraints of a 5 ms sample time for the articulatory block of the model, the amplitude immediately decays exponentially. The sample time for the acoustic sources in the acoustic filtering block is 28 microseconds.

When there is voicing, a voice-synchronous component is added to the articulatory component of glottal area AG, to give a total glottal area AGC. The mean value of AGC is made to match AG and a base offset is permitted to simulate breathy phonation conditions. AGC is allowed to reach a maximum value as high as the largest AG value in the run. The oscillatory component of AGC is symmetrical in time whereas the acoustic component of glottal flow, the voice source waveform, is usually skewed to the right. AGC is treated as a section of length 0.5 cm in the filter and the subglottal resonators are included as a uniform tube. The fraction of energy reflected back to the glottis from the subglottal tube is controlled by a parameter. Increased acoustic loss from the vocal tract when the vocal folds are abducted is an important feature of articulatory synthesis. If a [VsV] sequence for example is simulated with fixed glottal loss, voicing energy trapped inside the nearly closed vocal tract for [s] gives unrealistic low frequency energy during the contoid phase. Addition of glottal area-dependent losses improves naturalness in two ways: not only by greatly reducing this extraneous voicing energy, but also by increasing bandwidths as seen through frication and aspiration noise for the [s].

2.1.6. *Acoustic filtering and radiation.* The acoustic filtering stage of the model is based on the reflected pressure wave method of Kelly and Lochbaum (1962). A transform from output volume flowrate to radiated sound pressure of +6dB/octave is used. These aspects of the modelling will not be described here, since the matching of the two speakers has focussed on the earlier stages of modelling.

2.2. ANALYSIS AND ARTICULATORY MODELLING FOR TWO WOMEN SPEAKERS

The aim is to differentiate, within a single acoustic signal of speech, between common linguistically determined signal features and acoustic pattern features which vary across speakers. The method used is to combine two different sets of speaker descriptions with the same linguistic string and the same accent of English in a simplified representation of the same physical processes. It is hoped that details of acoustic pattern features may have an internal consistency that can be traced back to speech production mechanisms. If it could be shown that many aspects of acoustic variation are rule-governed in this particular way, then it might prove possible, eventually, to characterise a speaker type by generalised rules for articulatory synthesis or by derived acoustic covariation rules applicable in acoustic based synthesis and in automatic speech recognition also.

2.2.1. *Choice of speakers.* From ten speakers recorded for a physiologically-oriented database of some R.P. English speech, two women, sr and gb, with apparently well contrasted voice types were selected for this study. Their aerodynamic traces, seen in Figure 11, differ greatly. As argued above, aerodynamic differences are likely to imply a multiplicity of acoustic pattern differences. It remains to be demonstrated which of their speech production parameters differ and which are similar, and what the implications are for

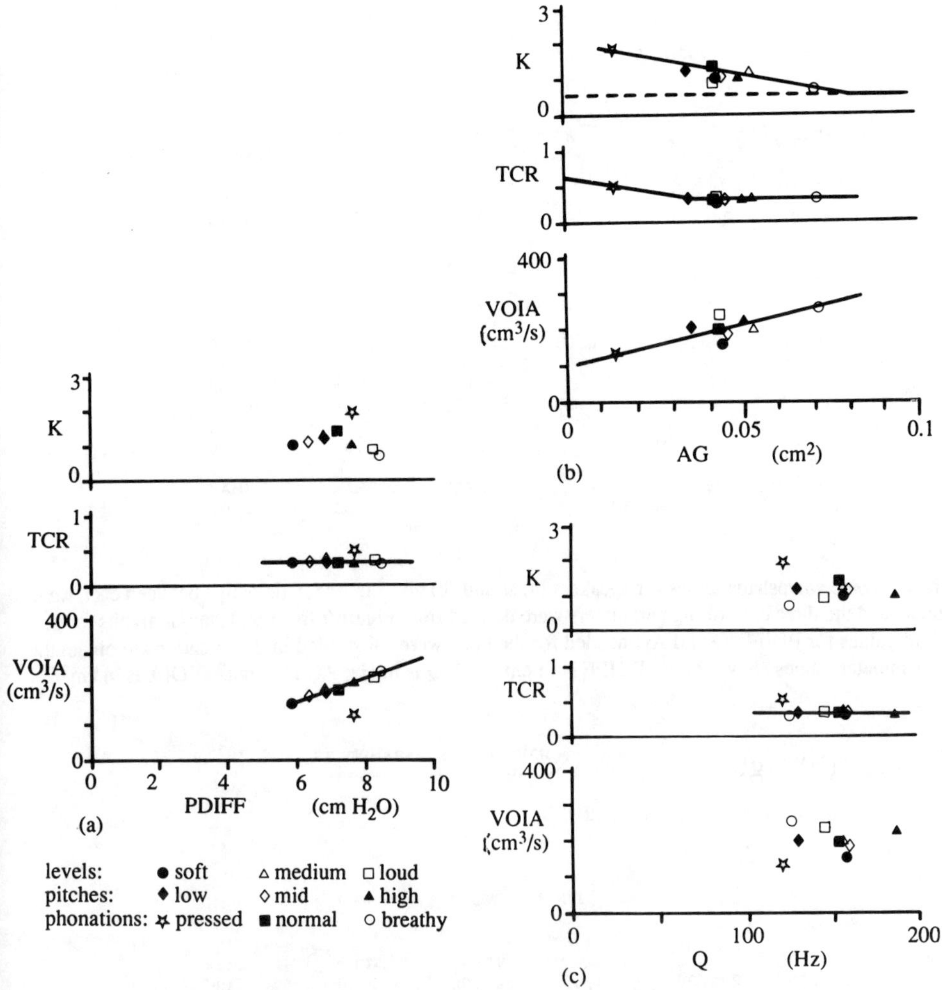

Figure 7 Real speech data for the voice parametric model from aerodynamic traces and inverse filtering for speaker sr. The three voice wave parameters K, TCR and VOIA, defined in Figure 5, are shown as functions of the three controlling parameters: (a) PDIFF the air pressure drop across the glottis, (b) AG the articulatory component of glottal area, (c) Q the effective mass and stiffness of the vocal folds.

(a) sr

WAVE SHAPE PARAMETERS

TKSW = 2 (VOIA, TCR, K)

PDIFF	Q	AG	VOIA	TCR	K
	Q1		VOIA1	TCR1	K1
10.00	200.00	0.05	308.10	0.22	1.10
			VOIA2	TCR2	K2
5.00	100.00	0.05	118.10	0.22	1.10
			VOIA3	TCR3	K3
10.00	100.00	0.05	228.10	0.22	1.10
			VOIA4	TCR4	K4
10.00	100.00	0.15	510.30	-0.64	-0.90
	MINIMUM		0.01	0.00	0.50
			VOIAMAX	TCRMAX	KMAX
	MAXIMUM		300.00	0.55	2.00

Figure 8 Voice waveshape tables for speakers (a) sr and (b) gb. Linear relationships between each wave parameter and the three controlling parameters were derived from Figure 7 for sr and similar graphs for gb. Standard values for PDIFF, Q and AG needed for the table were substituted in the equations to obtain the wave parameter values shown here. PDIFF is in cm H_2O, Q is in Hz, AG is in cm^2, VOIA is in cm^3/s.

(b) gb

WAVE SHAPE PARAMETERS

TKSW = 2 (VOIA, TCR, K)

PDIFF	Q	AG	VOIA	TCR	K
	Q1		VOIA1	TCR1	K1
10.00	250.00	0.05	-47.00	0.50	1.38
			VOIA2	TCR2	K2
5.00	100.00	0.05	-206.00	0.50	-0.63
			VOIA3	TCR3	K3
10.00	100.00	0.05	321.00	0.50	0.23
			VOIA4	TCR4	K4
10.00	100.00	0.15	688.00	0.20	0.23
	MINIMUM		0.01	0.00	0.50
			VOIAMAX	TCRMAX	KMAX
	MAXIMUM		250.00	0.55	2.00

VOIA, VOIF, ASPA AND FRICA PARAMETERS

DPMON	4.500000	3.000000	Minimum pressure difference for voice onset
DPMOFF	1.500000	1.500000	Minimum pressure difference for voice offset
AEXP	2.000000	2.000000	AREA EXPONENT USED FOR CALCULATING VOIA
VATON	0.100000	0.100000	Voice amplitude threshold (onset)
VATOFF	0.100000	0.100000	Voice amplitude threshold (offset)
AGMAX	0.120000	0.040000	EMPIRICAL CONSTANT FOR VOIA
AGMIN	0.030000	0.010000	0.99 REFERENCE LEVEL FOR VOIA AND VOIF
KF	3.080000	9.850001	Empirical constant for voif
KFX	0.016000	0.070000	Fractional constant for voif
FX	2	2	Voif calc. switch: 1 old, 2 proportional
VFT	2.000000	2.000000	VOICE FREQUENCY THRESHOLD
vtol	0.300000	0.300000	VOIA THRESHOLD LEVEL FOR REDUCING ASPA & FRICA
vfac	1.000000	0.500000	REDUCTION FACTOR FOR ASPA & FRICA
AFSW	2	2	ASPA/FRICA CALCULATION METHOD SWITCH
AREF	0.000200	0.000200	REFERENCE AMPLITUDE FOR CALCULATING ASPA/FRICA
KN	1.000000	1.000000	EMPIRICAL CONSTANT FOR ASPA/FRICA

(b) gb (a) sr

Figure 9 Additional parameters for the voice wave model. Values for speakers (a) sr and (b) gb are shown. The definitions of the parameters and the methods of estimating their values are described in the text. KF and KFX were estimated by the method shown in Figure 10.

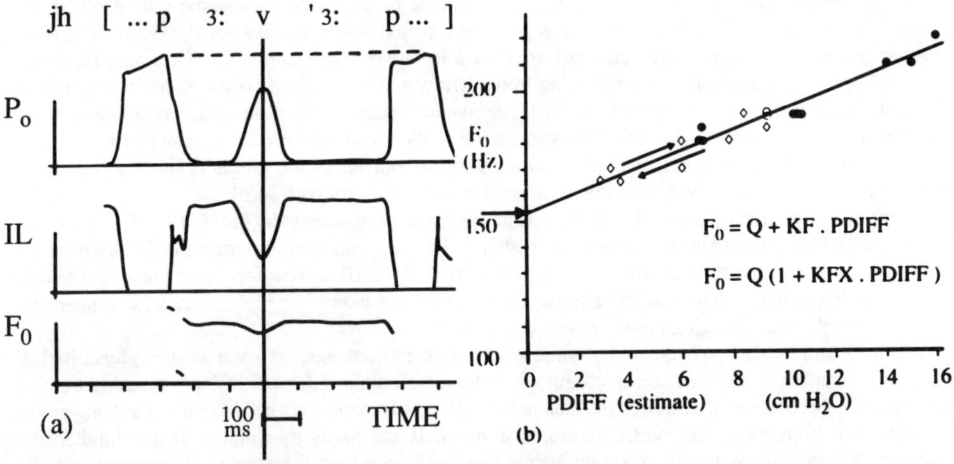

Figure 10 (a) Traces for speaker jh illustrating some of the effects of a severe vocal tract constriction for [v]. P_O is oral air pressure, IL is intensity level high pass filtered at 500 Hz, F_0 is fundamental frequency. The dashed line shows estimated subglottal air pressure P_{sg}. (b) Linear relationship derived from (a): filled circles for mid-vocoid in [...p3:p3:...] at soft, medium and loud levels, open circles for mid-vocoid in [...p3:v'3:...], diamonds for transitions into and out of the region of high oral pressure for [v]. The arrow shows Q, estimated at 154 Hz.

acoustic pattern features of their speech.

2.2.2. *Experimental methods for the natural speech.* Total output volume flow rate U_O was measured with a Rothenberg mask (Glottal Enterprises). Oral air pressure P_O was measured with a polyethylene tube as described above, with a Gaeltec pressure transducer system. A trace proportional to constriction cross-section area was derived from the flow and pressure signals with an Aerodynamic Speech Analyser (R. Caley) which gave an output signal $A = U_O / P_O^{0.5}$. These three signals, low-pass filtered at 50 Hz to reduce the voicing ripple, were displayed as hard copy traces on a mingograph 803. Additional traces were the sound pressure signal unprocessed, a laryngograph signal (Laryngograph Ltd.) and F_0 from the laryngograph signal, using a fundamental frequency meter (F.J. Electronics). Intensity level traces high-pass filtered at 500 Hz and at 3.9 kHz (intensity meters and audio filter, F.J. Electronics) were displayed also. Some of these traces are shown in Figure 11. Four channels of data were recorded on FM tape (Racal Store 4), as follows: sound pressure from a B&K condenser microphone type 4134 placed outside the mask, with sound level meter type 2231, the laryngograph signal, volume flow rate of air unfiltered and oral pressure unfiltered. The microphone and inverted laryngograph signals were recorded also onto videocassette via PCM (Sony PCM-F1 and SL-F1E) with Oros modification to remove the spectral shaping when recording (the standard SPAR format).

2.2.3. *Voicing models for the two speakers.* Each speaker produced, on one expiratory breath, a series of [...p3:p3:...] with three different levels - soft, medium and loud, with three different pitches - low, mid and high, and with three different phonation types - pressed (or glottalised), normal and breathy. Although the microphone signal contained some audible background noise from the mingograph, the audio recordings were of high quality and could be used for detailed acoustic analyses, including inverse filtering, taking as input both sound pressure and total volume flow rate. In each case the laryngograph signal located the start of the closed phase of voicing; after the first estimate of the voice waveshape was obtained, the analysis window was adjusted to give as flat as possible a closed phase (Brookes et al., 1989). The voice waveshapes for the different phonation types are shown in Figure 6 as examples. Generally there was quite good agreement between inverse filtering from airflow and from sound pressure. Where the results disagreed, the airflow method was used, and this is shown in Figure 6, with peak values of glottal flow and the constant d.c. flow marked. Smooth lines were drawn as shown and voice wave parameters VOIA, TCR and K measured. A continuously open chink in the posterior portion of the glottis is common in normal voices (Hirano, pers. comm. 1985) and this is inferred from the data for both speakers.

From the mingograph traces subglottal pressure, pressure drop across the glottis PDIFF and volume flow rate of air through the glottis UG were estimated in the same mid-vocoid portions of [3:] used for the inverse filtering, and glottal area AG was calculated with the orifice equation. For these mid-vocoid portions the discrepancies between UG and the U_O trace are not likely to be large, since PC is near zero and VC is probably not changing rapidly here.

Figure 7 shows each of the voice wave parameters plotted against each of the physiological controlling parameters, for speaker sr. Similar graphs were obtained for gb. Where a strong trend is apparent a line has been drawn through the data points. The twelve constants needed for our voicing model were obtained from these data, partly by graphical methods and partly by solution of the simultaneous equations. Likely minimum and maximum values were estimated from the graphs. Parameter values for the voice waveshape tables were calculated, with values as shown in Figure 8. The tables were originally set up to simulate men's voices and this may partly explain why there are some negative values shown. These are from extrapolation of straight line relationships like those seen in Figure 7; all the values were positive at the operating points of the two speakers as simulated in the model. The Q1 parameter values were set, arbitrarily, above the highest Q value in Figure 7.

The remaining parameters for the voicing models are shown in Figure 9. Their values for the two speakers had to be roughly estimated where possible, or left at standard values. AGMAX and AGMIN, defining the 'plateau' region of AG for full voicing, were matched to the range of values for AG estimated from mingograph traces for each speaker. Inverse filtering showed a higher d.c. glottal flow and

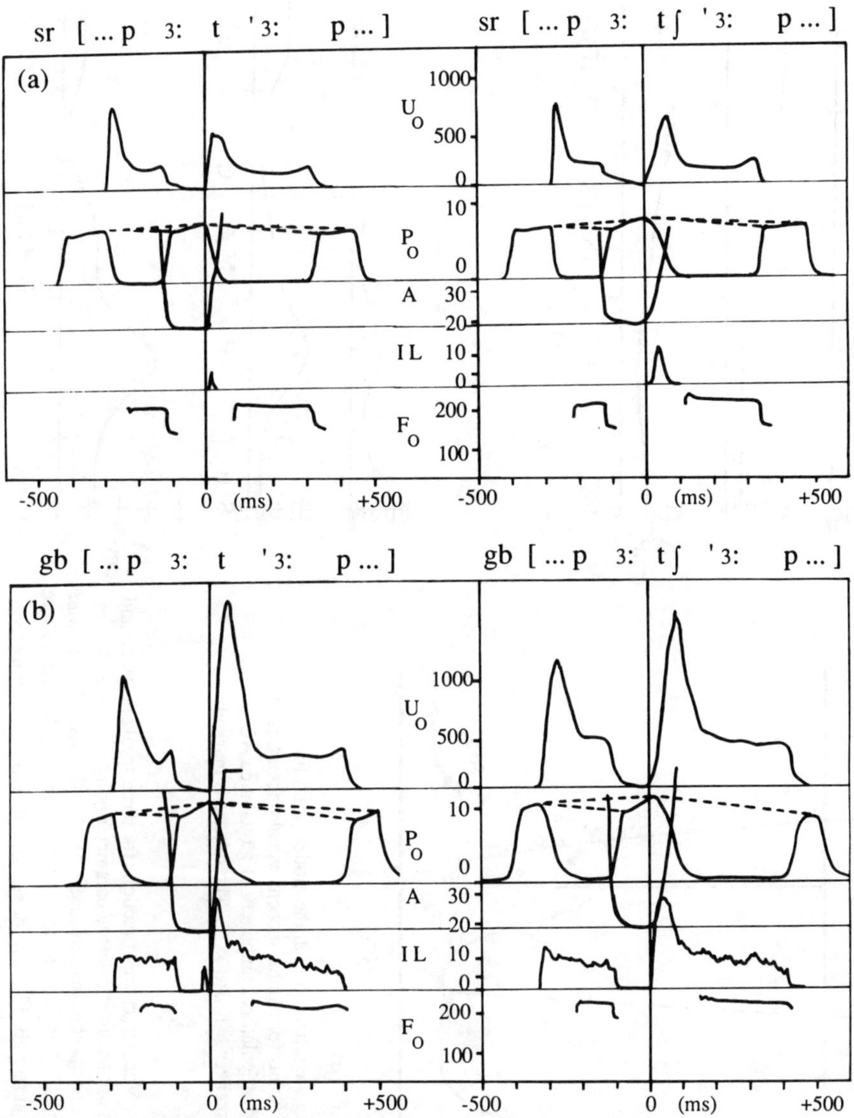

Figure 11 Natural speech data for (a) sr and (b) gb. From the top down the traces are: total volume flowrate of air through mouth and nose combined U_O (cm^3/s), oral air pressure P_O (cm H_2O), area trace for vocal tract constriction A (cm^2, calibration not shown), IL intensity level high pass filtered at 3.9 kHz, from the microphone signal (dB), F_0 from the laryngograph signal (Hz). Time is in ms with 0 located at the consonant release. Estimates for subglottal pressure P_{sg} are shown as dashed lines. On the left [...p3:t'3:p...]; on the right [...p3:tʃ'3:p...].

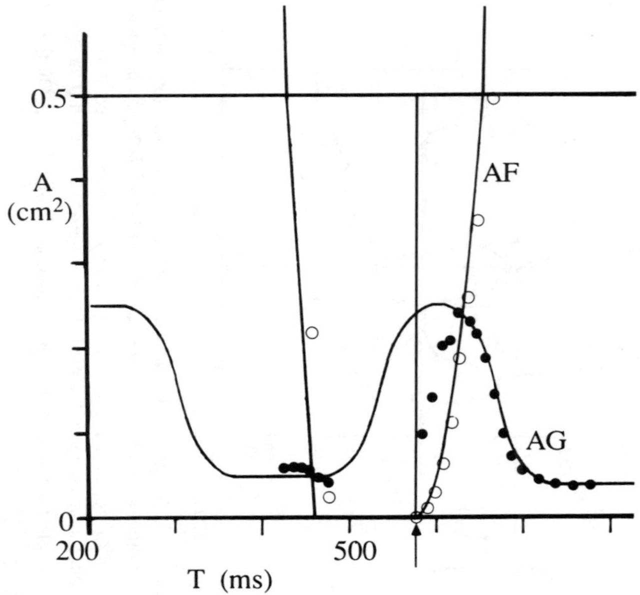

Figure 12 Matching the articulation in the model (solid lines) to the estimated data points for natural speech: vocal tract articulator AF (open circles) and articulatory component of glottal area AG (closed circles), for speaker sr [...3:tʃ3:...]. The release is marked by an arrow.

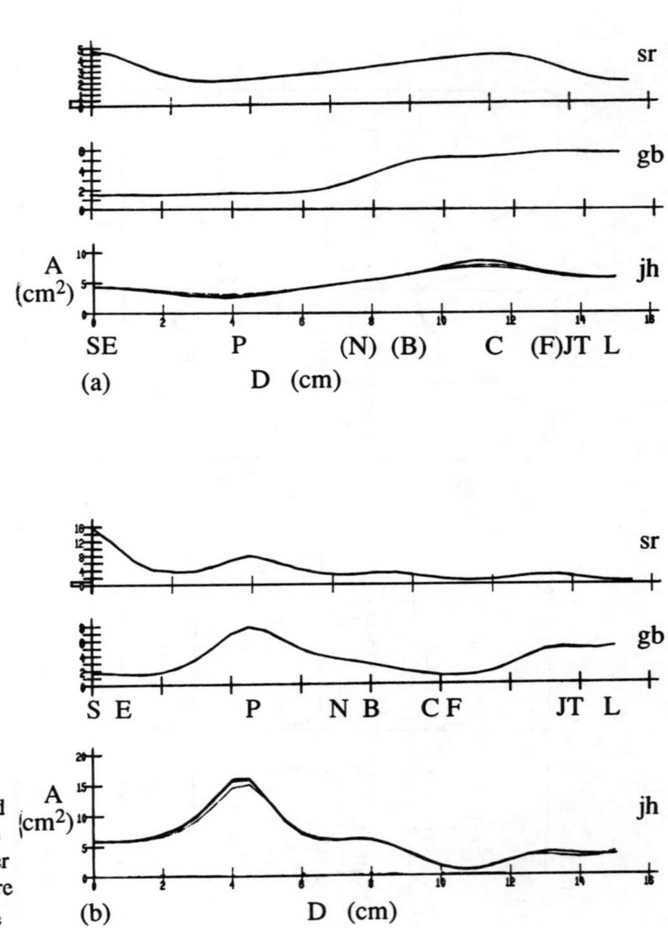

Figure 13 Vocal tract area functions for three speakers, sr, gb and jh. The positions of articulatory parameter points in the model are shown for one speaker in each case. (a) [3:], (b) about 70 ms after the release for [tʃ] in [...p3:tʃ3:p...]. Vocal tract lengths selected are 15 cm for jh and gb, 13.5 cm for sr, but with a 0.5 cm larynx tube added to give a total length of 14 cm.

aerodynamic traces indicated a larger articulatory component AG for gb than for sr. It seemed reasonable therefore to suppose that, when the vocal folds were vibrating, turbulence noise sources would be reduced less for gb than for sr. For this reason VFAC was set at a 0.5 reduction factor for sr but noise sources were unaltered with VFAC at 1 for gb.

2.2.4. *Control of F_0 for the two speakers.* A dip in F_0 is usually seen when oral air pressure rises, as shown in Figure 10(a) for speaker jh, the first of the women speakers analysed. Subglottal pressure PSG and transglottal pressure drop PDIFF were estimated and F_0 measured as oral pressure rose, to give the F_0-PDIFF relationship shown in Figure 10(b). Similar methods were used for speakers sr and gb. The KF and KFX values thus estimated are shown in Figure 9.

2.2.5. *Simulations of a plosive and an affricate for the two speakers.* The sequences chosen for analysis and modelling are shown in Figure 11. These are examples chosen from the series produced by each speaker. [t] and [tʃ] seem likely to include contrasts of noise sources during their release phases as well as different formant transitions. The A traces proportional to $U_O / P_O^{0.5}$ indicate a more sudden release for [t] and a more gradual one for [tʃ] for both speakers. Vocal tract constriction cross-section areas were calculated at 10 ms intervals, from the calibrated U_O and P_O traces. PDIFF was estimated and used with U_O to give a first estimate for AG. Figure 12 shows the data points obtained in this way for speaker sr [3:tʃ'3:]. The real speech data points, shown as open circles, for the release of the vocal tract tongue tip-blade alveolar ridge constriction, to be represented in the model by AF, could be well matched by a parabolic path, the solid line in Figure 12. The data points for AG estimated from the natural speech are even less reliable than those for AF, but there is at least an indication of the vocal fold adduction gesture following the vocal tract closure release. The estimated AG is judged to be least reliable immediately after the release, since just here subglottal pressure is likely to fall and estimates of PDIFF may be very inaccurate. In the modelling, the vocal fold abduction phase with AG increasing was matched to the path for the adduction phase, not to these unreliable data points. It is an advantage of the flow-dependent resistances of constrictions that area varies as the square root of pressure drop; errors in PDIFF are not as important as errors in UG.

Aerodynamic traces give indications of articulation where there is a severe vocal tract constriction but give no information on the overall vocal tract shape. This is needed for simulation of vowel-like configurations, and for the aerodynamic modelling specifically the enclosed volume must be estimated. An analytical tool called FORMFIND was developed for estimating some aspects of vocal tract shape from formant frequencies, based on the horn equation methods developed by Kuhn (1983). This method has the uncertainties associated with lossless models for inverse transforms to derive vocal tract geometry from a speech sound pressure signal, but it does at least give a possible vocal tract shape from which formant patterns similar to those of a particular speaker can be generated. This was confirmed by modelling [i:s'i:] for speaker jh. A uniform single tube resonator, closed at one end, the glottis, and open at the other, the lip, was defined for each of the three women speakers. Vocal tract average cross-section area and length had to be specified. Several lengths were tried for each speaker and the one giving smoothest shapes was selected (Atal et al., 1978). This initial 'average' cross-section area was chosen arbitrarily to be 3 cm^2 for all three speakers. It would be very helpful if purely acoustic methods could be devised to estimate vocal tract length and volume for approximations to uniform tubes for real speakers. Some results are shown in Figure 13. In the case of sr and gb the formant frequencies were measured from SPAR sfs spectrograms, like those shown in Figure 20. For jh, a hardware spectrograph (700, Voice Identification) was used. It was encouraging that the rather smooth area functions obtained for [3:] vocoids remained smooth even when four formants had to be matched. The results seem likely to be less reliable for Figure 13(b) shortly after the release for [tʃ] for several reasons: F1 cannot be measured reliably here, the assumption of an acoustic tube closed at one end and open at the other is probably not valid here where the glottis is large, and the assumption of no internal reflections for wave propagation along the vocal tract is probably violated here beacuse of the fairly severe constriction. If it could be demonstrated that the location of the constriction was appropriate, that would be valuable information.

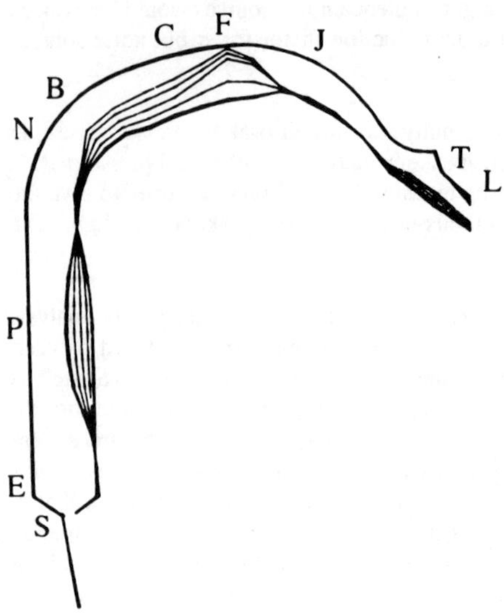

Figure 14 Vocal tract side view display for simulating sr [#3:tʃ'3:#]. The post-release phase for [tʃ] is shown at 25 ms (5 time units) intervals.

Figure 15 Complete articulatory time plan for simulating sr [V1C2V2] in the sequence [#p3:tʃ'3:p#] [#1C1V1C2V2C3#2]. The initial and final plosives [C1] and [C3] were removed but the timing of their actions appear here.

IDENTIFIER	VALUE	EXPRESSION
#1.EPLUX	60	#1.EPLUX:= C1.EALX - #1C1.D
#1C1.D	-36	#1C1.D:= -36
#2.EPLU	191	#2.EPLU:= C3.EAL + C3#2.D
C1.D1	0	C1.D1:= 0
C1.D2	-12	C1.D2:= -C1.D4/2
C1.D4	24	C1.D4:= 24
C1.EAG	48	C1.EAG:= C1.EAL + C1.D1
C1.EAL	48	C1.EAL:= C1.EALX + C1.D4
C1.EALX	24	C1.EALX:= ET0 + 24
C1.ETB	36	C1.ETB:= C1.EAL + C1.D2
C1V1.D	20	C1V1.D:= 20
C2.D1	8	C2.D1:= 8
C2.D2	5	C2.D2:= 5
C2.D4	23	C2.D4:= 23
C2.D6	0	C2.D6:= 0
C2.D8	-5	C2.D8:= -5
C2.D9	3	C2.D9:= 3
C2.EAF	114	C2.EAF:= C2.EAFX + C2.D4
C2.EAFX	91	C2.EAFX:= V1.EAF + V1C2.D
C2.EAG	122	C2.EAG:= C2.EAF + C2.D1
C2.EAGX	119	C2.EAGX:= C2.EAG - C2.D9
C2.EDF	109	C2.EDF:= C2.EAF + C2.D8
C2.EQ	92	C2.EQ:= C2.EAF - 22
C2.ETB	119	C2.ETB:= C2.EAF + C2.D2
C2.ETBX	91	C2.ETBX:= C2.EAFX + C2.D6
C2V2.D	26	C2V2.D:= 26
C3#2.D	-12	C3#2.D:= -C3.D4/2
C3.D4	24	C3.D4:= 24
C3.D5	19	C3.D5:= C3.D4 - 5
C3.D6	12	C3.D6:= C3.D4/2
C3.EAGX	198	C3.EAGX:= C3.EALX + C3.D5
C3.EAL	203	C3.EAL:= C3.EALX + C3.D4
C3.EALX	179	C3.EALX:= V2.EAL + V2C3.D
C3.ETBX	191	C3.ETBX:= C3.EALX + C3.D6
ET0	0	ET0:= 0
ETMAX	215	ETMAX:= #2.EPLU + 24
V1.D4	7	V1.D4:= 7
V1.EAF	75	V1.EAF:= V1.EALX + V1.D4
V1.EAG	95	V1.EAG:= C2.EAGX - 24
V1.EALX	68	V1.EALX:= C1.EAL + C1V1.D
V1C2.D	16	V1C2.D:= 16
V2.D1	-5	V2.D1:= -5
V2.D4	15	V2.D4:= 15
V2.EAFX	140	V2.EAFX:= C2.EAF + C2V2.D
V2.EAG	146	V2.EAG:= C2.EAG + 24
V2.EAGX	170	V2.EAGX:= V2.EAG + 24
V2.EAL	155	V2.EAL:= V2.EAFX + V2.D4
V2.EQ	135	V2.EQ:= V2.EAFX + V2.D1
V2C3.D	24	V2C3.D:= 24
V3.EQX	223	V3.EQX:= C3.EAL + 20

The model's parameter point positions and their area values were chosen with reference to the area functions obtained. Two examples of the choice of locations for the parameter points are shown in Figure 13. Vocal tract cross-section area is the articulatory dimension used in our articulatory modelling, but it is sometimes helpful to see whether the vocal tract side view looks plausible. An example of this for the release portion of sr [tʃ] is shown in Figure 14. Of course, mapping from area function to mid-sagittal view of the vocal tract or back is an important region of uncertainty in articulatory modelling.

Figure 15 shows the complete articulatory time plan for simulating the sequence containing [tʃ] as produced by speaker sr. The labels refer to the events E and durations D as shown in Figure 3 and 4. The initial and final [p] plosives appear in this time plan, but they were removed from the simulations, for simplification, by making a few simple changes to articulatory instructions and coordination. [tʃ] is consonant [C2], with [V1] before it and [V2] after it.

Figure 16 shows some of the graphs for the modelling of sr [#3:tʃ'3:#]. After the first run of the model, the aerodynamic traces from the modelling were compared with those for the real speech. Parameter values shown in Figure 4 were altered, notably the AF-AG coordination duration C2.D1 and the maximum value of AG for the consonant. When the constriction was made to move back during the release phase for [tʃ], the resultant cavity volume changes prevented a good match between real and synthetic flow rate traces here, across a wide range of back cavity volume changes made to allow for the uncertainties in this volume as estimated from FORMFIND. The position of the constriction was therefore kept constant. It seems that some aspects of the articulation simulation can be improved through the modelling itself, an analysis-by-synthesis approach to speech production. The graphs of Figure 16 are the result of about four runs of the model to get improved matching, specifically between the value and timing of the airflow peak for UC as well as the value and general shape of PC (Figure 16(b)) and the timing of the frication noise peak relative to the release (Figure 16 (c)).

Figure 17 shows some of the corresponding stages of modelling for gb. Figure 18 compares simulation of sr and gb in detail, with all time samples shown on the graphs. The similarities of the patterns for the different acoustic sources and the total glottal area in the two cases are clear. The actual magnitudes differ, with higher amplitudes of noise sources for gb than for sr. Voice wave amplitudes are not very different for the two speakers.

As a first attempt at closer matching of the voicing model's characteristics to the real speaker, the threshold for voice onset VATON was adjusted to make the first few cycles of voicing at a breathy onset grow in the model in the same sort of way as in the real speech. For Figure 19 VATON had been increased from its previous value of 0.1 cm^3/s. More direct comparisons need to be and could be made with inverse filtering on the actual speech sequence modelled instead of, as here, on the voice onset following a [p] release.

2.2.6. *Internal consistency of the modelling.* Some consideration was given to consistency across different stages of the modelling. For example, a value for UG, the aerodynamic component of glottal flow, was estimated from the inverse filtering results based on output flow. Assuming a triangular form for the wave, UG = d.c. flow + 0.5 (1 - TCR0) . VOIA and this gave UG values in mostly very good agreement with those measured from the mingograph traces. Thus the aerodynamic and voice source models are consistent here, which was not the case previously in our modelling. The d.c. flow of air through the glottis needs to be more explicitly included in the acoustic source generation models however. The articulatory and aerodynamic models are internally consistent to some extent, but they would be more strongly linked if some vocal tract dimensions for vowels could be measured with confidence. There is a mis-match of acoustic models between our analysis of natural speech and the filtering stage of our articulatory synthesis: in analysis a vocal tract area function is derived from formant frequencies on the assumption that there are no internal reflections, but in synthesis reflections at section boundaries form the basis for simulation of the filtering process.

2.2.7. *Comparison of the two speaker types and internal consistency of acoustic pattern features.*
gb operates with a larger glottal area than sr and, probably for this reason mainly, she generates higher

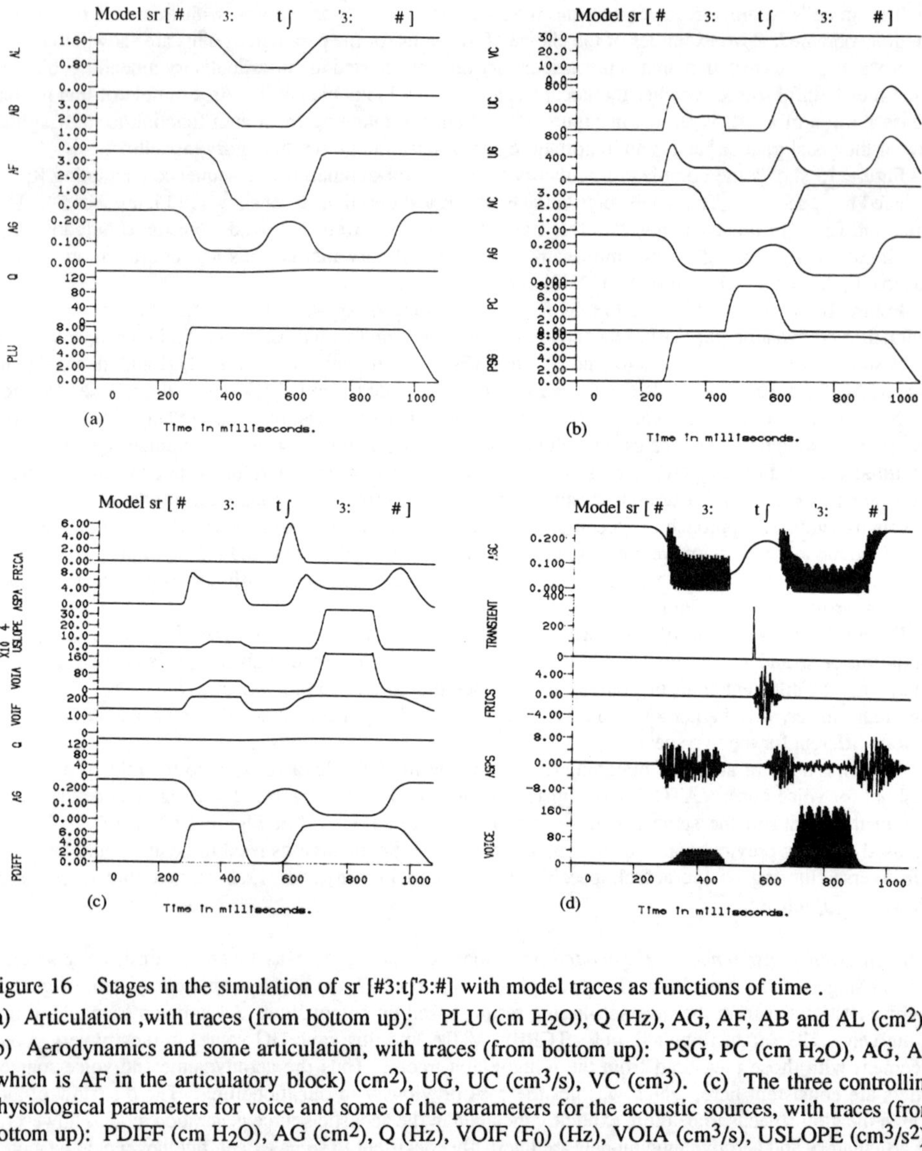

Figure 16 Stages in the simulation of sr [#3:tʃ'3:#] with model traces as functions of time . (a) Articulation ,with traces (from bottom up): PLU (cm H_2O), Q (Hz), AG, AF, AB and AL (cm^2) . (b) Aerodynamics and some articulation, with traces (from bottom up): PSG, PC (cm H_2O), AG, AC (which is AF in the articulatory block) (cm^2), UG, UC (cm^3/s), VC (cm^3). (c) The three controlling physiological parameters for voice and some of the parameters for the acoustic sources, with traces (from bottom up): PDIFF (cm H_2O), AG (cm^2), Q (Hz), VOIF (F_0) (Hz), VOIA (cm^3/s), USLOPE (cm^3/s^2), ASPA (amplitude envelope of aspiration noise source), FRICA (amplitude envelope of frication noise source (arbitrary units, the same for both noise source types). (d) Waveforms for all the acoustic sources and total glottal area, with traces (from bottom up): VOICE (cm^3/s), ASPS, FRICS (aspiration noise source and frication noise source respectively) (the same arbitrary units for both), TRANSIENT (arbitrary units), AGC (total glottal area, with a voice-synchronous component added to the articulatory component AG, for glottal losses in the filtering stage (cm^2). Irregularities in the waveshapes are due to sampling for this graphic display. Time markers are at 200 ms intervals.

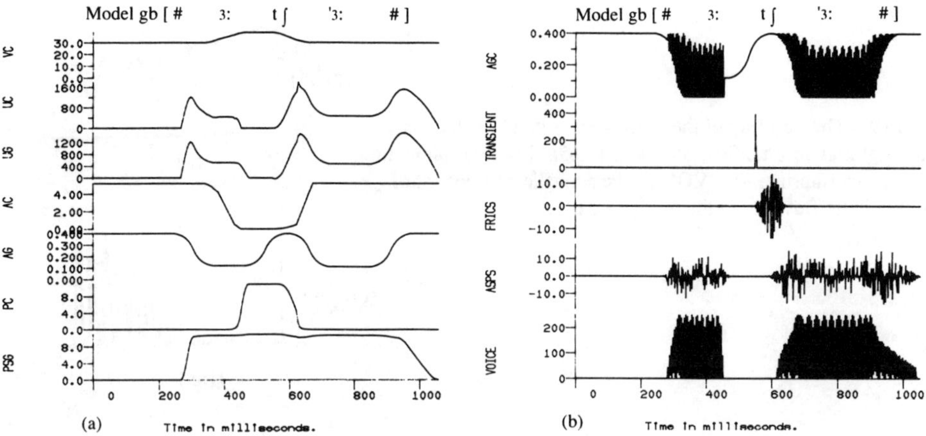

Figure 17 Some stages in the simulation of gb [#3:tʃ'3:#] with model traces, units and time markers as in Figure 16. (a) PSG, PC, AG, AC, UG, UC, VC. (b) VOICE, ASPS, FRICS, TRANSIENT, AGC.

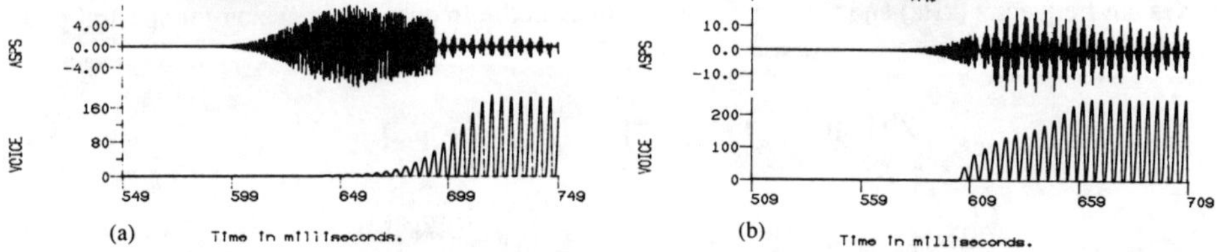

Figure 18 Simulation of (a) sr and (b) gb [#3:tʃ'3:#] showing details of acoustic sources and total glottal area for the release and post-release phase as functions of time. Traces (from bottom up): VOICE, ASPS, FRICS, TRANSIENT, AGC. Units as in Figure 16. All 28 microsecond samples are plotted here. Time markers are at 50 ms intervals.

(a)

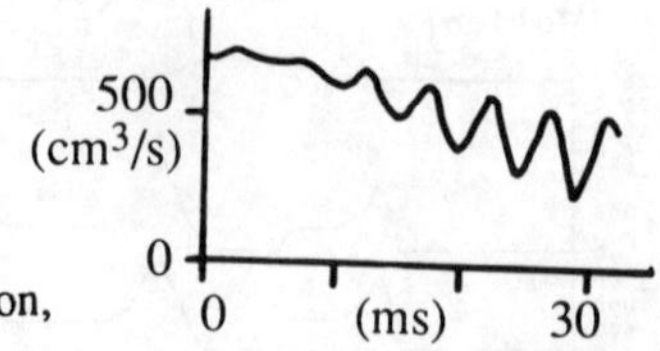

Figure 19 The build-up of the voice source in a breathy onset. (a) total glottal volume flowrate for sr [...p3:...] normal phonation, from inverse filtering. (b) VOICE, the acoustic component of glottal volume flowrate in the modelling of sr [#3:tʃ'3:#].

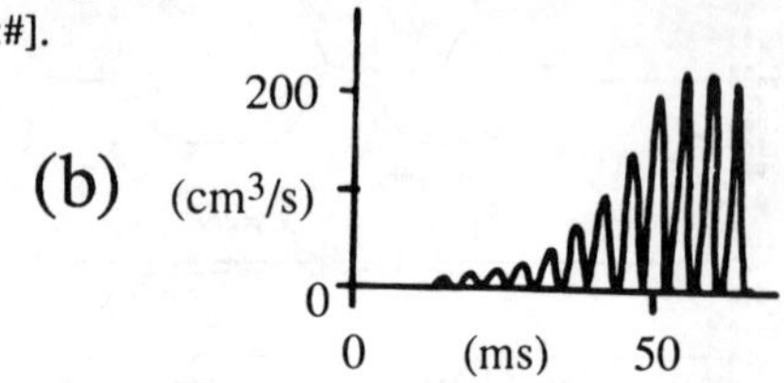

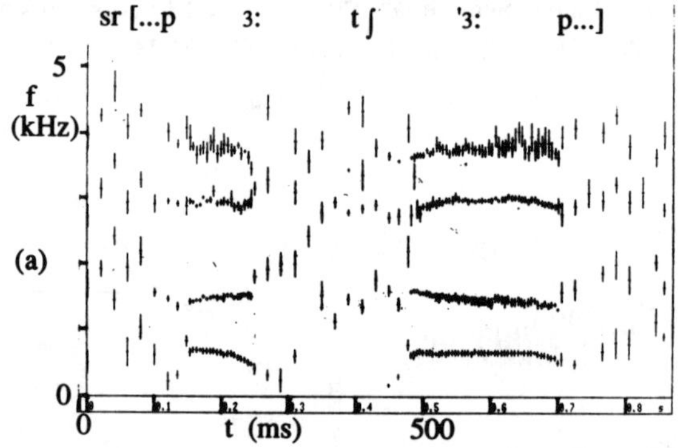

Figure 20 Excitation-synchronous LPC (SPAR sfs) spectrograms for (a) sr and (b) gb [...p3:tʃ'3:p...]. The axes are frequency (kHz) and time (ms). Wider lines in the frequency dimension for spectral peaks show wider bandwidths.

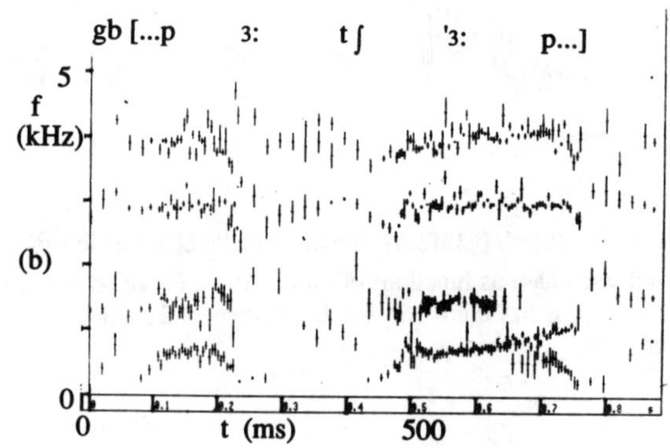

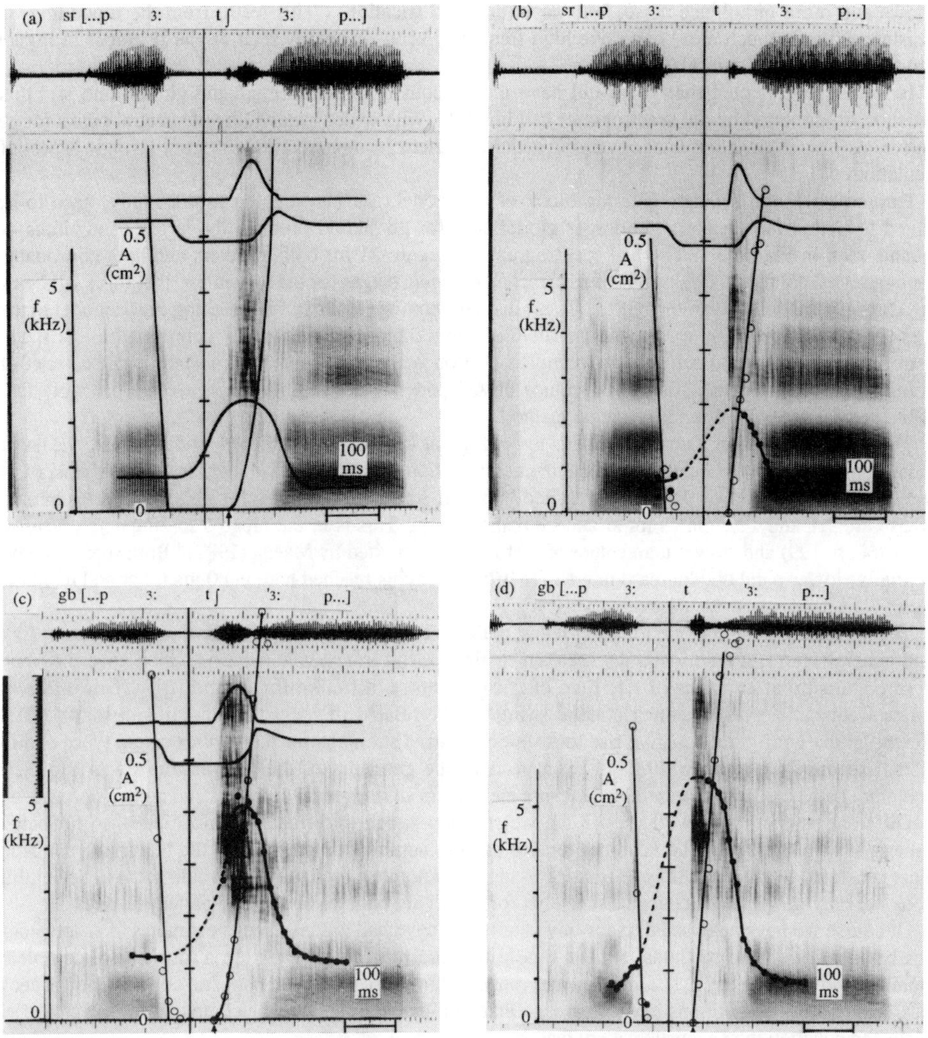

Figure 21 Comparisons of patterning for plosives and affricates produced by two speakers sr and gb. Spectrograms (LSI) have input spectral shaping flat, analysing bandwidth 400 Hz, with spectral shaping +6dB/oct. Axes are frequency with markers at 1 kHz intervals and time with a 100 ms marker. Overlays show movement paths for the main vocal tract constriction AF and the articulatory component of glottal area AG: circles for natural speech estimates, solid lines for simulation in the model. Cross-section area markers are at 0.1 cm^2 intervals. At the top of the spectrogram is shown the time course for, below ASPA and, above FRICA, the amplitude envelopes of aspiration noise and frication noise sources respectively in the modelling before reduction and modulation in the presence of voicing.

airflows and more turbulence noise, both aspiration and frication. This result from the modelling is consistent with the much more pervasive high frequency acoustic energy seen on the IL traces in Figure 11(b) for gb than in Figure 11(a) for sr.

It would be expected that gb would have more acoustic losses through the glottis than sr; this prediction is supported by the spectrograms of Figure 20, with wider bandwidths of spectral peaks for gb than for sr. This is an example of covarying acoustic pattern features which arise from a single aspect of articulation.

Parameters in the acoustic filtering block of the model, not investigated in this study, need to be adapted to match the speaker; the larger glottal area for gb should produce the larger bandwidths of formants seen in Figure 20. The LSI spectrograms in Figure 21 for both speakers and both consonants have overlays from the modelling. This has reached different stages for the four items, that for sr [tʃ] being more carefully modelled than the others. These displays confirm that careful matching of the model to the traces for natural speech up to constriction area of about 0.5 cm^2 is reasonably appropriate; this spans most of the important and acoustically complex portion between the release transient and the onset of voicing mixed with noise. Estimated values of AC and AG above about 0.5 cm^2 are probably less reliable in any case, with this aerodynamic method.

The very early part of the release seems to be of great importance for the /t/-/tʃ/ distinction. [t] has a rapid approximately linear increase in vocal tract constriction area, almost identical for the two speakers, in these examples: an area of 0.5 cm^2 is reached in 55 ms for sr and in 50 ms for gb. Initial rates of area increase are 9.1 and 12.5 cm^2/s for sr and gb respectively. This is in the range 5 to 20 cm^2/s given by Fant (1960, p. 199) and slower than values of 40 to 200 cm^2/s cited by Maeda (1987). Both speakers have an initially slower parabolic increase in area for [tʃ]: 0.5 cm^2 is reached here in 80 ms for sr and in 70 ms for gb.

The modelling gives a guide for interpreting noise on the spectrograms as aspiration or frication. As a broad generalisation, it seems that the speakers' actions result in acoustic patterns which have the same ordering and similar contrasts of rise time of frication noise, differing for [t] and [tʃ]. Their different formant frequencies for the same phonetic string can be related to vocal tract length and shape. Their different larynx configurations give rise to different relative strengths for the various sound types and to different spectral peak bandwidths. Other aspects of the geometry of their respiratory tracts and their articulatory paths could be assessed by the many methods of investigation currently under development.

Different listeners may well vary in the weight they attach to different acoustic cues, so for better synthesis it would probably be advantageous to capture as many as possible of the covarying acoustic effects generated in the modelling. For speaker-adapted automatic recognition it would seem to be helpful to look not only for isolated acoustic effects, but for covarying bundles of acoustic pattern features.

Of course we have analysed data for tiny details and have had to make rough estimates of articulatory paths by laborious hand methods. But it should become possible to build up a much more complete picture for a given speaker, over a much wider range of linguistic elements, phonetic contexts, and speech styles, with multiple repetitions in each case, so that confidence limits may be computed for the presence, domain and covariation of acoustic pattern features.

ACKNOWLEDGEMENTS

Thanks are due to Eric Brearley for running the experiments on natural speech, to Ted Allwood and Gustav Clark for the development of the computer model of speech production, to colleagues in the SPAR project for their analysis software, to David Pearson and Marion Shirt for their help with data gathering and analysis, and to all the ten speakers who gave their time and phonetic expertise, in particular to Jill House, Ginny Ball and Susan Ramsaran. Figures 2 and 3 are taken from Scully (1987) Speech Communication 6, 77-142, by permission of Elsevier Science Publishing. The work was supported by the S.E.R.C. and the D.T.I. as part of Alvey Project MMI 009, Speech Pattern Algorithmic Representation (SPAR).

REFERENCES

Allwood, E. and Scully, C. (1982) 'A composite model of speech production', Proc. Intl. Congr. on Acoustics, Speech and Signal Processing, IEEE, ICASSP 82, Vol. 2, 932-935.

Atal, B.S., Chang, J.J., Mathews, M.V. and Tukey, J.W. (1978) 'Inversion of articulatory-to-acoustic transformation in the vocal tract by a computer-sorting technique', J.Acoust. Soc. Amer., 63, 1535-1555.

Badin, P. and Böe, L.-J. (1987) 'Vocal tract vocalic nomograms: acoustic considerations. A crucial problem: formant convergence', Proc. Eleventh Intl. Congr. of Phonetic Sciences, 1-7 Aug 1989, Tallin, Estonia, USSR, paper Se 35.4, 352-355.

Berg, J. van den and Tan, T.S. (1959) 'Results of experiments with human larynxes', Pract. Oto-Rhino-Laryng., 21, 425-450.

Bickley, C.A. and Stevens, K.N. (1986) 'Effects of a vocal-tract constriction on the glottal source: experimental and modelling studies', Conf. on Voice Acoustics and Dysphonia, Gotland, Sweden, Aug, 1985, J. Phonetics, 14, 373-382.

Bocchieri, E.L. (1983) 'An articulatory speech synthesizer', Unpublished Ph.D. Diss., Univ. of Florida.

Bridle, J.S. and Ralls, M.P. (1984) 'An approach to speech recognition using synthesis-by-rule', in F. Fallside (ed.), Computer Speech Processing, C.U.P., Cambridge.

Broecke, M.P.R. and Cohen, A. (eds.) (1984) Proc. Tenth Intl. Congr. of Phonetic Sciences, Foris, Dordrecht, Holland.

Brookes, D.M., Howard, D.M. and Chan, D.S.F. (1989) 'Dynamic excitation control in parallel formant speech synthesis', FASE 88, Edinburgh 1988, Vol. 3, 1123-1130.

Browman, C.P. and Goldstein, L.M. (1986) 'Towards an articulatory phonology', Phonology Yearbook 3, 219-252.

Carré, R. (1981) 'Vocal tract modelling and speech synthesis: its use for improving knowledge on speech production', in T. Myers, J. Laver and J. Anderson (eds.) The Cognitive Representation of Speech, Advances in Psychology, 7, North-Holland, Amsterdam, 403-413.

Chiba, T. and Kajiyama, M. (1941) The Vowel, its Nature and Structure, Tokyo-Kaiseikan, Tokyo.

Childers, D.G., Yea, J.J. and Bocchieri, E.L. (1985) 'Source/vocal-tract interaction in speech and singing synthesis', Proc. Stockholm Music Acoustics Conf. SMAC83, Royal Swedish Academy of Music, No. 46:1, 125-141.

Coker, C. H. (1976) 'A model of articulatory dynamics and control', Proc. IEEE, 64, 452-460.

Cross, T.E., Johns, P.B. and Loasby, J.M. (1986) 'Transmission line modelling of the vocal tract and its application to the problem of speech synthesis', IEE Conf. Publication no. 258, Speech Input/Output; Techniques and Applications, IEE, London, 71-76.

Dunn, H.K. (1950) 'The calculation of vowel resonances and an electrical vocal tract', J. Acoust. Soc. Amer., 22, 740-753.

Fant, G. (1960) Second Edition 1970, Acoustic Theory of Speech Production, Mouton, The Hague.

Fant, G. (1980a) 'The relationships between area functions and the acoustic signal', Phonetica, 37, 55-86.

Fant, G. (1980b) 'Voice source dynamics', Proc. 10th Intl. Congr. of Acoustics, Sydney, and STL-QPSR, Stockholm, 2-3/1980, 17-37.

Fant, G., Liljencrants, J. and Lin, Q. (1985) 'A four-parameter model of glottal flow', STL-QPSR, Stockholm, 4/1985, 1-13.

Flanagan, J.L. (1972) Speech Analysis, Synthesis and Perception, 2nd. Ed., Springer-Verlag, Berlin.

Flanagan, J.L., Ishizaka, K. and Shipley, K.L. (1975) 'Synthesis of speech from a dynamic model of the vocal cords and vocal tract', Bell System Tech. J., 54, 485-506.

Flanagan, J.L. and Ishizaka, K. (1976) 'Automatic generation of voiceless excitation in a vocal cord - vocal tract speech synthesizer', IEEE Trans. on Acoustics, Speech and Signal Processing, ASSP-24, 163-170.

Flanagan, J.L., Ishizaka, K. and Shipley, K.L. (1980) 'Signal models for low bit-rate coding of speech', J.Acoust. Soc. Amer., 68, 780-791.
Fowler, C.A., Rubin, R., Remez, R.E. and Turvey, M.T. (1980) 'Implications for speech production of a generalised theory of action', in B. Butterworth (ed.), Language Production, Academic Press, London.
Fujimura, O. (1981) 'Temporal organisation of articulatory movements as a multidimensional phrase structure, Phonetics, 38, 66-83.
Fujisaki, H. and Ljungqvist, M. (1986) 'Proposals and evaluation of models for the glottal source waveform, Proc. Intl. Congr. on Acoustics, Speech and Signal Processing ICASSP 86, Tokyo, 1605-1608.
Gauffin, J. and Sundberg, J. (1980) 'Data on the glottal voice source behaviour in vowel production', Proc. 10th Intl. Congr. of Acoustics, Sydney, and STL-QPSR, 2-3/1980, 61-70.
Gracco, V.L. and Abbs, J.H. (1988) 'Central patterning of speech movements', Exper. Brain Res., 71, 515-526.
Guérin, B. (1983) 'Effects of the source-tract interaction using vocal fold models', in Titze, I.R. and Scherer, R.C. (eds.), Vocal Fold Physiology, 3rd. Intl. Conf. on Physiology and Biophysics of the Voice, 1983, Iowa City, Denver Center for the Performing Arts, Denver, Colarado, 482-499.
Guérin, B., Mryati, M. and Carré, R. (1976) 'A voice source taking account of coupling with the supraglottal cavities', Proc. IEEE Intl. Conf. on Acoustics, Speech and Signal Processing ICASSP, 47-50.
Haggard, M. (1979) 'Experience and perspectives in articulatory synthesis', in B. Lindblom and S. Öhman (eds.) Frontiers of Speech Communication Research, Festschrift for Gunnar Fant, Academic Press, London, 259-274.
Harshman, R., Ladefoged, P. and Goldstein, L. (1977) 'Factor analysis of tongue shapes', J. Acoust. Soc. Amer., 62, 693-707.
Holmes, J.N., Mattingley, I.G. and Shearme, J.N. (1964) 'Speech synthesis by rule', Language and Speech, 7, 127-143.
Holmes, J.N. (1981) 'Requirements for speech synthesis in the frequency range 3-4 kHz', Proc. F.A.S.E. Symposium, Venice April 1981, 169-172.
Holmes, J.N. (1988) Speech Synthesis and Recognition, Van Nostrand Reinhold, Wokingham.
Honda, K., Hibi, S.R., Kiritani, S., Niimi, S. and Hirose, H. (1980) 'Measurement of the laryngeal structures during phonation by use of a stereoendoscope', Ann. Bull., Res. Inst. of Logopedics and Phoniatrics, Univ. of Tokyo, 14, 73-79.
Ishizaka, K. and Flanagan, J.L. (1972) 'Synthesis of voiced sounds from a two-mass model of the vocal cords', Bell Syst. Tech. J., 51, 1233-1268.
Karlsson, I. (1986) 'Glottal wave forms for normal female speakers', J. of Phonetics, 14, 415-419.
Kelly, J.L. and Lochbaum, C.C. (1962) 'Speech synthesis', Proc. 4th. Intl. Congr. of Acoustics, Copenhagen, paper G 42, 1-4.
Kempelen, von (1791) Le Mécanisme de la Parole, Suivi de la Description d'une Machine Parlante, Vienna, J.V. Degan.
Kiritani, S. (1977) 'Articulatory studies by the X-ray microbeam system', in M. Sawashima and F. S. Cooper (eds.) Dynamic Aspects of Speech Production, Univ. of Tokyo, 171-190.
Kiritani, S., Imagawa, H., Takahashi, T., Masaki, S. and Shirai, K. (1982) 'Temporal characteristics of the jaw movements in the production of connected vowels', Ann. Bull., Res. Inst. of Logopedics and Phoniatrics, Univ. of Tokyo, 16, 1-10.
Kuhn, G.M. (1983) 'From acoustic tube to acoustic cues', Expository Report 23, Institute for Defence Analyses, Princeton, New Jersey.
Ladefoged, P. (1979) 'Articulatory parameters', in E. Fischer-Jorgensen, J. Rischel and N. Thorsen (eds.), Proc. Ninth Intl. Congr. of Phonetic Sciences, 6-11 Aug, 1979, Copenhagen, Institute of Phonetics, Univ. of Copenhagen, Vol. 1, 41-47.
Laine, U.K. (1989) 'Studies on modelling of vocal tract acoustics with applications to speech synthesis', Thesis, Helsinki Univ. of Technology.

Laine, U.K. and Vilkman, E. (1987) 'Acoustic-mechanical feedback in vocal source - tract interaction', Proc. Eleventh Intl. Congr. of Phonetic Sciences, Aug 1-7, 1987, Tallin, Estonia, USSR, 5, 19-22.
Liljencrants, J. (1985) 'Speech synthesis with a reflection-type line analog', Diss. Doctor of Science, Royal Institute of Technology, Stockholm, Dept. of Speech Communication and Music Acoustics.
Lindblom, B.E.F. and Sundberg, J. E. F. (1971) 'Acoustical consequences of lip, tongue, jaw and larynx movements', J. Acoust. Soc. Amer., 50, 1166-1179.
Lindqvist-Gauffin, J, (1970) 'The voice source studied by means of inverse filtering', STL-QPSR Stockholm, 1/1970, 3-9.
Linggard, R. (1985) Electronic Synthesis of Speech, C.U.P., Cambridge.
Macklem, P.T. and Mead, J. (1967) 'Resistance of central and peripheral airways measured by a retrograde catheter', J. Applied Physiology, 22, 395-401.
Maeda, S. (1975) 'An articulatory model of the tongue based on a statistical analysis', paper presented at a meeting of the Acoust. Soc. Amer., Cambridge, Mass, paper 2.
Maeda, S. (1982) 'A digital simulation method of the vocal-tract system', Speech Communication, 1, 199-229.
Maeda, S. (1987) 'On the generation of sound in stop consonants', RLE, M.I.T., Speech Communication Group Working Papers, 5, 1-14.
McGlone, R.E., Proffit, W.R. and Christiansen, R.L. (1967) 'Lingual pressures associated with alveolar consonants', J. of Speech and Hearing Research, 10, 606-615.
Mermelstein, P. (1972) 'Speech synthesis with the aid of a recursive filter approximating the transfer function of the nasalized vocal tract', Proc. 1972 Intl. Conf. on Speech Communication and Processing, Boston, Mass., 152-155.
Monsen, R.B., Engebretson, A.M. and Vemula, N.R. (1978) 'Indirect assessment of the contribution of subglottal air pressure and vocal-fold tension to changes of fundamental frequency in English', J. Acoust. Soc. Amer., 64, 65-80.
Mrayati, M., Carré, R. and Guérin, B. (1988) 'Distinctive regions and modes: A new theory of speech production', Speech Communication, 7, 257-286.
Muller, E.M. (1980) 'Variations in the supraglottal air pressure waveform and their articulatory interpretation', in N. Lass (ed.), Speech and Language: Advances in Basic Research and Practice, Vol. 4, Academic Press.
NAG (1987) 'The NAG Fortran Mini Manual - Mark 12', Numerical Algorithms Group, Oxford.
Nittrouer, S., Munhall, K., Scott Kelso, J.A., Tuller, B. and Harris, K.S. (1988) 'Patterns of interarticulator phasing and their relation to linguistic structure', Haskins Labs. SR-95, 1-15.
Ohala, J.J. (1975) 'A mathematical model of speech aerodynamics', in G. Fant (ed.) Progress in Speech Communication, Wiley, Stockholm, Vol. 2, 65-72.
Ostry, D.J. and Munhall, K.G. (1985) 'Control of rate and duration of speech movements', J. Acoust. Soc. Amer., 77, 640-648.
Pedley, J., Schroter, R.C. and Sudlow, M.F. (1970) 'The prediction of pressure drop and variation of resistance within the human bronchial airways', Respiratory Physiology, 9, 387-405.
Regnault (1868) Comptes Rendus, 66, 209.
Rice, D.A. (1980) 'Sound speed in the upper airways', J. of Applied Physiology, 49, 326-336.
Richardson, E.G. (1940) Sound, 3rd. ed., Edward Arnold, London.
Rosenberg, A.E. (1971) 'Effect of glottal pulse shape on the quality of natural vowels', J. Acoust. Soc. Amer., 49, 583-590.
Rothenberg, M. (1968) The Breath-Stream Dynamics of Simple-Released-Plosive Production, Bibliotheca Phonetica, No. 6, Karger, Basel.
Rothenberg, M. (1973) 'A new inverse-filtering technique for deriving the glottal airflow waveform during voicing', J. Acoust. Soc. Amer., 53, 1632-1645.
Rothenberg, M. and Mahshie, J. (1986) 'Induced transglottal pressure variations during voicing', J of Phonetics, 14, 365-371.
Sawashima, M., Hirose, H., Ushijima, T. and Niimi, S. (1975) 'Laryngeal control in Japanese consonants

with special reference to those in utterance-initial position', Ann. Bull., Res. Inst. of Logopedics and Phoniatrics, univ. of Tokyo, 9, 21-26.

Scully, C. (1975) 'A synthesizer study of aerodynamic factors in speech segment durations', in G. Fant (ed.), Progress in Speech Communication, Wiley, Stockholm, Vol. 2, 227-234.

Scully, C. (1986) 'Speech production simulated with a functional model of the larynx and the vocal tract', Proc. Conf. on Voice Acoustics and Dysphonia, Gotland, Sweden, Aug. 1985, J. of Phonetics, 14, 407-414.

Scully, C. (1987) 'Linguistic units and units of speech production', Speech Communication, 6, 77-142.

Scully, C. and Allwood, E. (1984) 'Lung and larynx coordination in a composite model of speech production , in M.P.R. van den Broecke and A Cohen (eds.), Proc. Tenth Intl. Congr. of Phonetic Sciences, Foris, Dordrecht, Holland, 372-377.

Scully, C. and Allwood, E. (1985a) 'Production and perception of an articulatory continuum for fricatives of English', Speech Communication, 4, 237-245.

Scully, C. and Allwood, E. (1985b) 'Simulation of singing with a composite model of speech production', Proc. Stockholm Music Acoustics Conf., July -Aug. 1983, A. Askenfelt, S Felicetti, E Jansson and J. Sundberg (eds.), Royal Swedish Academy of Music, Stockholm, Vol. 1, 247-259.

Scully, C. and Clark, G.C.F. (1986) 'Analysis of speech signal variation by articulatory synthesis', in Speech Input/ Output: Techniques and Applications, IEE Conf. Publication no. 258, IEE, London, 83-87.

Shadle, C.H. (1986) 'Models of turbulent noise sources in the vocal tract', Proc. Inst. of Acoustics, 18, 213-220.

Shirai, K. and Masaki, S. (1983) 'An estimation of the production process for fricative consonants', Speech Communication, 2, 111-114.

Sondhi, M.M. and Schroeter, J. (1987) 'A hybrid time-frequency domain articulatory synthesizer', IEEE Trans. on Acoustics, Speech and Signal Processing, ASSP-35, 955-967.

Stevens, K.N. (1971) 'Airflow and turbulence noise for fricative and stop consonants: static considerations', J. Acoust. Soc. Amer., 50, 1180-1192.

Stevens, K.N. (1972) 'The quantal nature of speech: evidence from articulatory-acoustic data', in P.B. Denes and E.E. David (eds.), Human Communication, a Unified View, McGraw-Hill.

Stevens, K.N., Kasowski, S. and Fant, C.G.M. (1953) 'An electrical analog of the vocal tract', J. Acoust. Soc. Amer., 25, 734-742.

Stevens, K.N. and House. A.S. (1955) 'Development of a quantitative description of vowel articulation', J. Acoust. Soc. Amer., 27, 484-493.

Sundberg, J. and Gauffin, J. (1979) 'Waveform and spectrum of the glottal voice source', in B. Lindblom and S. Öhman (eds.) Frontiers of Speech Communication Research, Festschrift for Gunnar Fant, Academic Press, 301-320.

Titze, I.R. (1980) 'Comments on the myoelastic-aerodynamic theory of phonation', J. of Speech and Hearing Research, 23, 495-510.

Titze, I.R. (1981) 'Biomechanics and distributed-mass models of vocal fold vibration', Chapter 18 in K.N. Stevens and H. Hirano (eds.), Vocal Fold Physiology, Univ. of Tokyo Press, 245-270.

Titze, I.R. (1986) 'Mean intraglottal pressure in vocal fold oscillation', in Proc. Conf. on Voice Acoustics and Dysphonia, Gotland, Sweden, Aug. 1985, J. of Phonetics, 14, 359-364.

Titze, I.R. (1989) 'On the relation between subglottal pressure and fundamental frequency in phonation', J. Acoust. Soc. Amer., 85, 901-906.

Vincent, N. J., Knudsen, R., Leith, D.E., Macklem, P.T. and Mead, J. (1970) 'Factors influencing pulmonary resistance', J. of Applied Physiology, 29, 236-243.

Warren, D.W. and DuBois, A.B. (1964) 'A pressure-flow technique for measuring velopharyngeal orifice area during continuous speech', Cleft Palate J., 1, 52-71.

Wu, H.Y., Badin, P., Cheng, Y.M. and Guérin, B. (1987) 'Continuous variation of the vocal tract length in a Kelly-Lochbaum type speech production model', Proc. Eleventh Intl. Congr. of Phonetic Sciences, 1-7 Aug. 1989, Tallin, Estonia, USSR, Vol. 2, 340-343.

ARTICULATORY-ACOUSTIC RELATIONSHIPS IN FRICATIVE CONSONANTS

CHRISTINE H. SHADLE
Department of Electronics and Computer Science
University of Southampton
Southampton SO9 5NH
England

ABSTRACT. The work described in this paper is concerned with improving fricative models by investigating the acoustic mechanisms involved in their production more closely. Previous studies reviewed fall into four categories: general aeroacoustics, mechanical models of speech, analysis of speech, and theoretical models, including synthesis. The work reported here combines these various approaches. Three levels of experiments with mechanical models of increasing realism are described. As a result, two source mechanisms are identified: the obstacle source, in which sound is generated at an abrupt obstacle to the airflow, such as the teeth, downstream of the constriction, corresponding to /s, ʃ/; and the wall source, in which sound is generated by striking a wall at an oblique angle to the flow, corresponding to /ç, x/. Source spectra differ, and are given in each case. Comparison to speech data from which the models were derived supports and extends the conclusions.

1. Introduction

A fricative consonant is produced when the vocal tract is constricted somewhere along its length enough to produce a noisy sound when air is forced through the constriction. As with vowels, the location of the constricion affects the timbre of the resulting sound. In addition, the vocal folds may vibrate simultaneously, generating a periodic sound at the glottis and modulating the airflow through the constriction. Thus we have, in English, the voiced-voiceless pairs of fricatives, arranged by constriction location, from most anterior to most posterior: /v,f; δ, θ; z,s; ž,ʃ/, and the voiceless glottal fricative, /h/ (these are, respectively, the italicized portions of *v*ine, *f*ine; *z*ip,*s*ip; a*z*ure, a*ss*ure; *h*eat). Figure 1 illustrates a range of constriction locations by showing schematized midsagittal cross-sections of the vocal tract for the three fricatives /f, ʃ, x/ (as in *f*in, *sh*in, lo*ch*); in each case, an arrow points to the region of greatest constriction.

Fricatives are primarily classified by presence of voicing and location of the constriction. Finer classifications have been suggested, dividing fricatives into sibilants and non-sibilants, or stridents and non-stridents. The lack of agreement regarding which feature is more useful, or indeed which fricatives ought to be in which group, reflects our poor understanding of the acoustic mechanism of fricatives.

W. J. Hardcastle and A. Marchal (eds.), Speech Production and Speech Modelling, 187–209.

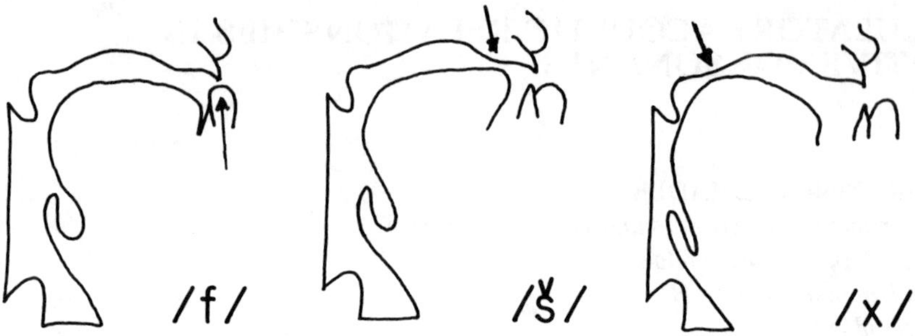

Figure 1: Diagram of the midsagittal cross-section of the vocal tract during the production of the fricatives /f, ʃ, x/.

This poor understanding of the acoustic mechanism of fricatives exists for three reasons. First, there is not a complete theoretical formulation of sound generation due to turbulence. Second, there is no mechanical vibration that is obviously correlated with the speech signal, as there is in the case of vowels; thus the primary sound generation process is more difficult to measure and to model physically. Third, the output signal – that is, the speech – is intrinsically noisy, and thus must be described statistically rather than analytically. Two signals might thus be produced by the same configuration under conditions considered identical in all practical terms, and yet appear quite dissimilar in the time domain. The process of comparing, coding, or analyzing speech signals is more complex; the models of the signals and the systems which produce them must also be more complex.

Past work on fricatives has been conducted using one or more of the following three approaches: 1) Mechanical model experiments; 2) Theoretical models, sometimes including synthesis and perception experiments; 3) Speech analysis, often including extensive articulatory measurements. We will take these up in turn, after reviewing relevant aeroacoustics studies.

2. Previous Work

2.1 GENERAL ASPECTS OF UNSTABLE AND TURBULENT JETS

Techniques such as flow visualization have established that as air exits from a constriction it forms a jet, which gradually mixes with the surrounding air. The Reynolds number (Re) characterizes the degree of turbulence generated as this mixing takes place. It is defined by

$$\mathrm{Re} = \frac{Vd}{\nu} \tag{1}$$

where V = a representative flow velocity, usually taken to be that in the center of the constriction exit, d = a representative dimension, usually the constriction diameter, and ν = the kinematic viscosity of the fluid, which for air is 0.15 $\mathrm{cm}^2/\mathrm{sec}$. As Re increases, an initially laminar flow will pass through an unstable region and finally become fully turbulent. Turbulent flow is distinguished by irregular, high-frequency fluctuations in velocity and pressure at a given point in space (Schlichting, 1979). The critical Reynolds

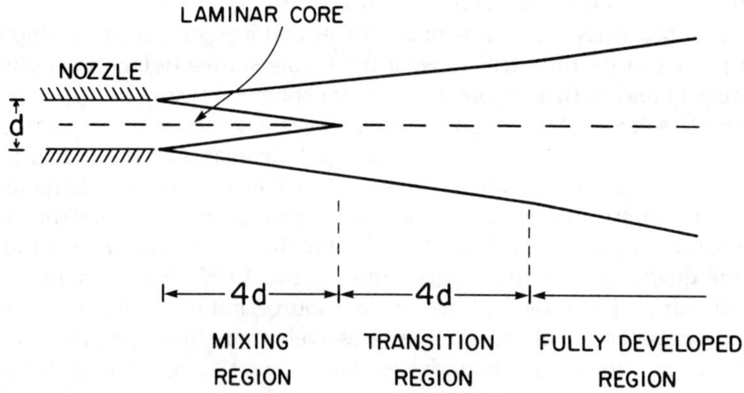

Figure 2: Diagram of the mixing, transition and fully developed regions of a fully turbulent jet. After Goldstein (1976).

numbers, Re_{crit}, separating these regions vary according to the geometry and degree of prior turbulence of the fluid. For a jet issuing from a circular hole, the unstable region would typically occur for $160 < Re < 1200$ (Goldstein, 1976).

The dimensions of a fully turbulent jet in the subsonic range depend only on the constriction diameter and shape; thus, visually, all jets can be scaled to look the same. Theoretical work by Lighthill (1954) and others established that the sound generated by jets scales as well, that is, that the spectral characteristics of the sound generated by a jet depend only on the jet velocity and diameter.

Sound is generated by the random pressure fluctuations of the turbulent fluid. A good summary of the theoretical and empirical efforts to describe this sound generation process may be found in Goldstein (1976). For our purposes, the essential facts are as follows. For a jet emerging from a constriction of diameter d at $Re > Re_{crit}$, three regions, the mixing, transition and fully developed regions, can be defined, as shown in Fig. 2. From both theory and experiment it appears that nearly all of the sound power is generated in the mixing and transition regions, possibly with most of it coming from the mixing region (Goldstein, 1976). If half of the sound power is assumed to be generated in the mixing region, the total power, P, generated by the jet is proportional to V^8 (where V is the flow velocity), which agrees with Lighthill's prediction (1952). The total sound power spectrum has a broad peak at about SV/d Hz, where V is the flow velocity in the center of the jet as it exits the constriction, d is the jet diameter, and S, the Strouhal number, defined by this equation, is equal to 0.15. (The frequency of the spectral peak depends on the type of spectrum chosen. The Strouhal number at the peak is $S = 0.15$ when the noise spectral density, an equal-bandwidth representation, is plotted; $S = 1.0$ for the third-octave spectrum.) The sound pressure measured at a particular point in the far field will have a similar spectrum, with a peak frequency dependent on the angle at which the measurement is made. Measurements within the jet itself show that the high-frequency sound originates closer to the nozzle than does the low-frequency sound (Fletcher and Thwaites, 1983).

Lighthill described three types of sound sources that are present to varying degrees in the sound produced by turbulent flow: a monopole source (which is equivalent to a sphere pulsing in and out), a dipole source (two spheres pulsing in opposite phase), and a quadrupole source. A monopole source obeys a V^4 power law, meaning that the total sound power generated by a flow monopole increases as the fourth power of the flow velocity V; the dipole source obeys a V^6 power law; the quadrupole source, V^8. An equivalent statement is that the efficiency of conversion of the kinetic energy of the flow into sound is proportional to $M = V/c$ for the monopole, M^3 for the dipole, and M^5 for the quadrupole source (Morse and Ingard, 1968). For subsonic velocities ($M < 1.0$) the quadrupole is thus the least efficient source, but its relative contribution to the total sound generated should become progressively more important as M increases. Dipole sources occur along rigid boundaries, which exert an alternating force on the fluid; quadrupole sources exist in free jets. Thus if a jet impinges on an obstacle such as a flat plate, the sound generation can be modeled by a combination of quadrupole and dipole sources. It is not well understood how the energy divides between quadrupole and dipole sources in such a case (Goldstein, 1976).

Two studies have combined theory and experiment to focus on the sound generated by a spoiler in a duct. Heller and Widnall (1970) showed that although flow dipoles in general have sound power proportional to V^6, when they are located well inside a duct, the radiated sound power is proportional to V^4 instead. This was borne out by a later study of Nelson and Morfey (1981), who were also able to show that the relationship was different above and below the cut-on frequency of the duct. They were able to collapse their source characteristics at a range of flowrates according to theory for various relative spoiler areas. These collapsed curves did not show the broad peak typical of free jet spectra; instead they fell steadily in amplitude with inceasing frequency.

2.2 MECHANICAL MODEL EXPERIMENTS

One of the earliest efforts to consider fricative production in terms of the aerodynamics involved was that of Meyer-Eppler in 1953. He compared sound-pressure vs. flow relationships for plastic tubes with different-sized elliptical openings. He derived effective-width formulas that allowed sound-pressure vs. Re plots for different ellipse sizes to coincide, and then used these derived formulas on data of human subjects uttering fricatives to infer articulatory parameters from pressure measurements.

It is not clear, however, that Re_{crit}, which Meyer-Eppler defined to be the lowest Reynolds number at which measurable sound was generated, should be the same for the plastic tubes and the three fricatives /s, ʃ, f/ that he studied. First, Re_{crit} is probably lower for an irregularly shaped vocal tract than for the plastic tubing. Second, sound is most likely being generated for these three fricatives both when the air passes through a constriction (over the tongue for /s, ʃ/ or between teeth and lower lip for /f/) and when the jet of air strikes an obstacle (the teeth for /s, ʃ/ or upper lip for /f/). The intensity of the sound generated may therefore be related more to the distance between the constriction and the obstacle and the physical properties of the obstacle than to the effective area of the constriction. Since the tubes he used had a constriction but no obstacle, application of the effective-width formulas developed for the tubes to the strident fricatives may give misleading results.

Heinz (1958) carried out experiments using a 17 cm tube imbedded in a wooden sphere to approximate the dimensions and radiation impedance of a vocal tract. By placing a cylindrical plug with a 0.2 cm diameter axial hole at the mouth and 4 cm back from the mouth, he approximated the articulatory configurations for fricatives such as /ϕ/ and /ç/, respectively. He obtained far-field directivity patterns and spectra for a variety of frequencies and flow rates. As expected, intensity rose with flow rate, except at the half-wavelength resonance of the constriction. Heinz ascribed this behavior at this resonance to effective movement of the source relative to the constriction. However, the source he used in calculating the system response was a localized pressure source that did not change position with flowrate. The source spectra derived by subtracting the system response from the measured sound spectrum were fairly flat, but with dips at resonance frequencies, which Heinz assumed were a consequence of greater than expected resonance bandwidths due to turbulence losses. He also calculated the incremental flow resistance from his data.

2.3 THEORETICAL MODELS

Fant (1960) used a distributed model of the vocal tract to investigate the effects of source location, source spectrum, and constriction resistance. Concluding that the theory of turbulence was too undeveloped to be useful, he judged the accuracy of his models by how closely the predicted specra matched the spectra measured from the speech of a single subject. Models for all fricatives used a series pressure source that generated either white noise for /s, š,ş,č/ or integrated white noise (i.e., with a -6 dB/oct slope) for /x,f/.

Fant did not attempt to make a physical argument relating the two types of source spectra to the distinguishing features of the fricatives. The location of the source – whether at tongue or teeth – is more clearly linked to the place of articulation. The model for /x/, which used a source located at the tongue, produced the best match of all of the fricatives. For /s/ and /š/, although sources at both tongue and teeth were used, it appeared that neither location by itself would provide a good match at all frequencies. Fant suggested that quite possibly /s/ was produced with sources at both locations, with a low spectral level below 1 kHz, but did not attempt a physical justification for this particular source characteristic. Fant was aware that changes in the source location would alter the frequencies of the zeros it produced in the output, but he wrote that this effect would probably prove to be perceptually unimportant.

Flanagan et al. (Flanagan, Ishizaka, and Shipley, 1975, 1980; Flanagan and Ishizaka, 1976) elaborated on this model by allowing for multiple noise sources, one for each section of the uniform tube model, with the strength of each source (the variance of the white noise pressure source) depending on Re of the section (computed from its area and the glottal volume velocity). A given pressure source was included only if its Reynolds number exceeded the value of Re_{crit} determined from Meyer-Eppler (1953). The model allows for a dependence of both intensity and acoustic resistance on flow rate. The sound sources are now distributed throughout the vocal tract, but the source due to a single section is still localized. Further, the spectral characteristic of the source is unchanged by flow rate (except for its overall amplitude), which contradicts findings of Heinz (1958) and Thwaites and Fletcher (1982), among others. Likewise, Re_{crit} is constant regardless of upstream conditions of the tract configuration. The method of computing Re based on the cross-sectional area is not sensitive to the turbulence generated when a jet of air impinges on an obstacle. This model, like Fant's, assumes linear elements, independence of source and filter functions, and one-dimensional sound propagation.

2.4 SPEECH ANALYSIS

Efforts to model fricatives, such as that of Fant (1960), and work with mechanical models, such as that of Heinz (1958), form two approaches to understanding the acoustic mechanisms involved in fricative production. A third approach is simply through acoustic analysis of spoken fricatives. The work of Hughes and Halle (1956), Strevens (1960), Heinz and Stevens (1961) on English, and others, detailed thoroughly by Ladefoged and Maddieson (1986), is useful, but perhaps the most important conclusion is that the considerable variation across speakers makes it difficult to draw any useful generalizations about the acoustic parameters used to differentiate the fricatives.

Some authors pair articulatory and acoustic data in an effort to overcome such problems. Ladefoged and Maddieson exploit not only X-rays, but electropalatography and other methods whenever possible. As a result, they are able to make fine articulatory distinctions. However, their descriptions of the spectra still rely on the same crude measures of minimum frequency of the high-energy region, steepness of the slope at cut-on, and the overall amplitude. We will discuss their conclusions regarding the acoustic mechanism for each fricative at a later point. For now, we will say only that by pointing out the full range of articulatory variation possible, they are guaranteeing that there will be a full range of acoustic variation, which we are unlikely to be able to explain at present.

Others are able to proceed further by narrowing the problem considerably. Cohen and Perkell (1986), for instance, used EPG data to demonstrate the effect of the sublingual cavity in distinguishing /s, ʃ/. Hardcastle (1984) has used similar data to study pathological fricatives, and to look at the effect of groove shape on /s, ʃ/.

The approach taken by the author, in light of the above work, has been to focus primarily on experimenting with mechanical models, to relate these occasionally to analysis of human speech, and by this means to test the assumptions on which theoretical models are based (e.g., that source and filter are independent) and determine needed parameter values (e.g. source characteristics). This work has proceeded by means of three sets of experiments with progressively more realistic mechanical models. We will now discuss these models, beginning at the lowest level.

3. Level I Models

3.1 METHOD

The first set of models was deliberately designed to include only the features thought to be important in fricative production. Thus, the constriction was included, but without significant taper; the vocal tract bend was eliminated, and the transverse cross-section was circular throughout. The models, shown in Fig. 3, consisted of a cylindrical plug forming an axial constriction in a tube of circular cross-section; optionally, a semi-circular obstacle was positioned transverse to the flow downstream of the constriction. Preliminary analysis showed that the articulatory parameters having the greatest acoustic effect were the flowrate, the length of the front cavity, and the presence (or absence) of the obstacle. Acoustic measurements of sound generated by air passing through the mechanical models showed that the obstacle generated relatively high-amplitude noise, which was localized at the obstacle. This led to a division of the models into the "obstacle" and "no-obstacle" cases. Both cases were then examined closely at a range of flowrates for three different front-cavity lengths. Further details of these experiments are given in Shadle (1985, 1986a).

3.2 RESULTS

Theoretical modelling of the obstacle case posited a dipole source at the surface of the obstacle, which has the transmission-line equivalent of a series pressure source. The source characteristic was derived from the experimental data, as shown in Fig. 4, and had a rather flat spectrum falling off with frequency; this shape was repeated at higher amplitude for higher flowrate, similar to spoiler-in-duct spectra generated by Nelson and Morfey (1981). When the derived source was used to excite a transmission line whose parameters were determined from the area function of the model according to classic theory, the predicted far-field spectrum matched the experimentally observed spectrum well. This demonstrated that the obstacle case was well-modeled by a series pressure source located at the obstacle, and that that source is independent of the filter. In other words, the source spectrum depends only on flow rate and distance from constriction to obstacle, not on location of the constriction-obstacle unit within the tract.

The no-obstacle case was not modeled theoretically, but it was established that the source should be spatially distributed. Also, if the source was assumed to be determined solely by the flowrate and the dimensions of the constriction, that source was dependent on the location of the constriction within the tract. Thus, altogether, the no-obstacle case is more problematic than the obstacle case.

3.3 COMPARISON WITH SPEECH

Comparison of these two models to actual sustained fricatives, as shown in Fig. 5, demonstrated that the obstacle case was equivalent to the fricatives /s, ʃ/, with the obstacle playing the role of the teeth. Extensive study of the transfer function of the mechanical model (possible when the articulatory configuration is both simple and completely known) revealed several facts about the mechanical models that could then be extended to the vocal tract for these fricatives. A pressure source located within the front cavity will excite all poles of the entire tract, and in addition a real zero near 0 Hz and a complex conjugate pair of zeros at a frequency related to the distance between the pressure source and the anterior end of the constriction. Since the distance is so short, a change in it of the order of a millimeter will change the frequency of the free zero by a few hundred Hertz. In addition, the pressure source generates zeros that nearly cancel all of the back cavity poles. Thus, the main features remaining in the transfer function are the front cavity poles, the free zero, and the zero near 0 Hz. Because of

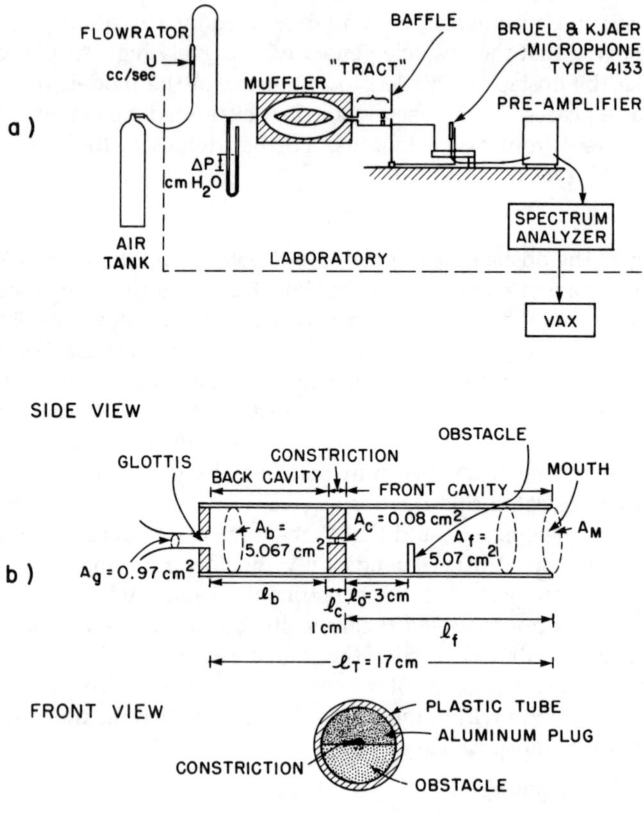

Figure 3: Cross-sectional diagram of Level I models, with and without obstacle.

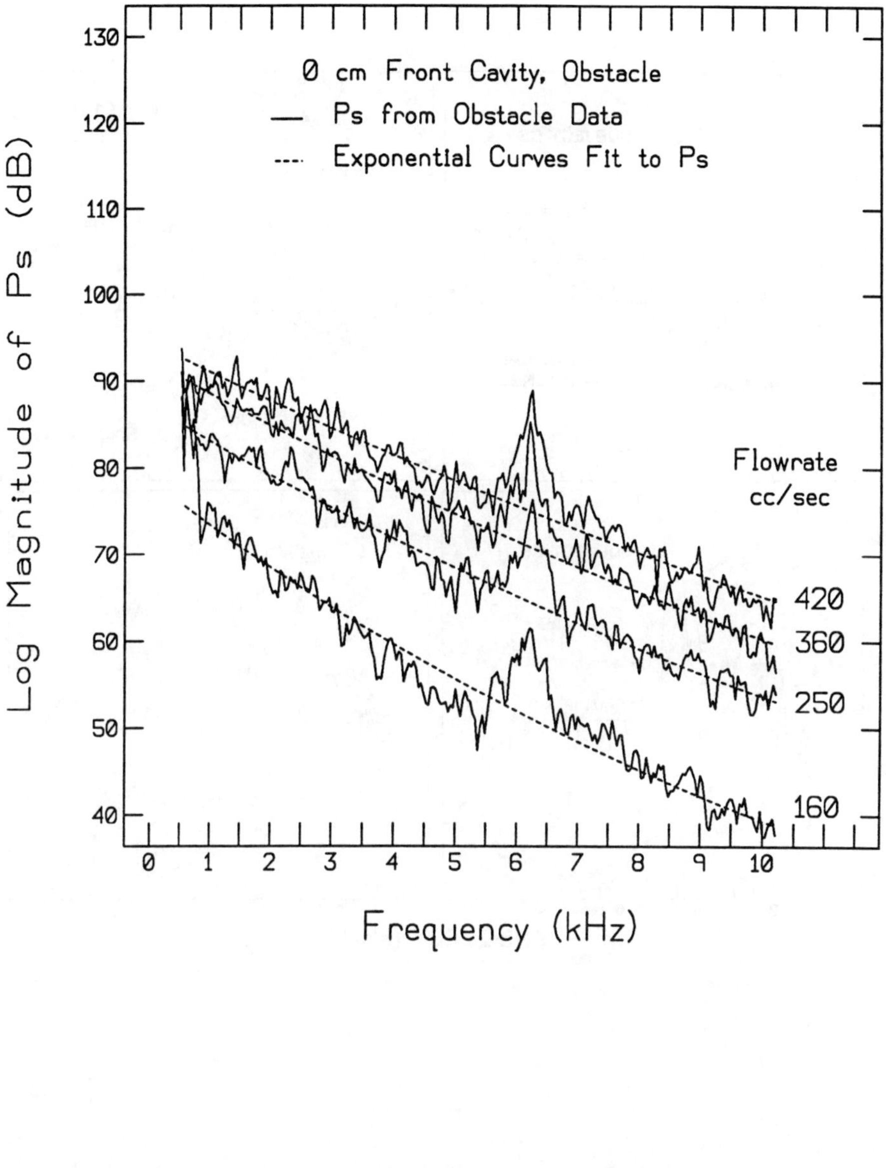

Figure 4: Equivalent pressure source p_s, derived from far-field sound pressure. Dashed lines represent an exponential line-fit. Length of front cavity is 0 cm, obstacle is 3 cm downstream of constriction.

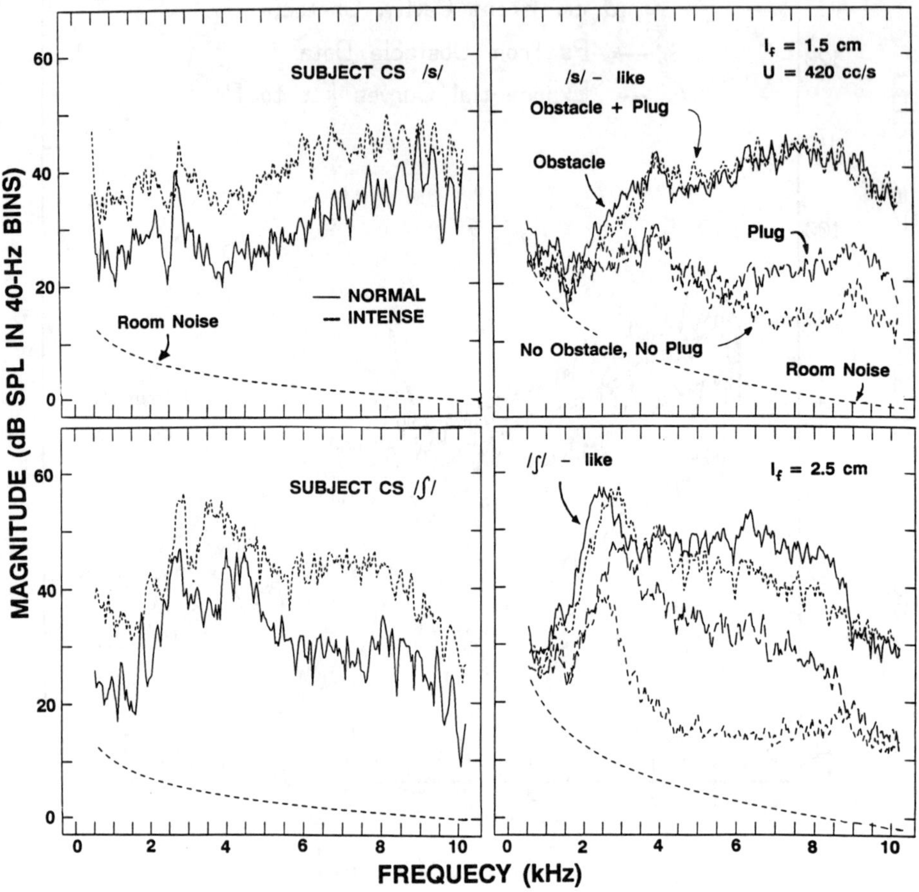

Figure 5: On left: Spectra of /s, ʃ/ by female speaker. On right: Spectra generated by the air flowing through obstacle-case models of Fig. 3.

the low-frequency zero, and the high damping due to the relatively short front cavity, the first formant may be quite low in amplitude.

The no-obstacle case was possibly equivalent to the weak fricatives /ϕ,f,θ/, but due to the high degree of inter-speaker variation, it was difficult to be certain about this claim. Certainly, Catford's data (1977) showing a major drop in spectral energy of /s, ʃ/ but not /θ/ following removal of false teeth support the classification of /s, ʃ/ as obstacle case, and /θ/ as *not* obstacle case. But, as we shall see, the no-obstacle Level I model is not the only alternative to the obstacle case. Thus, the classic theory was shown to be correct if modified in this manner: for /s,ʃ/, a series pressure source located *at the teeth,* with spectral characteristic flat and falling, and having higher amplitude for higher flowrate, is valid. For other fricatives, neither case matched adequately, indicating that the models were perhaps over-idealized.

4. Level II Models

4.1 METHOD

Fricatives other than /s, ʃ/ were investigated using Level II models, which featured constriction shapes that were non-circular in the transverse plane, as shown in Fig. 6. These were used in the Level I setup shown in Fig. 3, and as before, spectra were computed at a variety of flowrates and constriction locations. Full details are given in Shadle (1986b).

4.2 RESULTS

Comparison of speech spectra and model output showed that /ç,x/ were best matched by a "flat-topped" constriction which forced the air to flow along the wall of the tube, as shown in Fig. 7. This generated more sound than a centered circular constriction of the same cross-sectional area and length, and showed even greater dependence on the location of the constriction within the tube. Thus, the demonstrated source-tract dependence of the no-obstacle case appeared to be due to sound generation along the walls of the tube downstream of the constriction. However, theoretical models of the flat-topped plug indicated that the similarity to speech might possibly be due to an artifact (Shadle, 1986b). Essentially, Level II models were so highly idealized that it was unclear whether the same physical mechanisms for sound production were involved.

The Level II models did, however, allow clarification of the problem as follows: Since the self-noise of turbulent air is much lower in amplitude than that generated by air impinging on a rigid object, what objects does the air impinge upon in its path through the vocal tract? How do these objects affect the sound generated?

5. Level III Models

5.1 METHOD

To answer these questions, more realistic models of the vocal tract were needed. These Level III fricative models were designed and built in the following manner. The mid-sagittal X-ray tracings of fricatives given by Fant (1970) were used as a pattern with which to cut "tracts" out of inch-thick (2.54 cm) plexiglass. The "tract" was then sandwiched between two layers of 1/4" thick plexiglass, and the entire sandwich bolted

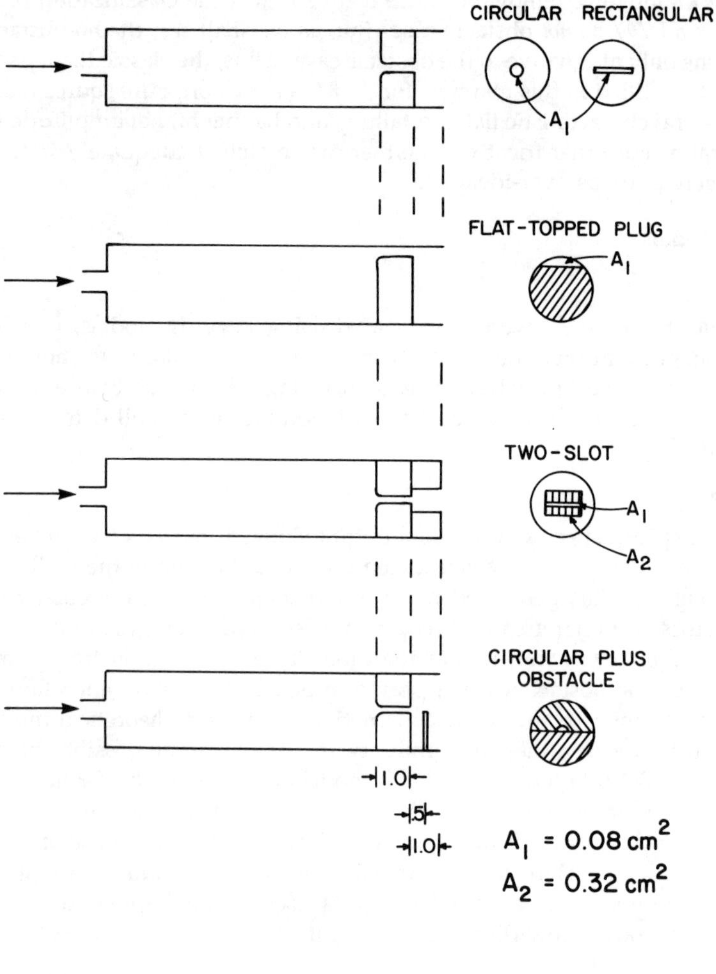

Figure 6: Cross-sectional and front view of Level II models.

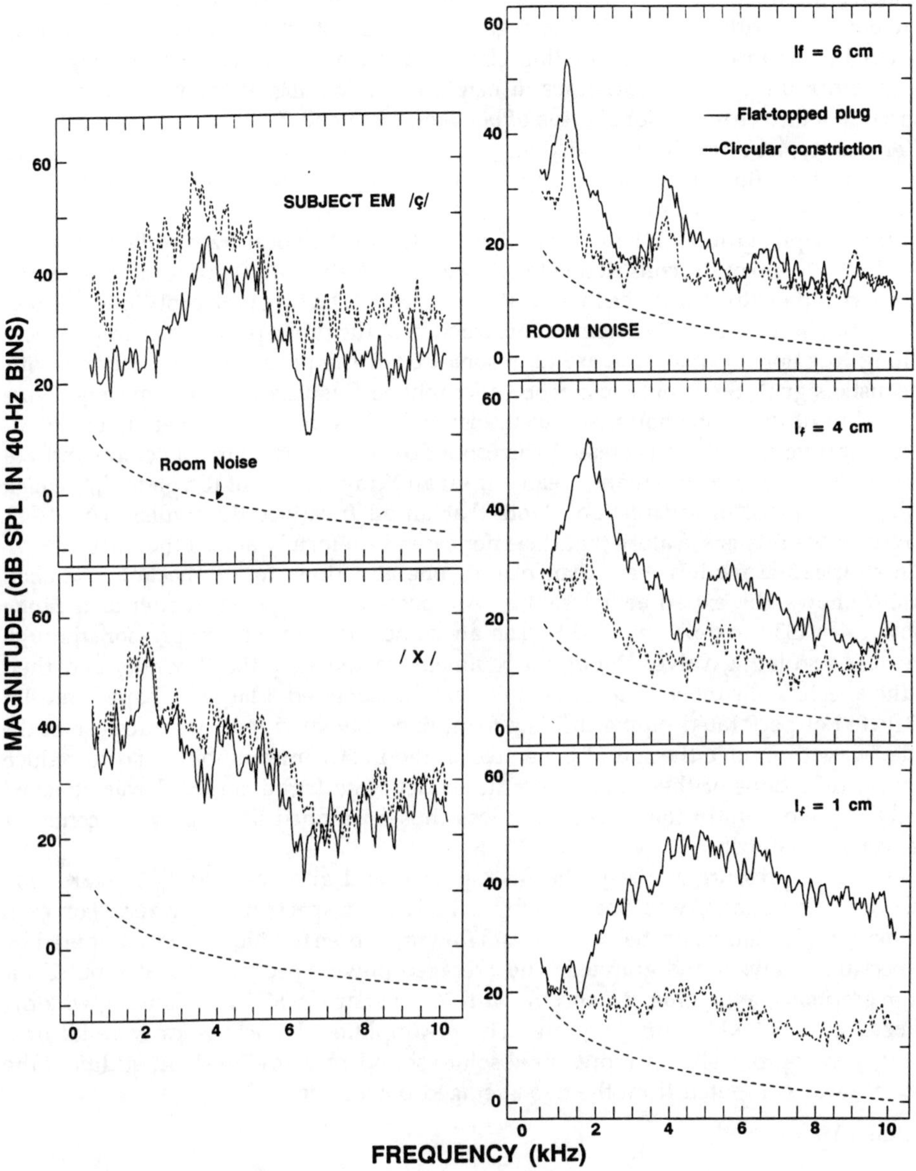

Figure 7: On left: Spectra of /ç, x/ by female speaker. Solid lines = normal, dotted lines = intense production. On right: Spectra produced by the indicated constrictions and lengths of front cavity, at flowrate 400 cc/sec.

together. The overall area function was thus on average correct, but at specific locations along the tract, incorrect, since there was a constant 1" width throughout. This was deemed unworkable only in the region of the constriction, where the area was corrected by the insertion of modelling clay to create a central channel of length 0.5 cm and width 0.1 cm. The variations in height of the models at the point of greatest constriction led to cross-sectional areas of between 0.015 and 0.03 cm^2, which is actually smaller than typical fricative constrictions. Effectively this would tend to generate more sound at a given flowrate; since a range of flowrates was investigated, the difference was not deemed problematic.

In the completed model, holes were then drilled in the vocal tract walls at various points downstream of the constriction to allow a Bruel & Kjaer 4170 probe microphone to be inserted and lie flush with the inner wall of the tract. The tip of the probe microphone was 1.25 mm in diameter. The tapered construction of the probe provides an impedance-matching function so that there are no resonances of the probe itself interfering with the acoustic signal. Since only one probe microphone was used, aluminum plugs were machined to fit the probe holes, and were inserted when those holes were not in use.

The fricative models thus created corresponded to the fricatives /ʃ, ç, x/, and are shown in Fig. 8. /ç/ was actually created from an X-ray tracing of the glide /j/, which has the same place of articulation. Note that on all fricatives the uvulas have been removed, since this is a feature that does not extend uniformly across the tract.

These fricative models were then placed, one at a time, in an anechoic chamber. Figure 9 shows the experimental setup. Air passed through the model at a range of flow rates (330 to 1670 cm^2/sec). The air came from an air compressor, through a pressure-regulating valve, through a rotameter to measure the flowrate, and then into the anechoic chamber. There, the 4 cm air duct entered a large L-shaped muffler constructed of perforated piping inside a foam-lined plywood box. The purpose of the muffler was to remove the sound generated in the upstream plumbing and to reduce upstream reflections of the sound from the model. The fricative model was attached directly to the output of the muffler, and was thus positioned in roughly the center of the anechoic chamber.

The sound pressure at the probe microphone, and at a far-field 1/2" B&K 4133 condenser microphone, was analyzed by a B&K 2032 spectrum analyzer. For each combination of fricative, probe location, and flowrate, three functions were computed by the spectrum analyzer and graphed. The averaged power spectrum was computed for each microphone, averaging 100 spectra with 0% overlap. A 62.5 ms Hanning window was used at a 25.6 kHz sampling rate. The preamplifier characteristics were factored in so that the graphs shown represent absolute sound pressure level. In addition, the coherence was computed from the two averaged power spectra.

5.2 RESULTS

5.2.1 *Source Location.* An example of the power spectra and coherence for two probe locations is shown in Fig. 10. When the two microphone signals were highly coherent (i.e. close to 1.0), sound at both microphones was similar, which indicates that in this case (probe point 3) the probe microphone was detecting the tract resonances only. When the signals were incoherent (i.e. close to 0.0), the sound at both microphones

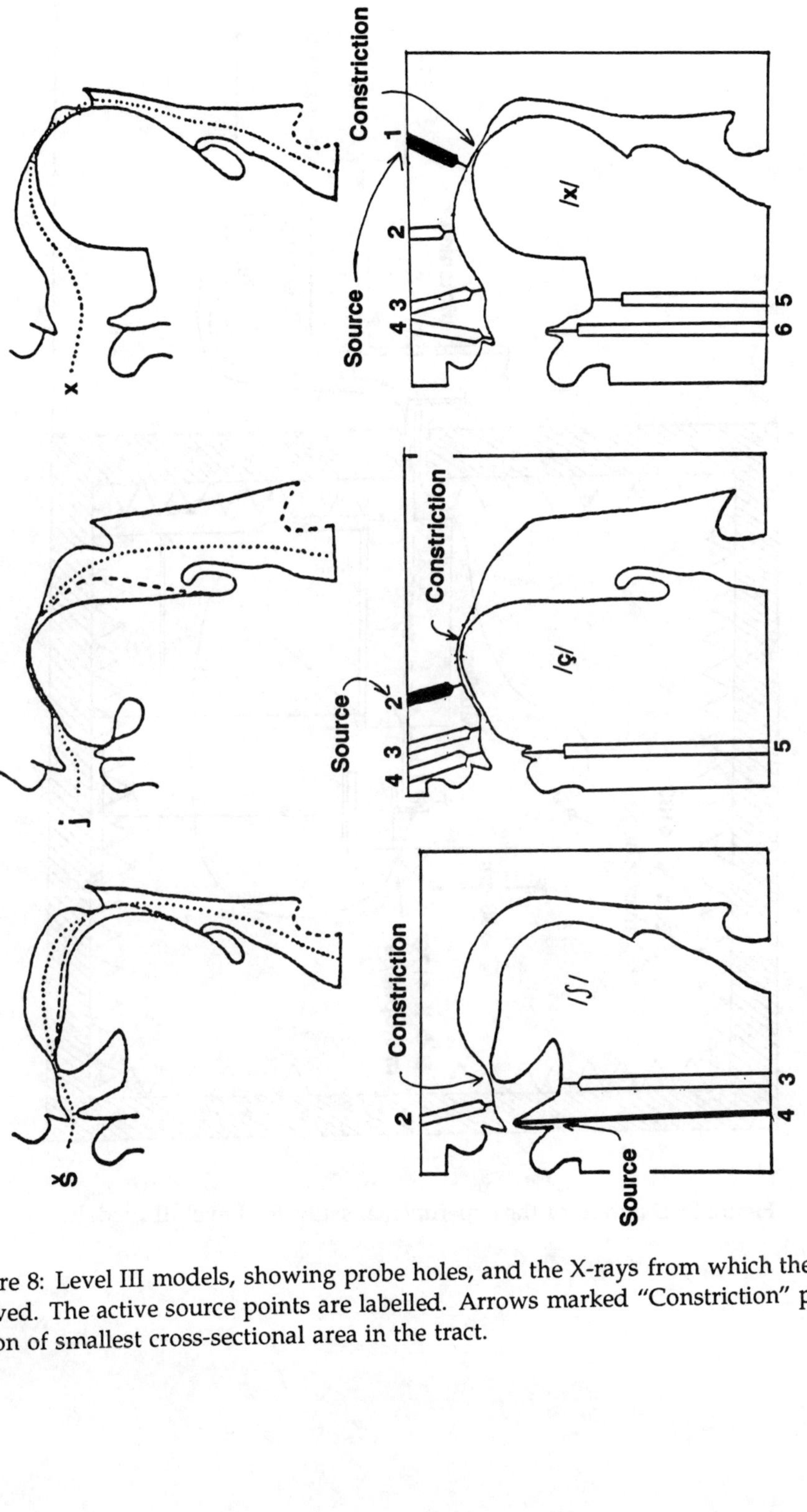

Figure 8: Level III models, showing probe holes, and the X-rays from which they were derived. The active source points are labelled. Arrows marked "Constriction" point to region of smallest cross-sectional area in the tract.

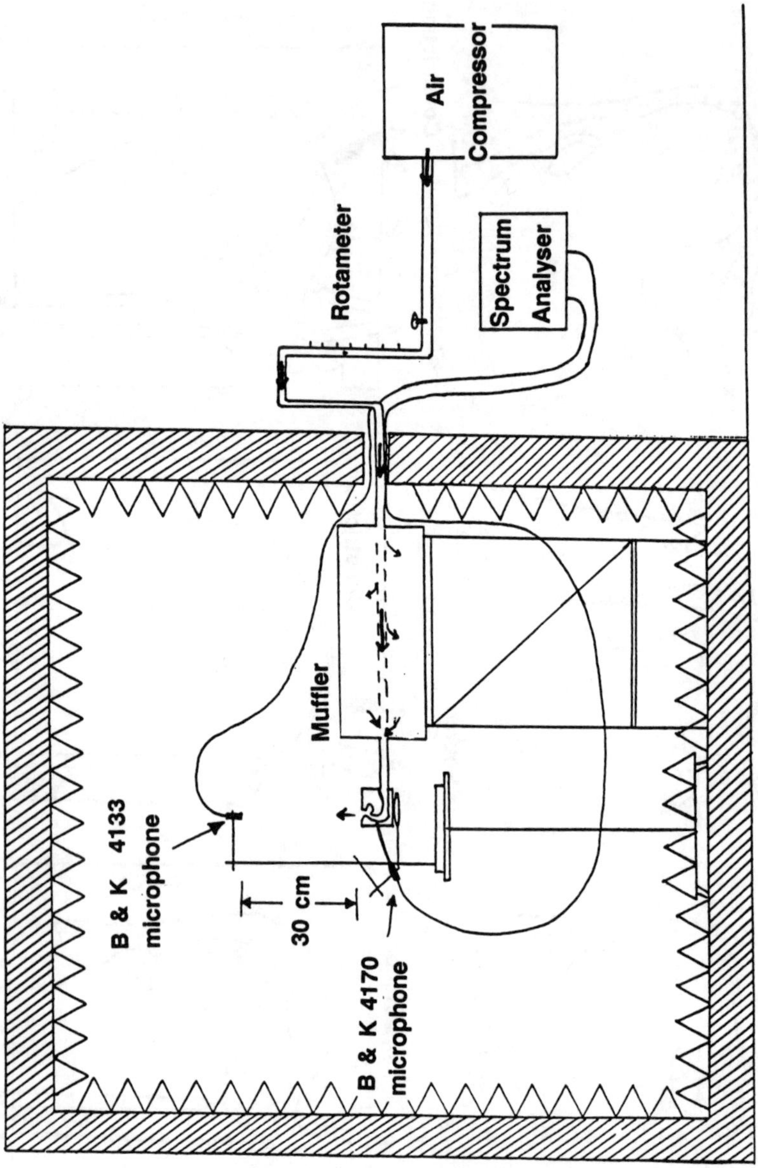

Figure 9: Diagram of the experimental setup for Level III models.

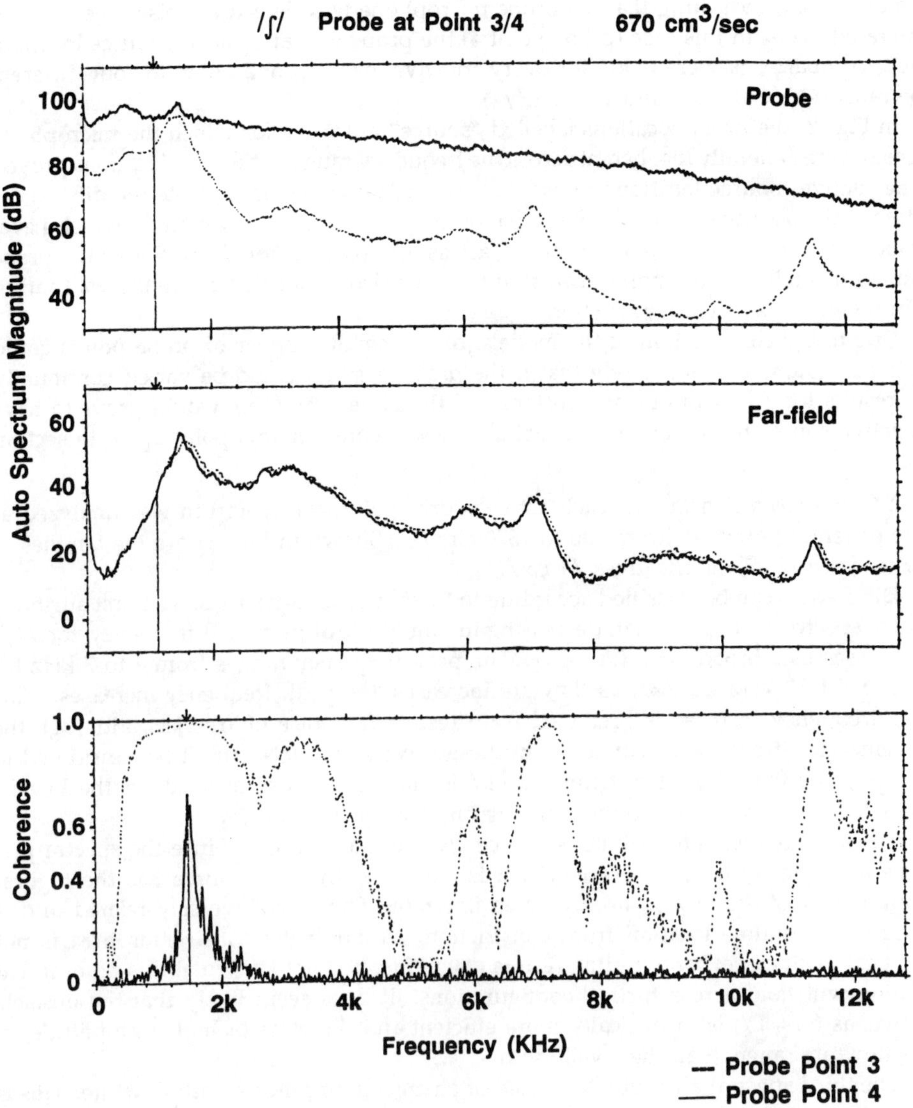

Figure 10: Example of the spectrum analyzer output for /ʃ/, for probe point 3 (floor of mouth, just posterior of lower teeth) and point 4 (tip of lower teeth).

was dissimilar, indicating that the probe microphone was detecting noise where it was generated. Thus in this case (probe point 4) the probe was at or near a source location. These comparisons were made for every fricative, every probe point, at four different flowrates (330, 670, 1000 and 1330 cm^2/s).

In Fig. 8, the probe locations labelled "source" are those for which the microphone signals were generally incoherent across the frequency range. The same locations proved to be "active" source locations across the range of flowrates used. Note the differences between the fricatives: for /ʃ/ the active probe point is at the lower teeth, which Level I model results had led us to expect to act as an obstacle; for /ç,x/ the active probe point is in each case the point nearest to the constriction, located along the wall rather than on any downstream projection.

Due to the construction of the models, only a small number of probe points could be investigated. It seems likely that if the probe position could be varied continually, the results for /ʃ/ would stay substantially the same, but /ç,x/ would prove to have an active source *region* rather than point. We will consider this point again in section 5.3.

5.2.2 *Source Characteristic.* At each active source point, the spectrum was analyzed at nine different flowrates (from 330 to 1670 cm^2/s). Shown in Fig. 11 are the families of curves for each of the fricatives /ʃ,ç,x/.

Differences can be classified according to three spectral attributes. The first attribute, source spectrum shape, is flat, decreasing in amplitude at increased frequency, for /ʃ/, but possesses a broad peak for /ç,x/. The peak frequency ranges from 4 to 7 kHz for /ç/, or 5 to 9 kHz for /x/; as flowrate increases, the peak frequency increases. This peak-frequency increase is predicted from results for noise of free jets, although the jets emerging from the constrictions in these fricative models cannot be considered to be free. The flat falling spectrum for /ʃ/ is similar to those derived for the Level I obstacle-case models, and to those derived by Nelson and Morfey.

The second spectral attribute is source spectrum amplitude. Since the spectra preserve absolute sound pressure levels, it is clear that the /ʃ/ source has the highest amplitude, /x/ the lowest, at any given flowrate. These are inversely related to distance of the source location from constriction, although the constriction area is not identical in all cases, confounding results. Therefore it is not known at this point if that relation will hold across further configurations. It does seem likely that the obstacle source, as for /ʃ/, is intrinsically more efficient and therefore of higher amplitude, in this flowrate range, than the "wall" source.

The third spectral attribute is the rate of change of amplitude with flowrate. This is not uniform across the entire frequency range, nor is it expected to be. If we consider the difference in amplitude from lowest to highest flowrate at 2 and at 12 kHz, we obtain a rough measure of the dependence of amplitude on flowrate. For /ʃ/, the difference is 17 dB at 2 kHz, 29 dB at 12 kHz; for /ç/, 13 and 18 dB; for /x/, 8 and 15 dB. Since the rate of change of amplitude is one of the prime characteristics differentiating flow noise source types, these differences are significant, and point once more to a different source mechanism at work.

In summary, the properties of source location and spectral shape group /ç,x/ together, distinguishing them from /ʃ/; spectral amplitude and rate of change of ampli-

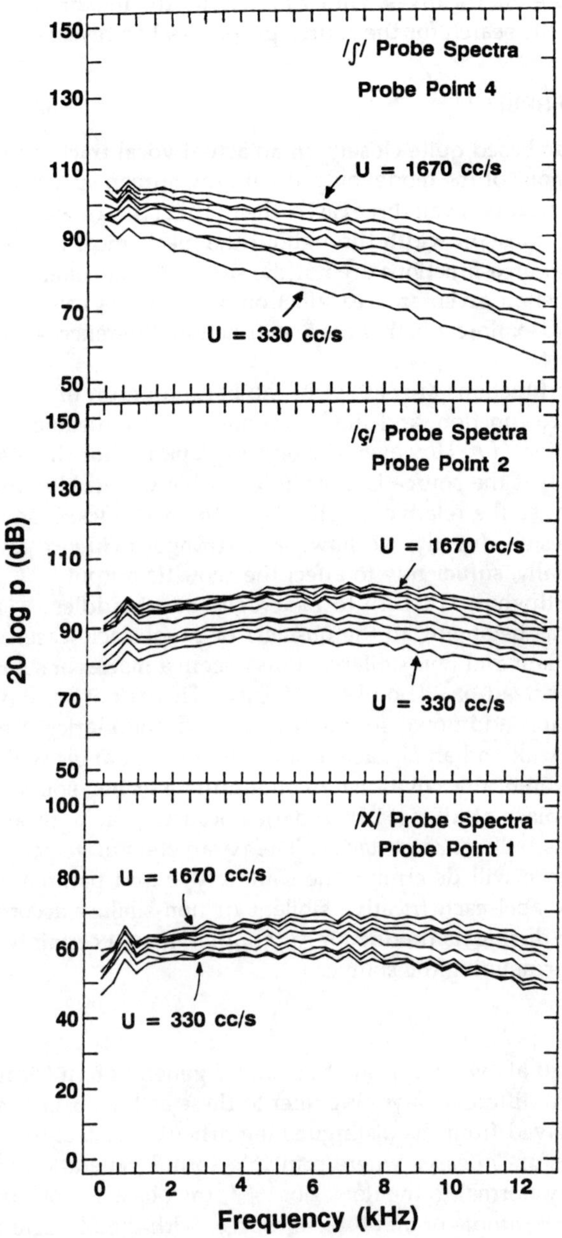

Figure 11: Families of source spectra for each of the three fricatives. These are output of the probe microphone while positioned in the active source point, for the indicated range of flow rates.

tude with flowrate separate the fricatives, with /ʃ/ having the maximum values, and /x/ the minimum. Thus, the search for the source properties has revealed distinguishing acoustic features.

5.3 COMPARISON WITH SPEECH

Because these models were based quite closely on an actual vocal tract, it was possible to make detailed comparisons of the model output with the human speech analyzed by Fant. Figure 12 shows the results: in all three cases, the output spectra were significantly dissimilar. Because of the constant width throughout the mechanical model, however, the human and mechanical area functions are not the same. A computer program that calculates the transfer function given an area function and various loss factors (Badin and Fant, 1984) was used to explore whether the area function differences would explain the acoustic differences.

The predicted transfer functions are shown in the bottom graph of each section of Fig. 12. Adjusting the area function, and the constriction-source distance with it, did explain the discrepancies for /ʃ/. However, the best explanation for the discrepancies in /ç, x/ appeared to be that the source is distributed. This would blur the free zero and, in the case of /x/, alter the relative amplitude of the first and second formants. Although this could be tested directly, we now have stronger evidence that the wall source is distributed spatially, sufficiently to affect the acoustic output.

At this point it is worthwhile to return to Ladefoged and Maddieson (1986), who identified every fricative as belonging to the obstacle or no-obstacle case, which they declared equivalent to sibilant and non-sibilant. It may seem a matter of semantics only to refer to a "wall" case instead of a "no-obstacle" case. This is not so, however. The turbulent jet of air generates additional sound at any rigid boundaries it encounters. The difference between a wall and an obstacle is a matter only of angle with respect to the jet axis, and could theoretically fall along a continuum. Further, sound generation will not necessarily take place at all rigid boundaries occurring at right angles to the flow; that is, not all objects will act as obstacles. The parameters of velocity, angle and area of the jet at the surface will determine the source type and parameters. It thus seems a bit premature to label each fricative sibilant or non-sibilant according to the position of the lower teeth during its production, although that will certainly eventually be an important parameter defining the source.

6. CONCLUSION

Based on the work reported above, it appears that sound generation in fricative consonants occurs in at least two different ways. We refer to these as the obstacle source and the wall source, terms derived from the distinguishing articulatory feature.

The obstacle source occurs in cases where principle sound generation is at a rigid body that is approximately normal to the flow. For /ʃ/, the obstacle appears to be the lower teeth; /s/ is also an example of an obstacle source, with the obstacle possibly at upper rather than lower teeth. The obstacle source is characterized by maximum source amplitude for a given flow velocity, by a spectral shape that is flat and falling off with increased frequency, and by a maximum rate of change of sound pressure with volume velocity. It can be modeled with a series pressure source at the obstacle. It appears

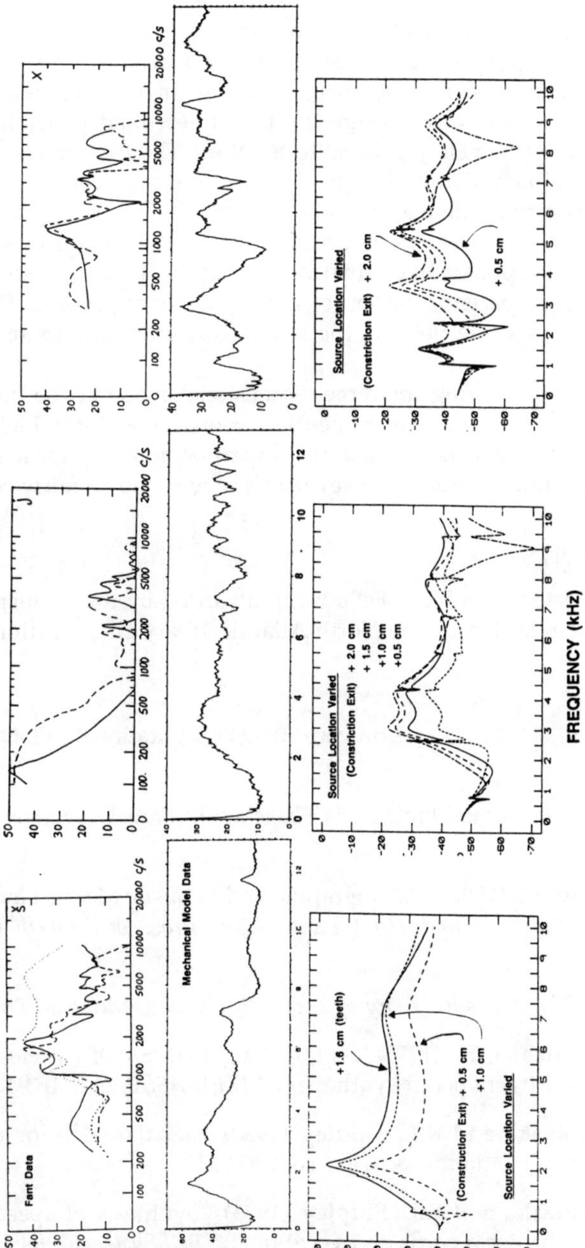

igure 12: For each fricative shown, the acoustic output measured by Fant (1960), that roduced by the mechanical model with the midsagittal profile of Fant's subject, and ansfer functions derived using VCTR on the area functions indicated.

that the air must be travelling above a certain critical velocity when it impinges on the obstacle, implying a certain relationship among the parameters constriction area, flowrate (or pressure drop across the constriction), and distance to the obstacle from the constriction. The details of this relationship are not currently known.

The wall source occurs in cases where the principle sound generation is along a relatively rigid wall, approximately parallel to the flow. The fricatives /ç,x/ use a wall source, and it appears likely that many other fricatives do as well. It is characterized by high, but not maximum, source amplitude for a given flowrate, by a spectral shape with a broad peak, and by a high, but not maximum, rate of change of sound pressure with volume velocity. It appears likely that it is in fact a distributed source; details of the coherence of different parts of the source are unknown at present. The distributed source should be located at the point in the model corresponding to just downstream of the constriction.

It appears that by considering the three-dimensional shape of the tract in order to characterize the source, we can obtain a one-dimensional model in which the source is independent of the tract. This allows us to use superposition and linear system theory, which is a substantial simplification. However, this may require a fairly complex model of the source.

ACKNOWLEDGMENTS

This work was supported by a Hunt Fellowship awarded by the Acoustical Society of America, and a grant awarded by the North Atlantic Treaty Organization, in 1984.

References

Badin, P. and Fant, G. (1984) "Notes on vocal tract computation," *STL-QPSR* 2-3/1984, 53-108.

Catford, J.C. (1977) *Fundamental Problems in Phonetics*, Indiana University Press, Bloomington.

Cohen, M. and J. Perkell (1986) "Palatographic and Acoustic Measurements of the Fricative Consonant Pair /s/ and /š/," Paper A3-5, *Proc. of the 12th Int'l Cong. on Acoust.*, Toronto.

Fant, C.G.M. (1960) *The Acoustic Theory of Speech Production*, Mouton, The Hague.

Flanagan, J.L. and Ishizaka, K. (1976) "Automatic generation of voiceless excitation in a vocal cord-vocal tract speech synthesizer," *IEEE-ASSP* 24:2, 163-170.

Flanagan, J.L., K. Ishizaka and K.L. Shipley (1980) "Signal models for low bit-rate encoding of speech," *J. Acoust. Soc. Am.* 68, 780-791.

Flanagan, J.L., K. Ishizaka and K.L. Shipley (1975) "Synthesis of speech from a dynamic model of the vocal cords and vocal tract," *Bell System Technical Journal* 54:3, 485-506.

Fletcher, N. and S. Thwaites (1983) "The physics of organ pipes," *Scientific American* 248(1), 94-103.

Goldstein, M.E. (1976) *Aeroacoustics*, New York, McGraw-Hill.

Hardcastle, W.J. (1984) "New methods of profiling lingual palatal contact patterns with electropalatography," *Univ. Reading Work in Progress* 4, 1-40.

Heinz, J.M. (1958) "Sound generation by turbulent flow in an acoustic resonator," Unpublished M.S. thesis, Dept. of Elec. Engr. and Comp. Sci., M.I.T., Cambridge, Mass.

Heinz, J.M. and Stevens, K.N. (1961) "On the properties of voiceless fricative consonants," *J. Acoust. Soc. Am.* 33:5, 589-596.

Heller, H.H. and Widnall, S.E. (1970) "Sound Radiation from Rigid Flow Spoilers Correlated with Fluctuating Forces," *J. Acoust. Soc. Am.* 47, 924-936.

Hughes, G. and Halle, M. (1956) "Spectral properties of fricative consonants," *J. Acoust. Soc. Am.* 28:2, 303-310.

Ladefoged, P. and I. Maddieson (1986) "Some of the sounds of the world's languages," *UCLA Working Papers in Phonetics* 64.

Lighthill, M.J. (1952) "On sound generated aerodynamically: I. General Theory," *Proc. Roy. Soc.* A211, 564.

Lighthill, M.J. (1954) "On sound generated aerodynamically: II. Turbulence as a source of sound," *Proc. Roy. Soc.* A222, 1.

Meyer-Eppler, W. (1953) "Zum Erzeugungsmechanismus der Gerauschlaute," *Z. Phonetik* 7, 196-212.

Morse, P.M. and K.U. Ingard (1968) *Theoretical Acoustics*, New York, McGraw-Hill.

Nelson, P.A. and Morfey, C.L. (1981) "Aerodynamic Sound Production in Low Speed Flow Ducts," *J. Sound and Vib.* 79, 263-289.

Schlichting, H. (1979) *Boundary Layer Theory*, 7th ed., New York, McGraw-Hill.

Shadle, C.H. (1985) *The Acoustics of Fricative Consonants*, Ph.D. thesis, Massachusetts Institute of Technology, Cambridge, Mass; released as R.L.E. Technical Report 506.

Shadle, C.H. (1986a) "Models of Turbulent Noise Sources in the Vocal Tract," *Proc. IOA* 8:3, 213-220.

Shadle, C.H. (1986b) "Models of fricative consonants involving sound generation along the wall of a tube," *Proc. of the Int'l Cong. on Acoust.*, Toronto, Canada, paper A3-4.

Strevens, P. (1960) "Spectra of fricative noise in human speech," *Language and Speech* 3, 32-49.

Thwaites, S. and Fletcher, N.H. (1982) "Wave propagation on turbulent jets: II. Growth," *Acustica* 51, 44-49.

ARTICULATORY-ACOUSTIC-PHONETIC RELATIONS AND MODELING, REGIONS AND MODES

R. Carré, M. Mrayati* Institut de la Communication Parlée, INPG-ENSERG, 46, Avenue Félix Viallet, 38031 GRENOBLE Cedex
***) Scientific Studies and Research Center, POB 4470, DAMASCUS**

ABSTRACT

An overview of theories in speech production is given, and classical vocal tract models are presented. These models have been mainly built to facilitate the understanding of articulatory, acoustic, phonetic relations. A new model of the vocal tract based on distinctive regions and modes will be described and its main characteristics presented.

INTRODUCTION

Speech production is a complicated non-linear phenomenon not yet fully discovered. Although a lot of progress has been achieved, much research is still needed. A great amount of the available data is just empirical observations, some of which are not yet fully understood.

In order to make progress in speech production research, we need to know the relations existing between different levels of description, such as : articulation, acoustics, phonetics, and perception. Some aspects of these relations have been formulated through certain basic theories. Figure 1 is a schematic representation of such relations.

This paper consists of two parts. Part I is a general review of relevant literature, which starts with a brief analysis of well known theories related to some aspects of speech production only. In part II, we present a new speech production model based on the concept of distinctive regions and modes (Mrayati, Carré and Guérin, 1988). This model provides new insights into some principles of articulatory-acoustic relations and its applications to the phonetic theory.

We would like to indicate that, for the sake of clarity in this presentation, we do not refer to nasal sound production and nasal tract modeling which are in general treated using a similar approach. Subjects such as motor control theories, etc. are not discussed here.

Principal data relevant to the relations studied are presented with emphasis on the important role they play in the process of formulating a speech production model and identifying its parameters.

A brief overview of important classically known articulatory-

W. J. Hardcastle and A. Marchal (eds.), Speech Production and Speech Modelling, 211–240.

acoustic relations and articulatory-phonetic relations are also given.

Several well known models are described. They can explain specific cases of the above mentioned relations. The main criteria to be satisfied by a speech production model are summarized.

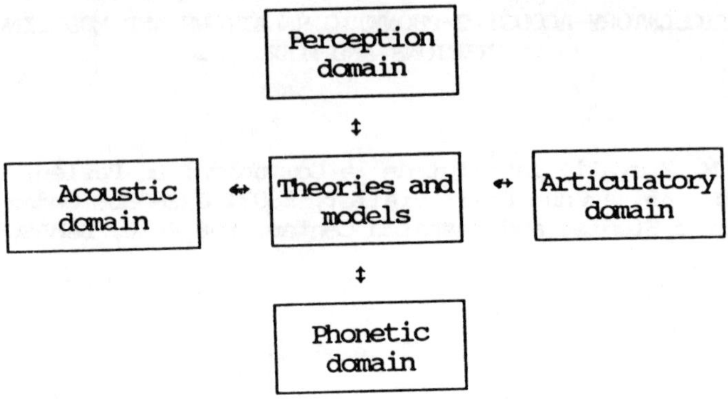

Figure 1. Schematic representation of certain speech domains and their relations.

PART I. GENERAL REVIEW OF RELEVANT LITERATURE

I.1. THEORIES OF SPEECH PRODUCTION

Over the years, we have witnessed important great theoretical achievements in the field of speech production. Chiba and Kajiyama (1941) proposed the wave propagation theory of speech production. They showed that the speech sound is produced by an uni-dimensional acoustic wave propagating in the vocal and nasal tracts. Standing waves of flow velocity and pressure exist in the vocal tract for several resonant modes. Figure 2 shows a basic case of the application of this theory, namely the neutral tube case. This theory is now well established and standing waves have been measured (see for example Firth, 1986).

The source filter theory has been a great step forward in the formulation of the speech production process and its models. Fant (1960), in his "acoustic theory of speech production" elaborated basic principles which explain many phenomena present in this process, and which have given rise to several models. Figure 3 provides a simplified illustration of the source filter composition of voiced sounds. The time domain representation of the source is a consequence of air pulses emitted through glottis, the spacing between adjacent pulses being T0. In the frequency domain, the source is a spectrum of harmonics of the voice fundamental frequency F0=1/T0. The effect of the vocal tract filtering is that the amplitude of each of the harmonics

S(f) is multiplied by the value of the transfer function T(f) at that frequency, the product being the spectrum of the radiated signal. The formants of the spectrum originate from the maxima of the vocal tract transfer function and are more precisely defined as the resonances of the vocal tract. The source filter theory is well adapted for studying static configurations or a succession of static configurations.

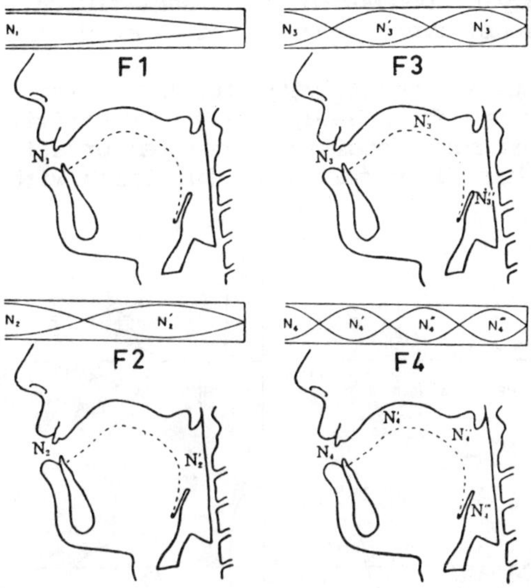

Figure 2. Standing waves in the vocal tract (Chiba and Kajiyama, 1941).

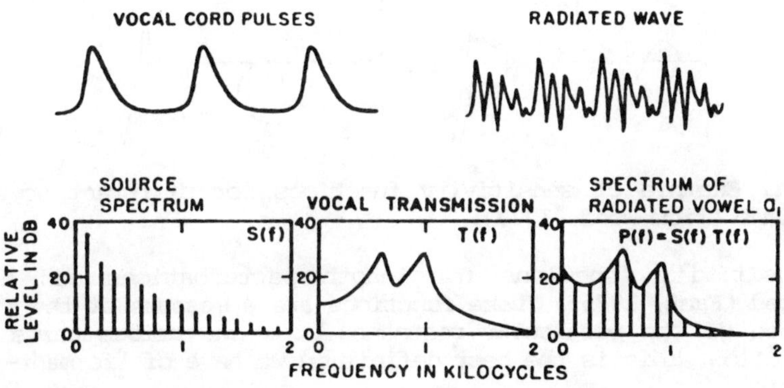

Figure 3. The source filter concept (Fant, 1960).

The perturbation theory has been successfully used to study the relation between a small area function variation and a corresponding acoustic parameter variation (Ungeheuer, 1962 ; Schroeder, 1967 ; Mermelstein, 1973). This concept is further developed leading to the concept of sensitivity function (Fant, 1967 ; Fant and Pauli, 1974 ; Fant, 1975 ; Mrayati and Carré, 1976 ; Wood, 1979 ; Fant, 1980). Sensitivity functions for local area perturbations of any area function A(x) relate, for a given formant, small local spatial variations to formant frequency or bandwidth. Such sensitivity functions provide information on "sensitive" parts of the vocal tract for changing formant frequencies.

Figure 4 displays sensitivity functions for different arabic vowel configurations (from Wood, 1979). Sensitivity functions are generally calculated for a positive local perturbation of A(x). As an example, we note that F1 of the vowel [u] rises with increasing area at the lips.

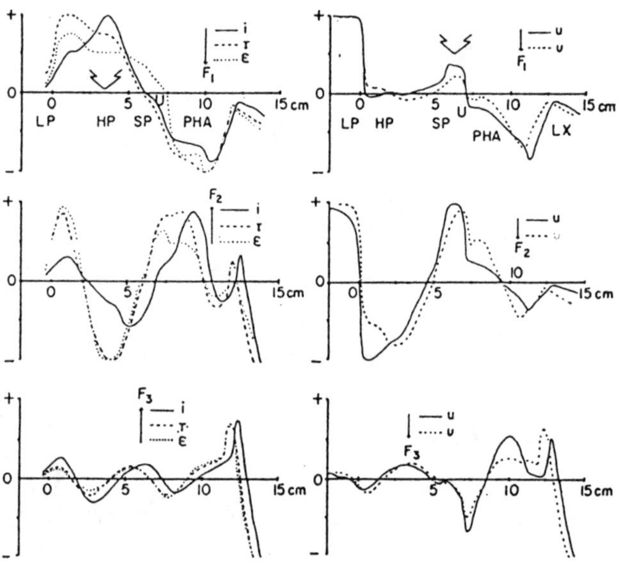

Figure 4. Example of sensitivity functions for different vowel configurations (from Wood, 1979).

Sensitivity functions for length perturbations can also be calculated (Fant, 1975). These functions are a measure of the relative dependance of the particular resonance mode on various parts of the area function. This is the best definition we have of "formant-cavity" affiliations.

Sensitivity functions relate formant frequency variations to small local area perturbations. Inversely, area perturbation functions can be calculated in order to obtain a specific formant frequency variation (or no variation at all).

Figure 5 shows the variation of the first formant obtained by perturbing the uniform area function in proportion to the sensitivity function curve corresponding to the first formant. The other formants remain stable (Schroeder, 1967). There exist also an infinite number of perturbation functions applied symmetrically with respect to the mid-point of the uniform tube, which will have almost no influence on the formant pattern. Heinz in 1967 generalized the work of Schroeder to the case of arbitrary area functions instead of a uniform tube.

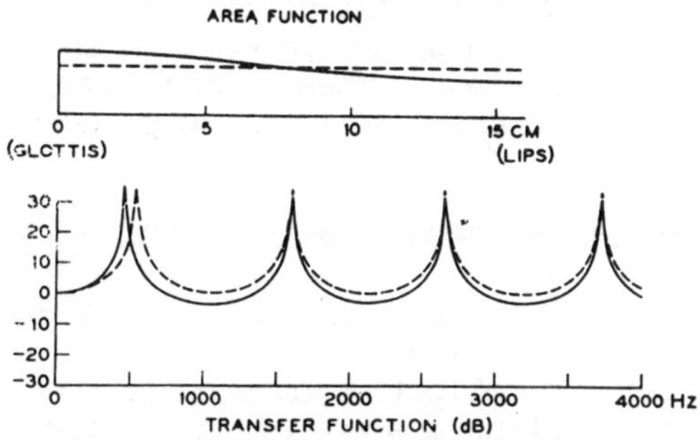

Figure 5. Example of the application of perturbation theory in three formant clamping case (F2, F3, F4) (Schroeder, 1967).

Finally, the <u>quantal theory</u> of speech, proposed by Stevens in 1972, has been another important achievement in the theoretical formulation of the speech production process. Stevens postulated that the articulatory-acoustic relations are quantal in the sense that the acoustic pattern may change from one quasi-steady state to another as the articulatory parameter is varied continuously.

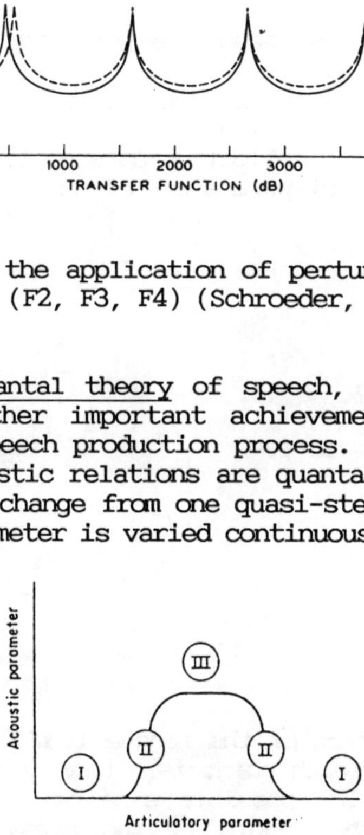

Figure 6. Schematized relation between an acoustic parameter of speech and a parameter that describes some aspect of the articulation (Stevens, 1972).

Figure 6 shows such a relation. In regions II, large acoustic changes happen for small shifts in articulation. In regions I and III, the acoustic parameter remains relatively stable when large variations are made in the articulation. It is suggested that "the existence of relations of this type between articulation and sound output has certain implications with regard to the inventory of articulatory and acoustic properties that are used as codes for signaling linguistic information" (Stevens 1972). Two examples of this quantal nature are, firstly, the acoustic variations when the degree of narrowing of a constriction is modified and secondly, certain acoustic stabilities present in vowel production.

In conclusion, the source filter theory and the perturbation theory are well adapted to study quasi-static configurations. But the vocal tract shape changes during speech production and the changes are larger than small perturbations. These large variations must be studied and formalized before creating a speech production model.

I.2. RELEVANT DATA FOR SPEECH PRODUCTION MODELING

The classical approach in control theory when creating a model for a given system is the following :

-a) observing the system and collecting data on it,

-b) identifying the pertinent parameters of the system,

-c) modeling the system.

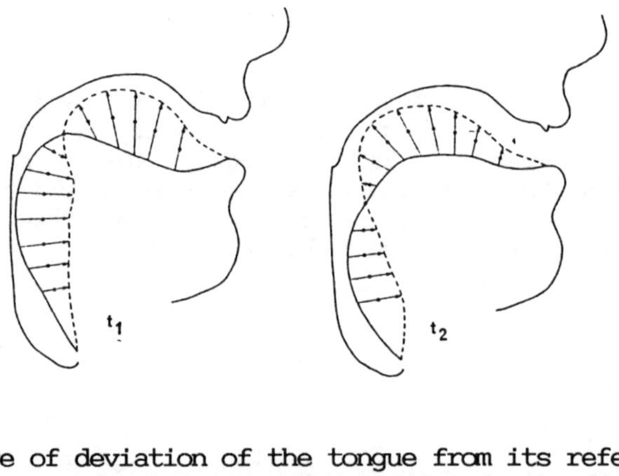

Figure 7. Range of deviation of the tongue from its reference neutral position. The movement of each point is either in the direction of the arrow or away from it. The possible size of the movement for each point is indicated by the length of the arrow through that point (Harshman et al., 1977).

In the field of speech production, the first stage of this process is well developed. The large amount of data already obtained have not yet yielded decisive identification results and, consequently, we do not have the final and comprehensive model of speech production. Ladefoged and Harshman (1979) reported that "no one has ever published a study in which they describe a vocal tract model that could generate formants that correspond accurately to the set of formant frequencies that have been observed in a wide range of vowels spoken by a number of different speakers, using only plausible methods of accounting for the anatomical differences between speakers".

We will now briefly review some of the important data relevant to the subject.

Harshman et al. (1977) using X-rays taken during the pronunciation of ten vowels by 5 English speakers, analyzed tongue movement by means of factor analysis. They show that tongue shapes can be described in terms of two underlying factors, or tongue gestures around the neutral position, as illustrated in figure 7. This gestures were labelled by Ladefoged : Front Rising (FR) and Back Rising (BR). Four main constrictions can be obtained from these two factors.

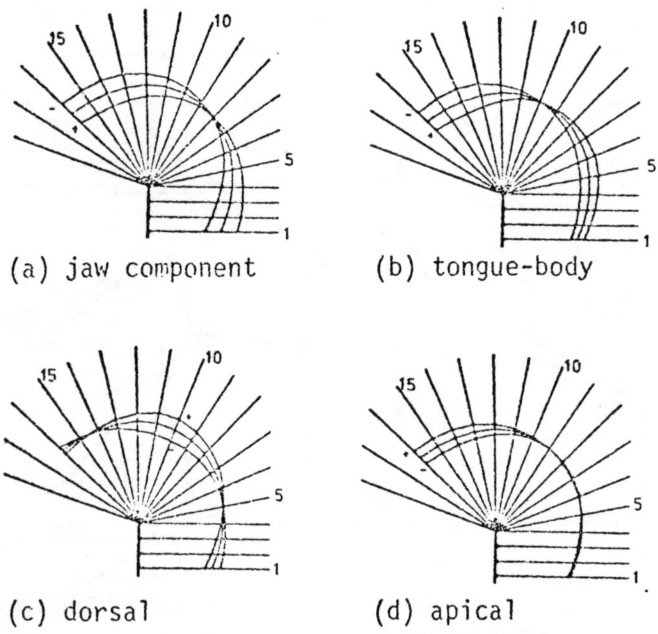

Figure 8. Decomposition of the tongue contour in 4 linear components (jaw, tongue body, tongue hump, and tongue tip) (Maeda, 1979).

It is interesting to note that the above results on natural speech suggest an anti-symmetrical movement to be an important tongue gesture.

Similar work done on French X-ray data reached a similar conclusion but differed in a better identification of parameters which are in this case four (Maeda, 1979) (see figure 8).

In the two studies, there are synergetic symmetrical movements present as well as synergetic non-symmetrical movements applied to certain regions along the tongue.

At this stage, we notice that the tongue movements are around a neutral position or a mean position. This remark, as we will see later, is very important.

Wood (1979) reported vocal tract area functions estimated from 38 sets of X-ray vowel articulations collected from the literature and from new X-ray motion films of English and Arabic speech. He has shown that four constriction locations exist which are situated along the hard palate, along the soft palate, in the upper pharynx, and in the lower pharynx. He concluded that each location is appropriate for a definable class of vowel qualities, confirming as such the quantal nature of at least this aspect of vowel articulation. He distinguished four quantal regions as shown in figure 9.

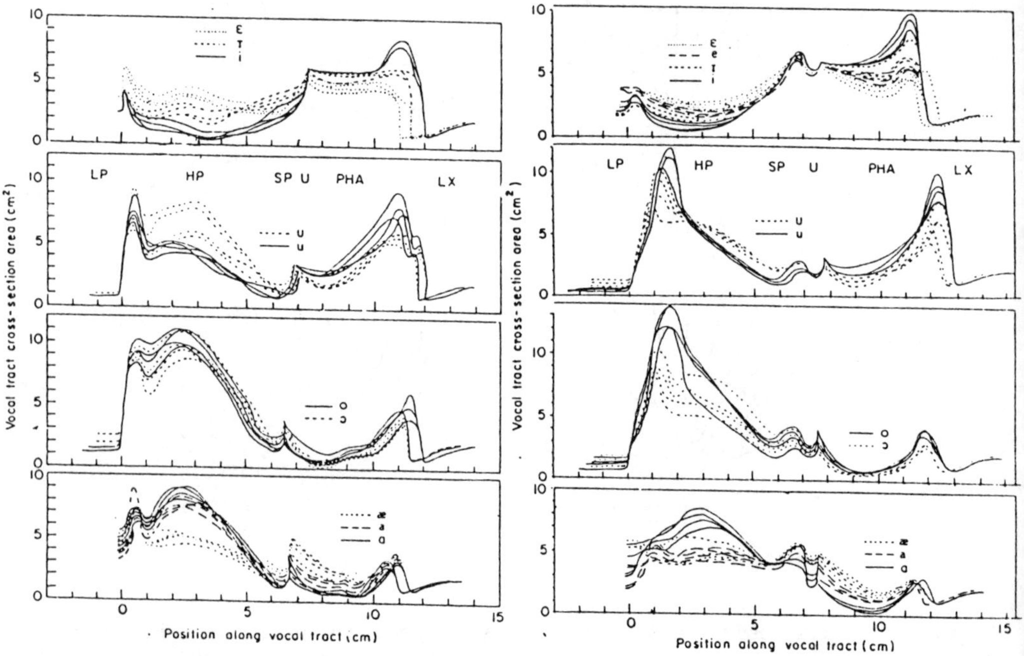

Figure 9. Area functions for English vowels (Wood, 1979)

Wood (1986) also reported other very interesting data derived from X-rays of natural speech. These data show that the movements of the lips and of the larynx are correlated. He hypothesizes that the reason for this synchronization is to prevent tongue gestures from having erratic spectral consequences. If we do not have larynx depression accompanying lip-rounding, the relative locations of the four above mentioned zones with respect to the tongue and its gesture change and, consequently, this gesture would have different acoustic outputs in the two cases.

In conclusion, we would like to point out that this underlying symmetrical and non-symmetrical synergy gesture of the tongue seems to be an important mechanism in speech production. Furthermore, the tongue musculature is well adapted for such gestures (Wood, 1979).

I.3. BRIEF REVIEW OF IMPORTANT CLASSICALLY KNOWN ARTICULATORY-ACOUSTIC RELATIONS.

In the frequency domain, the acoustic signal is the product of a source and a filtering process. The frequency response of the vocal tract is the simplest way of describing the main characteristics of the filter (formant frequencies and bandwidths) (Fant, 1960). The area function is generally defined from sagittal plane X-ray data.

Recall that two types of sensitivity functions exist and relate articulatory to acoustic parameters. The first which deals with transversal local area variations is proportional to the difference between kinetic and potential energy along the vocal tract. The second which deals with local length variations is proportional to the sum of the two energies.

Sensitivity functions as defined above can help to define formant stability as a function of local perturbations.

The distribution of the kinetic plus potential energy along the vocal tract is also a measure of the relative dependence of the particular resonance mode on various parts of the area function. This is the best definition we have of 'formant-cavity' affiliation (Fant, 1980).

From articulatory data and articulatory modeling, articulator movements and corresponding acoustic trajectories can be compared. All other parameters being constant, mandible movement alone causes considerable shifts in F1.

Changing the location of the tongue body downward along the tract can be an extremely effective means for lowering F2.

A formant frequency lowering is associated with decreases in lip opening area.

A larynx lowering decreases the frequencies of all formants.

The inverse relation, i.e. acoustic-articulatory relation, generates pseudo-area functions which remain fictional in the sense that they do not necessarily resemble natural area functions. An infinity of different shapes have the same F-pattern (Atal et al., 1978). Articulatory constraints have to be used to select the right shape. Dynamic movements are also very useful in determining the continuous area variation underlying the formant frequency variations.

I.4. BRIEF REVIEW OF IMPORTANT CLASSICALLY KNOWN ARTICULATORY-PHONETIC RELATIONS.

It is generally accepted that the most powerful criteria that

relate the articulatory and phonetic domains are the following :
-a) place of articulation which can vary between the glottis and the lips,
-b) the manner of articulation which distinguishes between stops, affricates, fricatives, etc. although they have the same place of articulation,
-c) the mode of articulation which distinguishes between voiced and unvoiced, emphatic and non emphatic, etc.

In changing the manner of articulation, non linear phenomena take place. On the other hand, when moving the place of articulation, we notice that we pass by regions having stable acoustic behaviour and other regions having large changes in acoustic behaviour. These two observations suggest strongly a non-linear quantal nature in speech production (Stevens, 1972). This observation can be seen from the nomograms to be introduced in the following section. This tendency for quantal relations among acoustic and articulatory parameters is an important factor shaping the inventory of acoustic and articulatory attributes that are used to signal distinctions in language (Stevens, 1972). As in static speech sound distinction, non-linear behaviour is also exploited in speech dynamics according to specific strategies. In natural and modeled speech production, inherent physiological constraints influence not only the static and target configurations but also the dynamic behaviour.

I.5. MODELS AND NOMOGRAMS IN SPEECH PRODUCTION

A specification of an area function as a more or less continuous graph of cross sectional area from the glottis to the lips allows detailed calculations of the acoustic response but is not practical for systematic descriptions. Preceding theories and data help to define simple relevant representations of vocal tract shapes. The schematic representations or models are described by a reduced number of parameters.

Several models of speech production have been proposed. We will review the main models with their nomograms. These models are those capable of providing insight into some area-function-acoustic relations or articulatory-acoustic relations. Nomograms which relate formant frequency variations to variations of model parameters one at a time, give the behaviour of the model.

First, the models consisted of 2, 3, 4, etc. cylindrical cavities approximating the vocal tract configuration. Then, a horn-shaped tongue constriction has been proposed giving a better approximation of the main constriction. Model parameters are generally the place and the degree of the main constriction. Finally, articulatory models are controlled by articulatory parameters such as tongue height, tongue body, tongue tip, etc.

I.5.1. Fant (1960) elaborated a very instructive model composed of four cavities representing the vocal tract. He provided excellent nomograms relating the five formant frequencies to the dimensions of these four cavities. These nomograms reflect the variations in the formant frequencies due to variations in place and area of the constriction. Figure 10 shows an example of the Fant model and figure 11 an example of its nomograms.

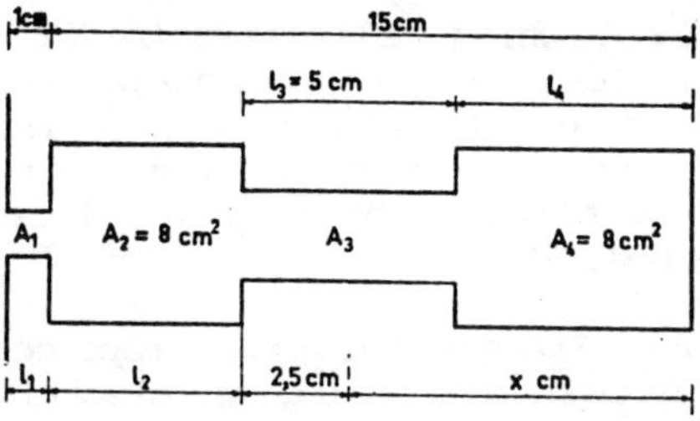

Figure 10. Three parameter vocal tract model based on four homogeneous tubes, simulating the lip section, a front cavity, the tongue section, and a back cavity (Fant, 1960).

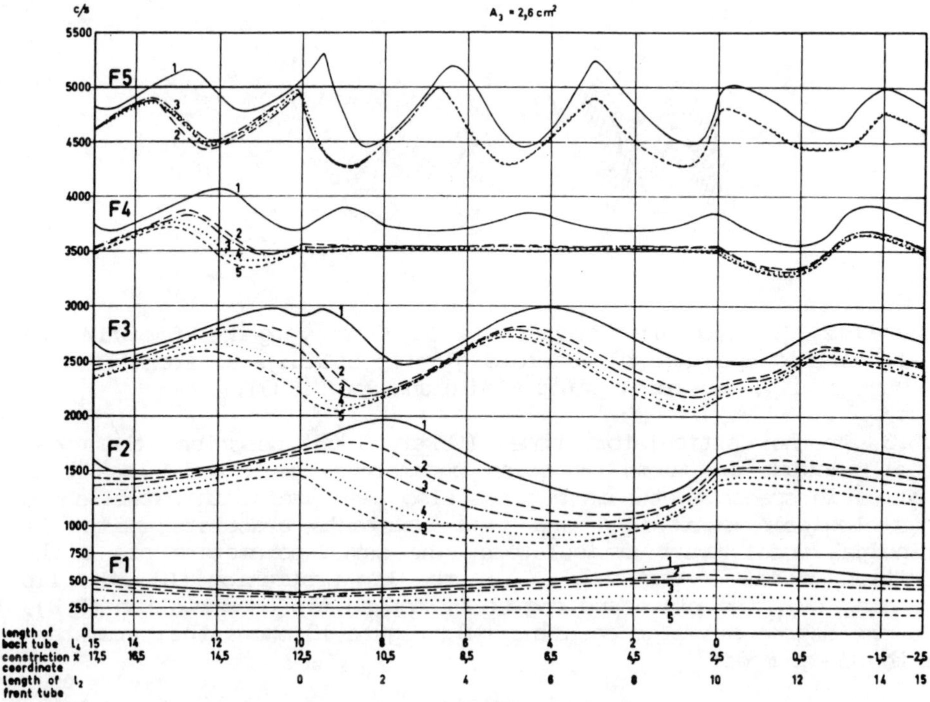

Figure 11. Nomograms relating F1, F2, F3, F4, and F5 of the three parameter model to the location of the constriction (Fant, 1960).

Ladefoged and Bladon (1982) validated certain aspects of these nomograms by attempts to produce them by human speakers.

Several interesting phenomena could be demonstrated. Cases of quantal speech production behaviour can be shown. When the constriction is at 10.5 cm from the glottal end, for example, we notice that we have a maximum for F2.

A second phenomenon, which is related to the first one, occurs when two formants come close to each other to produce a "focal point". In this convergence region, the affiliation of the lower and upper resonances switches from one cavity to the other when the tongue constriction is shifted between these back and front cavities (see figure 12) (Badin and Boë, 1986).

I.5.2. As early as 1955, Stevens and House proposed a three-parameter model for vowel production, and using an electrical analog of the vocal tract, calculated the variations in the three formant frequencies against variations in the model parameters. They presented several nomograms relating acoustic parameters (F1, F2, F3) to articulatory parameters (place of constriction, degree of constriction, and lip parameter). Fant's three parameter model (1960) differs from this model only in the details since both have the same set of parameters.

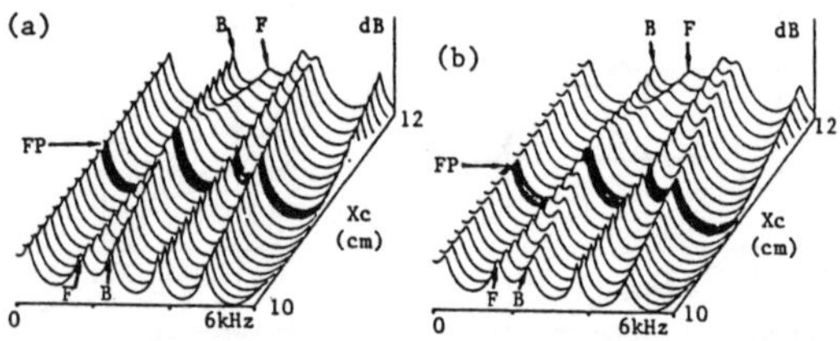

Figure 12. 3-D nomogram around the F2/F3 convergence for /i/, a) small glottis opening, b) moderate glottis opening. F: front cavity, B: back cavity, FP: focal point (Badin and Boë, 1986).

I.5.3. Another articulatory model (Coker, 1968) describes the vocal area function in terms of seven parameters. This model has been applied in speech synthesis but has also been useful in establishing physiological constraints and time constants associated with the principal vocal tract articulators. The seven parameters are : the position of the tongue body (X,Y), the lip protrusion (L), the lip rounding (W), the place and degree of tongue tip constriction (R,B), and the degree of velar coupling (N). Figure 13 shows the parameters of the Coker model.

I.5.4. Lindblom and Sundberg (1971) presented an articulatory model controlled by 6 parameters : labial height, labial width, jaw, tongue tip, tongue body, and larynx height. The model is used to reinterpret the classical notion of "tongue height".

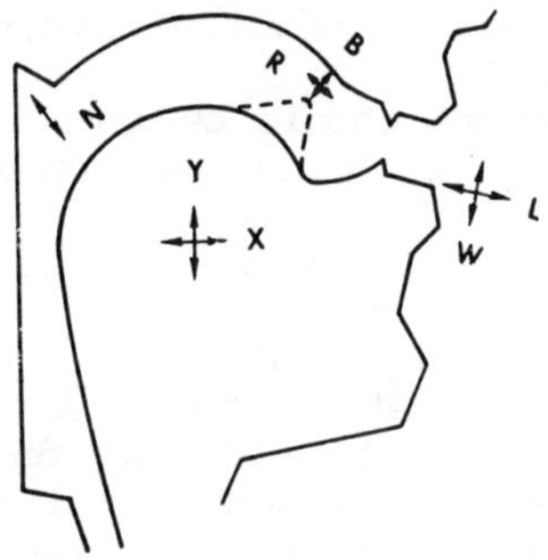

Figure 13. The parameters of the Coker model (Coker, 1968).

I.5.5. Mermelstein (1973) devised another articulatory model which represents the vocal-tract outline by means of variables specifying the position of the jaw, tongue body, tongue tip, lips, velum, and hyoid. This model is primarily used for speech synthesis.

I.5.6. Ladefoged and Lindau (1979) and Maeda (1979) proposed an articulatory model based on the results of statistical analysis of the tongue shape (see figure 7 and 8 above). For the description of a speech event, the tongue is regarded either as a composite of two independent parameters (Front Rising and Back Rising, Ladefoged and Lindau (1979) or as a composite of independently controllable systems such as the jaw and the tongue-body, tongue dorsal, and tongue tip (Maeda, 1979).

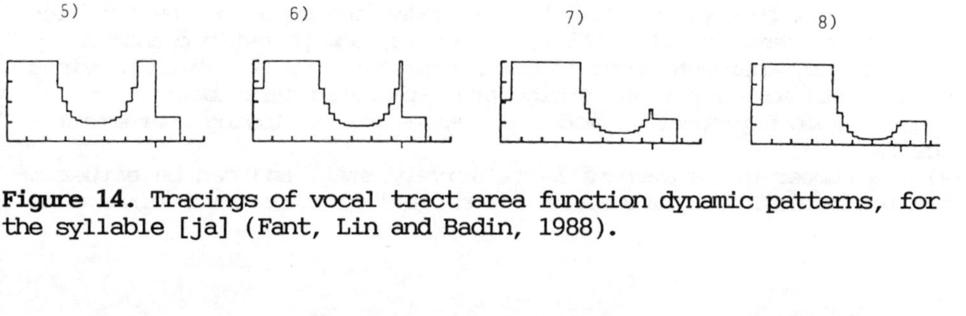

Figure 14. Tracings of vocal tract area function dynamic patterns, for the syllable [ja] (Fant, Lin and Badin, 1988).

I.5.7. Conclusions on the models.

When comparing the models mentioned above, one should look for the following criteria :

1) the capability of the model to describe natural phenomena as close as possible (vowel and consonant production, compensation, coarticulation, vowel reduction, etc.),
2) the parameters of the model should be identified on the basis of observations of the system. They should be orthogonal and capable of producing all possible states of the modeled system,
3) the number of the model parameters and the simplicity and significance of their command strategies,
4) the nomograms of the model are better if they have simple and easily interpretable relations between the articulatory domain and the acoustic domain, and between the articulatory domain and the phonetic domain ; these criteria render the model much more instructive and useful in explaining the speech production process,
5) the capability of exhibiting nonlinear phenomena inherent in natural speech production,
6) the capability of predicting existing phenomena.

It is interesting to mention, at this stage, that most of the known models imply a control strategy which consists of continuous longitudinal movement of the constriction. One of the best examples of such a strategy is given in figure 14. Notice that the constriction is moving upwards at the same time as it moves forward. The constriction reaches its highest place at the mid point of the vocal tract where it starts moving downward.

PART II. A DISTINCTIVE REGION MODEL BASED ON UPDATED RESULTS.

II.1. INTRODUCTION

In the following paragraphs, we are going to introduce and develop a new concept (Mrayati, Carré and Guérin, 1988 ; Carré and Mrayati, 1989; Mrayati, Carré and Castelli, 1989) which shows the existence of distinctive spatial regions along the vocal tract between the glottis and the lips. It also defines three distinctive modes which are related to the degree of narrowing at a constriction in the vocal tract. This new concept is founded on and derived from the above mentioned theories of speech production namely perturbation theory, wave propagation theory, source-filter theory, and quantal theory.

A quasi-articulatory model of speech production based on this new concept is proposed together with its nomograms and control strategy.

This new concept and the derived model are supported by the rich inventory of data on articulatory, acoustic, and phonetic phenomena.

This new approach seems to offer explanatory power which provides insight into some important articulatory-phonetic relations.

Our speech production model possesses the following characteristics :

-a) The number of parameters is relatively small and can be either 5 or 9 parameters. The parameters are related to transversal dimensions of the vocal tract.

-b) Our proposed model achieves the place of articulation displacement by means of vertical movements of region areas rather than by transversal movements of a constriction.

-c) The command strategy of these parameters is extremely simple and could be formulated using a limited number of simple rules and target values for articulators.

-d) This model has the advantage of very simple nomograms. These nomograms have at most one maximum. This, as we will see, renders the model very simple to exploit and offers a tool that explains in a simple manner many speech production phenomena and the relation between area function and acoustic parameters.

-e) The model is capable of producing consonants and vowels and other phenomena observed in natural speech production.

-f) Important principles supported by a lot of data, are inherently incorporated in the model. These principles are the anti-symmetrical behaviour of the vocal tract around its mid-point ; the synergy and compensatory principles which are a consequence of the antisymmetry characteristic.

-g) The model incorporates inherently several aspects of the quantal nature of speech : it has distinctive spatial regions along the vocal tract quantizing it in pseudo-orthogonal acoustic behaviour, each region is related to an optimal place of articulation ; the transversal movement corresponding to a narrowing is quantized and described using three modes ; these modes or a combination of them correspond to the manner of articulation. When changing the degree of narrowing or opening in one region, acoustic parameters change non-linearly and pass by a maximum either during narrowing or opening ; this maximum correlates, from our point of view, with the stable states defined by Stevens in his quantal theory.

-h) Thanks to the symmetrical aspects of our model, it can incorporate the influence of all parts of the vocal tract. One consequence of this aspect is that what is important in some vowel production for example is not only the degree of constriction but also the degree of opening of the cavity in symmetry around the mid-point with the constriction. As a matter of fact, if the vocal tract area function is expressed using a logarithmic scale, equal logarithmic area variation at the back constriction or front cavity could have the same acoustic effect.

II.2. REGIONS AND MODES.

In this paragraph, the Distinctive Regions and Modes theory (DRM) of speech production is summarized. For an acoustic tube, closed at one end and open at the other, there exist distinctive spatial regions (R) having specific Formant Transition Behaviour (FTB). These FTBs are monotonic as long as the variations of the cross section area (S) of the different regions (R) are within specified limits (between approximately 1 and 15 cm2, if the neutral tube is 5 cm2) defining a mode denoted One-Tract Mode (OTM). Two other modes can also be defined depending on region cross sectional areas. These are the Transition Mode (TM) corresponding to narrow S(R) (between around 0.05 and 1 cm2); and the Two-Tract Mode (TTM) corresponding to practically closed S(R) (between closure and around 0.05 cm2). It is important to emphasize that regions are defined for the uniform tube. These regions and modes are deduced from sign changes of sensitivity functions with distance from the glottis x and area S changes respectively.

These spatial regions are bounded by zero crossings of the sensitivity functions of the uniform tube (figure 14). If one is interested in the first three resonant modes only (the three formants), eight regions can be distinguished. They are called A, B, C, D, $\overline{D}$, $\overline{C}$, $\overline{B}$, $\overline{A}$.

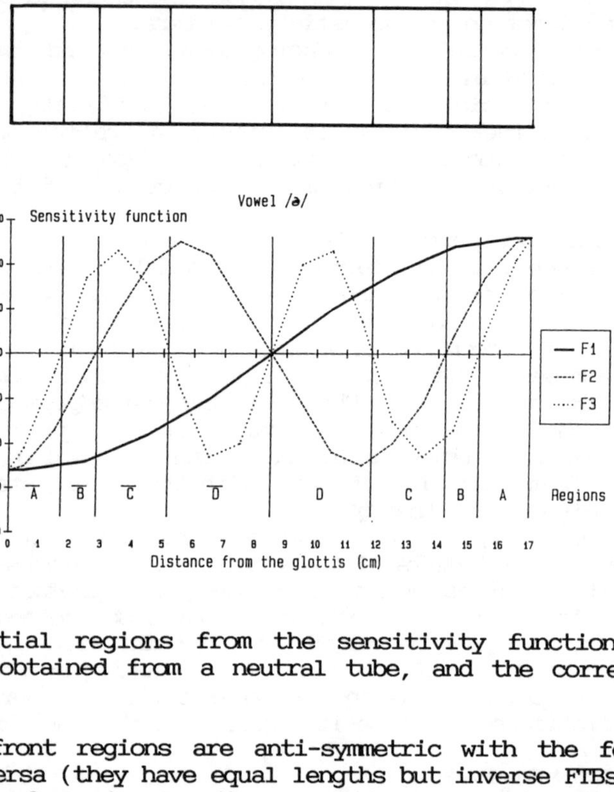

Figure 15. Spatial regions from the sensitivity functions for 3 resonant modes obtained from a neutral tube, and the corresponding region model.

The four front regions are anti-symmetric with the four back ones and vice versa (they have equal lengths but inverse FTBs). These regions have the following lengths respectively : L/10, L/15, 2L/15, L/5, L/5, 2L/15, L/15, L/10.

It is clear from the sensitivity functions that, for each of these regions, an increase or a decrease of its cross sectional area leads to a well defined variation sign of the three resonant modes. This aspect distinguishes the eight possible combination of variation signs (three formants with 2 possible variation signs for each (+ and -) result in 8 combinations). If one considers one resonant mode only, there exist two regions. When only the first two resonant modes are considered, there exist four regions. In general, the number of regions N is related to the number of resonant modes n by the following formula :

$$N = 2 * (1 +(1 + 2 + ... + n-1)).$$

We have already mentioned the anti-symmetrical structure of the regions of the vocal tract. As a consequence, it can be easily seen that there exist two major behavioral characteristics for the regions. The first is that an area increase in one of the front regions could acoustically compensate a similar increase in the corresponding back region. The second which is related to the first, occurs when the two corresponding regions change area in an opposite direction involving a maximal formant change. These **compensatory** and **synergy** aspects can be used to control formant frequencies.

The above mentioned regions and modes concept leads directly to the introduction of a new speech production model and a new efficient coding of the area function A(x,t). The model is composed of eight non equal uniform tubes of lengths L/10, L/15, 2L/15, L/5, and their symmetrical counterparts. The parameters of the model are the cross-sectional area of the eight regions S(R,t), and the total length of the vocal tract L(t). The important and new aspects of this model are its non equal tube lengths, and the transversal variation strategy of the area of these tubes. This new model has the advantages mentioned in paragraph II.1. above. Figure 15 shows a schematized form of this model.

Due to the pseudo-orthogonality of the model parameters around the neutral, the model is capable of producing the vocalic space in the F1-F2 plane with lowest parameter variations (lowest energy). In other words, it can been shown that, for the same dimensional ratio between cavities of the model, this model produces a vocalic space larger than any other model. It is important to mention that regions have the properties of including places where modifications of the area around the neutral gives optimal variations of the three formants at the same time. In other words, if we look for places in the vocal tract having the best modulation of the three formants together, and the largest dynamics of these modulations, these places are around the mid-points of the regions and around the neutral. Non linearities and saturations occur near the borders of the vocalic space.

II.3. ACOUSTICS-REGIONS RELATIONS

II.3.1. Uniform region model. Large variations.

The acoustics-regions relations are presented here in the case of 8 regions : A, B, C, D, $\bar{D}$, $\bar{C}$, $\bar{B}$, $\bar{A}$. Thus, three formants are taken into account.

Figure 16 gives the region nomograms calculated for a lossless tube closed at one end and open at the other : they relate numerically, for each of the eight regions, the three formant frequencies F1, F2, and F3 to variations of the region cross sectional areas S(R). One region area is varied logarithmically between 0.005 to 14.14 cm2 while the other ones are kept at 5 cm2. The OTM prevails up to a constriction of about 1 cm2, while the TM prevails from about 1 to about 0.05 cm2 where the TTM starts.

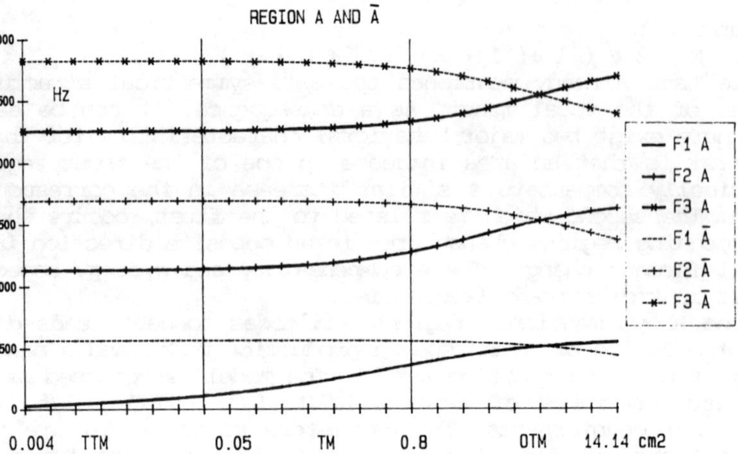
REGION A AND Ā
Hz
3000
2500
2000
1500
1000
500
0
0.004 TTM 0.05 TM 0.8 OTM 14.14 cm2
F1 A
F2 A
F3 A
F1 Ā
F2 Ā
F3 Ā

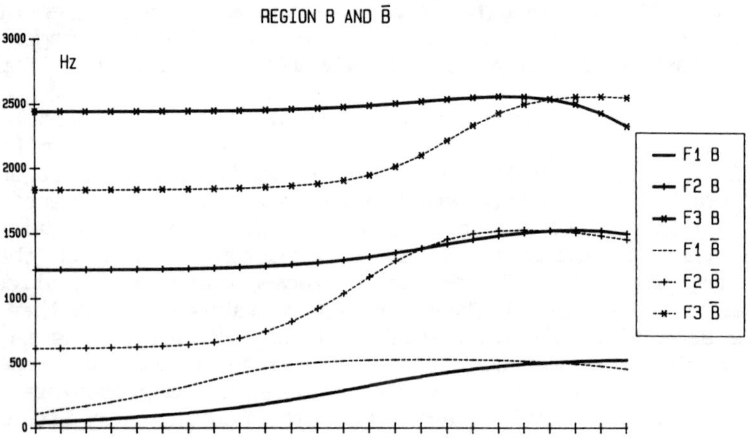
REGION B AND B̄
Hz
3000
2500
2000
1500
1000
500
0
F1 B
F2 B
F3 B
F1 B̄
F2 B̄
F3 B̄

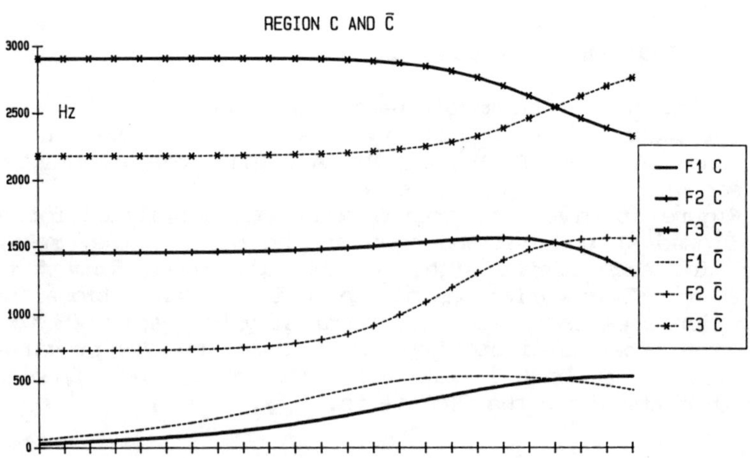
REGION C AND C̄
Hz
3000
2500
2000
1500
1000
500
0
F1 C
F2 C
F3 C
F1 C̄
F2 C̄
F3 C̄

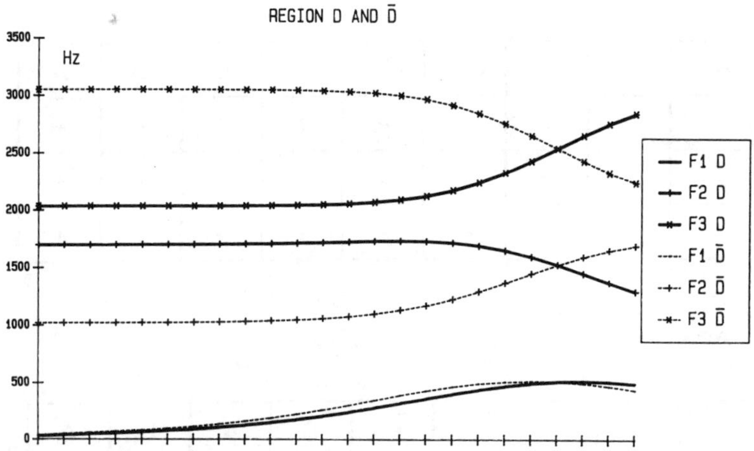

Figure 16. Nomograms obtained for a uniform tube when varying the cross sectional area of each of the regions, one at a time. The area is varied logarithmically from 0.005 to 14.14 cm2. The OTM prevails up to a constriction of about 1 cm2, while the TM prevails from about 1 to about 0.05 cm2 where the TTM starts. Eight regions are here defined.

In that last case, the three formant frequencies are quasi-stable. The values of the OTM boundaries as given above are mean values : in the example given on figure 16, a change of first formant transition behaviour takes place at 2.5 cm2 for region $\overline{D}$ and at 10 cm2 for D. Such nomograms can also be obtained for formant frequencies and bandwidths when including losses in the vocal tract (heat, viscosity, radiation, wall vibration losses).

Figure 17 shows the Formant Transition Behaviour matrices for the OTM, TM, and TTM modes. In the OTM case, an increase in the area at the lips (A region) will increase the three formants while at the glottis ($\overline{A}$ region), it will decrease the three formants. An increase of the area at region C will increase F1 and decrease F2 and F3. The OTM matrix is strictly true when the area is varied between 2.5 to 10 cm2.

The nomogram data can also be represented in the F1-F2, F2-F3, F1-F3 planes. Figure 18 shows the F1-F2 formant variations in the case of varying two regions together (case of a 4 region model). The starting tube is a uniform tube. Each two region area is varied from 1.78 to 14.14 cm2 logarithmically. Because of the anti-symmetrical behaviour of the vocal tract, the F1-F2 trajectories for BA and $\overline{AB}$ on the one hand, DC and $\overline{CD}$ on the other hand, strictly coincide.

OTM case :

Regions:	$\bar{A}$	$\bar{B}$	$\bar{C}$	$\bar{D}$	D	C	B	A
F1	–	–	–	–	+	+	+	+
F2	–	–	+	+	–	–	+	+
F3	–	+	+	–	+	–	–	+

TM and TTM cases :

Regions:	$\bar{A}$	$\bar{B}$	$\bar{C}$	$\bar{D}$	D	C	B	A
F1	–	+	+	+	+	+	+	+
F2	–	+	+	+	+	+	+	+
F3	–	+	+	–	+	–	+	+

Figure 17. Formant Transition Behaviour Matrices for the OTM case and for the TM and TTM cases. Eight regions are defined here.

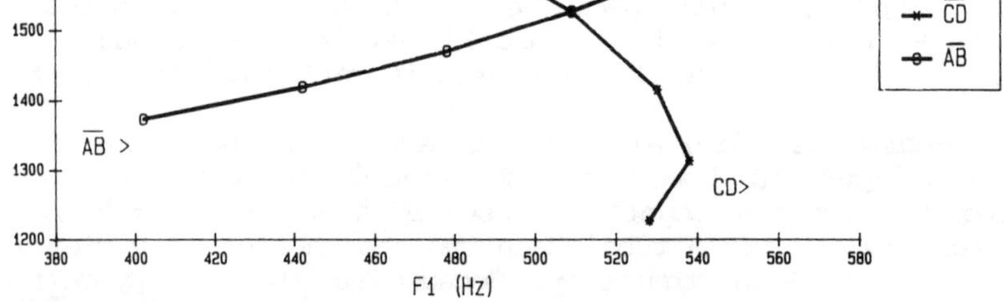

Figure 18. Nomograms obtained for a uniform tube when varying the cross sectional area of two regions together. The area is varied logarithmically from 1.78 to 14.14 cm2. The formant variations are

represented in the F1-F2 plane.

Furthermore, F1 is almost stable when DC area is more than 5 cm2 and when $\overline{CD}$ is less than 5 cm2. As will be seen later on, such a property can be used to explain articulatory-phonetic relations.

II.3.2. Non-uniform region model.

The nomograms obtained for a uniform region model and their corresponding representations in the F1-F2 plane can also be calculated for a non-uniform region model (vowel configuration).

Figure 19 displays the F1-F2 trajectories for small cross-sectional area variations of each of the eight regions in the case of 11 French vowels. One region is varied around its starting value by a factor of $\sqrt{2}$ or $\sqrt{2}/2$.

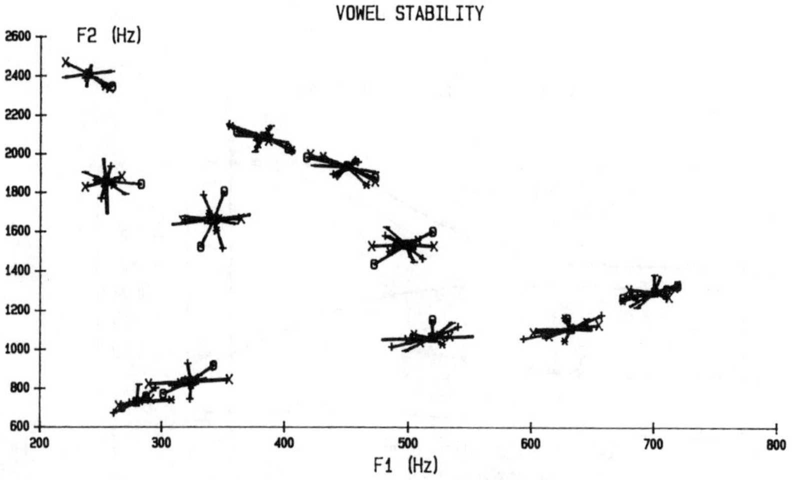

Figure 19. Formant frequency variations obtained when the cross-sectional area of each of the eight regions is multiplied by a factor of $\sqrt{2}$ or $\sqrt{2}/2$. The region configurations correspond to 11 vowels.

In the case of non-uniform configurations corresponding to vowels, the formant transition behaviour is generally preserved. With eleven configurations corresponding to eleven French vowels, about 5 % of the transitions do not follow the behaviour obtained for a uniform region model. As an example, figure 20 shows nomograms for two vowels [e] and [o]. For vowel [e], the formant transition behaviour related to regions D and $\overline{D}$, is clearly preserved and the synergy and compensation principles can be applied. For vowel [o], looking at regions C and $\overline{C}$, the nomograms show formant stability ranges : F3 for region C, F2 for $\overline{C}$, etc. the formant transition behaviour should always be considered taking into account the stability of certain formants. This important quantal behaviour will be treated in a forthcoming study.

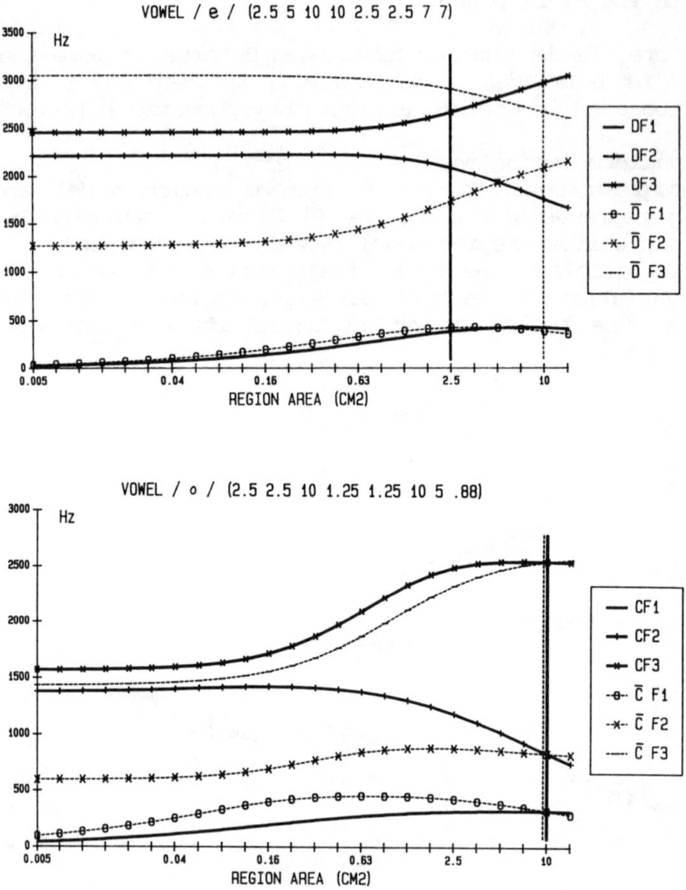

Figure 20. Nomograms for vowels [e] and [o]. The region model configurations are given on the top of the figures (glottis to lips).

II.4. RELATIONS BETWEEN REGION MODEL AND VOCAL TRACT.

Regions C, D and $\overline{C}$, $\overline{D}$ are well correlated with the tongue location whereas the regions A and B correlate with the lips and the tongue tip and $\overline{A}$ and $\overline{B}$ correlated with the larynx and the beginning of the pharynx.

Radiation effect, lip rounding and spreading on the one hand, larynx movement on the other hand, have a combined effect which leaves the effective vocal tract length practically unchanged and, consequently, the region boundaries are stable. It is well known that there is a certain coordination between the movements at the level of the lips and those at the level of the larynx. Larynx depression takes place as a countermeasure to the result of rounding and spreading in order to keep the axis of symmetry of the vocal tract unchanged (Wood, 1986). The modeling of these effects will be treated in a forthcoming study.

Dynamic aspects of the region model.

Representation of the vocal tract in regions with specific formant transition behaviour (FTB) leads to a specific command strategy : namely, transversal change of region areas instead of longitudinal constriction displacement. The morphology of the vocal tract and the tongue musculature seem adapted to such a strategy (Wood, 1979). The transversal region command strategy provides an inherent quantal longitudinal displacement of a constriction (figure 21). When two successive regions (D and $\overline{D}$ for example) are changed in order to produce a constriction displacement, the shape of the equivalent constriction is changed passing from low cross-sectional area (position 1) to higher value (position 2) before returning to low values (position 3).

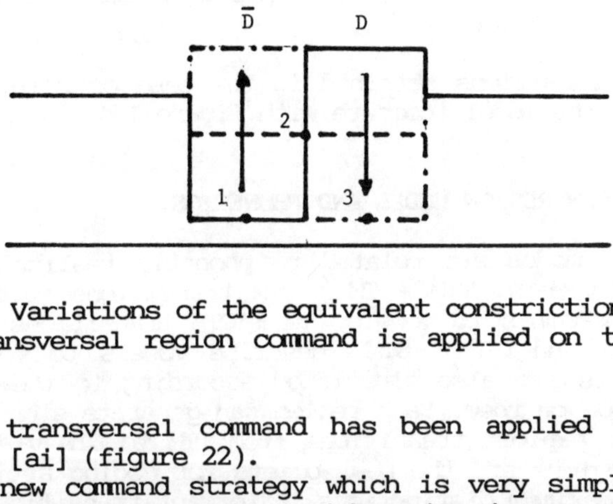

Figure 21. Variations of the equivalent constriction degree and place when a transversal region command is applied on two successive regions.

Such transversal command has been applied to obtain the VV transition [ai] (figure 22).

This new command strategy which is very simple to use in controlling the model parameters, inherently incorporates the classical strategy. Namely, transversal movement of regions around the neutral is equivalent to a certain longitudinal movement of a constriction.

The anti-symmetry property of the vocal tract is also well adapted for maximum acoustic effect obtained by synergetic movements: a tongue constriction in the front part of the vocal tract leads to a large cavity in the back part (tongue volume preservation) which induces a synergetic effect on formant frequency variations.

A jaw movement affects A, B, C, D region areas. Such a movement mainly involves a first formant variation.

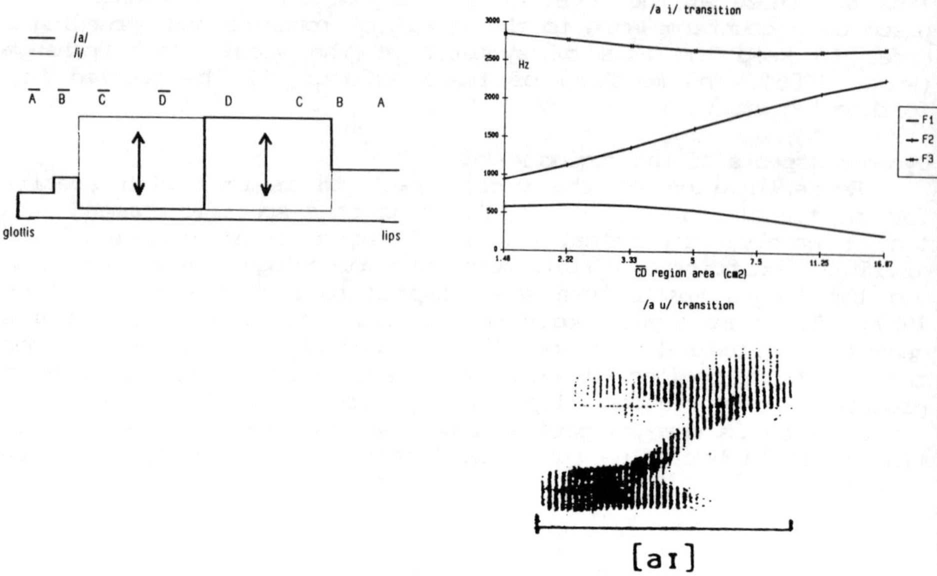

Figure 22. Formant transitions obtained in the case of natural diphthong [ai], and with the model (compare with figure 14).

II.5. RELATIONS BETWEEN REGION MODEL AND PHONETICS.

The regions and modes are related to phonetic realizations for both consonants and vowels. TTM + TM + OTM transitions successively describe plosive consonants to vowels, TM + OTM transitions describe fricatives to vowels and OTM + OTM describe vowels to vowels and diphthongs. Consonants are also classified according to the distinctive regions that produce them. Each region can generate stops (voiced or unvoiced) when we exploit transitions from TTM to OTM via the TM. The mid-points of regions are the best points for coding articulation movements producing maximum distinctions in formant transitions around the neutral (figure 23). These mid-points are the well known closure points for producing plosives. As such, we have [p,b] at the A region, [t,d] at the B region, etc. When the vowel is other than 'schwa' (neutral position), movements of more than one region are involved. Even before closure is achieved for a stop consonant, the tongue has begun movement towards a position appropriate for the following vowel. Consequently, several regions change area together, resulting in superposition of the effects of all region changes.

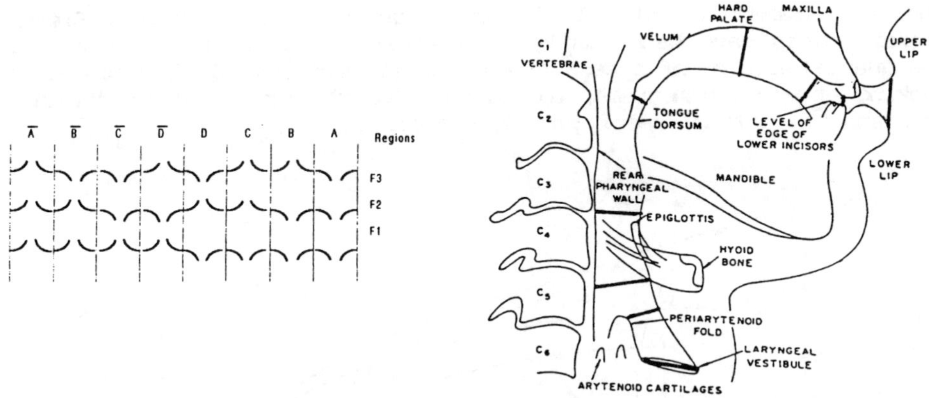

Figure 23. The schematized eight possible three formant transitions and their regions of production when closing and opening the region from a neutral configuration. On the right, the eight region boundaries drawn on a schematized vocal tract (from Perkell, 1969).

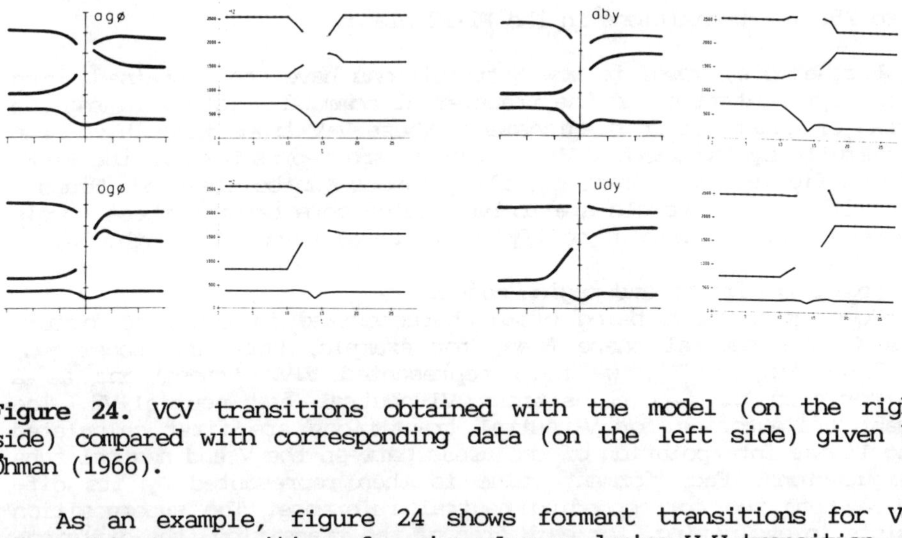

Figure 24. VCV transitions obtained with the model (on the right side) compared with corresponding data (on the left side) given by Öhman (1966).

As an example, figure 24 shows formant transitions for VCV obtained by superposition of region closure during V-V transition. In this example, the beginning of the closure corresponds to the beginning of the vowel transition. The involved regions are moved at the same time. For comparison, we have also shown the transitions obtained by Öhman (1966) on natural speech.

The vowel configurations can be represented using regions. Regions C, D, $\bar{D}$, and $\bar{C}$ are well correlated to the tongue constriction locations observed by Wood (1979) for different languages and also by Harshman et al. (1977) (see figures 9 and 7 above). Eleven French vowels have been easily represented using the region concept. The quantal nature of speech production (Stevens, 1972) is noticed in looking at vowel nomograms, for example for the vowel [o] in the case of C and $\overline{C}$ regions (see figure 20 above).

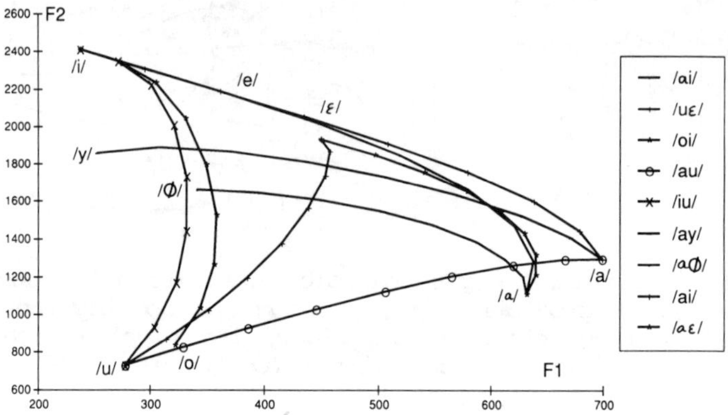

Figure 25. V-V transitions in the F1-F2 plane.

Furthermore, vowel to vowel transitions have been obtained using region representation and the transversal command strategy in regions instead of constriction displacement. These V-V transitions have been successfully synthesized. V-V transitions are represented in the F1-F2 plane on figure 25. Formant stability ranges can be observed. Transitions with rounded vowels are to be studied more because these vowels have been obtained without modifying the model physical length.

Superposition effects and region model.

Vowel production being often characterized in terms of distorsions from a neutral shape (see, for example, Harshman, Ladefoged, and Goldstein, 1977), we have represented V1V2 transitions as a superposition of two transitions V1-neutral and neutral-V2. The formant trajectories for V-neutral transitions are first calculated using linear interpolation of the areas between the V and neutral tube configurations. Each formant value is then represented by its difference from the corresponding neutral reference. The superposition consists in evaluating, for each step of the transition, the differential value of the formant as the sum of the differential values of V1-neutral and neutral-V2 formants. Figure 26 shows the results obtained in the F1-F2 plane for [ai], [iu], and [ua]. Vowel-neutral transitions [aə], [iə], and [uə] are also represented. For comparison, transitions obtained by linear interpolation of areas between V1 and V2 configurations are shown.

In the case of CV transitions, the superposition involves C-neutral and neutral-V transitions. Figure 27 shows an example for [Ca] where the consonant C is obtained by closing one of the regions A, B,

C, or D.

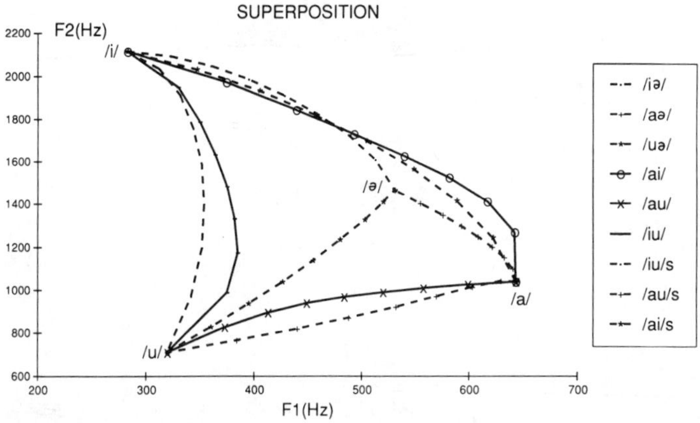

Figure 26. V1V2 transitions in the F1-F2 plane. Dotted lines : V-neutral transitions and V1V2 transitions obtained by superposition. Thick lines : transitions obtained by direct calculation.

CONCLUSIONS

A new model for the study of acoustic-articulatory-phonetic relations has been proposed which makes clear relationship between dynamic vocal tract configurations and their acoustic correlates. A new efficient coding of the area function has also been introduced, together with a speech production model based on regions and modes. The simple control strategy of the region model is believed to be capable of leading to productive research in speech dynamics. Examples of synthesized speech sounds exploiting the above mentioned relations have been successfully obtained. New tools have been proposed which are explanatory in education and speech research. The problem of synthesis-by-rules is simplified : formant transitions can be described in terms of region parameters. We hope that the new concept offers a step towards articulatory speech synthesis.

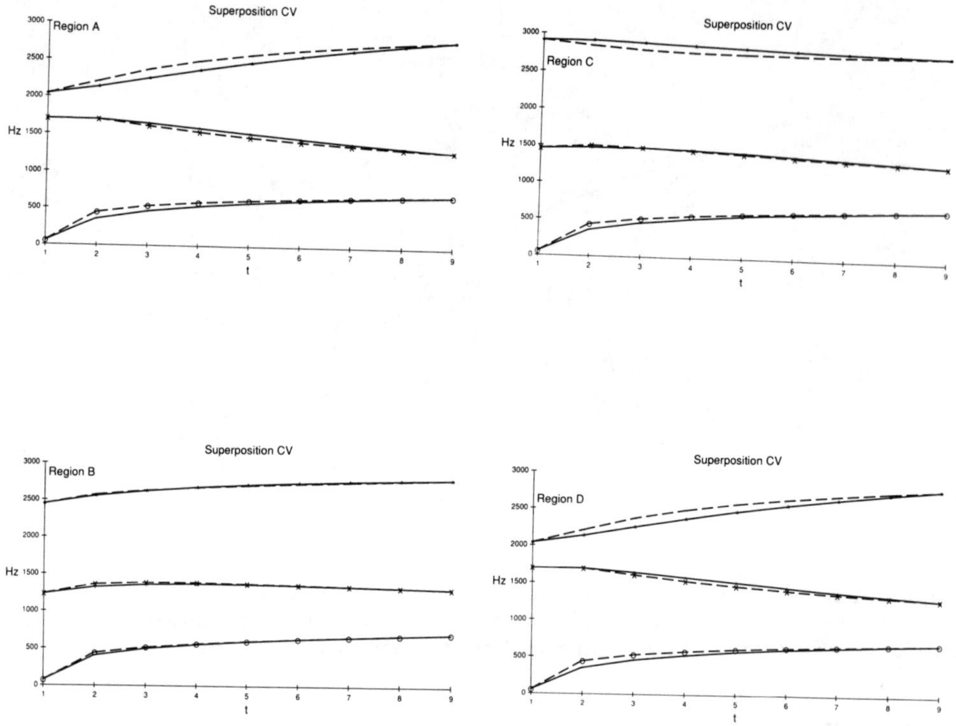

Figure 27. CV transitions in the time domain. Dotted lines : CV obtained by superposition. Thick lines : transitions obtained by direct calculation.

BIBLIOGRAPHY

Atal, B.S., Chang, J.J., Mathews, M.V., and Tukey, J.W., (1978) 'Inversion of articulatory-to-acoustic transformation in the vocal tract by a computer-sorting technique', J. Acoust. Soc. Am., 63, 1535-1555.

Badin, P., and Boë, L.J., (1986) 'Vocal tract vocalic nomograms : acoustic considerations. A crucial problem : formant convergence', Proc. of the XIth ICPhS, 2, 352-355.

Carré, R., and Mrayati, M., (1989) 'New concepts in articulatory-acoustic-phonetic relations. Perspectives and applications', Proc. of the Intern. Cong. on Acoustics, Speech and Signal Processing, Glassgow, 231-234.

Chiba, T., and Kajiyama, M., (1941) 'The vowel. Its nature and structure', (Tokyo).

Coker, C., (1968) 'Speech Synthesis with a parametric articulatory model,' Proc. Kyoto Speech Symposium.

Fant, G., (1960) 'Acoustic theory of speech production', Mouton, The Hague.

Fant, G., (1967) 'Kompendium i talöver föring', KTH Inst. för Talöverföring (in Swedish).

Fant, G., and Pauli, S., (1974) 'Spatial characteristics of vocal tract resonance modes', Speech Communication Seminar, Stockholm.

Fant, G., (1975) 'Vocal tract area and length perturbations', STL-QPSR, 4, 1-13.

Fant, G., (1980) 'The relations between area functions and the acoustic signal', Phonetica, 37, 55-86.

Fant, G., Lin, Q., and Badin, P., (1988) 'Speech production Models : Constraints and control strategy', 2d Joint Meeting of the Acoustical Societies of America and Japan. Poster NN6.

Firth, I.A., (1986) 'Modal analysis of a vocal tract,' STL-QPSR, 2-3, 1-12.

Harshman, R., Ladefoged, P., and Goldstein, L., (1977) 'Factor analysis of tongue shapes', J. of Acoust. Soc. of Am., 62, 693-707.

Heinz, J.M., (1967) 'Perturbation functions for the determination of vocal tract arrea functions from vocal tract eigenvalues', STL-QPSR, 1, 1-14.

Ladefoged, P., and Bladon, A., (1972) 'Attempts by human speakers to reproduce Fant's nomograms', Speech Communication, 1, 185-198.

Ladefoged, P., and Lindau, M., (1979) 'Where does the vocal tract end?', UCLA WWP, 45, 32-38.

Ladefoged, P., and Harshman, R., (1979) 'Formant frequencies and movements of the tongue', UCLA WWP, 45, 39-52.

Lindblom, B., and Sundberg, J., (1971) 'Acoustic consequences of lips, tongue, jaw and larynx movements', J. of Acoust. Soc. Am., 50, 1166-1179.

Maeda, S., (1979) 'Un modèle articulatoire base sur une étude acoustique', Bulletin de l'Institut de Phonétique de Grenoble, 8, 35-55.

Mermelstein, P., (1973) 'Articulatory model for the study of speech production', J. Acoust. Soc. Am., 53, 1070-1082.

Mrayati, M., and Carré, R., (1976) 'Relations entre la forme du conduit vocal et les caractéristiques acoustiques des voyelles françaises', Phonetica, 33, 285-306.

Mrayati, M., Carré, R., and Guérin, B., (1988) 'Distinctive regions and modes : a new theory of speech production', Speech Communication, 7, 257-286.

Mrayati, M., Carré, R., and Castelli, E., (1989) 'A new approach for speech dynamic studies', J. of Acoust. Soc. of Am., 85, S144.

Öhman, S.E.G., (1966) 'Coarticulation in VCV utterances : spectrographic measurements', J. Acoust. Soc. Am., 39, 151-168.

Perkell, J., (1969) 'Physiology of speech production. results and implications of a quantitative cineradiographic study', The MIT Press (Cambridge).

Schroeder, M.R., (1967) 'Determination of the geometry of the human vocal tract by acoustic measurements', J. Acoust. Soc. Am., 41, 1002-1010.

Stevens, K.N., and House, A.S., (1955) 'Development of a quantitative description of vowel articulation', J. Acoust. Soc. Am., 27, 484-493.

Stevens, K.N., (1972) 'The quantal nature of speech: evidence from articulatory-acoustic data', Human Communication: a Unified View (Mc Graw-Hill), 51-66.

Ungeheuer, G., (1962) 'Elemente einer akustischen Theory der Vokalarticulation', Springer-Verlag, Berlin.

Wood, S., (1979) 'A radiographic analysis of constriction locations for vowels', J. of Phonetics, 7, 25-43.

Wood, S., (1986) 'The acoustical significance of tongue, lip, and larynx maneuvers in rounded palatal vowels', J. of Acoust. Soc. Am., 80, 391-401.

EVIDENCE FOR NONLINEAR SOUND PRODUCTION MECHANISMS IN THE VOCAL TRACT

H. M. Teager, Sc.D. and S. M. Teager

Biomedical Engineering
Boston University School of Medicine
University Hospital
75 East Newton Street
Boston, Massachusetts 02118

ABSTRACT. Much of what speech scientists believe about the mechanisms of speech production and hearing rests less on an experimental base than on a centuries-old faith in linear mathematics. Based on experimental evidence we believe that the momentum waves, or the interactions of the inertia-laden flows leading to various modes of oscillation, within the vocal tract are neither passive nor acoustic. Measurements of flow within the vocal tract indicate that acoustic impedance, or the pressure-flow ratio, is violated. The pressure across any cross section of the tract is constant and does not exhibit the differentials expected from the markedly different separated flows across that same cross section. There has been little proof that the ear is primarily a frequency analyzer, or any solid explanation given for its extraordinary sensitivity. Finally, nonlinear processing techniques, that are less prey to Fourier artifacts, are described.

Introduction

For a number of years the scientific community has been looking at the vocal tract, hearing, and signal processing from a largely mathematical and parochial point of view. This narrow approach has resulted in some fairly significant errors. The objective of this paper is to reveal some of those errors, the experimental evidence that led to their discovery, and some of the conclusions based on our observations. This paper divides the subject into three parts. Part I discusses the vocal tract, Part II considers hearing, and Part III attempts to tie them together and discusses signal processing as it applies to speech.

W. J. Hardcastle and A. Marchal (eds.), Speech Production and Speech Modelling, 241–261.

Part I – The Vocal Tract

As background, there is substantial evidence that the operation of the vocal tract is neither linear nor passive, nor even acoustic. The data to support this assertion were first noted in experiments we ran over 25 years ago. In the early 1960's, when I was still a professor at MIT, we were working on machine aided cognition and man-machine interaction. I had worked on graphical input devices, a handwriting recognizing input tablet to be specific, but circumstances changed and keyboards were accepted as the ideal input device. I left visual models and decided to work on speech. The generally accepted opinion at that time was that the problem of speech recognition by machine would be solved in 20 years. As a starting point I began to look at speaker generated signals. Hours were spent at an oscilloscope as I made what my children called "woo-woo" sounds. After extensive observations it became clear that the signals I was examining could not possibly have come from a passive linear system, but rather were generated by a far more complex mechanism; but more on that later.

At that time there was also a flurry of activity in a number of laboratories to explore the possibility that divers, using helium, could produce recognizable speech. Groups at Bell Labs and in Sweden did some research in this area, but there was something disturbing in their reported results. Investigators had conducted a series of experiments with mixtures of gasses at high pressure, but there were no reports of anyone having used pure helium at normal atmospheric pressure just to see what would happen. The experiment is simple enough. Many a university physics instructor has inhaled a lung full of helium and generated "Donald Duck" noises to demonstrate the effects that result from the use of this much lighter gas, usually to the great delight of the freshmen. The velocity of sound in helium is three times greater than the velocity of sound in air. In our experiments however, we found that the formants, that is to say the speech frequencies, that were produced with nearly pure helium at atmospheric pressure, were shifted by a factor closer to 1.6 rather than the expected factor of 3.0. Moreover, we also found that the pitch of the speaker's voice also rose approximately by the same factor of 1.6.

The linear acoustic model would argue that all the resonances of a passive system should increase as the velocity increases. However what is not stated is that if the system were a passive mechanical one, such as the glottis, and analyzed as a mass and springs, the increase in resonance should *not* depend on the velocity of sound in the medium. Yet, as was just stated, the pitch of pure helium speech went up by a factor of 1.6. About that time, as we were pondering the problem, a salesman from DISA came and tried to sell us a hot-wire anemometer. For the pure fun of it, I stuck the hot-wire in my mouth and phonated. I had thought that the actual flows that occurred in the mouth during speech were relatively small and that they would be quite uniform. Much to my surprise, I found the following: if I took a single probe and measured the flow close to my tongue, using a storage scope to make sure that I was producing the same sound wave, I could observe one flow wave. If I repeated the measurement near the roof of my mouth, I saw an entirely different flow wave. Thus, with the vowel sound held constant two quite different flow waves were observed.

An example of this is illustrated in Figure 1 which shows three time traces for a vocalized 'ah' over one pitch period of approximately 6.1 ms duration. The top trace represents the output signal from a microphone placed 5" in front of the lips (axial) along with its baseline. The second trace, A in Figure 1, represents the flow measured 0.25" below the palate and 1.5" posterior to the lips with a single hot-wire normal to the flow. The third trace, B in Figure 1, represents the flow measured 0.25" above the tongue and 1.5" posterior to the lips (the separation between the tongue and hard palate being 0.75"). This experimental evidence indicated that something odd was going on.

First, the observed flows, basically two formants, looked as if they were 180 degrees out of phase which implied that the actual flow was switching back and forth between the tongue and the hard palate.

Second, the actual flow-to-pressure fluctuation measured inside the mouth was only 1% of the acoustic impedance. In other words the air flow was not traveling as an acoustic wave. The actual flow was some 100 times greater than the pressure would indicate by impedance considerations. To repeat, a factor of 100, and that is *not* negligible.

Third, the formant flows essentially stuck to the walls of the vocal tract. Most of the flow in the vocal tract, as far as it could be measured without doing violence to one's physiognomy, was along the roof of the mouth and along the surface of the tongue. Very little of interest was happening in between. What is actually happening is that the flow is switching between the roof of the mouth and the tongue. It is the same flow, moving up and down at the formant rate, switching about a thousand times a second. This first experiment however, was made with a single flow probe. A constant vowel sound was maintained and the probe was held in two positions within the mouth, a quarter of an inch apart. The data from this experiment was included in a proposal made to NIH in 1974. It was our first rude experience with the establishment. The proposal was rejected on the grounds that 100 years ago Helmholtz had proved that the vocal tract was a passive linear acoustic system and that this data obviously had to be nonsense.

Someone could obviously argue "Well, what about your experimental technique? These are not actually simultaneous flows; you are taking them at two different times." But our data has held up. We were able to obtain additional hot-wire anemometers and put them together in an array. Later as the arrays got fancier, the signals were recorded digitally and processed on a computer. Figure 2 illustrates the results of one of those experiments.

Figure 2 has two parts, 2-a and 2-b. The first part, Figure 2-a, presents four groups of consecutive time plots corresponding to approximately 240 pitch periods of the sustained voiced sound 'eh'. The time traces consist basically of sixty groups of four adjacent pitch periods, being averaged after alignment of their starting times, and then stacked one above the other to give a complete time history of that trace. The four 'stacks' of traces correspond respectively from left to right to 1) top left flow, 2) top right flow, 3) bottom flow, and 4) external sound. The scale for the top axial flows ranged from 0.2 to 10 feet per second and the bottom flow from 0.2 to 5 feet per second, in each pitch period. The flows were always positive and outwards, i.e. towards the front of the mouth. In order to observe what differences there might be in the flow-time traces the probe holding the three hot-wire anemometers was positioned in the rear of the mouth and moved slowly from left to right during the 240 pitch periods recorded.

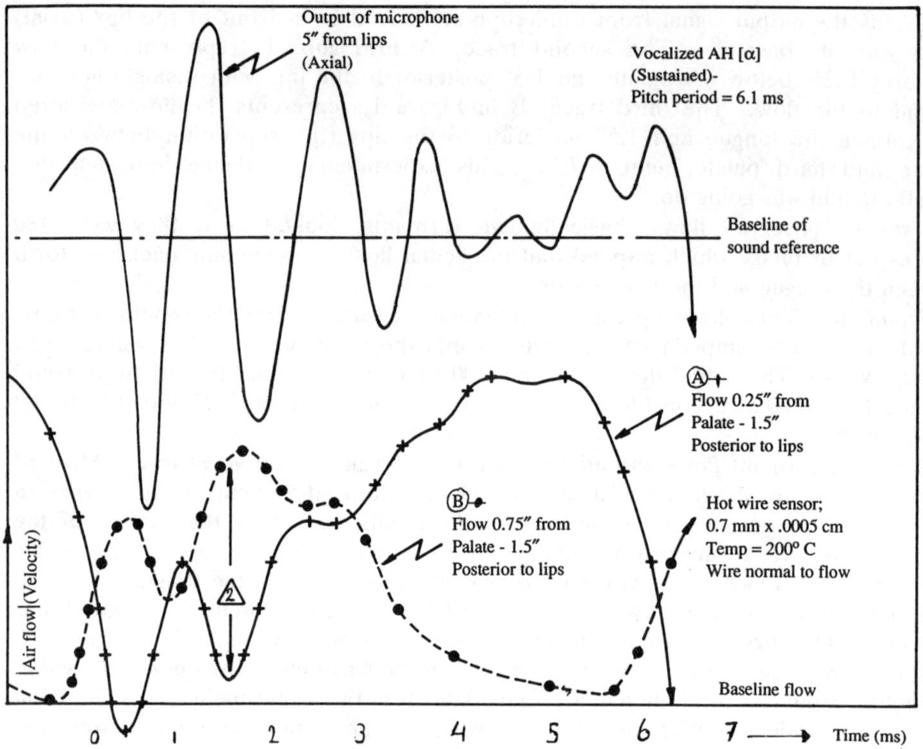

-Experimental Verification of Separated Flow

1. Simultaneous flows Ⓐ and Ⓑ, (Synchronized with sound wave) are non-similar.

2. Format perturbations of Ⓐ and Ⓑ are exactly 180° out of phase.

3. Flow at exit of nose and mouth (not shown) are similar to, and in phase with Ⓐ, but format perturbations are far smaller (10% as opposed to 50% of peak flow deviation).

Interpretation – The data are consistent with a pulsed jet whose average flow axis is close to the palate but whose direction is perturbed at the formant frequency.

Figure 1. Three time traces for a vocalized vowel ‘ah’ produced by a male speaker. The traces provide experimental verification of separated flow; original data and figure taken from a 1974 NIH proposal.

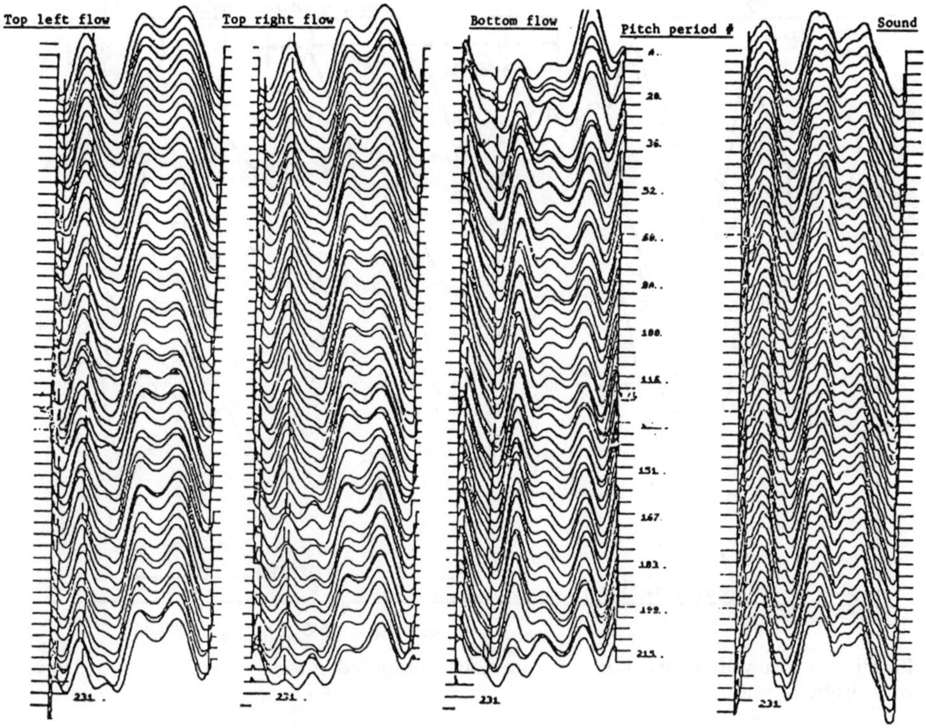

Figure 2-a. Internal mouth flows and generated sound for 'eh'. Back of mouth, left to right trajectory. Complete data array of flows and sound.

The second part, Figure 2-b, shows five time traces taken simultaneously. The two pitch periods illustrated were selected from the same data on 'eh' presented in Figure 2a. These pitch periods were selected from the data at the point where the probe was located in the center of the rear of the mouth. The traces from top to bottom are 1) the raw pressure data, 2) the computed integral of the raw pressure data, which would correspond directly to the flow for the linear acoustic model, 3) the top left flow, 4) the top right flow, and 5) the bottom flow. It is important to note that there is no evidence of the high frequency formant in the bottom flow. In this illustration we have an example of a flow within the mouth that does not contain a formant found outside the mouth. In other cases we have found the converse to be true, namely formants in the inside flows that do not exist outside the mouth. For this sound, which we believe is generated from collisions at the front of the mouth, the first formant is 660 Hz and the second formant is 1800 Hz.

Data collected later revealed other interesting phenomena. First, when the hot-wire sensors that could measure both normal flows as well as radial and axial flows were used, it was found that in the case of some formants, there was a rotation superimposed on the velocity out, i.e. towards the mouth. *The velocity out was always*

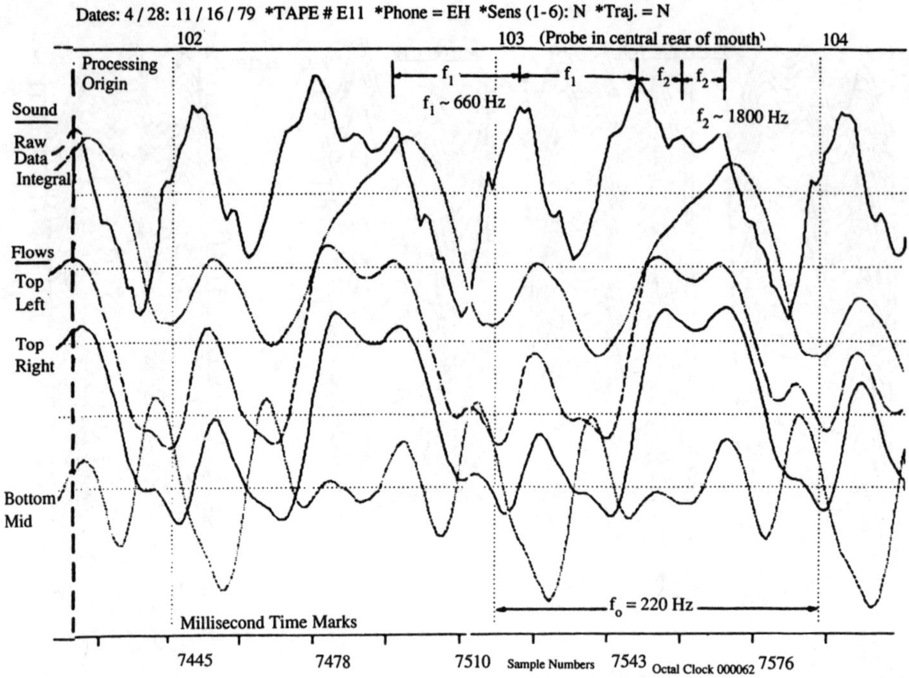

Figure 2-b. Simultaneous plots of sound, its computed integral and measured flows with probe in mid trajectory.

positive. In other words we have found that large vortexes of an axial or radial type can, and do, exist in the vocal tract. These swirling flows within the mouth were of the same order of magnitude as the flows that were coming out of the mouth. Similar experimental data for the voiced sound 'ih' were reported in [1].

Here, we should perhaps include a few words about hot-wire anemometers since their accuracy is always questioned first. Hot-wire anemometers have been used by aerodynamicists to make measurements of flow amplitude within a tenth of a percent out to frequencies of 100 kHz. For our work, we were willing to settle for 5% accuracies and a bandwidth out to 5 kHz. Both of those limits are conservative in terms of the frequencies and amplitudes that the technique is capable of measuring.

Where then is the anomaly? In the beginning of this paper we asserted that obviously errors have been made in basic assumptions. We could not possibly be dealing with an acoustic wave. Let me remind you of the factor of 100 discrepancy. For a typical pulsating flow in the mouth, we measured flows on the order of 10 feet per second. These flows ranged from 0.1 to 10.0 feet per second within a single pitch period. According to the impedance relationship, the pressures corresponding to those flows would be of the order of 140 decibels! The actual measured pressures were about 100 times smaller than that, which means that we have a wave that is not traveling by compression.

Everyone who is familiar with Professor P. M. Morse's book on theoretical acoustics [2] will know that it contains a wave equation describing acoustics, namely

$$\frac{\partial^2 p}{\partial x^2} = \frac{1}{c^2}\frac{\partial^2 p}{\partial t^2} \quad \text{where } c^2 = \frac{1}{k\rho}$$

Obviously this is a one-dimensional equation. If you wish you can include your ∇^2 gradient operator term, but fundamentally this is the wave equation for acoustics and it is very similar to the general wave equation. In fact it is identical in form to Laplace's wave equation for electromagnetic theory; therefore the same machinery can be used for its solution.

The electromagnetic equation however is derived from two other equations. The first is the continuity equation that says you cannot create material, $k\,\partial p/\partial t = -\partial u/\partial x$. The second, namely, $\partial p/\partial x = -\rho\,\partial u/\partial t$, is the dynamic equation derivable from the familiar $f = ma$ which states that force equals mass times acceleration. There was an error in the second equation that eluded Morse [2] when he wrote his first book. In his second book, Morse and Ingard [3], Ingard caught the error, namely that the two equations are actually these:

$$k\frac{\partial p}{\partial t} = -\frac{\partial u}{\partial x} - ku\frac{\partial p}{\partial x}$$

where the last term is a convection term and

$$\frac{\partial p}{\partial x} = -\rho\left(\frac{\partial u}{\partial t} + u\frac{\partial u}{\partial x}\right)$$

That last convection term, $u\,\partial u/\partial x$, however is very important. $f = ma$ is not a partial derivative; it must be a total derivative, which means that there should be a convection term in both equations. These two equations hold in the presence of an average flow, as is the case in speech. Thus we are dealing with a nonlinear set of equations.

Two observations in our data were covered too quickly; one is that the pressure fluctuations were 1% of what we thought they should be. The other is that there was no differential pressure across any section where the flow was separated and switching rapidly. One would expect to see a difference in pressure at such a cross section but we observed none, or at least there was none that we could detect within the accuracy of our measurements. *This means then that there is no* $\partial p/\partial x$ *driving the* $\partial u/\partial t$. This is not so surprising. An air stream with lots of velocity possesses a considerable amount of energy. It does not need to store up energy by compression; it can propagate just by giving up some of its available kinetic energy.

We suggest the following regarding the solution of these equations. They predict four kinds of waves, positive and negative waves going with and against the flow. That implies four combinations, depending on the size of the $\partial u/\partial x$ term. We have not solved this set of nonlinear partial differential equations; instead we have taken an experimental approach. It has been very, very difficult to get any support whatsoever for this work. When you tell people your acoustic equations should be reexamined, they say nonsense and they wave Morse's physics book at you. Without resources one cannot run experiments. However, even with our limited number of experiments on vowel sounds we found that the flow pattern for each vowel sound was unique. There were different flow patterns for 'ee', 'ih', 'eh', 'ooh', 'oh'; each pattern was distinctive. We also found combinations of separated flows, axial vortexes, radial vortexes and a variety of interactions between them.

Let us consider the possible dynamics inside a system with a large amount of available kinetic energy. Let us also accept that in speech production, we are dealing with an aerodynamic system. Viewed experimentally, from the pressure to flow aspect, it is not the same as an acoustic system, but rather has much in common with the dynamics of flight and aerodynamic instabilities.

Some three years ago, as I learned how to fly, I found that one encounters many aerodynamic instabilities when one is flying an airplane. The most obvious one is if the air separates from the wing, the plane stalls and goes into a spin. Other things can also happen; fly too fast in the wrong regime and an aileron or a tail member will start fluttering. My aerodynamics friends all admit that these things do happen. But when I ask "do you understand what is going on?" they shake their heads and say "no, we do not have to." "What we have to do is to produce an airplane where that will not happen." The flutter instability problem has never been really studied .

So far we have determined five major instabilities or modes of oscillation in the course of our work in speech generation. The first is the whistle, a jet that is tangentially exciting the cavity. The second is a jet that is along the inside wall of the cavity. The third is a jet with a swirl inside a cavity. The fourth is a jet with a radial vortex, and the fifth is the old Aeolean instability. Here we discuss only two of the five instabilities.

The first model we studied very carefully was an ordinary policeman's whistle, with a tangential air jet. It has a little cavity usually with a pea, or ball, inside, but the pea is not necessary. We analyzed the physics from measurements with hot-wire anemometers, pressure sensors, etc., to find out what was going on. It turned out that these whistles were relaxation oscillators. There was a vortex inside, but that was of second-order interest. What was really happening was that when the pressure inside the chamber decreased, the jet was deflected into the cavity. Cavity pressure then built up, and blew the jet out again. The pressure increased and decreased at unequal rates. Thus a policeman's whistle is a relaxation oscillator with a significant second harmonic. Furthermore the vortex, inside the whistle, amplitude modulates the sound at a rate determined by the precession rate of the vortex inside. The pea, if included, changes the precession rate in a rather random fashion that makes the whistle sound unique.

When we took our whistle model and put in the equations for what would happen when it was blown with helium, or with a high density gas, we calculated that in the case of helium the pitch of that whistle should go up by a factor of about 1.6, and indeed the physical experiment showed that it did. We never did run the experiment for high density gasses. Unfortunately this is the only model we analyzed in detail.

An Aeolian instability is generated by air flowing past a single wire and producing a periodic wave of pressure and flow downstream. Dr. Theodore von Kármán looked at that problem about 40 years ago, but avoided its dynamics. What he did instead was to say, "we will obtain, in the case of a large enough air flow, a system that is shedding vortexes downstream." He then predicted what the optimum, or stable, spacing was for those vortexes that were shed. He knew very well that he was avoiding a very difficult dynamics problem, namely the time behavior of the vortexes bound behind the wire. The bound vortexes are not shed for low flow velocities. At low frequencies, however, sine wave pressure and flow fluctuations are generated.

Each one of the five major instabilities represents a different set of dynamic equations. They lead to regenerative oscillations. With a jet of air flowing through a cavity one cannot have a passive system. Such a system is not passive, but rather is an oscillator. The reason is that a jet is very, very sensitive to the surrounding pressure. People have known this since Tyndall, in the mid 1800's, when this effect was utilized as a "flame manometer". They used a jet of gas as a sound detector, because the jet was very sensitive to the ambient pressure. Moreover, a jet sweeping through a cavity entrains a great deal of air and exhausts that cavity. When it exhausts the air on one side of the cavity, the jet switches to the other side of that cavity. There was a lore on switching amplifiers using jets and cavities called "fluidics." It died about the time I was a graduate student at MIT because integrated circuits came along. In every case, very little was understood about either these instabilities or the 'Coanda' effect.

Professor Henri Pierre Bouasse of the University of Toulouse, not too far from here, is still remembered for his superb series of classical physics texts. He spent most of his productive professional life cataloguing and studying fluid-acoustic effects in wind instruments. His experiments and conclusions appear in his remarkable, but now neglected, books. The fundamental conclusions and the framework for our work on airflow were presaged by Bouasse, and his works merit much more careful study than they have received [4-8].

A few additional comments on some of the characteristic inertial effects of jets and axial vortexes are in order. Sound waves are assumed to be able to travel freely in any direction. Jets and vortexes cannot. A jet of air inside a cavity with an inlet and an outlet, such as the mouth, acts as a barrier to that cavity's outlet. An axial vortex in a similar cavity can also act as a barrier to that cavity's outlet, but in a different manner than the jet. The swirling axial vortex acts as a nonlinear plug. When the pressure inside the cavity is increased, the vortex is compressed cutting off the flow. When the pressure inside the cavity is decreased, the vortex expands allowing more flow. This is exactly the description of a positive feedback system which will oscillate under almost any circumstance, and indeed does.

Sound waves are interesting in their own right, but it is the sense of hearing which allows us to correlate the nonlinear production mechanisms with perception. Let us now discuss hearing in general terms.

Part II – Hearing

If you are willing to grant that there is solid evidence that the vocal tract is neither passive, nor linear, and not even acoustic, we can continue to talk about hearing. Where is the evidence that the ear is a Fourier frequency analyzer? The answer is that there is none. It was stated as an assertion by Ohm a long time ago [9]; Ohm's law of acoustics states that you could take any periodic wave and break it down into its frequency components. That is in the realm of mathematics. There is no argument that such an analysis can be done. But is that what the ear is doing? Seebeck, who worked with sirens about the same time as Ohm, demonstrated that he could generate a complex tone which is missing its pitch but nevertheless the pitch could still be heard. He published a paper in German [10] that I've been trying to have translated for several years. It turns out that it is in archaic German and very few people can translate it. The substance of this paper however was that one hears periodicities rather than multiple pure tones. Seebeck was very strongly criticized at the time.

Later on, a marvelous combination of skills in mathematics, physics, anatomy, and music came together in the person of Hermann Helmholtz. He laid down the principles that have lasted for more than one hundred years. Helmholtz said the ear is a frequency analyzer, except for that part of it which obviously must be different in order to be able to work with transient sounds. Helmholtz recognized that there had to be something other than pure tones if the ear was going to be able to do anything with a transient sound since a short transient contains all frequencies. He then proceeded to write a book where he accepted the existence of tonal analyzers. The only remaining question was, where were they? He said that the hair cells were the tonal analyzers. Each hair cell is a resonator that resonates at a different frequency. His book on this topic, *On the Sensations of Tone* [11], is a classic.

Back in the twenties and thirties, von Békésy had to take account of the fact that nerve response was too slow. There was an additional difficulty in that a hair in a fluid cannot overcome damping and resonate. He came up with the "place theory" of where the ear measured frequencies, but again he accepted the fundamental premise that the ear is, in fact, only interested in pure-tone frequencies. Let me repeat that I do not know of anyone who has ever run a definitive experiment that proved that the ear was only a tonal analyzer.

Let us now look at other evidence. More recently, in the late 1960's, a very competent professor of anatomy at the Boston University School of Medicine, the late Dr. Elizabeth Moyer, helped us run some experiments on the ear. We placed a piece of temporal bone, which includes the middle and inner ear, in liquid nitrogen at -85 degrees F; *the bone did not crack.* The bone around the cochlea is very hard; it is filled with water along with something akin to antifreeze. We know that water expands about 10% when it freezes, but in this experiment, the round window in the temporal bone did not even bulge. To us, this implied that the spiral ligament, or the cochlea, had compressed.

The cochlea is shaped like a directional antenna. It is at 45 degrees and has three turns, yet it is always analyzed as if it were stretched out in a straight line; it is generally accepted that this assumption does not make a difference. However, the cochlea is *not* stretched out; it is coiled. And there is a good reason why it should be coiled, especially if it is compressible. There are other examples in nature of a tube whose membrane does not split when it is frozen and undergoes a change in volume, the needles of many conifers for instance.

It has always been assumed that the energy in the ear is carried as an acoustic wave in the bulk motion of the fluid. We believe however that most of the energy is traveling as a wave along the inside cochlear surface. Moreover, an examination of the energy relationships indicates that there must be an amplification mechanism there that we do not understand. There is no other way of accounting for the amount of energy inside the cochlea at the threshold of hearing. An interesting observation from the field of medicine should be included here. If a patient is overdosed with kanamycin (sulfate) three things stop functioning simultaneously: 1) the cilia that clean the lungs, 2) the tubules in the kidneys, and 3) the sense of hearing, the patient becomes deaf. The only thing those three organs have in common is that they all have ciliated motile cells. Our hunch is that the two or more rows of outer hair cells in the cochlea are pumping fluid. Recently Brownell *et al.* discussed similar observations [12]. Those outer hair cells might be setting up their own vortex which would then act as an amplifier. Nature needs a low-noise amplifier if it is to measure deflections on the order of one hundredth of an angstrom unit. We are dealing with deflections of that order at the threshold of hearing. Once again, we may have hunches but we do not really understand the dynamics of what is going on inside a real cochlea.

From a design standpoint, it is obvious that ears, or the sense of hearing, exist for survival. Every mammal, for example, has a very good pair of ears and the object is to extract information about threats, opportunities, and neutral sounds from the environment or from pressure fluctuations. That means that there has to be a fair amount of analysis. Nature does not give rewards for being inefficient. Any system that does not work very well has its fate sealed. It will be replaced by another system that works a little bit better.

Something else that should be mentioned à propos of hearing is the case of a bird that can hear better than we can in terms of its 2–40,000 Hz range of hearing, yet it does not even have a cochlea. It has perhaps one square millimeter of sensory surface in each ear. There is *no* way that that system can be doing a mechanical frequency selection. Yet birds sing, and they obviously hear what they sing because they learn from each other and can produce modulations in pitch and amplitude that are far greater than anything we can even hear. This is better observed when a bird song is recorded and then slowed down.

Let us discuss the types of noises that exist in nature. We repeat, there are *no* pure tones in nature; the tuning fork and the vibrating string are very rare things that man invented. They did not exist until man made them. What does exist in nature? We have things falling, things breaking, things rubbing, things hitting, transients every one. Some of these transients are interesting in that they are repetitive. If a repetitive transient sound is broken down with a Fourier analysis, little that is recognizable will be obtained. Fourier analysis makes sense only when one has stationary, periodic signals which is definitely not the case for speech signals with all their inherent variability and modulations.

To extract the information contained in the average transient noise, one does not need to break it down into an infinite number of frequency components or harmonics. What does one really want to know? First, one would like to know how much energy was involved in the process that *produced* that noise. A loud crash gets more attention than a soft tap. The soft tap might be more interesting from a frequency point of view but the crash was produced by more energy. For its survival value, as living systems we are more interested in the energy that produced the loud crash than we are in the actual energy that reached our ears in the sound wave. This means that we are interested in *both* the square of the frequency *and* the square of the amplitude of the sound wave. In any physical process which produces sound, the energy that is tied up in that production process goes as the square of both these quantities [13]. We are also interested in the mode of oscillation. It turns out that most transient sounds, examined closely enough, have a structure. They have both a modal structure as well as an amplitude structure. The amplitude envelope and modal frequency structures tell us a great deal about the processes that produced the transients. Therefore, we would assert that the ear is not primarily a Fourier analyzer. It does not care about an infinite number of frequency components. What the ear is concerned with however, are the energy modulations that were involved in producing the transient.

These energy modulations can be extracted with a very simple system, if one could do three things. First, filter the sound, so that one is not dealing with widely ranging frequency components. Second, demodulate the result. Third, do correlations to find out what is going on. We would submit that those are the three things that are going on in the ear and the brain. It is not necessary, nor even desirable, to take a periodic or an aperiodic wave, multiply it by all the possible sines and cosines, and then integrate and average the result as required for a Fourier analysis. One would be smearing and destroying the very information that one is trying to extract. One loses the modulations in a Fourier analysis; they are gone.

Recall what we said earlier in the case of the vocal tract, namely that we are producing sounds with a nonlinear oscillatory system. Nonlinear systems have some very interesting aspects. First, they mode-lock. You can not have all possible frequencies being produced at the same time. The modes will tend to space themselves apart by at least a factor of two in frequency. If they are closer than that they will coalesce. Second, energy can be transferred. Assume we have a tightly stretched string and we pluck it. The partial differential equation describing the vibrations of that string is a familiar one. In the case of constant tension, the differential equation can be easily solved. However, it turns out that a very important term has been neglected. Namely, as the string deflects, its tension is changing so that its position at any time couples into all modes of oscillation. What is the result? The high frequency oscillation damps out quickly, because it is responsible for stretching the string the most for a given amplitude. On every cycle that energy couples back into the lower frequency modes. This happens with a drum, it happens with a stretched string, and it happens with any arbitrary natural sound. Third, and perhaps the most important point of all, is that there is no way we know of where people are able to look at a Fourier analysis of a signal and say whether that signal was produced by an active system or by a passive one. Fourier analysis is oblivious to that kind of distinction. Take spectra, and you will see spectra, but you will not know whether those spectra were produced by an active or by a passive system.

Part III – Speech, Hearing, and Related Signal Processing

Let us now return to discussing speech and tie speech and hearing together. Although the human vocal apparatus is capable of producing a wealth of sounds, it is interesting to note which sounds are actually selected for speech communication. Other than some Canary Islanders who use whistling to communicate over long distances, the phonemic sounds that are used by mankind in different cultures around the world seem to be pretty much the same. This amazing fact must mean that there has to be something unique and distinctive about the sounds that are actually chosen. After all, we do have a choice of using anything we please. Why then do we particularly pick the ones we use?

Let us review a few definitions. Figure 3-a is taken from the *Handbook of Speech Pathology and Audiology* edited by L. E. Travis [14]. It is a good example of a speech waveform and represents three pitch periods of a 'zuh', the voiced sibilant (z). It illustrates the high frequency s-type of sound associated with the 'zuh'; the pitch period is obvious.

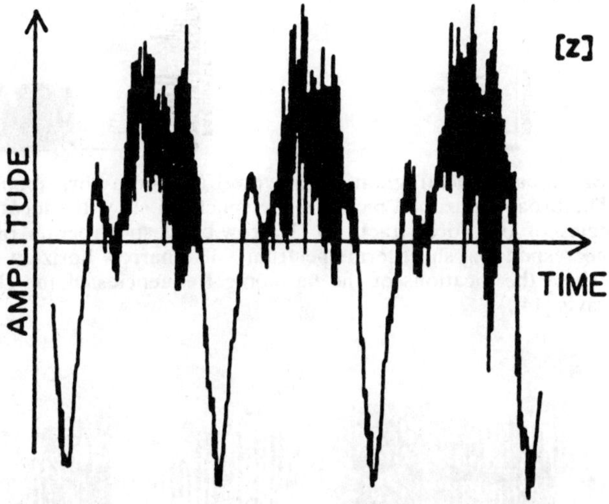

Figure 3-a. The acoustic speech waveform of 3 periods of a production of the voiced sibilant (z). The positive portions of the cycles tend to display more friction than the negative portions. (From Travis [14].)

In Figure 3-b, A and B are spectrograms for the vowel (a). The frequency bands where the energy appears to be concentrated for this, and other sounds, can be found by breaking the sound wave down in the conventional manner, namely into its Fourier components. If the decomposition is done with a set of wide bandpass filters, then wide frequency bands of energy will be found. On the other hand if a set of narrow bandpass filters is used, then energy bands all across the spectrum will be found. This can clearly be seen in Figures 3-b A and 3-b B. Anyone who has taken a 1024 point Fourier transform of a speech wave learns very quickly that all the frequencies and harmonics will be present.

Figure 3-c illustrates a broad-band sound spectrogram of the word "primed". The numerous horizontal bands are defined as the formants while the vertical striations correspond to the pitch periods. This example very clearly shows that sometimes the definition of what is a formant and what is a frequency band becomes very hazy. For example, what is going on near the upper end of the spectrum above the diphthong 'ai' shown in Figure 3-c? Is it one very broad formant or are there 1, 2, 3, 4, or 5 frequency bands? In the same figure, does the area above 'r' illustrate two formants or 4 frequency bands?

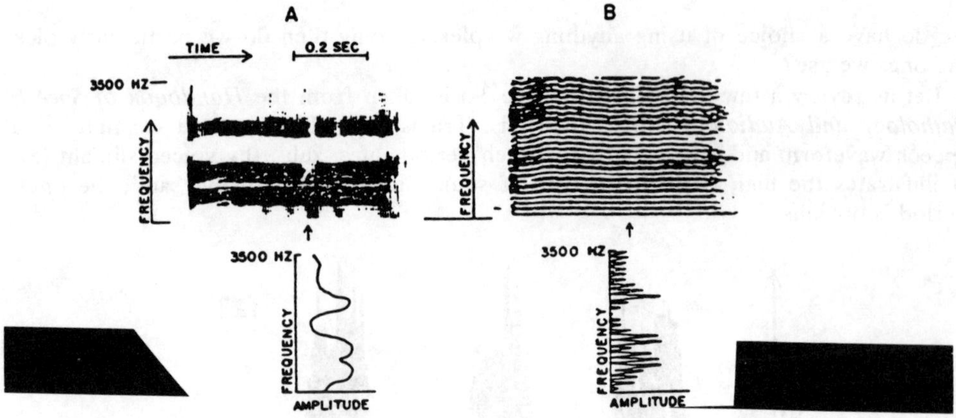

Figure 3-b. A. Broad-band spectrogram of the vowel (a) and a corresponding short-term spectrum. The broad horizontal bands in the spectrum show the locations of the resonant frequencies of the vocal tract. B. Narrow-band spectrogram of the same vowel (a) and a corresponding short-term spectrum. The narrow horizontal curves in the spectrogram show the locations of the harmonic frequencies of the glottal voice source. (From Travis [14].)

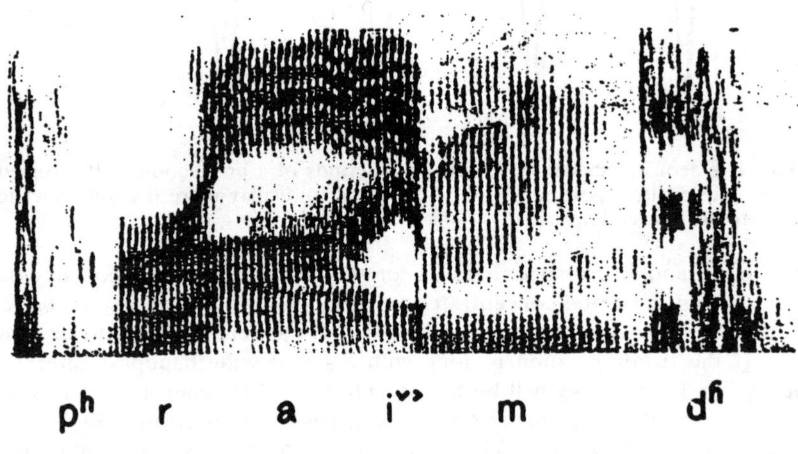

Figure 3-c. Broad-band sound spectrogram of the word "primed". (From Travis [14].) Note width and uncertain edges of the formant bands. Parsing is partial. Transition sounds are not indicated. Onset, peak, and end time are needed for each sound.

One fact became very clear in our early research namely that no pitch periods were ever the same twice. Pitch periods vary from being slightly different to being very different, but they are always different.

Since the time of Helmholtz, the assumption has been that vowels are a combination of two pure tones. Given two pure tones, a vowel is produced. Peterson and Barney [15] published their classic work on English vowels in 1952. They had many voices enunciate each of ten different vowels. Various formant frequencies were estimated for each of the chosen examples and the selected raw data was plotted in the formant-1-versus-formant-2 plane. Each vowel sound seemed to fall into a different region. Figure 4-a, which shows this data clearly, is also from Travis [14]. Had Peterson and Barney plotted all the continuants, the delineation between vowels would have been lost. The 'm', 'n', 'l', 'r', 'ai' and other sounds were not plotted; these and the rest of the omissions are listed in Table I along with the ten vowels included.

Table I

Glossary:		Missing:	
i	heed	ai	nine
I	hid	oi	boy
ϵ	head	o	boat
æ	had	ay	hay
α	hod	l	L
⊃	hawed	m	M
U	hood	n	N
u	who'd	r	force
Λ	hud		
ɝ	heard		

If each speaker's vowels are averaged over a period of time and then plotted in a formant-1-vs-formant-2 plane they should lie in a loop. Peterson noted this [16] and it is illustrated in Figure 4-b again from Travis [14].

In 1948 Licklider [17] showed that if one only listened to either the higher (above 1900 Hz) or the lower (below 1900 Hz) frequencies of a sound, most of the vowel sounds could be identified with only *one* formant, if that formant had a wide enough band. In other words two formants are not necessary. We have repeated the experiment many times and it has held true. The front vowels 'e', 'ih', 'eh', etc. can be identified with one formant only. This leads to the interesting question, namely, where is the information?

Around the early 1970's, some researchers examined the vowel recognition problem and found that when an actual word was sounded, the ratio of formants was not what the vowel loop model would predict. Therefore the conclusion was drawn that phonemes do not really exist and that recognition takes place at the word or syllable level. If a word composed of vowels and continuants is listened to piece by piece, almost every sound within that word can be distinguished. When one puts those sounds back together, they will make up the word spoken. The actual word may be perceived differently since every speaker pronounces words in a different manner from every other speaker in much the same way that everyone has different handwriting. People make the same speech sounds; however, in pronouncing a particular word they might not use the same sequence of phonemes. The

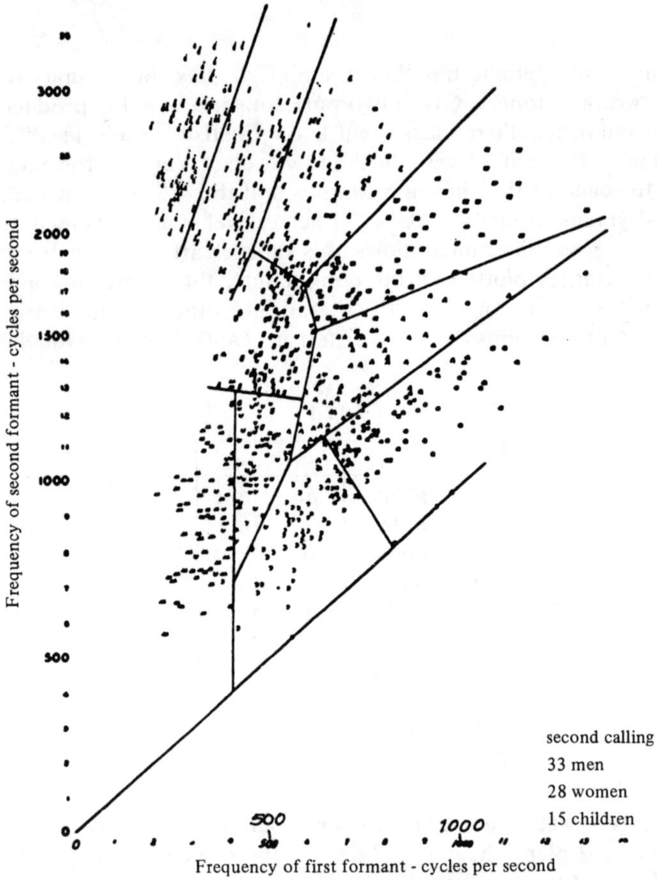

Figure 4-a. The frequency of the second formant versus the frequency of the first formant for the ten vowels produced in *hVd* series by 76 speakers. The lines have been drawn to provide approximately optimum separation among the vowels according to the two formants, when the vowels are classified according to the lists from which the speakers read. *Peterson and Barney*. (From Travis[14].)

phenomenon that distinguishes the sound of a word or of a phoneme is not the pure tones in that word or phoneme, but rather it is the modulations of those tones. These modulations can be both tracked and quantified.

Our approach to processing the speech waveform is unconventional when compared to the traditional spectral methods that employ the Fourier transform. Since they require integrating, or effectively averaging, over a long period of time, Fourier transforms are incapable of recovering the rapid modulations of speech. As was stated earlier in the discussion on hearing, our approach is to first locate the modes of oscillation, then to adaptively bandpass filter the speech, and finally to

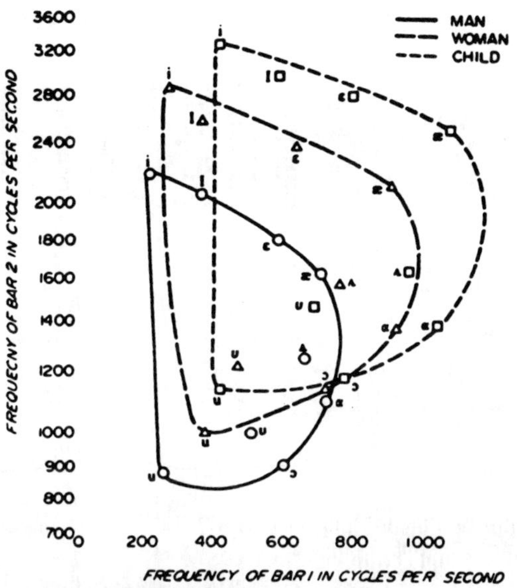

Figure 4-b. Loops constructed with the frequency of the second formant plotted against the frequency of the first formant for vowels produced by a man, a woman, and a child. *Peterson, 1951b*. (From Travis [14].)

demodulate the results in order to characterize and identify the speech sound(s). A detailed description of the algorithms used to perform these different filtering and demodulation operations is beyond the scope of this paper. However a typical example of the application of these methods to speech is shown and described in Figure 5.

Figure 5 illustrates a sound wave of about six consecutive pitch periods of the mid 'ai' part of the word "nine". It is shown along with its analysis. This figure, which is the direct computer output from one of our analysis programs, requires some description. The five traces shown on the left-hand side represent filtered time traces of pressure. From top to bottom, they are 1) the full band sound wave, 2) the lowest frequency bandpass filtered wave, 3) the mid-frequency bandpass filtered wave, 4) the high frequency bandpass filtered wave, and 5) the noise residue. The filters used to obtain this output were 'soft' in the sense that they were highly damped and did not produce any lasting oscillations. Although these filters are linear, they are unconventional. If one uses very sharp narrow-band filters to separate the modes of oscillation prior to demodulation, then the response of those filters to a pulse of energy will be dominated by the transient ringing, or lasting oscillation, of the filters. Instead, it is best to use wide-band filters that are as narrow as possible without destroying or rearranging the energy in the original wave. Filters with Gaussian-like responses work very well and are the types of filters that were actually used.

To illustrate that the different modes of oscillation have different modulation patterns, a nonlinear demodulation algorithm was applied to the output of each of the

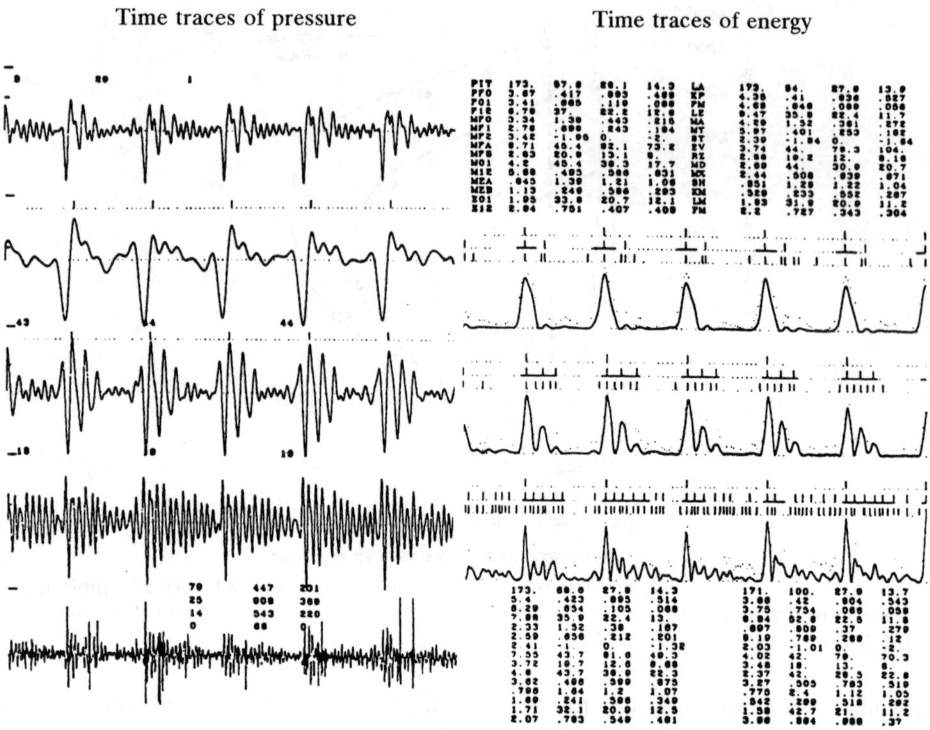

Figure 5. Direct computer output from an analysis program showing pressure and energy time traces and other data for the mid 'ai' part of the word "nine" produced by a male speaker.

bandpass filters. The three time traces on the right-hand side of Figure 5 are the results of that processing. They represent the energy that was required to produce the three pressure traces respectively shown to their immediate left. In these three energy traces it is important to note that if formants were primarily damped sine waves, as the linear acoustic theory presumes, then the energy traces found on demodulation should be exponentially decaying in shape over the entire pitch period with no additional 'bumps' or concentrations of energy. We believe that the presence of these 'bumps' as the dominant behavior, illustrates that the acoustic components, or formants, may well be the result of pulsatile flow interactions.

The sets of three sequences of short vertical bars above each of the energy traces on the right are related to the maxima and minima of the computed energy for each of the filtered bands immediately below them. The energy trace for the fifth pressure trace, or noise residue, was not computed. The columns of numbers above the three energy traces are tabulated values for the energy and amplitude parameters.

The processed output from the lowest frequency bandpass filter, the top right or first energy trace, shows a very large single pulse of energy occurring once each pitch period. It may be the single pulse coming out of the glottis. In other words it may be the representation of the energy of the glottal wave. For the demodulated output of the second bandpass filter, one finds a mode of oscillation which is very heavily modulated as evidenced by the three successively decreasing pulses nearly evenly spaced in each pitch period. At higher frequencies one obtains rough sounds which one could also listen to alone and still be able to distinguish characteristics of the word "nine".

If the vocal tract were simply a passive linear system as source-filter theory presumes, we would expect a formant to be excited and then damp out once per each pitch period. There is no mechanism in that theory for the modulation of energy. Yet the energy time traces indicate that *the formants are modulated*.

From these results and from a knowledge of the dynamics of air flow in confined geometries, one can see that sound generation in the vocal tract is very much an active distributed process. It probably depends on the physical locations and types of vortexes present in the entire vocal tract. It would appear that the three energy traces are produced in three separate parts of the vocal tract, working from the glottis to the tongue. The random, noisy fifth time trace, which was not converted to energy, is produced by the teeth and lips.

Conventional wisdom says that in a passive linear system, one can not tell where the sound is being produced. In an actual human voice however, one can measure and deduce the location where the sound is being produced since the sound is being actively generated in the vocal tract. One hears it coming from the mouth because that is where it was produced. One hears another component coming from the back of the throat because that is where that component was produced. The high frequency components generated in the glottis do not make it through the oral cavity to the outside of the mouth. The pulsatile sheet jet coming out of the narrow slit of the vocal folds during phonation, generates a considerable amount of high frequency noise which is inherent in the process. The pulsatile jet proceeds through the vocal tract and drives or excites everything downstream from it. If you think of it another way, what do you remember most about the water going over Niagara Falls, the froth or the falling water? The sound that is generated from the second-order processes is the "froth"; the main source of energy is the glottal jet.

The vocal tract is indeed very much an active sound generating system [18–20]. The sounds that human beings almost universally utilize for speech are in fact completely distinguishable on the basis of the amplitude and frequency modulations of their energy envelopes. Each vowel sound has a unique modulation that is generally tied in with its high frequency second formant. This unique characteristic of the selected speech sounds and the fact that they are not difficult to generate might well account for their universality.

Conclusions

Experimental evidence, concomitant theory, and discussion have been presented showing the presence of nonlinear processes as the primary sound producing mechanisms in the vocal tract during phonation. These processes arise from the nonlinear interaction of the sheet jet flows and the generated flow vortexes within the confined geometry of the vocal tract, with the vortex probably playing the role of the active oscillator in effecting modulations. Following a critical examination of the hearing mechanism and recognizing that the ear is set up to recognize these modulation effects, a corresponding comprehensive methodology for analyzing speech signals was described that incorporates the new knowledge on the fundamental role of modulation in the speech production process. An example was shown of the application of these new methods to speech.

Acknowledgements

This paper was presented by Dr. J. F. Kaiser of Bellcore. The authors wish to take this opportunity to express their gratitude to Dr. Kaiser, who is their long-time friend and esteemed colleague. His graciousness, generosity of spirit, and moral support have been of inestimable value in making the writing and delivery of this work possible. They remain in his debt.

References

1. Teager, H. M. (1980) 'Some Observations on Oral Air Flow During Phonation', *IEEE Trans. Acoust., Speech, Signal Processing*, **ASSP-28**, 599-601.
2. Morse, P. M. (1948) *Vibration and Sound*, McGraw-Hill Book Co., New York, NY.
3. Morse, P. M. and Ingard, U. (1968) *Theoretical Acoustics*, McGraw-Hill Book Co., New York, NY.
4. Bouasse, H. P. M. (1923) *Jets, Tuyaux et Canaux*, Librairie Delagrave, Paris, France.
5. Bouasse, H. P. M. (1926) *Acoustique Générale*, Librairie Delagrave, Paris, France.
6. Bouasse, H. P. M. (1929) *Instruments à Vent*, 2 volumes, Librairie Delagrave, Paris, France.
7. Bouasse, H. P. M. (1929) *Tuyaux et Résonateurs*, Librairie Delagrave, Paris, France.
8. Bouasse, H. P. M. (1931 & 1932) *Tourbillons, Forces Acoustique, Circulation Diverses*, 2 Volumes, Librairie Delagrave, Paris, France.
9. Ohm, G. S. (1843) 'Uber die Definition des Tones, nebst daran geknüpfter Theorie der Sirene und ähnlicher tonbildender Vorrichtungen', *Annalen der Physik*, **LIX**, 513 ff; ibid **LXII**, 1 ff.

10. Seebeck, A. (1843) 'Uber die Sirene', *Annalen der Physik*, **LX**, 449 ff; ibid **LXIII**, 353 ff and 368 ff.
11. Helmholtz, H. L. F. v. (1954) *On the Sensations of Tone*, Dover Publications, Inc., New York, NY.
12. Brownell, W. E., Bader, C. R., Bertrand, D. and Ribaupierre, Y. d. (1985) 'Evoked Mechanical Responses of Isolated Cochlear Outer Hair Cells', *Science* **227**, 194-196.
13. Miller, D. C. (1937) *Sound Waves, Their Shape and Speed*, Macmillan Company, New York, NY, p. 49.
14. Travis, L. E. (ed.),(1971) *Handbook of Speech Pathology and Audiology*, Chapter 6, D. J. Broad and G. E. Peterson, Appleton-Century-Crofts, New York, NY.
15. Peterson, G. E. and Barney, H. L. (1952) 'Control Methods Used in a Study of the Vowels', *J. Acoust. Soc. Amer.*, **24**, 175-184.
16. Peterson, G. E. (1951) 'Vocal gestures', *Bell Laboratories Record*, **29**, 500-503.
17. Licklider, J. C. R. and Pollack, I. (1948) 'Effects of Differentiation, Integration, and Infinite Peak Clipping Upon the Intelligibility of Speech', *J. Acoust. Soc. Amer.*, **20**, 42-51.
18. Teager, H. M. and Teager, S. M. (1981) 'The Effect of Separated Air Flow on Vocalization', in *Vocal Fold Physiology: Contemporary Research and Clinical Issues*, Bless, D. M. and Abbs, J. H. (eds.), College Hill Press, San Diego, CA 92120, 124-143.
19. Teager, H. M. and Teager, S. M. (1983) 'A Phenomenological Model for Vowel Production in the Vocal Tract', in R. G. Daniloff (ed.) *Speech Sciences: Recent Advances*, College Hill Press, San Diego, CA.
20. Teager, H. M. and Teager, S. M. (1983) 'Active Fluid Dynamic Voice Production Models or "There is a Unicorn in the Garden"', Chapter 32 in *Vocal Fold Physiology*, I. Titze and R. Scherer (eds.), Denver Center for the Performing Arts Press, Denver, Colorado.

TESTING THEORIES OF SPEECH PRODUCTION: IMPLICATIONS OF SOME DETAILED ANALYSES OF VARIABLE ARTICULATORY DATA

J.S. PERKELL
Research Laboratory of Electronics
Massachusetts Institute of Technology
Room 36-543, 50 Vassar Street
Cambridge, Massachusetts 02139
U. S. A.

ABSTRACT. Because of indirect relationships between theories and data in speech production, most theories are difficult to test. As an example, a general theoretical overview is outlined which suggests that speech production consists of sequences of movement ensembles from one set of articulatory goals to the next. A number of factors influence the nature of the goals; some of the most important are quantal or non-linear relationships among properties of the production mechanism, the acoustic characteristics of the resulting sounds and capabilities of the perception mechanism. The programming of articulatory movements is a function of: the nature and sequencing of articulatory goals, the moment-to-moment state of the vocal tract and the particular acoustic requirements of individual utterances. Three examples of articulatory data are presented, each of which bears in some way on these ideas and also contains a great deal of variability. One set of data suggests a mechanism for the definition of articulatory goals; the second suggests possible influences on the strategy of anticipatory coarticulation; and the third implies that variability in articulatory data may obscure potentially interesting principles. In spite of the fact that such a general theory cannot be tested directly, each example shows that by making a detailed examination of variable articulatory data, we may be able to advance our thinking about theoretical issues.

1. Introduction

Relationships between theory and data in speech production are rarely straightforward; consequently, the field of speech research is represented by a body of literature which contains conflicting theories and results and a large proportion of discussion to data.

There are valid reasons for this state of affairs. Speech is the result of the most complicated human motor act. It is controlled by neural processes about which we have almost no knowledge, and it is expressed in the coordinated actions of several neuromuscular systems, each with a complex set of neural and biomechanical properties. While phonological or conceptual models of speech production can be tantalizingly straightforward, actual speech production is highly variable and is governed by constraints and "rules" which are poorly understood. To compound matters, there are major problems with studying such a com-

W. J. Hardcastle and A. Marchal (eds.), Speech Production and Speech Modelling, 263–288.

plicated and variable process, including the fact that much of what we want to study is difficult to access physically. Transducing and analyzing the ensemble of relevant parameters is challenging, and the potentially most interesting parameters, those of the underlying neural processes, are mostly beyond our reach.

The main point of this paper is that even with these difficulties, variable articulatory data can help us understand theoretical issues. This point will be illustrated by outlining a theoretical position and presenting examples of data which may reveal something about aspects of the theory. Different theories are possible; one alternative is cited below to illustrate possible contrasts in certain areas. Given our current state of knowledge, no theory can be tested rigorously; data can be interpreted as providing only circumstantial support for any particular theoretical position. Thus the data may be interpreted differently from the way it will be done here. Hopefully, this presentation will stimulate discussion on how these and other data might support alternative theories, how this theoretical position might be supported by additional data and what methods can be used to sharpen our formulation and testing of theoretical issues.

2. Theoretical overview

Our theoretical overview suggests that speech production consists of sequences of ensembles of discrete movements towards articulatory goals. The articulatory goals have static and dynamic components. The precise nature of the goals is determined by several kinds of influences. The most important of those influences include quantal or non-linear relationships among properties of the production mechanism, the acoustic characteristics of the resulting sounds and capabilities of the perception mechanism (cf. Stevens, 1972; 1989).[1] The programming of articulatory movements is a function of the sequence of articulatory goals, the moment-to-moment state of the vocal tract and the particular acoustic requirements of individual utterances.

Figure 1 is a schematic illustration of the basic principle of what Stevens (1972; 1989) calls the "quantal nature of speech" (QNS). The curve illustrates the change in a perceptually salient acoustic parameter in response to a continuous change in an articulatory parameter. In regions labeled I and III, the acoustic parameter is relatively stable, and in region II the acoustic parameter undergoes an abrupt transition. Stevens hypothesizes that the stable acoustic cues resulting from such non-linear relationships in production are detected or selected out by quantal mechanisms in perception.[2]

As an example, a family of these relationships may be found between gradually-changing glottal aperture and the resulting sound, when the respiratory system is programmed to provide a steady expiratory flow of air. At the widest glottal opening no sound is generated. As the aperture is gradually narrowed, stable regions are traversed in which there is successively aspiration noise, vocal-fold vibration and again, no sound (i.e., a glottal stop). Between each stable region, there is an abrupt transition. Another example is in the relationship between vowel formant frequencies and location of the major vocal-tract

[1]It has been hypothesized (Stevens, 1972; 1989; Perkell, 1980) that these relationships and the goals that they define are correlates of distinctive features; that topic is beyond the scope of this discussion.

[2]Thus, the production and perception of a particular acoustic cue would theoretically involve two non-linearities, one in production and one in perception.

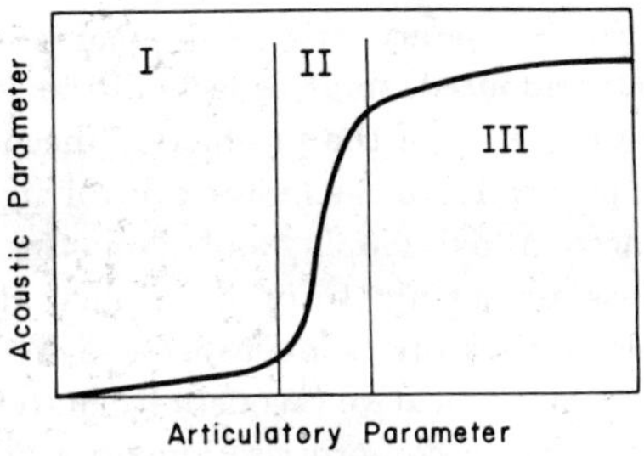

Figure 1: Schematization of a relationship showing the change in a perceptually salient acoustic parameter as a function of change in some articulatory parameter. The regions labeled I, II and III are explained in the text (Stevens, 1972; 1989). The same type of function can conceivably illustrate the relationship between two parameters at a number of stages in the sequence of transformations that takes place in production and perception.

constriction. Stevens (1972; 1989) argues that there are regions of constriction location in which vowel formants are relatively stable with respect to some change in constriction location, and others in which they are unstable. (See also, Wood, 1979, Mrayati, Carré and Guerin, 1988). We hypothesize that such relationships help to define a set of articulatory goals which are utilized in speech production.

An alternative theory claims that speech production consists of the concatenation and elaboration of dynamically-specified articulatory gestures (cf. Fowler, 1980; Saltzman and Munhall, in press: Browman and Goldstein, 1986). From our perspective, there appear to be some differences in emphasis between this notion and our goal-based theory. In characterizing the basic underlying units of speech, the QNS relies equally on relationships between: a) production and the acoustics and b) the acoustics and perception of speech. The gesture-based theory may be less concerned about the origin of the underlying units, but it would appear to lean more heavily on inherent dynamic properties of the production mechanism. Our goal-based approach claims that to the extent that underlying units have dynamic components, those components exist mainly because they help to produce perceptually-salient acoustic cues.[3] In the gesture-based theory, timing of articulatory events seems to be a consequence of relatively straightforward scaling of gestural dynamics. In our goal-based framework, certain aspects of timing are determined in a more complex manner.

When discrete elements are concatenated in speech production, whether they be phonetic segments, syllables, morphs, words or even larger units, adjustments have to be made at the boundaries of the units to create the seamless transitions that are observed in measurements of the acoustic signal and of articulation. The gesture-based theory claims that when dynamically-specified units are concatenated, a weighted summation of conflicting influences on each articulator takes place to produce the observed kinematics (Munhall and Saltzman, in press). To the extent that adjustments might have to be made to account

[3]Quantal-based arguments have been made about dynamic components of fundamental units as well, but such ideas have not yet been explored in much depth (Stevens, 1989).

for modifications of parameters such as rate and stress, those adjustments apparently do not refer to phonetic context (or to the acoustic requirements of concatenated units). In the theoretical approach we have adopted, some aspects of the kinematics of speech movements result from underlying properties of the production mechanism (i.e., biomechanics, physical boundary conditions, properties of the motor control apparatus), and adjustments have to be made to take into account overlapping and competing influences on articulators (see Perkell, 1980). These ideas appear not to be in conflict with the gesture-based theory. However, our approach also claims that some aspects of speech kinematics have to be determined on the basis of the communicative (acoustic) requirements of sound sequences. Thus the resulting kinematics reflect a compromise among a number of conflicting influences, which include acoustics, context effects and variations in a number of factors such as speaking style and rate.[4]

3. Variation and the study of speech production

As mentioned above, there is a large amount of variation in speech. This variation is expressed in a number of ways, within and across speakers, and it has a number of sources. There is obvious within-speaker variation in speaking style and rate. There are also idiosyncratic, dialectal (cf. Labov, 1986) and cross-language (cf. Lindblom, 1975) differences in sound patterns. A major source of within-speaker variation is probably the inherently-variable nature of the motor control system in combination with the fact that speakers adjust for listeners' varying need for clarity of articulation (sometimes producing extremely reduced utterances). Some of this variation can convey information, and some of it (such as token-to-token variation of the same utterance) may be considered to be a form of noise.

The kinds of very general theoretical ideas we are considering here do not attempt to account for most sources of variability. Thus variability can hinder attempts to explore such theories. However, it can also be exploited. Studies of motor equivalence in speech have examined variation in the contributions of individual articulators to achieving multi-articulator goals (cf. Hughes and Abbs, 1976; Edwards, 1985; Abbs, 1986). The idea behind this work is that a goal may be relatively invariant, while the coordinated contributions of individual articulators toward reaching that goal may vary in a reciprocal way. Thus, in achieving a relatively invariant small lip aperture when moving from an open vowel, the lower lip may move upward a lot with respect to the mandible, while the mandible moves up a little, or vice-versa. Results of such studies have been mixed, some supporting the concept of motor equivalence, and others questioning its validity (cf. Gentil, *et al.*, 1983; Smith and McLean-Muse, 1987; Sussman, 1980; Folkins and Brown, 1987). Nevertheless, the general idea of looking for revealing patterns in variable behavior remains as a source of potentially useful experimental paradigms. In using such paradigms, it is important to control as much as possible for different sources of variability and consider the potential

[4]Some of the differences between the two theoretical approaches may have to do with the fact that the gesture-based theory is being driven in part by an ambitious effort to incorporate it into a simulation of speech production. Formulating such a complicated a theory in the form of a working model places severe constraints on its potential generality, but has the very large benefit of imposing discipline on the development of the theory.

consequences of sources of variation that are beyond control.

4. Three Experimental Results

The following examples explore variability in articulatory data. Our conclusion from these examples will be that by looking in detail at individual (un-averaged) articulatory trajectories and events and examining the resulting patterns of variability, we can investigate predictions made from very general models of speech production. In each example, variable articulatory data are considered in terms of their possible interaction with the resulting acoustics.[5] The first example examines articulatory data on tongue body locations for the vowels /i/, /a/ and /u/ in light of hypotheses based on the theory of the quantal nature of speech. The second example presents data on anticipatory coarticulation of lip protrusion for the vowel /u/. It examines temporal aspects of lip protrusion movements with respect to hypotheses derived from the two theoretical overviews described above. The third example uses some of the same data as the first to examine context effects on tongue body location for the vowels /i/,/a/ and /u/. It emphasizes the importance of looking at variation by suggesting that certain context effects may be obscured by the variable nature of speech motor control. Thus, the first and third examples explore "static" aspects of articulatory "goals" or "targets", and the second explores a mechanism of articulatory timing.

In each example, there is more than one source of variation. The examples differ according to which source of variation is considered to be the parameter of interest and which are considered to be sources of error or "noise". The examples are presented as preliminary demonstrations of an experimental approach; by themselves, they are insufficient to test the basic theoretical notions. In order to fully test such theories, a very large amount of work has to be done in which many sources of variation are explored systematically, and the theories have to be further translated into a number of more detailed, testable experimental hypotheses.

4.1. AN INDIRECT TEST OF THE QUANTAL NATURE OF SPEECH IN THE PRODUCTION OF THE VOWELS /i/, /a/ and /u/

We have found Stevens' ideas about the quantal nature of speech to be rather compelling, but in need of experimental support. (Many of Stevens' examples are based on modeling.) Thus, we have searched for ways to test ideas about the QNS experimentally, and have begun by trying to explore ideas based on the relationship between vowel acoustics and place of articulation. These results have been published previously (Perkell and Nelson, 1985; Perkell and Cohen, 1989) and will be summarized here as an example of how we may begin to find evidence that supports the kind of framework outlined above.

4.1.1. *Hypothesis.* Our paradigm attempts to exploit the variability that is found in multiple tokens of the vowels for individual speakers. Figure 2-a is a schematic diagram of midsagittal tongue contours for the vowels /i/,/u/ and /a/. Variation in constriction location and degree of constriction are indicated schematically by arrows labeled "L" and "C".

[5] As of yet, only limited acoustic analysis has been done, although it is considered to be an important component of future research along these lines.

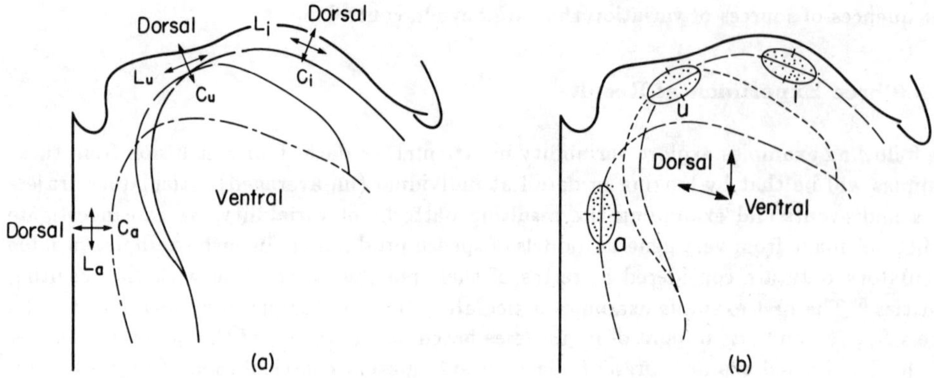

Figure 2: Part (a): A schematic diagram of midsagittal tongue contours for the vowels /i/,/u/ and /a/. Variation in constriction location and degree of constriction are indicated schematically by arrows labeled "L" and "C". Part (b): Hypothetical scatter plots in the midsagittal plane of the target location of one point on the tongue surface at each constriction location, corresponding to multiple repetitions of /i/, /a/ and /u/.

We have examined data which reflect constriction location and degree of constriction, for multiple tokens of these vowels in several speakers of American English (Perkell and Nelson, 1982; 1985; Perkell and Cohen, 1989). That examination was based on two premises raised in Stevens' discussion of the QNS (see also Wood, 1979; Fant, 1960; Mrayati, Carré and Guerin, 1989): 1) the acoustics for these vowels are insensitive to some variation in constriction location, and 2) because of the small cross-sectional vocal-tract area at the point of maximum constriction for these vowels, small dorsal-ventral tongue displacements produce large percentage changes in cross-sectional area and large changes in the vowel acoustics. Thus, the acoustics of the peripheral vowels should be relatively ***insensitive*** to ***variation in constriction location*** and ***sensitive*** to ***variation in dorsal-ventral tongue-body displacements*** (perpendicular to the vocal-tract midline and tongue surface) in the regions of maximum constriction. (At locations away from the region of maximum constriction, limited dorsal-ventral displacement variation produces smaller percentage variation in the area function and much less effect on the acoustics.)

These ideas led to the hypothesis that for multiple tokens of these vowels positioning of points on the tongue dorsum should be more precise in the dorsal-ventral direction than tangent to the vocal-tract midline, at or near the place of maximum constriction. Our criterion for relative precision is illustrated in Fig. 2-b, which shows hypothetical scatter plots in the midsagittal plane of the target location of one point on the tongue surface at each constriction location, corresponding to multiple repetitions of /i/, /a/ and /u/. The ellipses have two-sigma major and minor axes and represent an 86% probability of containment for a bivariate normal distribution. The positioning of points within an ellipse can be considered to be more precise in the direction of its minor axis.

In the following discussion, we will speculate that results such as these will be due to a combination of quantal effects that involve more than one kind of non-linear relationship: motor command-to-displacement, displacement-to-area function and area function-to-acoustic. These relationships should help to define articulatory targets for the peripheral vowels, and also to minimize the requirement for precision in the underlying motor control.

4.1.2. *Results and Discussion.* Midsagittal-plane X-ray microbeam data for two speakers are shown in parts A and B of Fig. 3. Data for the vowel /i/ are shown in the upper panel of each part of the figure, and data for the vowel /a/ are shown in the lower panel. The points in the figure represent the vowel target locations of pellets fixed to the tongue dorsum at three locations: rear, mid and blade. The "vowel target location" was defined to occur at the time the points reached minimum velocity and extreme displacement during the voiced interval of the vowel. N indicates the number of points in each grouping. A two-sigma ellipse is drawn around each grouping, with its major axis passing through the group mean and aligned with the principal component of variance for the group. Anterior is to the right. For these two subjects, the data are from a variety of stress and context conditions, and there were insufficient numbers of the vowel /u/ to include in the analysis.

Figure 4 shows results of a similar experiment. For this experiment, a third speaker pronounced about 15 (randomized) repetitions of a number of nonsense utterances of the form /bV'(C)Vbʌ/, in which the vowels were /i/, /u/, and /a/, and the consonant, when present, was /b/,/ʔ/ or /h/.[6] Stress was placed on the second syllable. The full corpus included all combinations of the vowels and consonants. The movement of one point on the dorsal tongue surface was transduced (with respect to two fixed points on the maxilla) using an alternating magnetic field movement transducer system (Perkell and Cohen, 1986). Using Fig. 3 as a reference, we estimate that the transduced tongue point for this subject was approximately between the the mid and rear locations for the first two subjects, near the region of maximum vocal-tract constriction for the vowel /u/. A subset of utterances was selected for this figure in order to minimize the effects of context: for each utterance, both vowels were the same, and the consonant, when present, was /h/ or /ʔ/. The figure shows scatter plots of the target locations of the tongue point for the examples of /u/, /i/ and /a/ that occurred in the stressed (second syllable – part A) and unstressed (first syllable – part B) positions.[7] Example utterances are /bu'hubʌ/ and /bi'ʔibʌ/. As in Fig. 3, each symbol represents the algorithmically-determined location of the tongue point at the first instance of a minimum in tangential velocity that occurred during the voiced portion of the vowel, and two-sigma ellipses are drawn in. The irregular line at the top of each plot is the outline of the posterior portion of the hard palate.

As implied by Stevens' discussion of the QNS (and according to Perkell and Nelson, 1982; 1985; Perkell and Cohen, 1989), these data may be accounted for in the following way. The vowel /i/ may be produced by a mechanism suggested by Fujimura and Kakita (1979). The tongue blade is stiffened and grooved, and the posterior genioglossus muscles pull the back of the tongue toward the genioid tubercle of the mandible, thus pushing the tongue body and blade upward and forward. When the stiffened, grooved tongue blade is being pushed against the hard palate, the area function at the place of maximum

[6] The subject was not instructed to produce an intervocalic glottal stop, but he often did so in utterances which specified no intervocalic consonant.

[7] In Fig. 4, tokens containing /h/ are indicated by "h", and tokens containing no consonant or glottal stop are indicated by the symbol for the surrounding vowels.

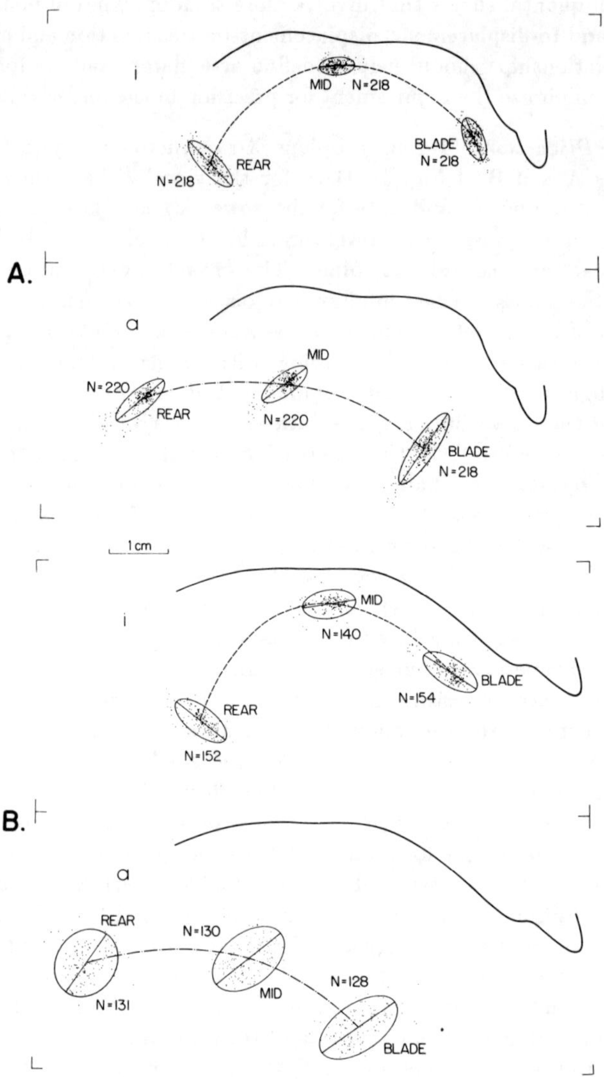

Figure 3: Results for two speakers, parts A and B, from Perkell and Nelson (1982;1985). Data for the vowel /i/ are shown in the upper panel in each part of the figure, and data for the vowel /a/ are shown in the lower panel. Each point in the figure represents the target location of a pellet fixed to the tongue dorsum for a repetition of the vowel. N indicates the number of points in each grouping. A two-sigma ellipse is drawn with its major axis passing through each group mean and aligned with the principal component of variance for the group. The approximate outline of the hard palate is shown for orientation purposes; it is not located accurately with respect to the data. Anterior is to the right.

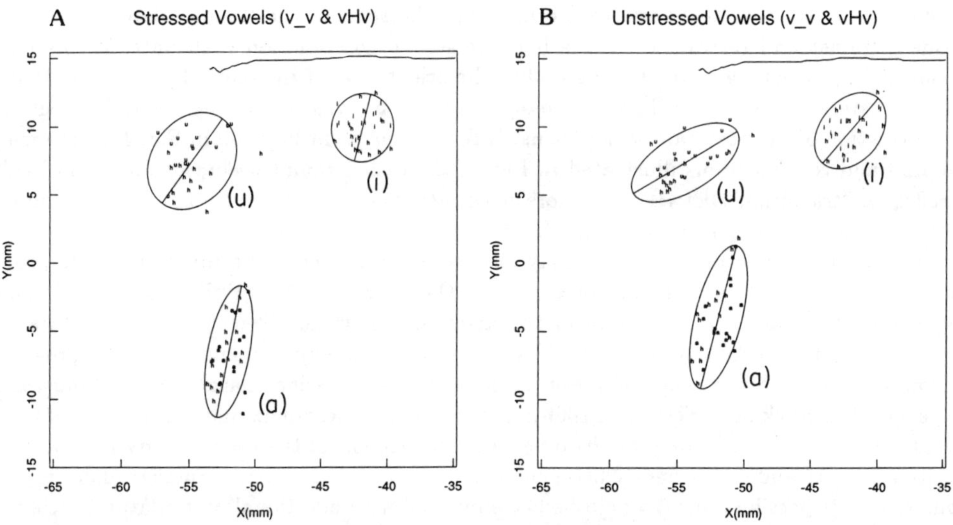

Figure 4: Scatter plots of the vowel target locations of a single dorsal tongue point for nonsense utterances of the form /bV'(C)Vbʌ/, in which both vowels were /i/, /u/, or /a/, and the consonant, when present, was /ʔ/ or /h/. Target locations for stressed (second syllable) vowels are shown in part A and for unstressed (first syllable) vowels are shown in part B. Each symbol represents the algorithmically-determined location of the tongue point at the first instance of a minimum in tangential velocity that occurred during the voiced portion of the vowel. Two-sigma ellipses are drawn as in Fig. 3. The irregular line at the top of each plot is the (accurately-located) outline of the posterior portion of the hard palate.

constriction and the resulting formants are stable over a relatively wide range of contraction of the posterior genioglossus (and accompanying movement of the back of the tongue). Fujimura and Kakita (1979) characterize this mechanism as showing a "saturation effect": over a range of variation of an important aspect of the "input" (genioglossus contraction), the "output" (area function in the front part of the oral cavity and the resulting vowel formants) is relatively stable. This saturation effect could be characterized by regions II and III in Fig. 1, except that the vertical axis would be cross-sectional area in the region of the hard palate, and the horizontal axis would correspond to contraction of the posterior genioglossus.

Thus, quantal effects are to be found not only in non-linear relationships between articulation and acoustics; they may also be manifested in mechanical interactions among articulators, boundary conditions and resulting area functions. (There are additional possibilities; as suggested below for the vowels /a/ and /u/, another useful quantal effect may be the non-linear relationship between motor commands and force of muscle contraction.)

The results for /i/ in Fig. 3 strongly support the articulatory component of this combination of hypotheses. The orientation of the ellipses for the front two (mid and blade) tongue points is most likely due to the fact that the sides of the tongue blade are being pushed against and restrained by the hard palate, in combination with antero-posterior variation in positioning of the tongue body. The orientation of the rear ellipse reflects that antero-posterior variation in the position of the tongue body, as caused by a variable degree of contraction of the posterior genioglossus. If the tongue point location in Fig. 4 is between the mid and rear locations illustrated in Fig. 3, the rather rounded ellipses for /i/ in Fig. 4 reflect a "transition" between the more horizontal ones for the mid point and the more vertical ones for the rear point in Fig. 3.

A non-linear or saturation effect may also be hypothesized to be involved in the production of the vowel /a/. To produce an /a/, the tongue body is pulled downward and "squeezed" backward by the action of the hyoglossus muscle. The amount of backward tongue body movement may be limited by the anatomical arrangement of the hyoglossus, in combination with the S-shaped (saturation-like) length tension characteristic of muscles in general (cf. Rack and Westbury, 1969): over a range of near-maximal contraction of the hyoglossus muscles, there may be little posterior movement of the tongue body beyond its position for /a/ and consequently little change in the vowel acoustics. (Further pharyngeal constriction is possible, but it would be accomplished with additional musculature, such as the pharyngeal constrictors.)

The results for /a/ in Figs. 3 and 4 are reasonably supportive of this mechanism and the prediction in Fig. 2.[8] Much of the vertical component of spread in the data for /a/ may be due variation in lowering of the mandible which carries the tongue body downward by a variable amount. This variation has a relatively minor effect on the area function in the pharynx and the acoustics of the vowel. (For details, see Perkell and Nelson, 1982.)

Figure 5 illustrates a hypothetical mechanism for the production of /u/ which is similar to those for /i/ and /a/. The figure shows dorsal tongue contours that were generated by an anatomically-based tongue model (Perkell, 1974). The dashed contour represents a "neutral" tongue configuration, and the solid contour represents the configuration that results from contraction of "muscles" whose actions were simulated by groups of contractile elements schematized by the heavy solid lines. Fig. 5-a suggests that because of the "squeezing" action that results from the anatomical arrangement of the styloglossus muscle fibers, the resulting vocal-tract constriction is about in the correct location for /u/ and not along the more posterior axis of contraction of the muscle. This mechanism follows an idea suggested by Wood (1979) and Perkell (1979): there is a coincidence between anatomical and articulatory-to-acoustic factors which helps to reinforce the definition of a basic articulatory target. Fig. 5-b suggests that the vowel /u/ is produced by a combination of contraction of the styloglossus and posterior genioglossus muscles (cf. Smith, 1971; Alfonso, Honda and Baer, 1984). It is hypothesized that a near-maximal contraction of the styloglossus results in a velo-palatal vocal-tract constriction with a relatively invariant cross-sectional area, while a variable degree of contraction of the posterior genioglossus results in some antero-posterior tongue body movement and variation of the constriction location.

The orientation of the distributions for /u/ in Fig. 4 is consistent with this mechanism and the prediction in Fig. 2. The rounding of the distribution for stressed /u/ (Part A)

[8] Perkell and Nelson (1982; 1985) suggest that the ellipses in Fig. 3 are less vertical than predicted because (for anatomical reasons) the actual point of measurement moves somewhat more horizontally than a point lower down in the pharynx, and the subjects were leaning somewhat forward in the microbeam apparatus.

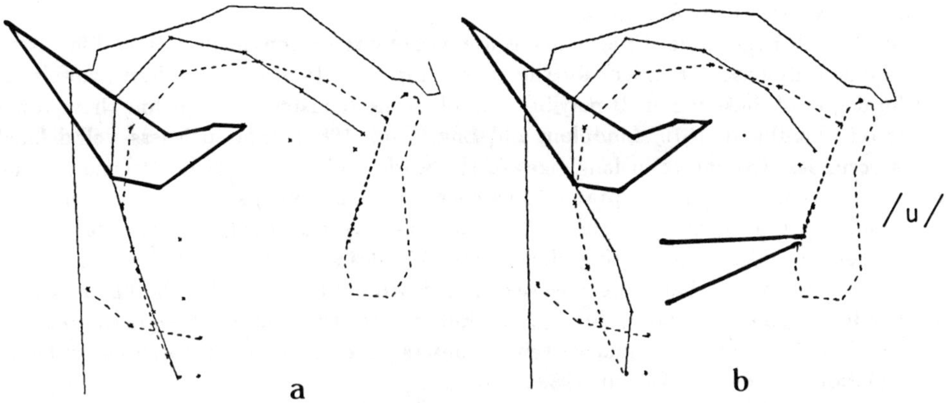

Figure 5: Dorsal tongue contours generated by an anatomically-based tongue model (Perkell, 1974). The dashed contour represents a "neutral" tongue configuration, and the solid contour represents the configuration that results from contraction of "muscles" whose actions were simulated by groups of contractile elements schematized by the heavy solid lines. Part a: simulation of contraction of the styloglossus muscles. Part b: simulation of the tongue shape for the vowel /u/, showing contraction of the styloglossus and posterior genioglossus muscles. (Contraction of the inferior longitudinal contributes to the shape in part b, but it is not shown.)

is due to the fact that it actually contains two separate, more elongated distributions: the tokens containing /h/ mostly occupy the ventral (lower) half of the ellipse and the tokens containing no consonant or glottal stop mostly occupy the dorsal (upper) half. It is not clear why such a separation takes place, but it may have to do with: a) creating a wide enough constriction for the /h/-containing utterances so that the glottal aspiration noise is transmitted with sufficient amplitude and/or b) avoiding the generation of turbulence noise at the vowel constriction location (Perkell and Cohen, 1989).

4.1.3. *Implications.* The preceding results are consistent with the idea that for each of the three peripheral vowels there is an articulatory-to-acoustic non-linearity which operates along the vocal-tract length and a muscle contraction-to-displacement non-linearity that operates in a direction normal to the vocal-tract midline. This combination of non-linear relationships results in relative acoustic stability for the vowels while allowing for imprecision in some aspects of the underlying motor control. These effects may contribute to the basic definition of articulatory targets for the peripheral vowels.[9]

Another possible implication of these results is that there is variation in the precision of control of different parts of production mechanism which depends on the requirements of individual sounds in a sequence. It may be that control is "concentrated" on those aspects

[9]There are other, more obvious non-linearities such as in consonant closures, in which some aspects of the closing movement should not have to be precisely programmed: the force of closure presumably may vary over some range with no appreciable effect on the acoustics.

of articulations which have critical acoustic consequences, and other less-critical aspects of articulation are relatively free to vary.

Not all such hypothesized mechanisms are as potentially convincing. It is difficult to make a quantally-based argument for the articulatory targets of non-peripheral vowels, at least for degree of constriction. It may be: that other mechanisms help to define those vowel targets (cf. Lindblom, 1975; Lindblom and Engstrand, 1989), that such less-well-defined vowels occur less frequently in languages of the world and/or that more variation in the acoustics of these vowels is acceptable to listeners. Stevens (1989) suggests that a quantal relationship may also influence the lip rounding gesture (for /u/), but considering the similar acoustic effects of lip closing movements, this idea seems somewhat less persuasive. In this case, it may be that more precise motor control is necessary for the lip target for /u/ than for tongue body targets for /i/, /a/ and /u/, but the additional required precision is made possible by more gradual movement onsets which accompany anticipatory labial coarticulation (Perkell and Cohen, 1989).[10]

4.2. A TEST OF CONCEPTUAL MODELS FOR ANTICIPATORY COARTICULATION OF LIP PROTRUSION

Coarticulation is the superposition of multiple influences on the movement of an articulator. Those influences can come from acoustic-phonetic context and from interactions with other articulators. In spite of the large number of experiments that have been done on coarticulation (cf. Öhman, 1966; Benguerel and Cowan, 1974; Lubker, 1981; Lubker and Gay, 1982; Bladon and Al-Bamerni, 1982; Recasens, 1984), there is not general agreement on a single model (cf. Henke, 1967; Öhman, 1967; Kent and Minifie, 1977; Bell-Berti and Harris, 1981; Perkell, 1980), and results from different experiments can appear to conflict with one another (cf. Bell-Berti and Harris 1979; Sussman and Westbury, 1981; Gelfer, *et al.*, 1982). We conducted a study of anticipatory coarticulation of lip protrusion in an attempt to explain some of the apparent conflicts.

4.2.1. *Hypothesis.* Figure 6 shows schematic diagrams of three pairs of traces of upper lip protrusion vs. time. Each pair illustrates specific predictions inferred from a relatively abstract "model" which has been proposed to account for experimental data.[11] In each panel, v1 represents an unprotruded vowel such as /i/; v2 represents a protruded vowel such as /u/, and c, c1 and c2 represents consonants that are presumably "neutral" (have no significant acoustic or biomechanical interaction) with respect to protrusion. The duration of the intervocalic interval is varied by choosing utterances with different numbers of intervocalic consonants. The solid vertical line represents onset of voicing for the /u/; the tic marks represent movement onset; and the dashed lines represent acoustic boundaries between phonetic segments (v1 and c(1), and c1 and c2). In the third panel the x's on the traces indicate the time of maximum acceleration (peak in the second derivative of protrusion vs. time).

[10] In general, there is a trading relationship between speed and accuracy in moment (cf. Wright and Meyer, 1983).

[11] Thus, the claims illustrated in Fig. 6 might be considered to be stronger than intended by originators of the underlying ideas.

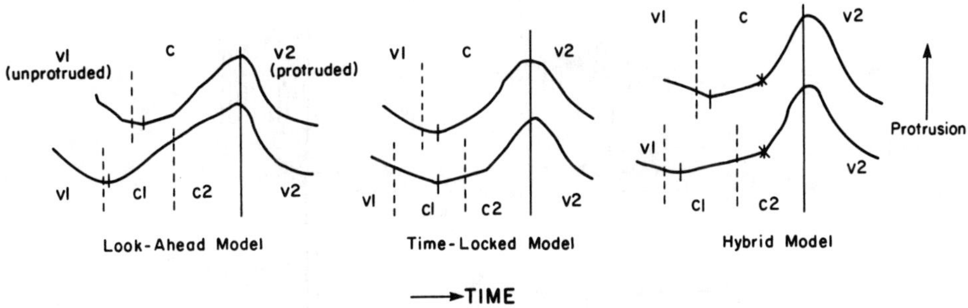

Figure 6: Schematic diagrams of pairs of traces of upper lip protrusion vs. time. In each panel: v1 represents an unprotruded vowel such as /i/; v2 represents a protruded vowel such as /u/; and c, c1 and c2 represent consonants that are presumably "neutral" with respect to protrusion. The solid vertical line represents onset of voicing for the /u/; the tic marks represent movement onset; and the dashed lines represent acoustic boundaries between phonetic segments (v1 and c(1), and c1 and c2). In the rightmost panel the x's on the traces indicate the time of maximum acceleration.

The first panel illustrates experimental findings which are characterized by the "look-ahead" model of anticipatory coarticulation (Henke, 1967; Öhman, 1967). In this case, movement onset occurs approximately at the acoustic offset of the /i/, when it is "allowed to" by relaxation of the acoustic constraint that the /i/ be unprotruded. For long intervocalic intervals the movement onset occurs earlier with respect to onset of voicing of the /u/ than it does for short intervocalic intervals. This model is consistent with the above-mentioned idea that certain (temporal) aspects of articulatory movements are determined with respect to (acoustic) characteristics of phonetic context.

The second panel illustrates experimental findings which are characterized by a so-called "frame theory" or "time-locked" model (cf. Bell-Berti and Harris, 1981). In this case, it does not matter when the preceding /i/ ends. Movement onset occurs at a fixed time before the onset of the /u/. This model is consistent with the gesture-based theory, in which most kinematic characteristics of speech articulations are a relatively straightforward consequence of the dynamic characteristic of gestural units.

The third panel illustrates the "hybrid model", which is a compromise between the first two. It is based on some preliminary observations made by Bladon and Al-Bamerni (1982) and Perkell and Chiang (1986). In this example, the lip protrusion gesture has two components, a gradual initial phase and a more rapid second phase. The phases are separated from one another by an inflection point which is a result of the underlying peak in acceleration. Here, the gradual movement onset occurs approximately at the time it is "allowed to" at the offset of the /i/, but the onset of the second, rapid phase is "time-locked" to the onset of the /u/. Thus as the intervocalic consonantal interval increases in duration, the duration of the gradual onset phase increases with it. Presumably the acoustics of the /i/ are not adversely affected by the small amount of protrusion that occurs during the gradual phase.

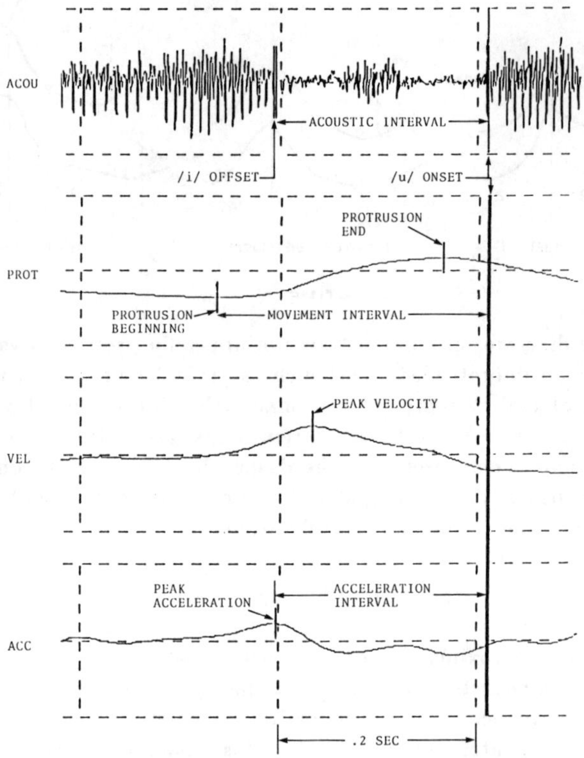

Figure 7: Illustration of data extraction: ACOU – a segment of the acoustic signal, PROT – the corresponding lip protrusion, VEL – lip velocity, and ACC – lip acceleration vs. time. See text for details.

4.2.2. *Methods and Results.* We recorded the acoustic signal and protrusion (horizontal) movements of the upper lip for four speakers of American English, using multiple repetitions of utterances such as "Its a *lee coot* again" and "Its a *leaked scoot* again". Control utterances such as "Its a *lee keet* again" and "Its a *leaked skeet* again" were included to look for protrusion-retraction movements associated with the intervocalic consonants. The first two derivatives of the displacement signal were computed to obtain velocity and acceleration vs. time. Data extraction is illustrated in Fig. 7, which shows from top to bottom: 1) a segment of the acoustic signal (ACOU), 2) lip protrusion (PROT), 3) lip velocity (VEL), and 4) lip acceleration (ACC) vs. time. Acoustic events in the time-expanded acoustic signal corresponding to offset of the /i/ and onset of the /u/ were labeled interactively. Events corresponding to protrusion beginning, protrusion end, peak velocity and peak acceleration were identified algorithmically.[12] Times and magnitudes of these events were recorded

[12]Protrusion onset time was defined as the time of the last onset of positive velocity before the peak in the protrusion gesture (protrusion end).

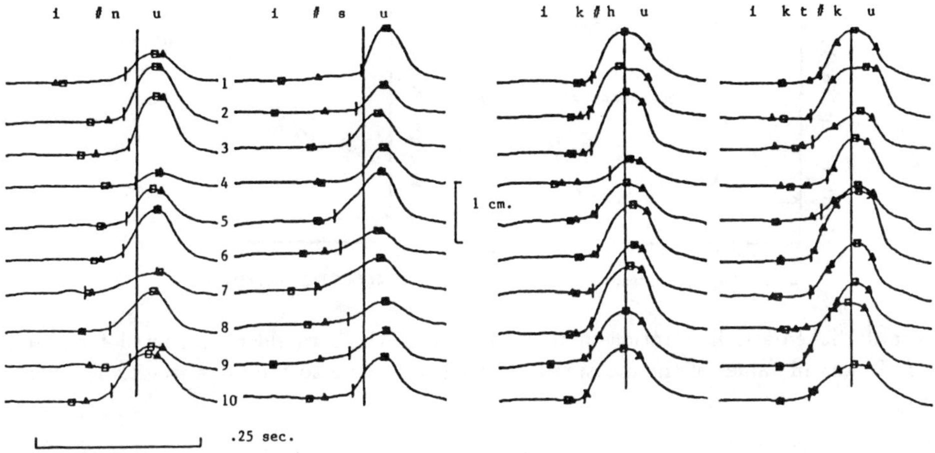

Figure 8: Sequence plots of lip protrusion vs. time for 10 repetitions each of the utterances "lee newt", "lee suit","leak hoot"and "leaked coot", for Subject 1. In each panel the vertical line indicates the "line-up time" of onset of voicing of the /u/. On each trace: the first box indicates the start of the protrusion gesture; the first triangle indicates the time of the acoustic end of the /i/; and the vertical tic mark indicates the time of maximum acceleration.

along with utterance identifiers, and a number of measures of magnitudes and durations were calculated.

Data were analyzed qualitatively and quantitatively. The first form of data analysis was to produce and examine so-called "sequence plots", each of which consists of a vertical series of plots of protrusion vs. time for all repetitions of an utterance. Four such sequence plots are shown in Fig. 8, from left to right for 10 repetitions each of the utterances "lee newt", "lee suit","leak hoot"and "leaked coot", for Subject 1. Each trace is marked with symbols corresponding to the times of events identified in the acoustic and movement signal streams. The heavy vertical line represents the "line-up" event of onset of voicing for the /u/. Observation of such plots enabled us to: eliminate grossly atypical tokens, gain a visual impression of token-to-token variation of displacement vs. time, verify that the control utterances showed very little evidence of consistent protrusion-retraction gestures and gain general impressions to guide quantitative analyses. Examination of these plots revealed a lot of variation in the shape and timing of protrusion trajectories.[13] However, it also verified our initial impression that the lip protrusion movement was almost always divided into two phases, an initial gradual phase and a second, more rapid phase.

The quantitative analysis done thus far has consisted of examining the relationships among the three time intervals indicated in Fig. 7. Those intervals are: a) time of acoustic

[13]One aspect of temporal variation apparent in Figs. 7 and 8 is in the timing of the peak in the protrusion gesture (movement end) with respect to the acoustic onset of the /u/. As shown in Fig. 7, movement end sometimes preceded /u/ onset.

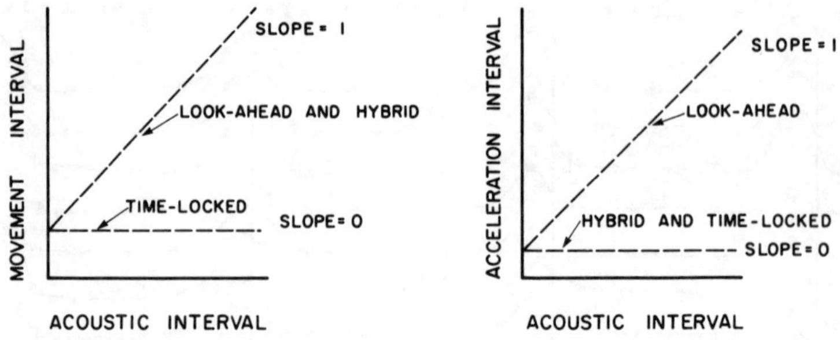

Figure 9: Schematic illustration of regression lines which would be expected for scatter plots of pairs of values of the measured intervals according to the three models illustrated in Fig. 6.

/i/ offset to time of acoustic /u/ onset (called the "acoustic interval"), b) time of movement beginning to time of acoustic /u/ onset (called the "movement interval") and c) time of maximum acceleration to acoustic /u/ onset (called the "acceleration interval"). Times of movement onset and maximum acceleration were chosen as model and measurement parameters because they should be physically related to the underlying forcing functions and thus to the motor control. The time of onset of voicing for the /u/ was chosen because we presume that the articulatory target for the /u/ should be realized accurately enough at that point to produce an appropriate percept.

Figure 9 illustrates schematically the regression lines which would be expected for scatter plots of pairs of values of these intervals according to the three "models". In both diagrams, the horizontal axis is the acoustic interval, time from acoustic /i/ offset to acoustic /u/ onset. The vertical axis in the left panel corresponds to values of the movement interval and in the right panel it corresponds to values of the acceleration interval. The look-ahead model predictsa regression line with a slope of one in both plots, and the time-locked model predicts a slope of zero in both plots. The hybrid model predicts a slope of one for the movement interval vs. acoustic interval and zero for the acceleration vs. acoustic interval.

Figure 10 shows plots of the movement interval (Part A) and acceleration interval (Part B) vs. acoustic interval for three of the speakers.[14] Each symbol on the plots represents values for a pair of intervals for one token. In each plot, a dashed line is drawn in which has a slope of 1.0 and passes through the origin. Thus points to the left of the dashed line represent instances in which the articulatory event (movement onset or maximum acceleration) precedes the offset of the /i/. For all 3 Subjects, and especially Subject 1, at any particular value of the acoustic interval, there is a great deal of variation in the movement or acceleration interval.[15] Because of this large amount of variation, linear least-square regressions had low multiple R^2 values; however their slopes were significantly

[14]Scatter plots from the fourth speaker contained two distinct groupings of points; therefore, those data could not be used in this analysis.

[15]In addition to variation which is inherent in the measured parameters, there are experimental sources of variation which include: the difficulty of establishing a uniform criterion for acoustic offset of the /i/ when it is followed by different kinds of consonants, and the fact that movement onset is a very gradual event.

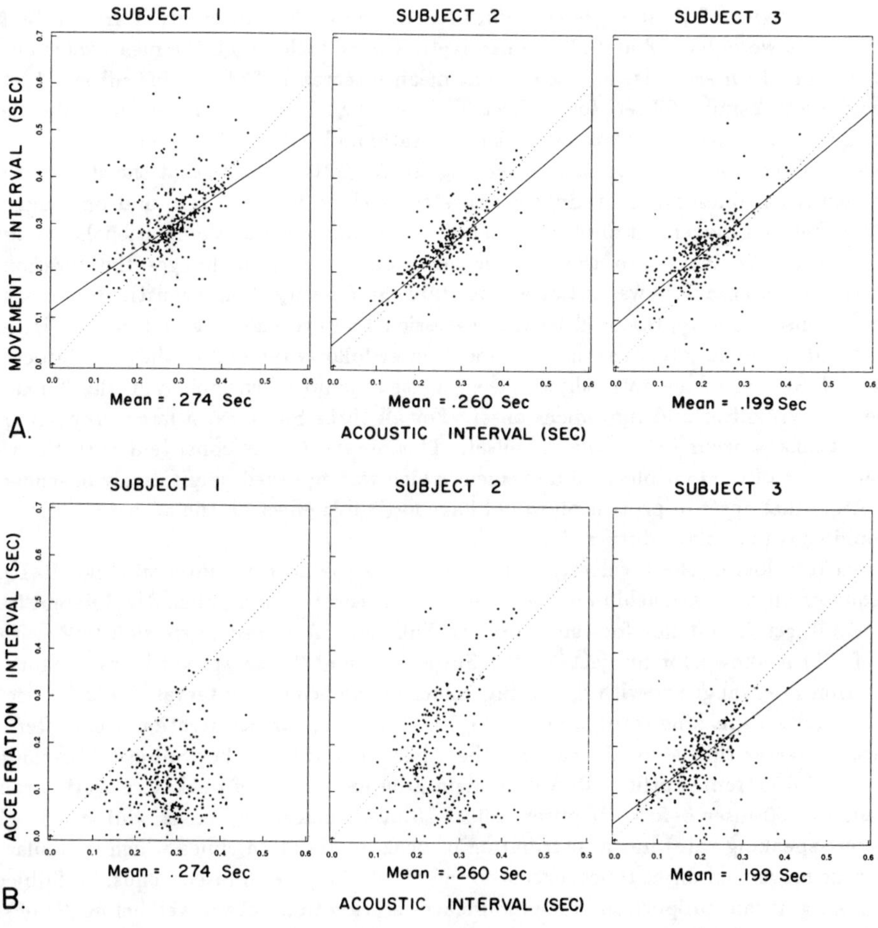

Figure 10: Scatter plots of pairs of values of movement interval (Part A) and acceleration interval (Part B) vs. acoustic intervals for Speakers 1, 2 and 3. The mean value of the acoustic interval is given below the plots for each Subject. When shown, in each panel, the solid line represents a robust regression and the dotted line has a slope of 1.0 and passes through the origin.

different from zero. The solid straight line through each distribution shows the result of a robust (least absolute values) regression, which de-emphasizes the influence of outlying points. The slopes of these regression lines for the movement intervals (Part A) for the three Subjects were .63, .78 and .77, respectively. Under each panel, the mean value of the acoustic interval (in seconds) is listed. That mean interval is .274 sec. for subject 1, .260 sec. for Subject 2 and .199 sec. for Subject 3. As a rough indicator of speaking rate, these values show that Subject 1 spoke at the slowest rate, and Subject 3, the fastest.

Several observations can be made from Fig. 10-A. To the extent that the slopes of the robust regression lines can be validly interpreted, a slope between zero and one suggests that /i/ offset and movement onset are related to one another, but the relationship is not a "rigid" one. As the duration of the acoustic (intervocalic) interval changes, the duration of the movement interval follows, but at a lower rate. So for very short acoustic intervals, the movement onset time represented by the regression line precedes the end of the /i/, and for very long acoustic intervals, the movement onset follows the end of the /i/. Subject 1, with the slowest speaking rate, showed the largest amount of variability in the "linkage" between the /i/ offset and movement onset. For all three Subjects, a large proportion of movement onsets occur before the /i/ offset. This observation is consistent with the very gradual nature of the first phase of movement which was observed qualitatively in sequence plots. A gradual onset of protrusion would have negligible effect on the acoustics of the /i/ and could thus take place during the /i/.[16]

In scatter plots of the acceleration interval versus the acoustic interval (Fig. 10-B), a least-squares linear relationship between these two intervals was significantly different from zero for Subject 3, but not for the other two Subjects. A robust regression line (with a slope of .71) is shown for Subject 3. For Subjects 1 and 2, the spread in values for the acceleration interval grows with increasing values of the acoustic interval.[17] For Subject 1 (with the slowest speaking rate), almost all of the times of peak acceleration occur after the /i/ offset, whereas this is not the case for the other two Subjects. For Subject 2, excluding the group of 6 extreme outliers above the main grouping, times of peak acceleration range from almost 100 msec before /i/ offset to just 20 msec. before /u/ onset. Subject 3 (with the fastest speaking rate) has a distribution of peak acceleration times which is similar to that for his movement onset times, except for an offset of about 70 msec. Thus, for Subjects 2 and 3, a significant proportion of maximum acceleration times occur well before /i/ offset, suggesting a potential influence on the acoustics of the /i/.[18]

4.2.3. *Discussion and Implications.* Further analyses of the data from this experiment are planned, so it is premature to say too much about the results at this stage. However, some general observations can be made which bear on the main point of this paper. In spite of the large amount of observed variability, there are relationships in the data which allow us to reject strong versions of all three models as means of accounting for the results of this

[16]These data do not speak to what underlies the movement onset, whether it is the result of a peak in a slight spreading gesture for the /i/, active onset of a protrusion gesture, or a combination of the two. Further work is needed to explore the nature of the movement onset.

[17]Thus the lack of a least-squares slope which was significantly different from zero for Subjects 1 and 2 could not be considered as support for the "frame theory".

[18]Of course, other articulators, especially the tongue body, are influencing the acoustics of the vowels, so an analysis like this one without additional articulatory and acoustic measurements has its limitations.

particular experiment. One possible interpretation of the results is that we are observing a compromise between two underlying principles. The timing of the articulatory event may be influenced somewhat by the timing of the related acoustic event, offset of the /i/. There could also be a tendency for the movement event to begin at a constant time before the onset of the /u/; that is, the lip protrusion gesture could have some "preferred" inherent dynamics. Variation in the relative strengths of these influences could account for differences observed across subjects and between type of articulatory event (movement onset or peak acceleration). The apparent cross-subject differences associated with their speaking rates tend to reinforce the very tentative notion that we are observing the overlapping influences of acoustic (phonetic) context and inherent dynamics of the lip protrusion gesture.

A number of additional preliminary observations and ideas need to be explored. For example, the sequence plots in Fig. 8 show differences in timing of movement end (peak of the protrusion gesture) with respect to acoustic onset of the /u/ (indicated by the heavy vertical line) across utterances for Subject 1. It is possible that such differences are due to influences of intervocalic consonants, which may not be "neutral" with respect to lip protrusion as assumed by the working hypothesis for this experiment (cf. Perkell and Chiang, 1986). This observation points to one of a number of additional relationships in the data that must be examined. The idea that movement onset is actually the peak in a retraction gesture for the /i/ (S. Boyce, personal communication) has implications for our interpretation of the data; it also needs to be examined in more detail. It is also important to note that because of the large amount of variation in the data, our tentative conclusions are not based on results which are "statistically significant". Even though further analyses of these and additional data are in order, these preliminary results indicate that there is value in such an exploratory approach.

4.3. TOKEN-TO-TOKEN VARIATION OF TONGUE BODY VOWEL TARGETS: THE EFFECT OF CONTEXT

This study used the tongue point displacement data from the third subject in the above-described vowel-target experiment (Section 4.1) to investigate the interaction between token-to-token variation and context effects (Perkell and Cohen, in press).

4.3.1. *Background, Methods and Results.* Recall that for this experiment, the speaker pronounced about 15 (randomized) repetitions of a number of nonsense utterances of the form /bV'(C)Vbʌ/, in which the vowels were all combinations of /i/, /u/, and /a/, and the consonant, when present, was /b/, /ʔ/ or /h/. The single transduced tongue point was close to the place of maximum vocal-tract constriction for /u/. Stress was placed on the second syllable.

Figure 11-A shows a plot of the target locations for all examples of /u/, /i/ and /a/ (minus a few spurious points) that occurred in the stressed (second) syllable position. As in the previous examples, the irregular line at the top of the plot is the outline of the posterior portion of the hard palate. Anterior is to the right. A "two-sigma" ellipse is drawn around each vowel grouping and is centered at its mean location. Part B is the same type of plot (from the same utterances) showing the articulatory target locations for the unstressed (first syllable) vowels.

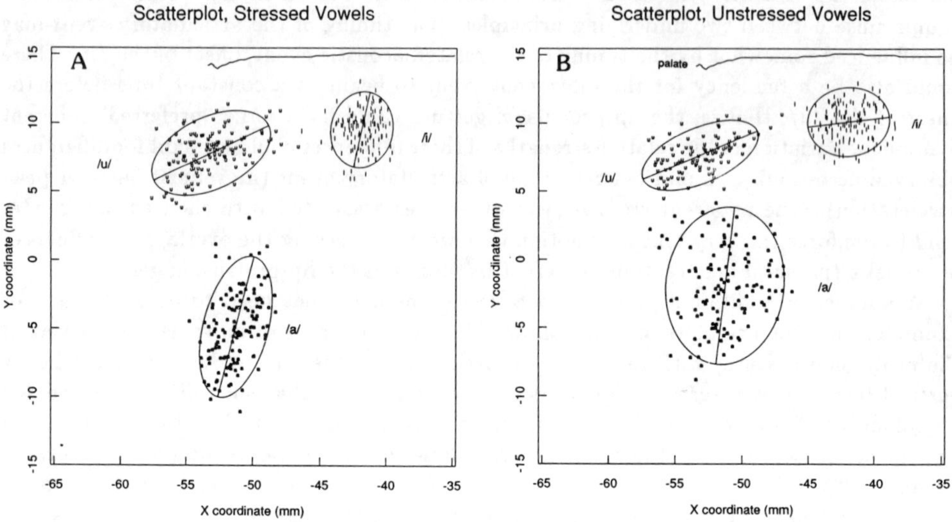

Figure 11: Plot of vowel target locations for all examples of /u/, /i/ and /a/ that occurred in stressed (second) syllables (Part A) and unstressed (first) syllables (Part B) from the second experiment described above in Section 4.1.

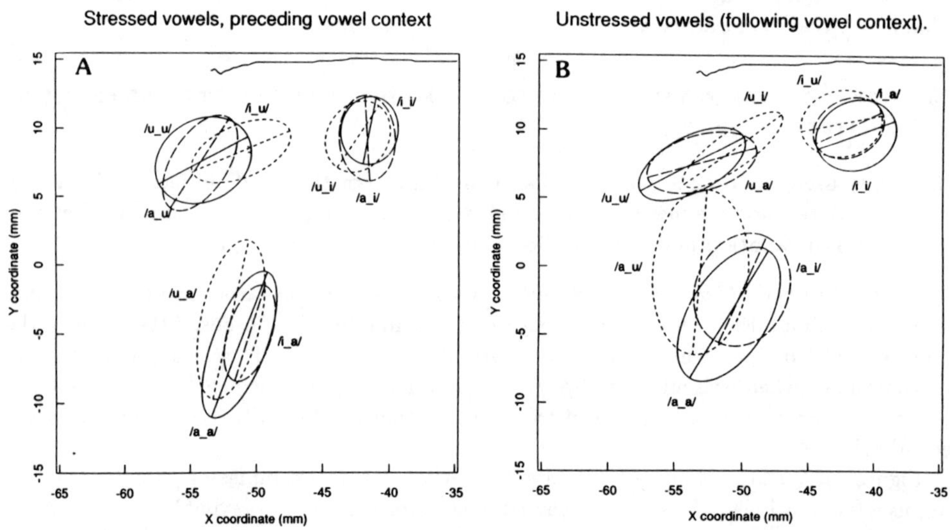

Figure 12: Two-sigma ellipses derived from the data illustrated in Fig. 11. Each ellipse represents the distribution of target points for a vowel, sorted according to context. Part A shows stressed vowels sorted by preceding (unstressed) vowel. Part B shows unstressed vowels, sorted by following (stressed) vowel. See text for details.

Figure 12 shows two-sigma ellipses derived from the same data, each representing the distribution of target points for a vowel, sorted according to context. Part A shows stressed vowels sorted by preceding (unstressed) vowel. Part B shows unstressed vowels, sorted by following (stressed) vowel. For example, in Part A, the ellipse labeled /i_u/ represents all instantiations of stressed /u/ preceded by unstressed /i/ (from the utterances /bi'ubʌ/, /bi'ʔubʌ/, /bi'bubʌ/, /bi'hubʌ/). In part B, /u_i/ represents all instantiations of unstressed /u/ followed by stressed /i/ (from the utterances /bu'ibʌ/, /bu'ʔibʌ/, /bu'bibʌ/, /bu'hibʌ/). In this figure, there is a general tendency for the principal component axes to be "rotated" somewhat in the direction of the context vowel target location, and for the ellipses to be displaced toward the context vowel target location. The strength of these effects varies, being less pronounced for the vowel /i/ than for /u/ or /a/. This difference for /i/ could be due to the fact that the maximum constriction (i.e., the acoustically most important aspect of the articulation) for /i/ is formed by the tongue blade, whereas the maximum constrictions for /u/ and /a/ are formed by the tongue body. The location of the transduced tongue point is more indicative of the tongue body position than tongue blade position, thus the acoustics of the /i/ should be somewhat insensitive to the displacement of this point.

Figure 13 shows the same vowel target positions, sorted according to vowel and consonant context. Part A, for the stressed vowels is derived from the same data as is Fig. 12-A, but the points contained by each ellipse in Fig. 13-A are further divided, according to intervocalic consonant. Thus, the top three ellipses in Fig. 13-A correspond to the stressed (second syllable) vowel targets for the utterances /bu'bubʌ/, /bu'hubʌ/ and /bu'ʔubʌ/, respectively (where ʔ represents no consonant or a glottal stop). Figure 13-B shows the same type of plots, for the unstressed vowel targets from the same utterances. Note that the scales differ among the cells. For the most part, the ellipses in each cell overlap. The most noticeable exception to this observation is for the consonant /b/, which is associated with some "displacement" of targets away from the center of the overall group. This effect is most likely due to anatomical and biomechanical interactions among the lips, mandible and tongue which should influence tongue body position for /b/, but not for /h/ or /ʔ/. The effect is expressed by the /b/ on both stressed (postconsonantal) vowel (Part A) and on the unstressed (preconsonantal) vowel (Part B). For the high vowels /u/ and /i/ (top and bottom rows respectively) the effect of /b/ is to displace the tongue body target for the vowel somewhat forward (to the right of the plot). For the vowel /a/ (middle rows), the effect of the /b/ is to displace the targets upward (Part A, stressed vowels) or forward (Part B, unstressed vowels). These displacements are consistent with the upward and forward positioning of the mandible required for the labial articulation of /b/.

4.3.2. *Discussion and Implications.* In these utterances for a single speaker, there is evidence of context effects of postconsonantal (stressed) vowel articulations on the targets for preconsonantal (unstressed) vowels and vice versa. There is also evidence for context effects of the intervocalic consonant on both the pre- and postconsonantal vowel targets. Thus the effects are overlapping. While the target displacement effects due to context are quite large, they are smaller than the effects of token-to-token variation. Since this evidence is derived from examination of multiple repetitions of the utterances, the effects could be "stochastic" in nature, and might not necessarily be evident in the same way in

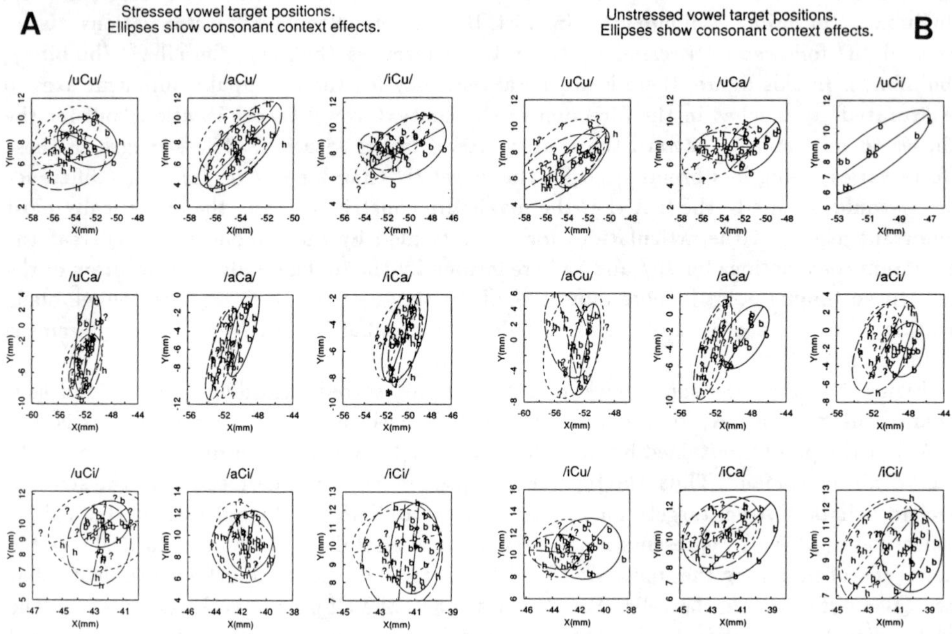

Figure 13: Vowel target positions, sorted according to vowel and consonant context. Parts A and B, are derived from the same data as in Fig. 11-A and 11-B, but the points contained by each ellipse in Fig. 13 are further divided, according to intervocalic consonant. See text for details.

any individual utterance. We conclude that the underlying mechanism for production of these vowels includes "bi-directional" (anticipatory and perseveratory) influences of context which are manifested in some variable or incomplete way in any single production of an utterance. Thus, exploration of the variation in articulatory data can reveal potentially important aspects of underlying mechanisms which might not otherwise be apparent (Perkell and Cohen, in press).

5. Conclusions

The three examples of data presented above have implications for different aspects of our theoretical overview.

Results of the study of variation in articulatory displacements for the vowels /i/, /u/ and /a/ are consistent with the notion that some articulatory targets are defined by quantal or non-linear relationships between parameters; the quantal relationships can be characterized as "saturation effects". A quantal relationship may be between an articulatory and an acoustic parameter or between two different articulatory parameters. But this very tentatively supported idea can only be part of a much bigger picture. An important component of the quantal theory of speech which was only mentioned above very briefly is the claim that the acoustic cues resulting from quantal relationships in speech production are selected out or detected by properties of the perceptual mechanism. The resulting acoustically-mediated combinations of quantal effects in production and perception may form at least part of the basis for definitions of distinctive features. Thus these effects would help to determine sound patterns of languages (Stevens, 1972; 1989). It is likely that a number of other principles also contribute to the nature of the underlying phonetic framework. Interactions which minimize articulatory effort while producing sufficient perceptual contrast seem to be able to predict certain articulatory targets on the basis of numbers of sounds in any particular language (Lindblom, 1975, Lindblom and Engstrand, 1989). These interactions are inherently continuous, as opposed to being quantal. As mentioned above, additional influences could include things as diverse as dynamic properties of articulatory gestures (cf. Saltzman and Munhall, in press) and sociological factors (cf. Labov, 1986).

The preliminary results on anticipatory coarticulation similarly contribute a very small amount of information to what is very likely a complicated aspect of speech production. The results suggest that multiple factors are likely to determine the timing of articulatory movements. Those factors probably include properties of the production mechanism and intended communicative (acoustic) characteristics of the speech signal. We have just begun to explore speech production strategies under the wide range of possible speaking behaviors, and as we proceed to do so, this component of the picture is also likely to appear more complicated.

The future discovery of increasing complexity is hinted at by the small study of context effects on vowel targets, with its demonstration of "stochastic" bidirectional and overlapping influences of articulatory targets on one another. Such effects may have implications for future experimental and theoretical work.

Considered together, the results suggest that as we are able to examine speech production in more detail, we will be confronted not only with increasing complexity, but also with a great deal of variability. However, by using appropriate techniques of data collection and analysis, it should be possible to systematically examine different sources of variability and look beyond that variability to further refine our understanding of underlying mechanisms.

Acknowledgements

Much of the work behind the data presented above was done by Marc Cohen (Sections 4.1 and 4.3) and Chih-ming Chiang (Section 4.2; see Chiang, 1987), to whom I am most grateful. I also thank Alex Samarov for help with the data analysis in the coarticulation experiment and Suzanne Boyce, Harlan Lane, Victor Sorokin and Mario Svirsky for their very helpful comments. This research was supported by N.I.H. Grant No. NS04332.

References

Abbs, J.H. (1986) 'Invariance and variability in speech production: A distinction between linguistic intent and its neuromotor implementation', in J.S. Perkell and D.H. Klatt (eds.), *Invariance and Variability in Speech Processes*, Lawrence Erlbaum Associates, Hillsdale, New Jersey, pp. 202-218.

Alfonso, P.J., Honda, K. and Baer, T. (1984) 'Coordinated tongue muscle activity during /əpVp/ utterances', Proceedings of the Tenth International Congress of Phonetic Sciences, IIB, pp. 390-394.

Bell-Berti and Harris (1979), 'Anticipatory coarticulation: Some implications from a study of lip rounding', **J. Acoust. Soc. Am., 65**, 1268-1270.

Bell-Berti and Harris (1981) 'A temporal model of speech production', **Phonetica, 38**, 9-20.

Benguerel, A.-P. and Cowan, H.A. (1974) 'Coarticulation of upper lip protrusion in French', **Phonetica, 30**, 41-55.

Bladon, A. and Al-Bamerni, A. (1982) 'One stage and two-stage temporal patterns of velar coarticulation', **J. Acoust. Soc. Am., 72**, S104(A).

Browman, C.P. and Goldstein, L.M. (1986) 'Towards an articulatory phonology', **Phonology Yearbook, 3**, 219-251.

Chiang, C-m. (1987) Models of Coarticulation, B.S. Thesis, Massachusetts Institute of Technology.

Edwards, J. (1985) 'Contextual effects on lingual-mandibular coordination', **J. Acoust. Soc. Am., 78**, 1944-1948.

Fant, G. (1960) *Acoustic theory of speech production*. The Hague: Mouton.

Folkins, J.W. and Brown, C.K. (1987) 'Upper lip, lower lip and jaw interactions in speech: comments on evidence from repetition-to-repetition variability', **J. Acoust. Soc. Am., 82**, 1919-1924.

Fowler, C. (1980) 'Coarticulation and theories of extrinsic timing', **J. Phonetics, 8**, 113-133.

Fujimura, O., and Kakita, Y. (1979) 'Remarks on the quantitative description of lingual articulation', in B. Lindblom and S. Öhman, (eds.), *Frontiers of Speech Communication Research*, Academic Press, London, pp. 17-24.

Gelfer, C.E., Bell-Berti, F. and Harris, K.S. (1982) 'Determining the extent of coarticulation', Expanded version of a paper presented at the 103rd meeting of the Acoustical Society of America, Chicago.

Gentil, M., Gracco, V.L. and Abbs, J.H. (1983) 'Multiple muscle contributions to labial closure during speech: Evidence for inter-muscle motor equivalence', Proc. Int. Cong. Acoust., **11**, 11-14.

Henke, W.L. (1967) 'Preliminaries to speech synthesis based on an articulatory model', Proc. 1967 IEEE Boston Speech Conference, 170-171.

Hughes, O.M. and Abbs, J.H. (1976) 'Labial-mandibular coordination in the production of speech', **Phonetica, 33**, 199-221.

Kent, R.D. and Minifie, F.D. (1977) 'Coarticulation in recent speech production models', **J. Phonetics, 5**, 115-135.

Labov, W. (1986) 'Sources of inherent variation in the speech process', in J.S. Perkell and D.H. Klatt (eds.), *Invariance and Variability in Speech Processes*, Lawrence Erlbaum Associates, Hillsdale, New Jersey, pp. 402-422.

Lindblom, B. (1975) 'Experiments in sound structure', Plenary address, Eighth International Congress of Phonetic Sciences, Leeds, England.

Lindblom, B. and Engstrand, O. (1989) 'In what sense is speech quantal?', **J. Phonetics, 17**, 107-122.

Lubker, J.F. (1981) ' Temporal aspects of speech production: Anticipatory labial coarticulation', **Phonetica, 38**, 51-65.

Lubker, J.F. and Gay, T. (1982) 'Anticipatory labial coarticulation: Experimental, biological and linguistic variables', **J. Acoust. Soc. Am., 71**, 437-448.

Mrayati, M., Carré, R. and Guerin, B. (1988) 'Distinctive regions and modes: A new theory of speech production', **Speech Communication, 7**, 257-286.

Öhman, S.E.G. (1966) 'Coarticulation in VCV utterances: Spectrographic measurements', **J. Acoust. Soc. Am., 39**, 151-168.

Öhman, S.E.G. (1967) 'Numerical models of coarticulation', **J. Acoust. Soc. Am., 41**, 310-320.

Perkell, J.S. (1974) A physiologically-oriented model of tongue activity during speech production, Ph.D. Thesis, Massachusetts Institute of Technology.

Perkell, J.S. (1979) 'On the nature of distinctive features: Implications of a preliminary vowel production study', in B. Lindblom and S. Ohman, (eds.), *Frontiers of Speech Communication Research*, Academic Press, London, pp. 365-380.

Perkell, J.S. (1980) 'Phonetic features and the physiology of speech production', in B. Butterworth (ed.), *Language Production*, Academic Press, London, pp. 337-372.

Perkell, J.S. and Chiang, C. (1986) 'Preliminary support for a "hybrid model" of anticipatory coarticulation', Proceedings of the 12th International Congress of Acoustics, A3-6.

Perkell, J.S. and Cohen, M.H. (1986) An alternating magnetic field system for tracking multiple speech articulatory movements in the midsagittal plane, Technical Report No. 512, Research Laboratory of Electronics, Massachusetts Institute of Technology.

Perkell, J.S. and Cohen. M.H. (1989) 'An indirect test of the quantal nature of speech for the vowels /i/, /a/ and /u/', **J. Phonetics, 17**, 123-134.

Perkell, J.S. and Cohen, M.H. (in press) 'Token-to-token variation of tongue-body vowel targets: the effect of context', Festschrift for Osamu Fujimura.

Perkell, J.S. and Nelson, W.L. (1982) 'Articulatory targets and speech motor control: A study of vowel production', in S. Grillner, A. Persson, B. Lindblom, and J. Lubker, (eds.), *Speech Motor Control*, pp. 187-204.

Perkell, J.S. and Nelson, W.L. (1985) 'Variability in production of the vowels /i/ and /a/', **J. Acoust. Soc. Am., 77**, 1889-1895.

Rack, P.M.H. and Westbury, D.R. (1969) 'The effect of length and stimulus rate on tension in the isometric cat soleus muscle', **J. Physiology, 204**, 443-460.

Recasens, D. (1984) 'Vowel-to-vowel coarticulation in Catalan VCV sequences', **J. Acoust. Soc. Am., 76**, 1624-1635.

Saltzman, E.L. and Munhall, K. (in press) 'A dynamical approach to gestural patterning in speech production', **Ecological Psychology**.

Stevens, K.N. (1972) 'The quantal nature of speech: Evidence from articulatory-acoustic data', in E.E. David and P.B. Denes, *Human Communication: A Unified View*, McGraw-Hill, New York, pp. 51-66.

Stevens, K.N. (1989) 'On the quantal nature of speech', **J. Phonetics, 17**, 3-46, 145-157.

Smith, B.L. and McLean-Muse, A. (1987) 'An investigation of motor equivalence in the speech of children and adults', **J. Acoust. Soc. Am., 82**, 837-842

Smith, T. (1971) A phonetic study of extrinsic tongue muscles. U.C.L.A. Working Papers in Phonetics, **18**.

Sussman, H.M. (1980) 'Methodological problems in evaluating lip/jaw reciprocity as an index of motor equivalence', **J. Speech Hear. Res., 23**, 699-704.

Sussman, H.M. and Westbury, J.R. (1981) 'The effects of antagonistic gestures on temporal and amplitude parameters of anticipatory labial coarticulation', **J. Speech Hear. Res., 16**, 16-24.

Wood, S. (1979) 'A radiographic examination of constriction location for vowels', **J. Phonetics, 7**, 25-43.

Wright, C.E. and Meyer, D.E. (1983) 'Conditions for a linear speed-accuracy trade-off in aimed movements', **Quart. J. Exp. Psych., 35A**, 279-296.

SPEECH AS AUDIBLE GESTURES

Anders Löfqvist
Department of Logopedics and Phoniatrics,
University Hospital
S-221 85 Lund,
Sweden
and
Haskins Laboratories,
New Haven, CT,
USA

"Speech is rather a set of movements made audible than a set of sounds produced by movements."
R. H. Stetson

ABSTRACT. Speech can be described as audible movements. By varying the positions and trajectories of the lips, the jaw, the tongue, the velum, and the glottis, a speaker creates variations in air pressure and airflow in the vocal tract. These variations in pressure and flow produce the acoustic signal that we hear when listening to speech. For this acoustic signal to be not only audible, but also structured in such a way that it can transmit linguistic information, the articulators have to be controlled and coordinated so that the acoustic variations in the produced signal conform to the phonetics and phonology of the language being spoken. This paper examines several aspects of the control and coordination of articulatory gestures during speech, with particular emphasis on gestural patterning, gestural coherence, and gestural aggregation

1. The Segment and its Structure

Preliminary to this discussion, it is instructive to present a view of the structure of the segment. Since there are several well known arguments against the usefulness of the segment in the analysis of speech, a brief examination of some of them is in order. Among the objections to the segment are its contextual variability at both the articulatory and acoustic levels, the problem of finding boundaries between successive segments in articulation and acoustics, and the variability between tokens of the "same" segment produced by different speakers. In spite of these arguments, more general considerations of speech communication (and my metaphysics) have convinced me that giving up the segment is bound to create even more serious problems (cf. Fowler, 1980; Fowler, Rubin, Remez and Turvey, 1980, for informed discussions of these complications and related matters). Note, incidentally, that the use of syllables instead of segments will not resolve any of the

W. J. Hardcastle and A. Marchal (eds.), Speech Production and Speech Modelling, 289–322.

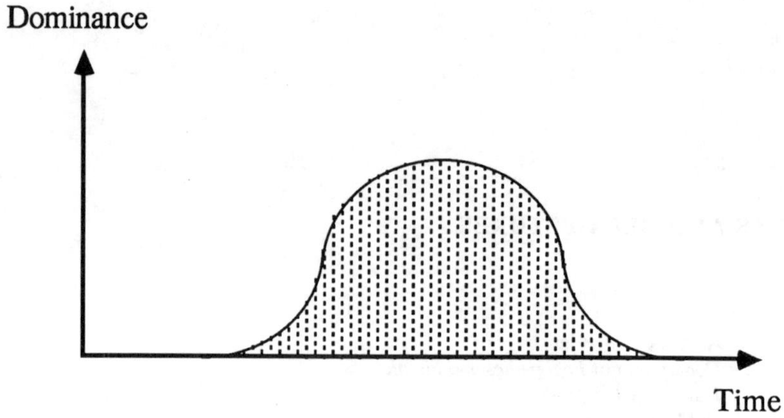

Figure 1. A representation of the segment in terms of its influence (dominance) on the vocal tract over time.

problems listed above, since, to the best of my knowledge, no one has been able to show unique articulatory and/or acoustic characteristics of syllables, or the existence of clear boundaries between syllables in speech.

The role of the segment can be defined as providing internal structure to words. We can think of a communication system, a language, without segments, where the primary units were words, or morphemes. Without an internal structure, they would be indivisible units. Each such unit would then have to be represented by a set of related articulatory/acoustic patterns. If, however, the lexicon of such a language increased, the units would presumably have to be rather long. Otherwise, it would have been difficult to keep them apart, acoustically, since they could not be built up by smaller building blocks, i.e. segments. Such a communication system would, furthermore, have required a large acoustic memory capable of retaining a broad band acoustic analog representation of the speech signal. The emergence of internal structure to words, i.e. of segments, obviates the need for such a memory. The internal structure can now be represented by discrete units, segments, rather than by the acoustic signal (cf. Mandelbrot, 1954, who also provides further justification for why segments exist in speech).[1]

We can view segments as an aspect of a more general phenomenon of perceptual and cognitive organization, i.e. categorization. The collapsing of different phenomena in the real world under the same category is a common capacity that is not limited to humans. At least some such categories are rooted in the biology of the perceptual system (see Brown, 1976; Rosch and Lloyd, 1978). Various non-human species have been trained to discriminate between the linguistic categories of human speech, but the proper interpretation of these studies remains debatable (see Remez, 1989). More

[1] "Pour qu'un message porté par une onde acoustique puisse etre relayé, et en particulier enseigné, il faut qu'il soit segmenté en morceaux discrets." (Mandelbrot, 1954, p. 6.) Briefly, his argument is that the use of an acoustic signal for the transmission of messages requires that the signal be decomposed into discrete units; otherwise a message would be difficult to transmit between people with different voice characteristics. Strictly speaking, this is not an argument for segments in particular but rather for some kind of discrete unit, such as segments or syllables. The problem of the static versus continuous nature of segments can be rationalized using Pattee's distinction between dynamic and linguistic modes of description (Pattee, 1977).

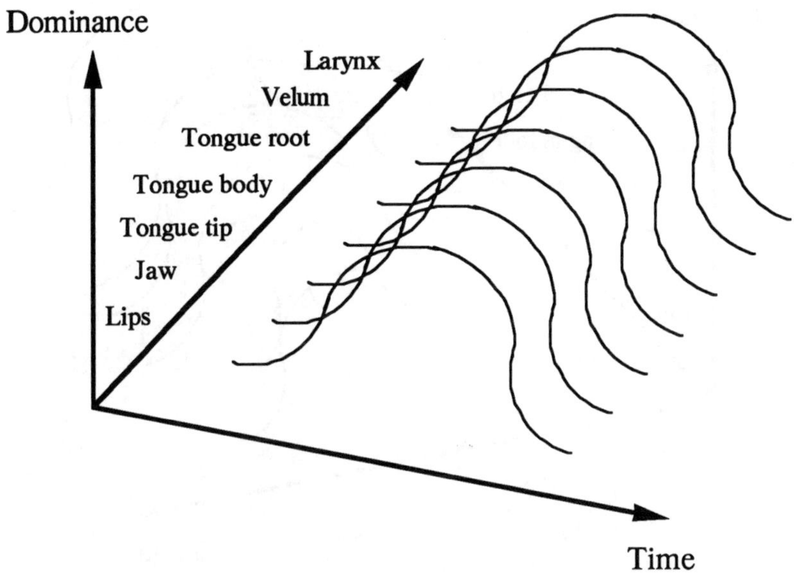

Figure 2. Same as Figure 1, but showing the dominance on different parts of the vocal tract.

importantly, however, similar categorization also occurs for ecological valid stimuli such as bird song (Nelson and Marler, 1989). The problem of contextual and interspeaker variability can thus be accounted for as far as acoustics and perception is concerned. Its rationalization in articulation will be taken up below. The search for absolute boundaries between segments in acoustics and articulation is misguided. All that is required is that there are detectable changes from one segment to another, irrespective of whether the exact boundary between them can be located or not. To take an example, if you are going down a river in a canoe approaching a waterfall, and happily make it through the fall, you can most definitely say that there was a transition, although you may not be able to tell exactly where the waterfall started. Moreover, some acoustic analysis techniques are in fact able to locate regions of major spectral changes in the speech signal that bear at least a superficial relation to boundaries between linguistic segments (cf. Atal, 1983; van Dijk-Kappers and Marcus, 1987; Bimbot, Ahlbom and Chollet, 1987; van Dijk-Kappers, 1988). These techniques also offer some possibilities of recovering the articulatory gestures underlying the acoustic signal.

After this digression, let us return to the structure of the segment. It is well known that the segments are not produced strictly sequentially but overlap with each other. This is referred to as coarticulation, coproduction, or blending. For example, a voiceless fricative /s/ preceding a rounded vowel is normally produced with rounded lips; however, when the fricative occurs before an unrounded vowel, no lip-rounding occurs. As a consequence of coarticulation, the vocal tract shape at any one time during speech represents an aggregate of gestures associated with different segments (cf. Fant, 1962). Coarticulation is pervasive in speech. Indeed, some researchers have claimed that it is a fundamental characteristic of speech and forms the basis for the rapidity with which information can be conveyed in speech (see Liberman, Cooper, Shankweiler, and Studdert-

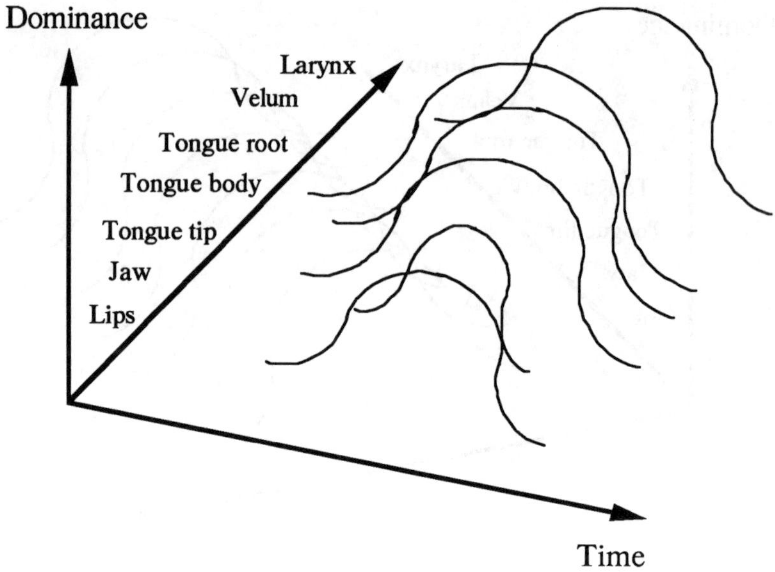

Figure 3. Same as Figure 2, but with different dominance relations for different parts of the vocal tract.

Kennedy, 1967).

One way of illustrating the structure of the segment is shown in Figure 1. Time runs along the x-axis, and the y-axis represents what can be called dominance.[2] It shows how much the segment influences the vocal tract at any one point in time, and thus represents the temporal evolution of the segment in the vocal tract. Moving along the time axis in Figure 1, we see that, initially, the segment has no influence on the vocal tract at all. At one point in time, it begins to affect the vocal tract, grabbing it, and influencing it more and more until it dominates it completely. Then, gradually, the segment releases its grip on the vocal tract until, finally, it has lost all influence.

The picture shown in Figure 1 is incomplete, since we have to acknowledge that the vocal tract consists of several parts that can be moved and controlled independently of each other; this is one of the bases of coarticulation. This concept is illustrated in Figure 2, where the segment is shown with an internal structure; each tract variable has its own temporal pattern of dominance. Here, all the dominance profiles are the same, i.e. the segment begins to influence all parts of the vocal tract at the same point in time and also to the same degree. This is a simplification. There is evidence, however, that such a temporal synchrony can be a correct representation under certain conditions, while in other cases there has to be an asynchrony between the dominance patterns for different parts of the vocal tract. Figure 3 elaborates this point by illustrating how a segment can

[2] This is similar to the "innervation waves" discussed by Joos (1948), and to the ideas about "prominence" (Fowler, 1983), and "gestural activation waves" (Saltzman and Munhall, in press). It may also be useful to make further distinctions between movements and gestures, and between tract variables and articulators, but these issues are beyond the scope of the present paper; a discussion of them can be found in Saltzman and Munhall (in press). While some more detail is added in Figure 2, it is still incomplete since the respiratory system has been left out.

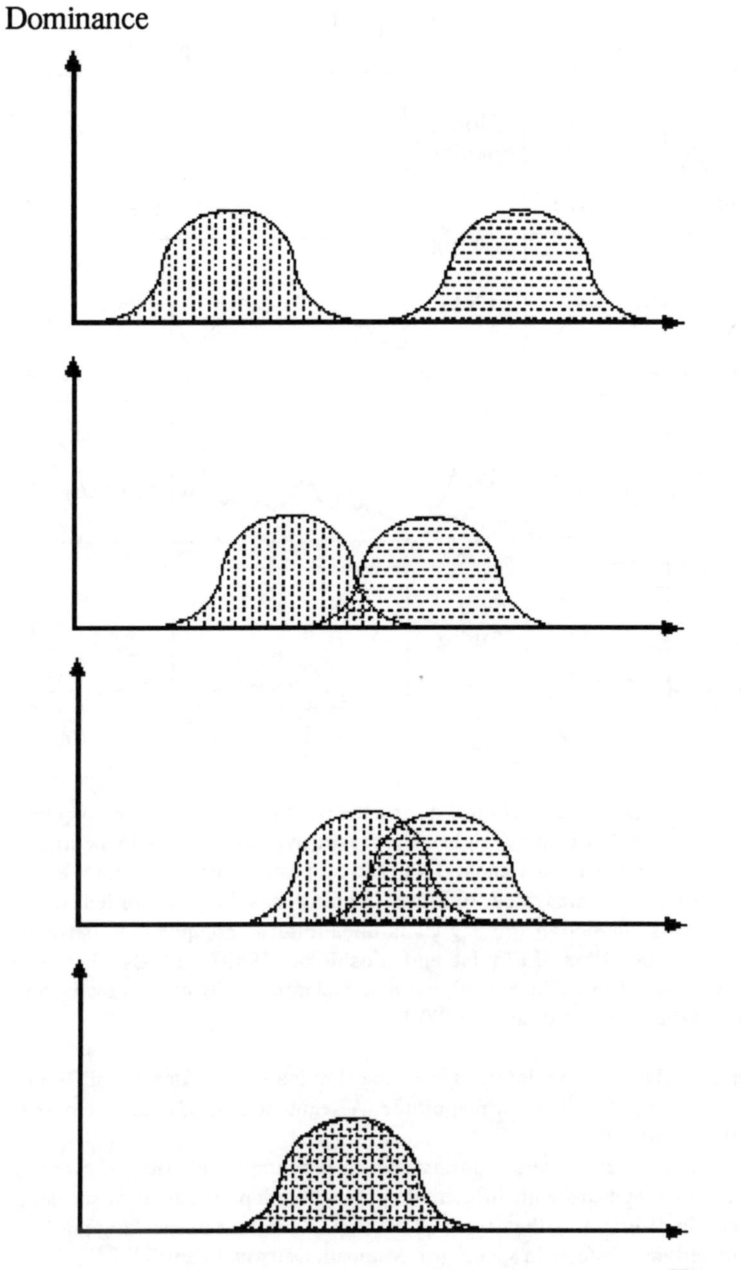

Figure 4. Representation of two segments with different degrees of overlap.

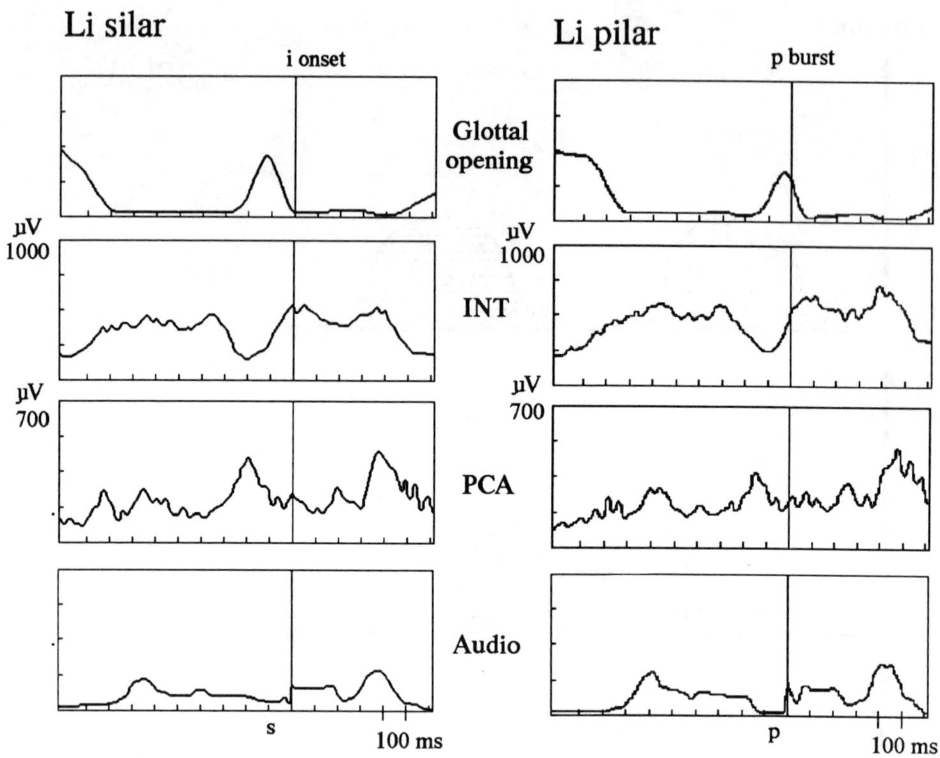

Figure 5. The glottal gesture and its motor control during the production of a single voiceless fricative /s/ (left) and a single voiceless stop /p/ (right). Curves are averages of 15-20 repetitions. The electromyographic signal from the interarytenoid (INT) and the posterior cricoarytenoid (PCA) muscles have been rectified and integrated. Glottal opening has been recorded using transillumination of the larynx. Although this signal is uncalibrated, comparisons between transillumination and fiberoptic films (Löfqvist and Yoshioka, 1980) and also between transillumination and high-speed films (Baer, Löfqvist and McGarr, 1983) have shown good agreement. (Modified from Löfqvist and Yoshioka, 1980.)

differ in its utilization of different articulators. Here, the dominance profiles for different articulators are offset in time and also differ in magnitude. A segment can, of course, have no influence at all on a specific articulator.

The gestures for different segments overlap during speech production. Figure 4 shows the simplified representation of two segments with different degrees of overlap. Presumably, speaking rate is one important factor in determining the degree of overlap. Variations in speaking rate can thus be used to uncover underlying gestures in speech (cf. Munhall, Ostry and Parush, 1985; Ostry and Munhall, 1985).

In the following, I will examine the internal structure of different speech segments, how the articulatory gestures within and across segments are organized, and some of the principles

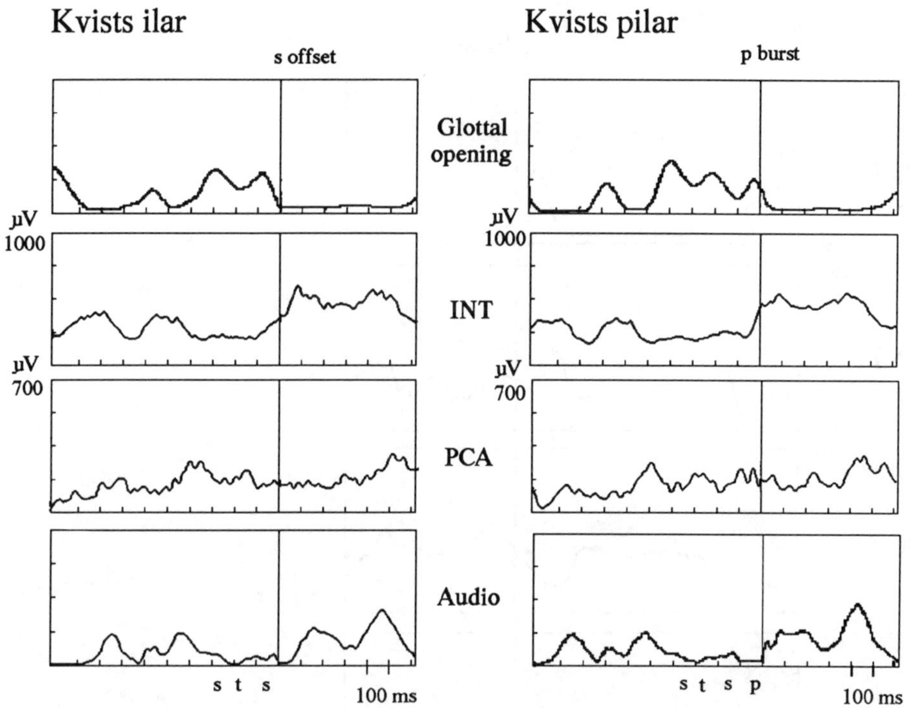

Figure 6. Glottal gestures and electromyographic activity of intrinsic laryngeal muscles during the production of two voiceless clusters /sts/ (left), and /sts#p/ (right). Curves as in Figure 5. (Modified from Löfqvist and Yoshioka, 1980.)

governing the aggregation of speech gestures. The examples will be drawn from different areas of speech production based on experimental investigations of articulation.

2. An Example - Voiceless Obstruents

Production of voiceless obstruents, stops and fricatives, requires cooordination of several articulatory systems. In the larynx, an abduction/adduction gesture of the glottis normally occurs to assist in the arrest of glottal vibrations and in the buildup of oral pressure. Movements of upper articulators are also necessary to produce a closure or constriction. These articulatory adjustments are made to produce the increase in oral pressure that drives the noise source associated with these consonants. The laryngeal and oral articulations must thus be coordinated within short temporal intervals.

The laryngeal articulatory gesture during voiceless consonant production has the form of a continuous opening and closing movement, cf. Figure 5 showing the glottal gesture and the activity of two intrinsic laryngeal muscles during the production of a single voiceless fricative and

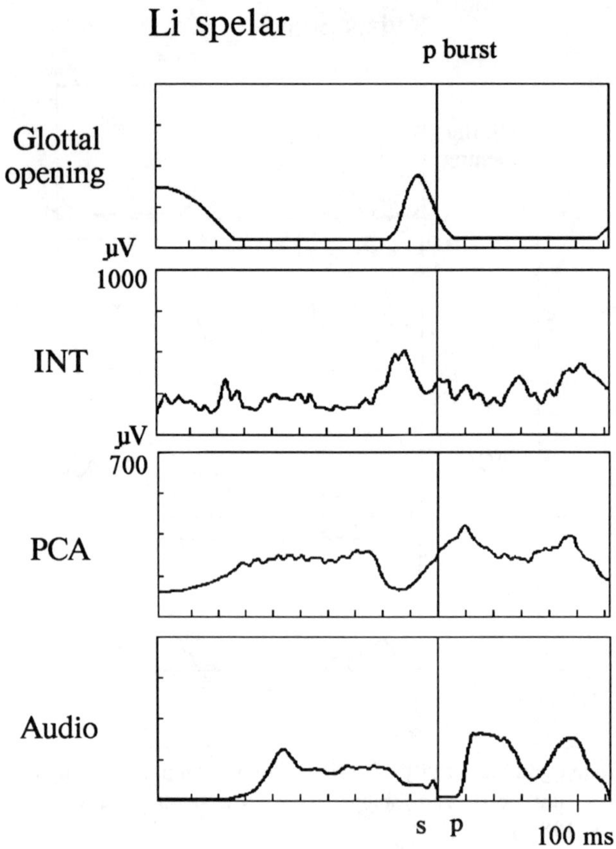

Figure 7. The glottal gesture and its motor control during the production of the voiceless cluster /sp/. Curves as in Figure 5. (Modified from Löfqvist and Yoshioka, 1980.)

stop, respectively. The glottis never appears to open and maintain a static open position during speech (see Löfqvist, 1980; Sawashima and Hirose, 1983). In clusters of voiceless consonants,one or more glottal gestures occur (Löfqvist and Yoshioka, 1980, 1981a,b; Yoshioka, Löfqvist, and Hirose, 1981; Yoshioka, Löfqvist, and Collier, 1982). This is illustrated in Figure6. The principle governing the occurrence of gestures in this case is that sounds requiring a high rate of airflow, such as fricatives and aspirated stops, are produced with a separate gesture. Thus, two, or even three, successive laryngeal gestures have been observed in voiceless clusters. In such clusters, variations in glottal opening occur although the glottis never closes completely during the cluster.

The clusters shown in Figure 6 exemplify situations where several gestures occur in succession. Here, a separate gesture occurs for the /st/ and /s/ in "Kvists ilar", and for the /st/, /s/, and /p/ in "Kvists pilar", respectively. We can also have a cluster where the gestures for the participating segments show greater overlap, as in Figure 7. Here, the sequence /sp/ has only one

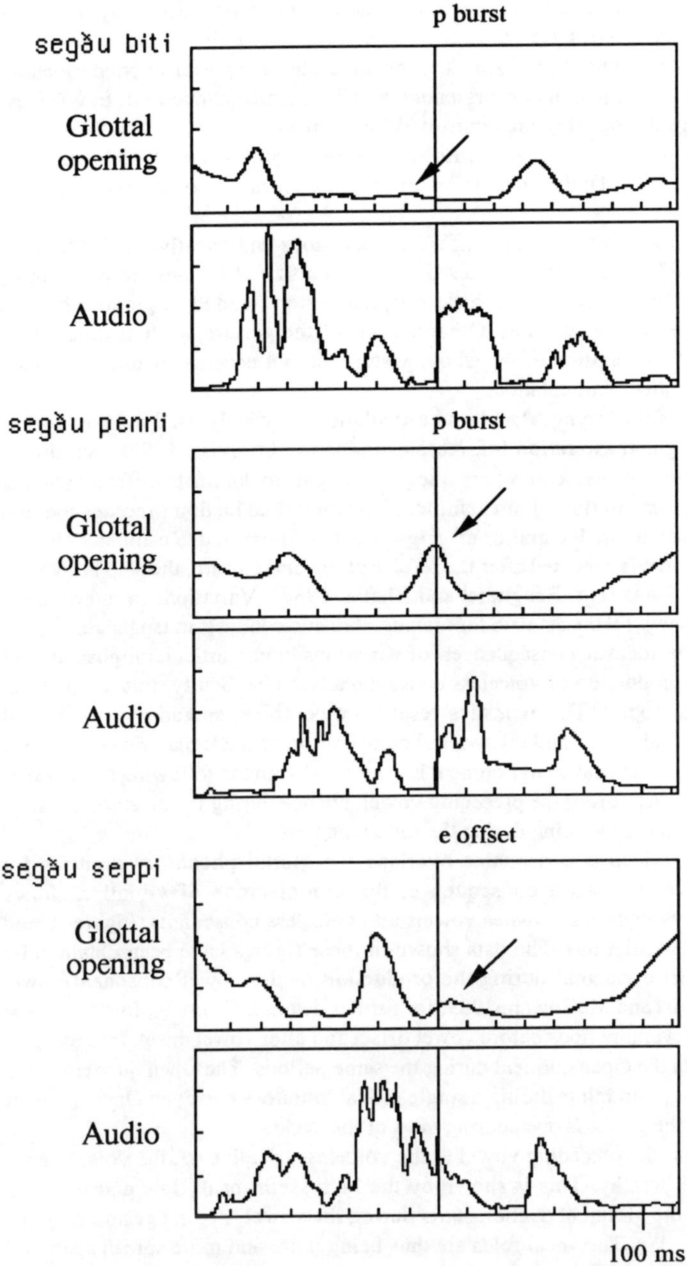

Figure 8. Glottal activity during the production of three different voiceless stops in Icelandic, unaspirated (top), postaspirated (middle), and preaspirated (bottom); the arrow points at the glottal gesture associated with the stop. (Modified from Löfqvist and Yoshioka, 1981a.)

glottal gesture. This may be regarded as the sum of two underlying gestures. Similarly, in Figure 6, the cluster /st/ is produced with a single gesture.

The glottal gesture is controlled by the intrinsic laryngeal muscles, the posterior cricoarytenoid (abduction), the interarytenoid, the lateral cricoarytenoid, and the vocalis (adduction). In voiceless clusters, the variations in glottal opening are controlled by the posterior cricoarytenoid and the vocalis muscles, which show a reciprocal pattern of activation (Löfqvist and Yoshioka, 1980; Löfqvist, McGarr, and Honda, 1984); the other adductor muscles appear to be suppressed in such clusters

Some aspects of the laryngeal gesture may differ between stops and fricatives (cf. Munhall, Ostry, and Parush, 1985; Löfqvist and McGarr, 1987). Thus, the size of the gesture is generally larger for fricatives than for stops. Similarly, voiceless aspirated stops tend to be produced with a larger gesture than their unaspirated cognates. Other aspects of the gesture, such as duration of abduction and adduction, or the duration of the whole gesture, do not necessarily differ between consonants with different manner of articulation.

Variations in the timing of the laryngeal and oral articulations is widely used in languages to produce contrasts of voicing and aspiration (cf. Abramson, 1977; Löfqvist, 1980). Within the current framework, this can be seen as a case where a segment begins to dominate different parts of the vocal tract at different points in time. For example, if the glottal adduction precedes the oral closure, preaspiration results as in Icelandic, cf. Figure 8 (Löfqvist and Yoshioka, 1981a). Conversely, if the glottal gesture is executed after the release of the oral closure, the stop is a voiced (or breathy voiced) aspirated one (e.g. Benguerel and Bhatia, 1980). Variations in intergestural timing is thus a simple and powerful mechanism for making phonetic contrasts in language.

Let us briefly examine the acoustic consequences of variations in interarticulator phasing and gestural overlap during the production of voiceless consonants (see also Scully, this volume, for additional discussion of this topic). The acoustics results are complex, spread over time, and involve differences in the sound source and the spectral composition of the signal. For example, a voiceless (post)aspirated stop in a vocalic environment is associated with the following sequence of source changes: periodic voicing during the preceding vowel, silence during the closure, transient noise, aspiration noise, and periodic voicing during the following vowel.

The glottal gesture for voiceless consonants overlaps the glottal phonatory gestures for preceding and following vowels. As a consequence, the glottal mode of vibration shows continuous variations at the boundaries between vowels and voiceless consonants. Figures 9 and 10 provide examples of such variations. The data shown in these figures have been obtained by inverse filtering the oral airflow signal during the production of the vowel-consonant-vowel sequences shown (see Löfqvist and McGowan, 1989, for further details). Figure 9 plots peak flow during the glottal pulse for several periods before vowel offset and after vowel onset, respectively. Figure 10 shows variations in the Open quotient during the same periods. The Open quotient is the ratio of the open phase to the period time during a single glottal vibratory cycle; an Open quotient close to 1 thus indicates that the glottis is open during most of the cycle.

During the transition from the preceding vowel to the voiceless fricative /s/, the voice source becomes successively more breathy. This is shown by the increase in peak flow and the open quotient. The reason is that the glottal abduction starts during the vowel, i.e. the gestures for the vowel and the consonant overlap. The vocal folds are thus being more and more spread apart and the vibrations continue with the folds being less and less in contact with each other during each vibratory cycle. For the voiceless stop, on the other hand, a similar pattern is observed for speaker RSS (American English) but not for AL (Swedish). This difference may reflect variations between

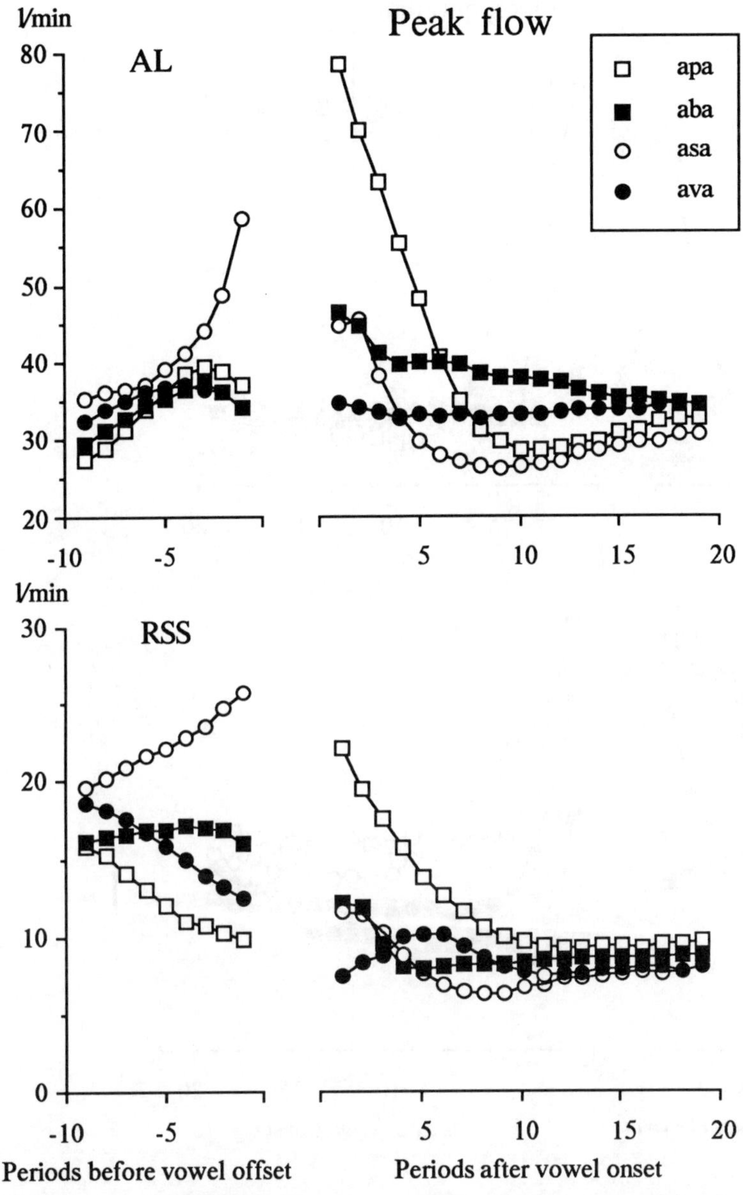

Figure 9. Peak flow of the glottal pulse during consonant-vowel and vowel-consonant transitions. The graphs plot peak flow during several cycles before vowel offset and after vowel onset for a male (AL) and a female (RSS) speaker. Each data point represents the mean of twelve measurements. The speaking fundamental frequency differs for the two subjects: 120 Hz for AL and 210 Hz for RSS, respectively. Hence, the same number of glottal cycles do not correspond to the same temporal interval. (From Löfqvist and McGowan, 1989.)

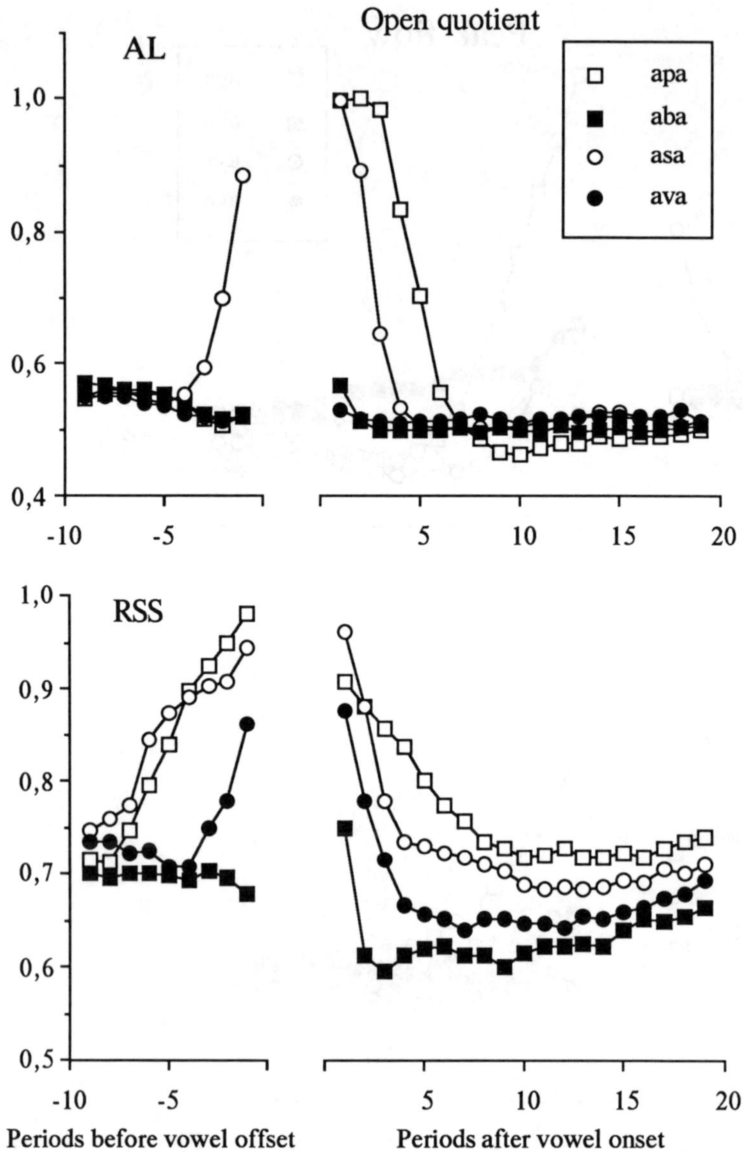

Figure 10. Open quotient of glottal vibrations during consonant-vowel and vowel-consonant transitions. (From Löfqvist and McGowan, 1989.)

languages (cf. Gobl and Ní Chasaide, 1988). As expected, no specific laryngeal adjustment appears to be made for the voiced consonants where peak flow decreases at the onset of the oral closure/constriction.

At the transition from the consonant to the vowel in Figures 9 and 10, the voice source is characterized by breathiness for both the voiceless stop and fricative. Here, the stop is aspirated, and the glottis is open at the oral release. The vocal folds are thus being adducted following the release while the rate of airflow is high. Note that for both peak flow and Open quotient, there is a pattern of decrease and increase after the vowel onset in the voiceless context. That is, the curves start at high values, drop to a minimum around periods 10-15, and then show a small rise. This can, again, be seen as a result of overlap between the gestures for the voiceless consonant and the following vowel. During glottal adduction at the transition between the consonant and the vowel, the vocal folds are forcefully brought together which results in a tight glottal closure evidenced by reduced values of peak flow and Open quotient. The degree of adduction then decreases to a setting appropriate for the vowel. Another factor that might also be involved (pointed out to me by John Ohala) is the reduction in subglottal pressure that commonly occurs during vowels following voiceless fricatives and aspirated stops (see Löfqvist, 1975, for measurements).

At the acoustic level, the change from modal to breathy phonation is accompanied by an increase in the amplitude of the fundamental and an increased tilt[3] of the source spectrum. Noise components also appear in the source signal due to turbulence in the open glottis, and additional resonances due to tracheal coupling may also be present (cf. Klatt and Klatt, in press).

In addition to these source changes, there are others that can also be accounted for by gestural overlap. For example, the vowel following a voiceless consonant usually has a high and falling fundamental frequency pattern (e.g. Ohde, 1984). This is most likely due to an increase in the activity of the cricothyroid muscle that is made to increase the longitudinal tension of the vocal folds and assist in suppressing glottal vibrations (Löfqvist, Baer, McGarr, and Seider Story, 1989). The rationale for this specific adjustment is that the glottal opening in itself may not necessarily arrest glottal vibrations but only do so in combination with aerodynamic and myodynamic factors. That such a suitable combination of aerodynamic and myodynamic factors is necessary can be inferred from the fact that the laryngeal fricative /h/ is often produced with a large glottal opening and yet the vocal folds may continue to vibrate, since the airflow through the glottis is not interrupted for this sound (cf. Yoshioka, 1981).

3. Gestural Coherence

It is reasonable to assume that the gestures forming a segment show internal stability in the form of coherence of patterns of muscle activity and/or movement. Following Lee (1984), we can describe gestural coherence at the spatial, temporal and scaling levels. Spatially, the same set of muscles should always be activated. Temporally, synchronicity, fixed temporal order, or fixed-phase relationships should occur among events. Finally, scaling should occur among elements once all elements are activated above threshold. In this section, I will review some proposed examples of gestural coherence. These examples are stable temporal relationships among gestures across variations in stress and speaking rate, and rapid compensations to induced mechanical perturbations during speech.

[3] The tilt of the source spectrum refers to the decrease in the amplitude of the higher harmonics, expressed in dB per octave.

3.1. TEMPORAL STABILITY

From the point of view of motor control it is attractive to assume that variations of a given movement pattern result from scaling between the individual parts making up the complex. That is, there is a generalized motor program that can be reparameterized (cf. Schmidt, 1975). For example, if a person writes a word on a piece of paper with a pencil or on a blackboard with a piece of chalk, different parts of the body are used - in the former case the hand and movements around the wrist, in the latter case the arm and movements around the shoulder joint. Since the written pattern on the blackboard can be regarded as a scaled up version of the pattern on the paper, it has generally been argued that there is but one underlying representation of the movement pattern that is instantiated by different parts of the body using a scaling relation. The alternative view that each pattern is somehow stored as a separate entity is at least intuitively implausible and inefficient. The generalized motor program predicts that when variations in speed and amplitude of a movement complex occur, the relationship between the individual movements should remain virtually unchanged. Many studies of different areas of motor control have purportedly shown this to be the case. For example, the model has been applied to locomotion (Grillner, 1975; Shik and Orlovsky, 1976; Shapiro, Zernicke, Gregor, and Diestel, 1981), handwriting (Viviani and Terzuolo, 1980), and typing (Terzuolo and Viviani, 1979).

Studies of speech motor control using the same experimental paradigm (i.e. varying stress and speaking rate) have argued that the model also applies to speech. At the intersegmental level, Tuller, Kelso, and Harris (1982), and Tuller and Kelso (1984) showed that the relationship between successive jaw lowerings for the vowels in a VCV sequence and lower lip raising for the medial (labial) consonant was stable across changes in stress and speaking rate. Similarly, at the intrasegmental level, Löfqvist and Yoshioka (1984) argued that the same was true for temporal relationships between oral and laryngeal gestures in voiceless consonant production.

The speech timing experiments not only extended the proportional timing model to speech but could possibly also rationalize some results on perceptual constancy in speech perception. In particular, studies by Port (1979), Fitch (1981), Miller and Grosjean (1981), and Summerfield (1981) have shown how variations in speaking rate affect the perceptual boundaries between segments. For example, it is well known that voice onset time (VOT, see Lisker and Abramson, 1964) is an important acoustic cue for the voiced-voiceless distinction in stop consonants (cf. Lisker, 1986, for a catalogue of other cues). Since the actual duration of VOT will change with speaking rate (e.g. Löfqvist and Yoshioka, 1984), the perceptual voiced-voiceless boundary should also shift, and this is indeed the case. While the production and perception studies would thus seem to converge, it is important to note that none of the perceptual studies have explored whether there is a constant proportionality between, say, the duration of a preceding vowel and the closure duration of a following stop that could account for the perceptual constancy. At most, the perceptual experiments have revealed boundary shifts in the expected directions.

The evidence in favor of the proportional duration model (i.e. the generalized motor program) has recently been re-examined by Gentner (1987). He argues that most of the studies showing support for the model have used improper analysis for assessing the model. When the proper analyses are applied, the model is not supported; at least the reparameterization is not linear.

Some of the methodological problems with the existing studies presenting evidence in favor of the proportional duration model are illustrated by the studies of speech production referred to above. One particular problem is the use of correlations to assess invariance and scaling (cf. Benoit, 1986). All experiments on speech have made part-whole correlations. For example, Löfqvist and Yoshioka (1984) analyzed the relation between closure/constriction duration and the

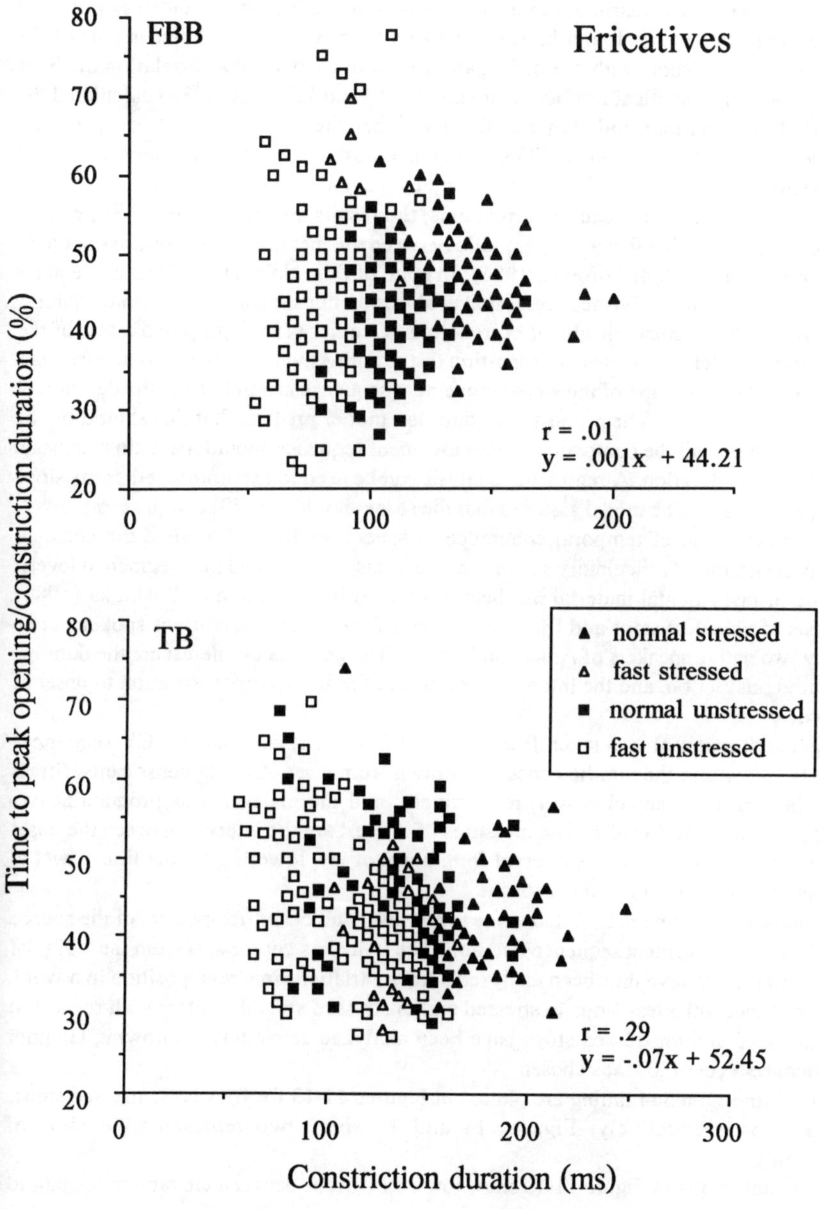

Figure 11. The duration of the oral constriction plotted against the ratio between the interval from constriction onset to onset of glottal adduction and constriction duration for fricatives spoken at different rates and under different stress conditions.

interval from onset of closure/constriction to the onset of glottal adduction in voiceless stops and fricatives. A high correlation was found between the two intervals, but the latter is part of the former, and correlating the whole with one of its parts would in itself yield a correlation of about 0.7. In order to avoid this statistical artifact, Munhall (1985) and Löfqvist (1986) calculated the expected part-whole correlations and then examined whether the obtained results significantly deviated from the expected ones. Munhall (1985) found that most of the effects could be accounted for by part-whole artifacts, whereas Löfqvist (1986) argued that the effects were real.[4]

If regression analysis is used instead of correlations, the results are more clear. The intercepts are generally non-zero, indicating that a constant ratio between the measured intervals does not hold (cf. Löfqvist and Yoshioka, 1984; Löfqvist, 1986). In spite of this, claims have been made about constant ratios even though the published regression analyses do not support such a strong claim.

In order to avoid the statistical pitfalls of correlations, Gentner (1987) proposed a test of the proportional duration model. The constant proportion test examines whether the ratio between one movement interval and the duration of the whole movement sequence is unrelated to the duration of the whole movement sequence. The proportional duration model predicts that this should be the case, since the durations of all the components of a movement sequence should maintain a constant proportion of the overall duration. A regression analysis can be used to examine whether the slope is different from zero or not. The model assumes that the slope should not differ from zero.

As a further examination of temporal coherence of speech gestures, I applied the constant proportion test to the timing of articulatory gestures at the intrasegmental and intersegmental levels, respectively. The intrasegmental material has been presented in Löfqvist and Yoshioka (1984). Briefly, it consists of voiceless stops and fricatives under different stress conditions spoken at two different rates by two native speakers of American English. The gestures of interest are the duration of the oral closure/constriction, and the interval from onset of oral closure/constriction to onset of glottal adduction.

The intersegmental material was taken from Löfqvist (1986). It consists of VCV sequences where the vowels are /a/ and the middle consonant drawn from a set of labial consonants. Stress was placed on the first and second vowel, respectively, and the material was produed at two different rates by a speaker of Swedish. The measures of interest are the interval between the onset of jaw lowering for the vowels, and the interval from onset of jaw lowering for the first vowel to onset of lower lip raising for the medial consonant.

The proportionality should be manifested at the token level and occur irrespective of the source of the change of overall movement sequence duration. The voiceless consonants from the Löfqvist and Yoshioka (1984) study have thus been analyzed across variations in stress, position in a word, and speaking rate. Since voiceless stops in stressed and unstressed syllables differ with respect to aspiration, the stressed and unstressed stops have been analyzed separately. Following Gentner (1987), a significance level of .05 was chosen.

The results for intrasegmental timing are plotted in Figures 11-13 for fricatives, stressed stops, and unstressed stops, respectively. Figures 14 and 15 show two representative plots of intersegmental timing.

The intrasegmental results in Figure 11-13 show low correlations between the ratio of the part to the whole and the whole. For the fricatives, the slope was significantly different from zero for subject TB ($t_{331} = 5.463$) but not for subject FBB ($t_{345} = .108$). A similar pattern holds for the

[4] However, the results published by Löfqvist (1986) only pertain to the row and column means. When the individual cells are analyzed, as they should more properly be, the results are less supportive of proportional timing.

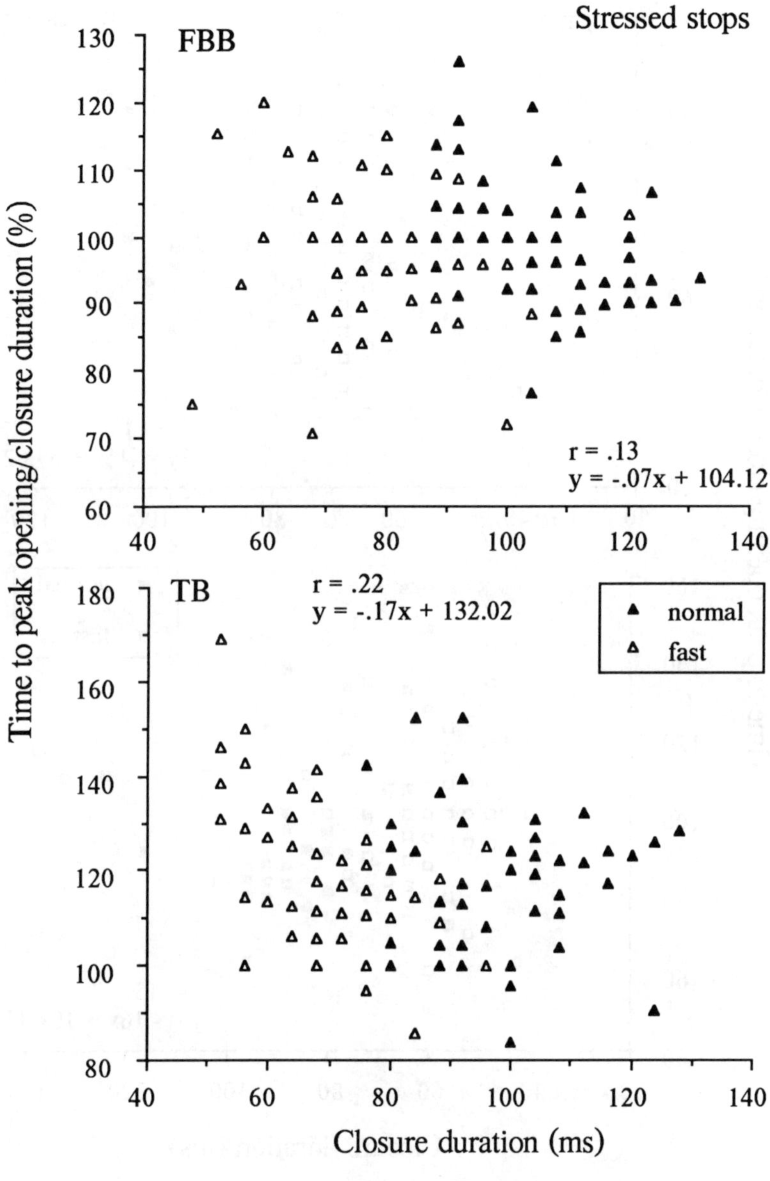

Figure 12. The duration of the oral closure plotted against the ratio between the interval from closure onset to onset of glottal adduction and closure duration for stressed stops spoken at two different rates.

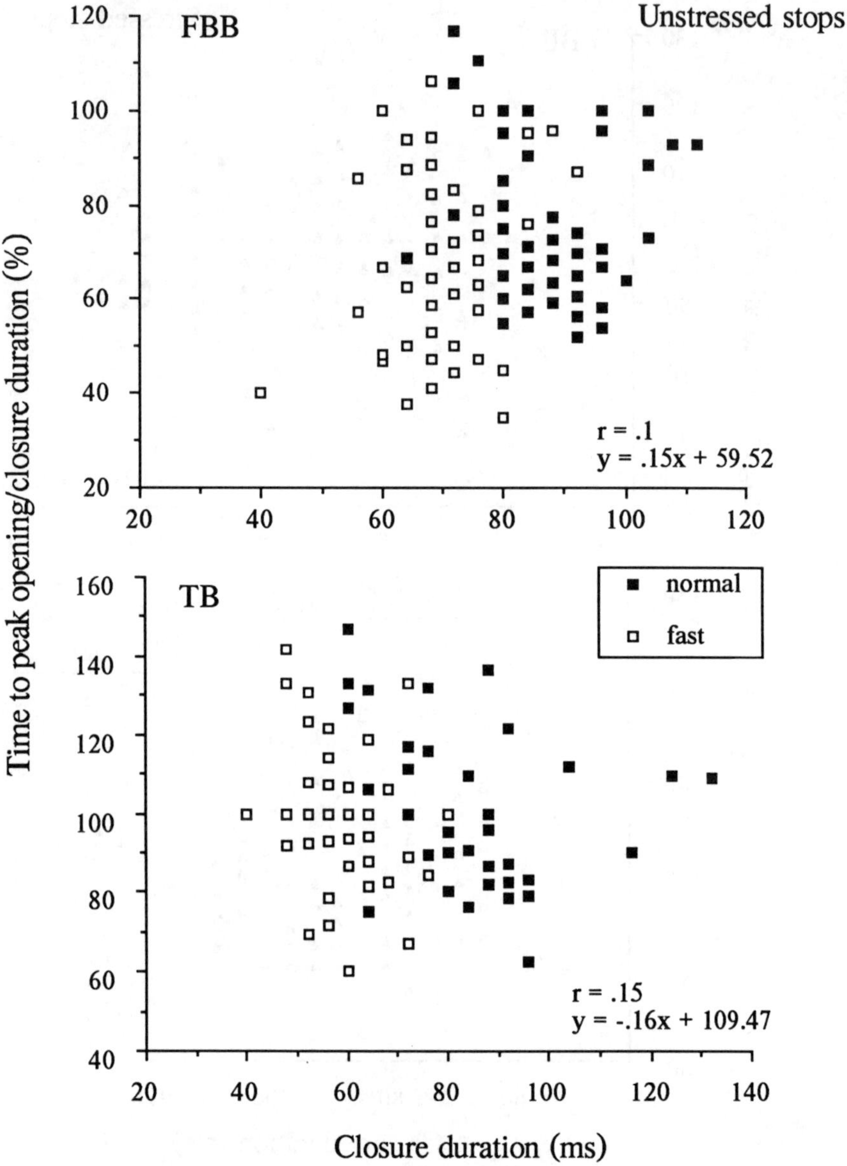

Figure 13. The duration of the oral closure plotted against the ratio between the interval from closure onset to onset of glottal adduction and closure duration for unstressed stops spoken at two different rates.

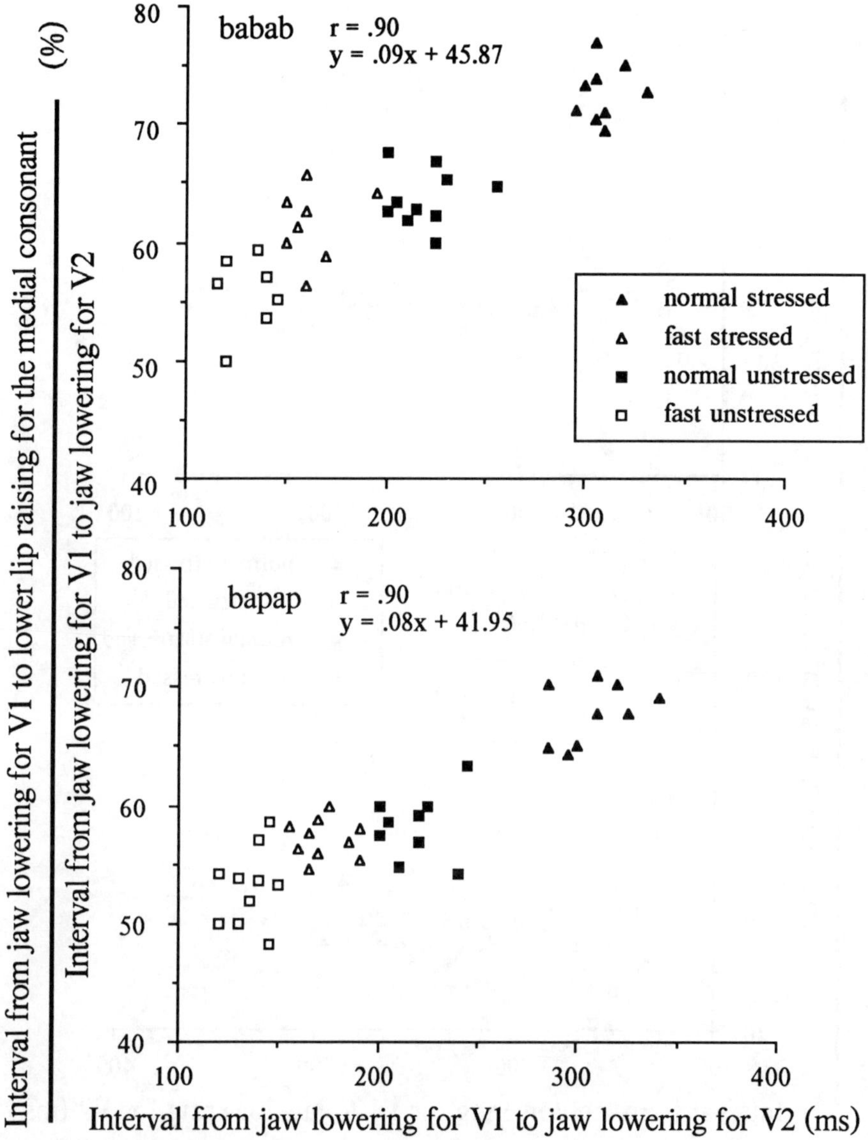

Figure 14. The interval between onsets of jaw lowering for the two vowels plotted against the ratio between the onset of jaw lowering for the first vowel to onset of lower lip raising for the medial consonant and the interval between jaw lowering onsets for the utterances /babab/ and /bapap/.

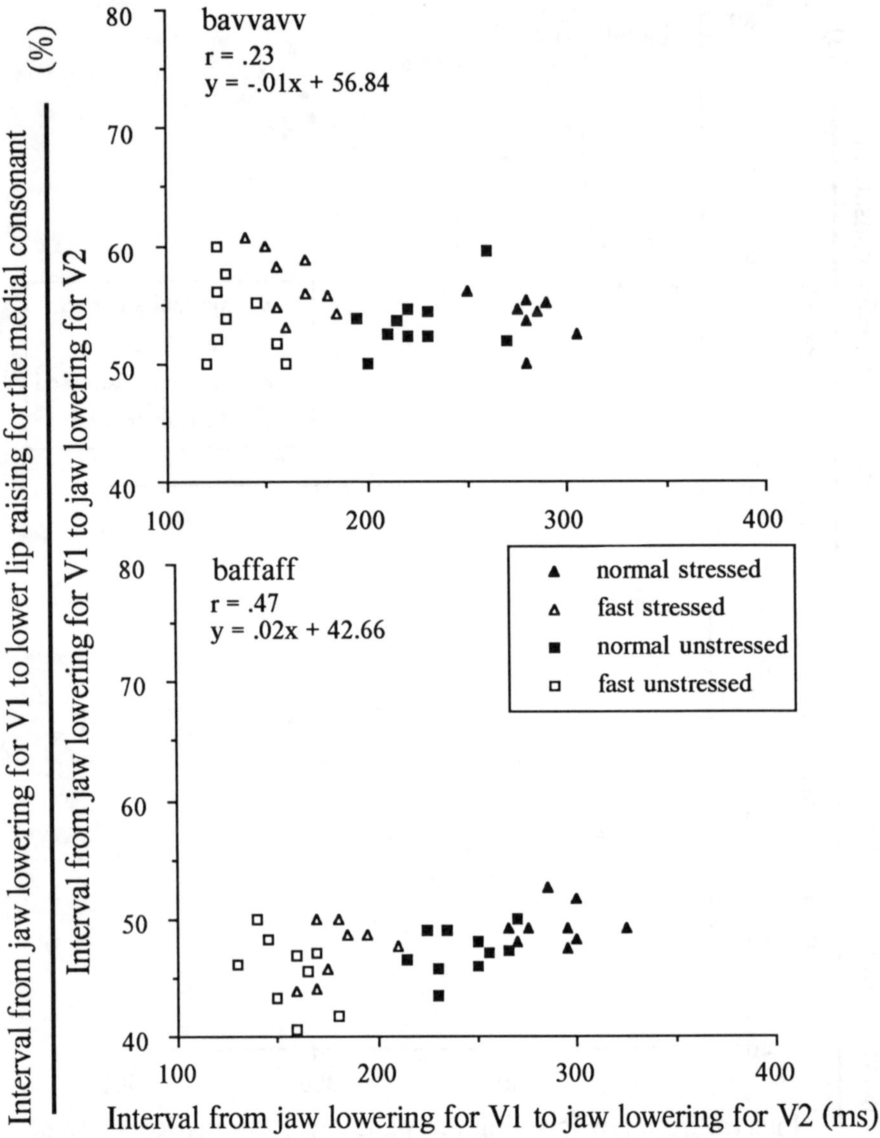

Figure 15. The interval between onsets of jaw lowering for the two vowels plotted against the ratio between the onset of jaw lowering for the first vowel to onset of lower lip raising for the medial consonant and the interval between jaw lowering onsets for the utterances /bavvavv/ and /baffaff/.

stressed stops, where subject TB has a slope different from zero (t_{137} = 2.598) but not FBB (t_{146} = 1.57). For the unstressed stops, the slope did not differ from zero for either subject with t_{134} = 1.176 and t_{91} = 1.43 for FBB and TB, respectively. For the intersegmental timing relations, all slopes differed from zero except for the sequence /bavvavv/ showed in Figure 15. Furthermore, correlations were uniformly high.

The outcome of the constant proportion test does not show any evidence of constancy at the intersegmental level, where the hypothesis was rejected in 90% of the cases examined. At the intrasegmental level, the idea of constant proportionality was only rejected in 33% of the cases. This could possibly indicate a difference in gestural cohesion within and across segments. However, the reason why the proportional duration model is not rejected here seems to be the great variability in the data. For example, the data for the fricatives in Figure 11 do not show any consistent ratio between time to peak opening and constriction duration. Instead, this ratio varies greatly for a given constriction duration.

Thus, this reanalysis does not show any support for the constant proportionality model in speech production. A suggestion that the proper metric is phase rather than temporal intervals was made by Kelso, Saltzman, and Tuller (1986) who also presented some evidence in support of this notion. Further investigations have, however, obtained conflicting results (Nittrouer, Munhall, Kelso, Tuller, and Harris, 1988).

There are some other examples of temporal coherence, in particular synchronicity, between the articulatory gestures for a given segment. For example, Kent and Moll (1975) found that closure for the /p/ and release of constriction for the /s/ in an /sp/ cluster occurred almost simultaneously irrespective of linguistic environment. Gay (1977) noted that the closing movements of tongue body, jaw, and primary articulator for the first vowel to the stop in a sequence of vowel-stop-vowel started almost simultaneously. On the other hand, Gracco and Abbs (1986) found asynchronous movement onsets for upper lip, lower lip, and jaw during the transition from a open vowel to a bilabial stop, but the order of movement onsets for these articulators was consistent. The same authors also noted a constant patterning of peak velocities for the same articulators. That is, the peak velocity of the upper lip preceded that of the lower lip, which in its turn occurred before that of the jaw. This pattern was maintained even when a perturbation was applied to the lower lip before closure (Gracco and Abbs, 1988). However, no such pattern was found at the release of the stop (Gracco, 1988). This would seem to reflect a tight coupling of the articulators in forming the labial closure, while they are less tightly coupled at the release.

One further aspect of segmental articulation that has been claimed to exhibit stable properties in the sense given above is the extent of anticipatory coarticulation. This problem has mostly been addressed in studies of labial coarticulation. For example, assume that a rounded vowel is preceded by a sequence of non-labial consonants. The relevant question here is: When does the rounding of the lips for the upcoming vowel start? There are two possibilities. Either the lip-rounding gesture starts at a time before the vowel that is independent of the number of preceding consonants, or the onset of lip-rounding is advanced in proportion to the duration of the preceding consonant sequence. At least two studies of lip-rounding (Bell-Berti and Harris, 1979, 1982) have obtained results supporting the first prediction. However, other studies have reported conflicting results. For example, Lubker (1981) and Sussman and Westbury (1981) both found that the onset of lip-rounding tended to precede the onset of a rounded vowel in proportion to the duration of the consonant string preceding it.

In this case of coarticulation, the results are conflicting. Possibly, language specific differences may be at play. This is strongly suggested by the results of a comparative study of anticipatory lip rounding in speakers of American English and Swedish (Lubker and Gay, 1982). This study showed that speakers of Swedish, in general, start to round their lips earlier than speakers of American English. A possible reason for this is that Swedish has more vowels and that the earlier lip-rounding is made to preserve vowel quality contrasts (cf. Boyce, 1988, for further examples of language specific patterns of coarticulation).

The material reviewed in this section, in particular the reanalysis of inter- and intrasegmental timing, does not show any strong evidence for proportional duration in the temporal relationships between speech gestures (see Sock, Ollila, Delattre, Zilliox and Zohair, 1989, for a similar conclusion based on acoustic data). There is, perhaps, no reason to expect this. The argument for constant proportionality is based on the idea that control is simplified if all components of a movement sequence are scaled, which would seem to rest on the assumption that the sequence is programmed as a whole. This may not necessarily be the case, however, since there are situations where it is useful to be able to control the individual components. Such cases will be discussed in the next section.

3.2. STABILITY IN THE FACE OF PERTURBATIONS

Normal motor control often requires rapid actions to cope with unexpected events such as stumbling during walking. From common, every-day experience it is well known that compensations for unexpected perturbations are in many cases rapid, functional, and successful. Controlled and systematic delivery of rapid and unexpected perturbations thus offers a valuable experimental paradigm for studying coordinated movements.[5]

While such unexpected mechanical disturbances may rarely occur during speech, static perturbations are sometimes found, e.g. when a person is speaking with a pipe in his/her mouth. Studies of vowel production using static perturbations, bite-blocks restricting jaw movements, generally indicate that speakers can, and do, rapidly compensate for them by adjusting tongue and/or lip movements (Lindblom, Lubker and Gay, 1979; Lubker, 1979; Fowler and Turvey, 1980; Gay, Lindblom and Lubker, 1981). On the other hand, the results of Hamlet and Stone (1976, 1978) suggest that speakers require some time to adjust to changes in oral morphology introduced by dental prostheses. Similarly, Flege, Fletcher and Homiedan (1988) investigated the production of dental consonants under bite-block conditions and found both articulatory, acoustic, and perceptual differences compared to normal productions.

Some studies have introduced dynamic perturbations to speech articulators and examined the compensations that speakers make (Folkins and Abbs, 1975; Folkins and Zimmermann, 1982; Abbs and Gracco, 1984; Kelso, Tuller, Vatikiotis-Bateson and Fowler, 1984; Gracco and Abbs, 1985, 1988; Shaiman, 1989). Most commonly, a load has been applied to the jaw and the compensatory responses in the lower lip, the upper lip, and the tongue have been examined using kinematic records and/or EMG. The received view from these studies can be briefly summarized as follows.

Compensations are rapid. If the jaw is loaded during the production of a dental fricative, increased activity in the genioglossus muscle occurs 20-30 ms after onset of the load (Kelso,

[5] A caveat is in order, however, concerning the proper interpretation of these experiments. For obvious reasons, a subject paticipating in such experiments cannot be unaware what is going to happen. Thus, the response strategy may differ from truly "normal" conditions. In spite of this, it is possible to draw some general conclusions. For example, by studying a person walking on ice and consequently adjusting his/her mode of locomotion and the strategy for coping with unexpected events we may still learn some general properties of locomotion even though we are sampling a limited part of the performance space.

Tuller, Vatikiotis-Bateson and Fowler, 1984). Similarly, an increase in upper lip activity can occur 35-70 ms after the jaw has been loaded during lip closure for a bilabial stop; the latency is not fixed, however, but depends on when the load was applied in relation to the onset of upper lip activity (Abbs, Gracco and Cole, 1984). Such short latencies imply that the responses are not due to reaction time processes; furthermore, subjects are usually not aware of making any compensations.

Compensations are functional. This means that the responses are not stereotypic but tailored to the needs of the ongoing motor act. For example, if the jaw is loaded during the transition from a vowel to a bilabial stop, compensatory responses are seen in the lower and upper lips. These responses thus assist in achieving the goal of the motor act, i.e. a lip closure. If, on the other hand, the jaw is loaded during the transition from a vowel to a dental fricative or a dental stop, a response is seen in the tongue which is involved in making the constriction/closure (Kelso, Tuller, Vatikiotis-Bateson and Fowler, 1984; Shaiman, 1989).

Compensations are effective. That is, the intended goal of the motor act is achieved. During speech, the goal of the motor activity is to produce an acoustic signal that can transmit information. In perturbed trials, no perceptually salient acoustic deviations have been reported (Abbs, Gracco and Cole, 1984).

These studies have examined articulators that are either mechanically linked (e.g. lower lip and tongue relative to the jaw) and/or functionally linked to produce a single constriction in the vocal tract (e.g. jaw-lips for bilabial constriction, jaw-tongue for tongue dorsum constriction). A reciprocal relationship has, furthermore, been suggested between jaw and lower lip movements in the control of oral aperture in both normal and bite-block speech (Hughes and Abbs, 1976; Folkins and Linville, 1983). These findings have, however, been criticised on methodological grounds (Sussman, 1988; Folkins and Brown, 1987).

If the perturbation paradigm is extended to examine coherence between remote articulators, the results are slightly different. Shaiman and Abbs (1987), and Munhall, Löfqvist and Kelso (1988) perturbed the lower lip during the transition from a vowel to a voiceless bilabial stop consonant and examined the laryngeal responses to the perturbation. As discussed above, the phasing of oral and laryngeal gestures is critical for maintaining the integrity of the segment. Both studies found an increase in the duration of the preceding vowel in the perturbed trials (cf. Folkins and Abbs, 1975, for a similar result). This increase was reflected by a delay in onset of glottal abduction. Munhall, Löfqvist and Kelso (1988) specifically monitored oral air pressure to make sure that this effect was not simply due to a delay in the build-up of oral air pressure; no evidence of venting was observed, however. More importantly, they also found that the normal coordination between the oral and laryngeal gestures was disrupted at the release of the stop closure in the perturbed condition. This was due to a decrease in the duration of the oral closure and an increase in the duration of the glottal devoicing gesture. The latter resulted from an increase in the adduction phase of the gesture, while the abduction phase remained unchanged. At the acoustic level, the disruption of the normal phasing manifested itself as a marked increase in VOT.

In relation to this particular acoustic change following the perturbation, we should note that the effectiveness of the observed compensations have generally been evaluated from kinematic records and auditory analysis of the produced speech. Changes in muscle activity inferred from electromyographic recordings have also been used, but such changes do not necessarily correlate with the kinematic adjustments (Shaiman, 1989). Since the goal of speech production is a time-varying acoustic signal, it is thus desirable to verify the success of compensations by acoustic analysis. While auditory analysis apparently showed compensations to be successful, some acoustic differences appear to occur between perturbed and control trials.[6]

The proper metric for evaluating the effectiveness should be auditory, however. There is no need for compensations to be absolute, or perfect. All that is required is that the compensations are sufficient to maintain the integrity of the acoustic signal. The reason is that categorical perception, i.e. poor within-category discrimination (Liberman, Harris, Hoffman and Griffith, 1957), takes care of acoustic variability under both normal and perturbed conditions when identification is required. On the other hand, judgements of acceptability seem to be lower for perturbed productions (Flege, Fletcher and Homiedan (1988).

The results of the perturbation studies show more evidence of gestural cohesion. That is, the integrity of a motor act is maintained when part of the system is mechanically perturbed. Short-time sensorimotor interactions are thus at play. Similar results have also been reported for other areas of motor control such as posture and arm movements (e.g. Nashner and McCollum, 1985; Soechting and Lacquaniti, 1989).

The disruption of normal articulatory phasing between oral and laryngeal gestures following lip perturbations suggests some limits of on-line compensations, however. The disruption may result either from the articulators being remote from each other or from a difference between spatial and temporal compensations. The critical factor in laryngeal-oral coordination is timing. Experiments on rhythmic finger movements suggest that a perturbation to the finger will not be compensated for immediately, but rather that the system requires one or two cycles to return to its initial state (Kay, 1986).

A useful concept for rationalizing the cohesion of gestures is that of coordinative structures (cf. Greene, 1971; Turvey, 1977; Kugler, Kelso and Turvey, 1980; Kugler and Turvey, 1987; Saltzman, 1986; Saltzman and Kelso, 1987; Saltzman and Munhall, in press). Designed to cope with the number of degrees of freedom to be directly controlled, a coordinative structure is a flexible linkage of muscles that has been set up temporarily for the execution of a specific action.

4. Gestural Aggregation

The view of the segment as a bundle of gestures predicts that overlapping gestures for successive segments should show blends or aggregates. Although coarticulation has received considerable attention over the years, most studies have focused on the temporal extent of coarticulation rather than on the resulting patterns. A few results suggest, however, that the proposed view of the segment can be useful in understanding observed patterns of coarticulation.

Löfqvist and Yoshioka (1981a) noted that a voiceless cluster of fricative and stop across a word boundary, /s#k/, could be produced with one or two laryngeal gestures depending on the speaking rate. That is, at slow rates, two gestures were observed, one for the fricative and one for the stop. At fast rates, however, only a single gesture occurred. More interestingly, at intermediate rates, a blend of two gestures was observed. These findings are reminiscent of Stetson's ideas on the "singling" and "doubling" of consonants as a function of speaking rate (cf. Stetson, 1951, p. 64).

The idea that these patterns could be explained as the blending of two underlying gestures was

[6] In addition to the increase in vowel duration and VOT discussed above, drastic temporal differences can be found in some recordings presented by Folkins and Zimmermann (1982). In their study, the perturbation consisted of electrical stimulation of the labial depressor muscles; the stimulation was applied during the transition from a vowel to a bilabial voiceless stop. Figure 3 in that study shows an increase in stop closure duration from approximately 300 ms to 500 ms in a normal and perturbed production, respectively.

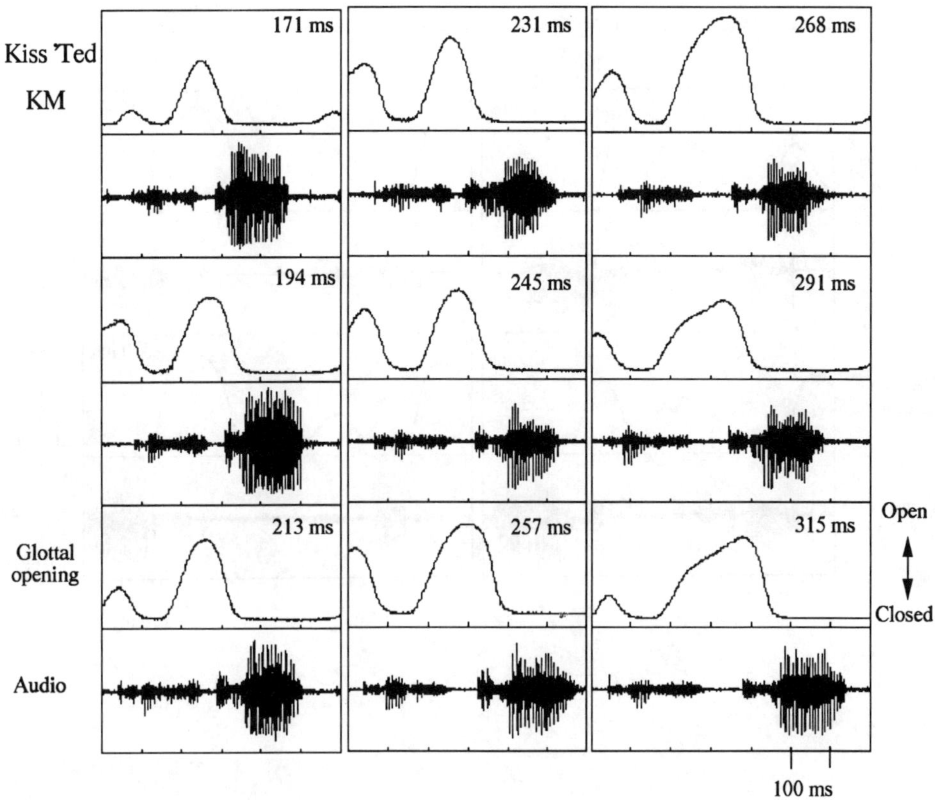

Figure 16a, b. Several productions of the utterance "Kiss 'Ted" arranged in order of decreasing speaking rate. The upper trace in each panel shows the transillumination signal with increasing glottal opening displayed upwards; the bottom trace shows the accompanying acoustic signal. The number in the upper right corner of each panel is the interval (in msec) from offset to onset of the vowels preceding and following the consonant cluster, respectively. (From Munhall and Löfqvist, submitted.)

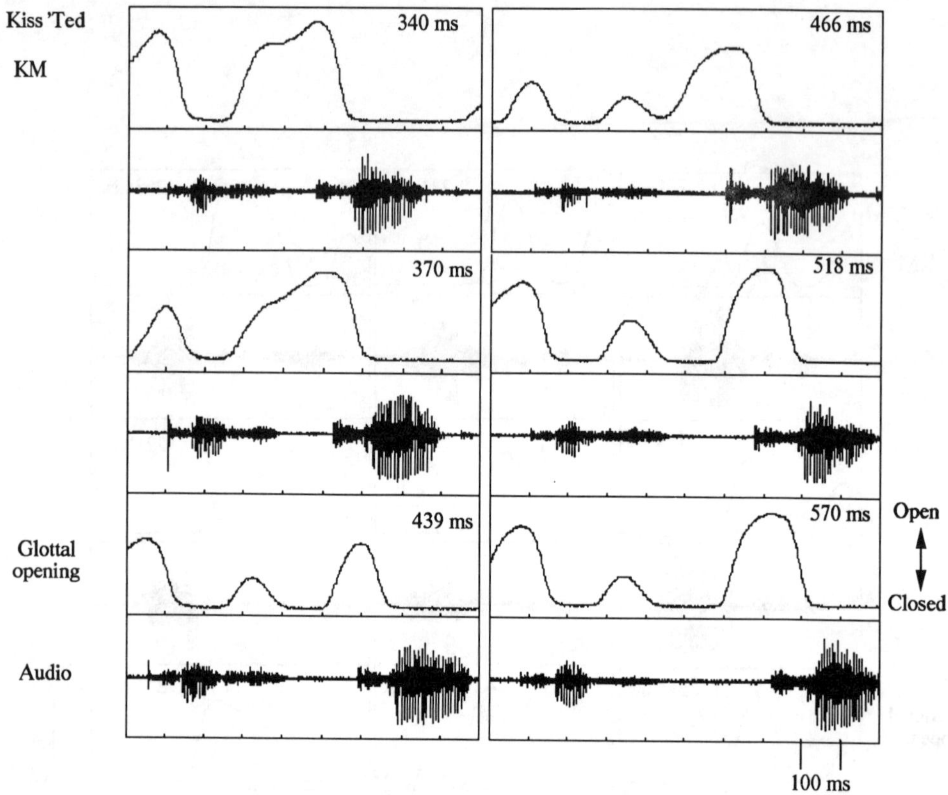
Kiss 'Ted
KM
340 ms
466 ms
370 ms
518 ms
439 ms
570 ms
Open
Closed
Glottal opening
Audio
100 ms

Figure 16b.

evaluated experimentally by Munhall and Löfqvist (submitted). They had two speakers produce the utterance "Kiss Ted", with stress on the verb or the noun, respectively, at several different speaking rates. Representative results are shown in Figure 16.

In this Figure, we see the following pattern. At fast rates, a single gesture is found with similar durations for the abduction and adduction phases. At slow rates, two separate gestures occur; here, the glottis is closed between the two gestures. At the intermediate rates, one gesture is made. This gesture does, however, show traces of the two individual gestures. Thus, it is not symmetrical but shows some extra adjustments as if two underlying gestures have been added together. We see, in other words, an example of an intra-articulator gestural aggregate.

As shown above in Figure 7, a cluster of /sp/ can be produced with a single glottal gesture. In this case, however, the amplitude and duration of the gesture is generally larger and longer than when the gesture is associated with only a single consonant. This can, again, be taken as the result of the blending of two underlying and overlapping gestures (see Saltzman and Munhall, in press, for a review of such examples).

Additional evidence for viewing speech production as gestural aggregation has been presented by Boyce (1988). She investigated the lip gestures associated with rounded vowels in English and Turkish. Boyce, first, found differences in the organization of the rounding gestures in the two languages. In particular, the English speakers produced a trough in the pattern of lip movements in utterances of the type VCV, where V is a rounded vowel and C represents one or more non-labial consonants. The trough manifests itself as an apparent retraction of the lips between the rounding gestures for the two vowels; it is seen in records of both movement and muscle activity (cf. Perkell, 1976, for a discussion of the trough effect). This difference between languages most likely reflects the fact that Turkish has vowel harmony.

Second, and of particular interest in the present context, Boyce showed how the pattern of lip movement for an utterance of the type /uCu/ could be decomposed into underlying gestures for each of the rounded vowels. That is, by adding the lip gestures for the rounded vowels in the utterances /uCi/ and /iCu/, the pattern observed for /uCu/ was obtained. Furthermore, by subtracting the lip movement for the utterance /iCi/ from that of /uCu/ it was possible to factor out the contribution of lip movements associated with the medial consonant(s). When this was done, the resulting movements for the utterance /uCu/ was remarkably similar for all subjects. Moreover, the created movement was more closely correlated with the activity pattern of the muscle associated with lip rounding in vowels, thus suggesting that some other mechanism was responsible for the consonant-related lip movements. Finally, Boyce also showed that the magnitude of the trough was at least partly related to the duration of the consonant(s) occurring between the two rounded vowels. This is the expected outcome if underlying gestures are being produced with varying degrees of overlap.

Blending of gestures is also evident in the x-ray tracings presented by Öhman (1966, 1967). In particular, when two flanking vowels and an intervening consonant share the same articulator, i. e., the tongue dorsum in the sequence /aga/ or /igi/, the tongue shape during the closure for the /g/ is a blend of the gestures for the vowel and the consonant (cf. Saltzman and Munhall, in press, for simulation of such patterns using gestural blending).

Coproduction has, furthermore, been observed for tongue gestures using electropalatography (Hardcastle, 1985; Marchal, 1988). That is, the gestures for two successive lingual consonants, such as /kt/ or /kl/, show different degrees of overlap. Again, the amount of overlap increases with speaking rate (Hardcastle, 1985).

The material reviewed in this section is consistent with the possibility that some articulatory patterns observed in fluent speech derive from a simple aggregation process. At the larynx, the process that combines overlapping gestures may be simply summation. For many years, one type of coarticulation, carryover, has been viewed in this way. Blending of overlapping gestures may, however, be a more general phenomenon accounting for commonly observed types of coarticulation. Such blending of gestures may have some possible advantages. Foremost of these is computational simplicity. In other biological phenomena, similar aggregation processes are evident. Neuronal excitation is one obvious example. The algebraic sum of excitatory inputs must exceed a threshold for a voltage spike to be generated. The behavior of reflex response fields in various lower vertebrates also shows some form of blending behavior (e.g. Berkinblit, Feldman and Fukson, 1986; Stein, Mortin and Robertson, 1986).

It is obvious, however, that not all blends will yield functional or coherent behavioural patterns. Simple algebraic combination of gestures could produce resultants that failed to maintain linguistic distinctions or that were so energetically inefficient that they must be avoided. This may be the reason why the decomposition of labial gestures carried out by Boyce (1988) was less successful when applied to speakers of Turkish.

5. Gestural Adaptation

The view of speech as audible gestures discussed here can rationalize several different facts of speech production. Similar ideas have also been used to account for phonological processes (Browman and Goldstein, 1986, in press) and also proved useful in the modelling of speech movements (Saltzman and Munhall, in press).

This view does not necessarily entail that all aspects of speech can be reduced to gestures. Rather, gestures are intentional, are made for a purpose, and have to be adapted to the environment. For example, the purpose of locomotion is to move an individual. In this process, the coordinated activities of the legs have to be adjusted to avoid objects in the environment and must also be adapted to the nature of the surface being traversed, i. e., if one moves uphill or downhill, on a hard or a soft surface, or on ice. The gestures of the vocal tract have to be adapted in a similar way. One important aspect of the environment during speech is the listener. Hence, a speaker normally adjusts his/her articulatory gestures to the listener. A prime example of such adjustments is the type of speech used by mothers addressing their young infants (cf. Jusczyk and Bertoncini, 1988, for a review).

Acknowledgments

I have benefited from discussions with several colleagues both before and during the conference, in particular Gérard Chollet, Vincent Gracco, Roger Moore, Kevin Munhall, John Ohala, and Elliot Saltzman. They do not necessarily subscribe to any particular views expressed here, however. This work was supported in part by NINCDS Grant NS-13617, and BRS Grant RR-05596 to Haskins Laboratories.

References

Abbs, J. and Gracco, V. (1984) 'Control of complex motor gestures: Orofacial muscle response to load perturbations of the lips during speech', Journal of Neurophysiology 51, 705-723.

Abbs, J., Gracco, V. and Cole, K. (1984) 'Control of multimovement coordination: Sensorimotor mechanisms in speech motor programming', Journal of Motor Behavior 16, 195-232.

Abramson, A. S. (1977) 'Laryngeal timing in consonant distinctions', Phonetica 34, 295-303.

Atal, B. (1983) 'Efficient coding of LPC parameters by temporal decomposition', in Proceedings ICASSP 2.6, 81-84.

Baer, T., Löfqvist, A. and McGarr, N. S. (1983) 'Laryngeal vibrations: A comparison between high-speed filming and glottographic techniques', Journal of the Acoustical Society of America 73, 1304-1308.

Bell-Berti, F. and Harris, K. S. (1979) 'Anticipatory coarticulation: Some implications from a study of lip rounding', Journal of the Acoustical Society of America 65, 1268-1270.

Bell-Berti, F. and Harris, K. S. (1982) 'Temporal patterns of coarticulation: Lip rounding', Journal of the Acoustical Society of America 71, 449-459.

Benguerel, A. P. and Bhatia, T. K. (1980) 'Hindi stop consonants: An acoustic and fiberscopic study', Phonetica, 37, 134-148.

Benoit, C. (1986) 'Note on the use of correlation in speech timing', Journal of the Acoustical Society of America 80, 1846-1849.

Berkinblit, M., Feldman, A. and Fukson, O. (1986) 'Adaptability of innate motor patterns and motor control', Behavioral and Brain Sciences 9, 585-638.

Bimbot, F., Ahlbom, G. and Chollet, G. (1987) 'From segmental synthesis to acoustic rules using temporal decomposition,' in Proceedings of the Eleventh International Congress of Phonetic Sciences 5, 31-34.

Boyce, S. (1988) The influence of phonological structure on articulatory organization in Turkish and English vowel harmony and coarticulation, unpublished doctoral dissertation, Yale University.

Browman, C. and Goldstein, L. (1986) 'Towards an articulatory phonology', Phonology Yearbook 3, 219-252.

Browman, C. and Goldstein, L. (in press) 'Tiers in articulatory phonology, with some implications for casual speech', in J. Kingston and M. Beckman (eds.), Papers in Laboratory Phonology I: Between the Grammar and the Physics of Speech, Cambridge University Press, Cambridge.

Brown, R. (1976) 'Reference - In memorial tribute to Eric Lenneberg', Cognition 4, 125-153.

Fant, G. (1962) 'Descriptive anlysis and the acoustic aspects of speech', Logos 5, 3-17 (Reprinted in Fant, G. (1973) Speech Sounds and Features, MIT Press, Cambridge, Mass., pp. 17-23).

Fitch, H. (1981) 'Distinguishing temporal information for speaking rate from temporal information for intervocalic stop consonant voicing', Haskins Laboratories Status Report on Speech Research SR-65, 1-32.

Flege, J., Fletcher, S. and Homiedan, A. (1988) 'Compensating for a bite block in /s/ and /t/ production: Palatographic, acoustic, and perceptual data', Journal of the Acoustical Society of America 83, 212-228.

Folkins, J. W. and Abbs, J. H. (1975) 'Lip and jaw motor control during speech: Responses to resistive loading of the jaw', Journal of Speech and Hearing Research 18, 207-220.

Folkins, J. and Brown, C. (1987) 'Upper lip, lower lip, and jaw interactions during speech: Comments on evidence from repetition-to-repetition variability', Journal of the Acoustical

Society of America 82, 1919-1924.

Folkins, J. and Linville, R. (1983) 'The effect of varying lower-lip displacement on upper-lip movements: Implications for the coordination of speech movements', Journal of Speech and Hearing Research 26, 209-217.

Folkins, J. and Zimmermann, G. (1982) 'Lip and jaw interaction during speech: Responses to perturbation of lower-lip movement prior to bilabial closure', Journal of the Acoustical Society of America 71, 1225-1233.

Fowler, C. (1980) 'Coarticulation and theories of extrinsic timing', Journal of Phonetics 8, 113-133.

Fowler, C. (1983) 'Converging sources of evidence on spoken and perceived rhythms of speech: Cyclic production of vowels in monosyllabic stress feet', Journal of Experimental Psychology: Human Perception and Performance 112, 386-412.

Fowler, C., Rubin, P., Remez, R. and Turvey, M. (1980) 'Implications for speech production of a general theory of action', in B. Butterworth (ed.), Language Production, Vol. 1: Speech and Talk, Academic Press, New York, pp. 373-420.

Fowler, C. and Turvey, M. (1980) 'Immediate compensation in bite-block speech', Phonetica 37, 306-326.

Gay, T. (1977) 'Articulatory movements in VCV sequences', Journal of the Acoustical Society of America 62, 183-193.

Gay, T., Lindblom, B. and Lubker, J. (1981) 'Production of bite-block vowels: Acoustic equivalence by selective compensation', Journal of the Acoustical Society of America 69, 802-810.

Gentner, D. (1987) 'Timing of skilled movements: Test of the proportional duration model', Psychological Review 94, 255-276.

Gobl, C. and Ní Chasaide, A. (1988) 'The effects of adjacent voiced/voiceless consonants on the vowel source: a cross language study', STL-QPSR (Department of Speech Transmission and Music Acoustics, Royal Institute of Technology, Stockholm) 2-3, 23-59.

Gracco, V. L. (1988) 'Timing factors in the coordination of speech movements', Journal of Neuroscience 8, 4628-4639.

Gracco, V. and Abbs, J. (1985) 'Dynamic control of the perioral dystem during speech: Kinematic analyses of autogenic and nonautogenic sensorimotor processes', Journal of Neurophysiology 54, 418-432.

Gracco, V. L. and Abbs, J. H. (1986) 'Variant and invariant characteristics of speech movements', Experimental Brain Research 65, 156-166.

Gracco, V. L. and Abbs, J. H. (1988) 'Central patterning of speech movements', Experimental Brain Research 71, 515-526.

Greene, P. (1971) 'Introduction', in M. Gelfand, V. Gurfinkel, S. Fomin and M. Tsetlin (eds.), Models of the Structural-Functional Organization of Certain Biological Systems, MIT Press, Cambridge, Mass., pp. xxi-xxxi.

Grillner, S. (1975) 'Locomotion in vertebrates: Central mechanisms and reflex interaction', Physiological Reviews 55, 247-304.

Hamlet, S. and Stone, M. (1976) 'Compensatory vowel characteristics resulting from the presence of different types of experimental dental prostheses', Journal of Phonetics 4, 199-218.

Hamlet, S. and Stone, M. (1978) 'Compensatory alveolar consonant production induced by wearing a dental prosthesis', Journal of Phonetics 6, 227-248.

Hardcastle, W. J. (1985) 'Some phonetic and syntactic constraints on lingual coarticulation during /kl/ sequences', Speech Communication 4, 247-263.

Hughes, O. and Abbs, J. (1976) 'Labial-mandibular coordination in the production of speech: Implications for the operation of motor equivalence', Phonetica 44, 199-221.

Joos, M. (1948) 'Acoustic phonetics,' Language 24, 1-136.

Jusczyk, P. and Bertoncini, J. (1988) 'Viewing the development of speech perception as an innately guided learning process', Language and Speech 31, 217-238.

Kay, B. (1986) Dynamic modeling of rhythmic limb movements: Converging on a description of the component oscillators, unpublished doctoral dissertation, The University of Connecticut.

Kelso, J. A. S., Saltzman, E. and Tuller, B. (1986) 'The dynamical perspective on speech production: Data and theory', Journal of Phonetics 14, 29-59.

Kelso, J. A. S., Tuller, B., Vatikiotis-Bateson, E. and Fowler, C. A. (1984) 'Functionally specific articulatory cooperation following jaw perturbations during speech: Evidence for coordinative structures', Journal of Experimental Psychology: Human Perception and Performance 10, 812-832.

Kent, R. and Moll, K. (1975) 'Articulatory timing in selected consonants', Brain and Language 2, 304-323.

Klatt, D. and Klatt, L (in press) 'Analysis, synthesis and perception of voice quality variations among female and male speakers', Journal of the Acoustical Society of America, in press.

Kugler, P., Kelso, J. A. S. and Turvey, M. (1980) 'On the concept of coordinative structures as dissipative structures: I. Theoretical lines of convergence', in G. Stelmach and J. Requin (eds.), Tutorials in Motor Behavior, North-Holland, Amsterdam, pp. 3-47.

Kugler, P. and Turvey, M. (1987) Information, Natural Law, and the Self-Assembly of Rhythmic Movement, Erlbaum, Hillsdale, NJ.

Lee, W. (1984) 'Neuromotor synergies as a basis for coordinated intentional action', Journal of Motor Behavior 16, 135-170.

Liberman, A., Cooper, F. S., Shankweiler, D. and Studdert-Kennedy, M. (1967) 'Perception of the speech code', Psychological Review, 74, 431-461.

Liberman, A., Harris, K. S., Hoffman, H. and Griffith, B. (1957) 'The discrimination of speech sounds within and across phoneme boundaries', Journal of Experimental Psychology 54, 358-368.

Lindblom, B., Lubker, J. and Gay, T. (1979) 'Formant frequencies of some fixed-mandible vowels and a model of speech-motor programming by predicitive simulation', Journal of Phonetics 7, 147-162,

Lisker, L. (1986) ' "Voicing" in English: A catalogue of acoustic features signaling /b/ versus /p/ in trochees', Language and Speech 29, 3-11.

Lisker, L. and Abramson, A. (1964) 'A cross-language study of voicing in initial stops: Acoustical measurements', Word 20, 384-422.

Lubker, J. (1979) 'The reorganization time of bite-block vowels', Phonetica 36, 273-293.

Lubker, J. (1981) 'Temporal aspects of speech production: Anticipatory labial coarticulation', Phonetica 38, 51-65,

Lubker, J. and Gay, T. (1982) 'Anticipatory labial coarticulation: Experimental, biological and linguistic variables', Journal of the Acoustical Society of America 71, 437-448.

Löfqvist, A. (1975) 'A study of subglottal pressure during the production of Swedish stops,' Journal of Phonetics 3, 175-189.

Löfqvist, A. (1980) 'Interarticulator programming in stop production', Journal of Phonetics 8, 475-490.

Löfqvist, A. (1986) 'Stability and change', Journal of Phonetics 14, 139-144.

Löfqvist, A., Baer, T., McGarr, N. S. and Story, R. S. (1989) 'The cricothyroid muscle in voicing

control', Journal of the Acoustical Society of America 85, 1314-1321.
Löfqvist, A. and McGarr, N. S. (1987) 'Laryngeal dynamics in voiceless consonant production', in T. Baer, C. Sasaki and K. S. Harris (eds.), Laryngeal Function in Phonation and Respiration, Little, Brown, and Co, Boston, pp. 391-402.
Löfqvist, A., McGarr, N. S. and Honda, K. (1984) 'Motor control of laryngeal articulation', Journal of the Acoustical Society of America 76, 951-954.
Löfqvist, A. and McGowan, R. S. (1989) 'Voice source variations in running speech', paper presented at the Vocal Fold Physiology Conference, Stockholm.
Löfqvist, A. and Yoshioka, H. (1980) 'Laryngeal activity in Swedish obstruent clusters', Journal of the Acoustical Society of America 68, 792-801.
Löfqvist, A. and Yoshioka, H. (1981a) 'Laryngeal activity in Icelandic obstruent production', Nordic Journal of Linguistics 4, 1-18.
Löfqvist, A. and Yoshioka, H. (1981b) 'Interarticulator programming in obstruent production', Phonetica 38, 21-34.
Löfqvist, A. and Yoshioka, H. (1984) 'Intrasegmental timing: Laryngeal-oral coordination in voiceless consonant production', Speech Communication 3, 279-289.
Mandelbrot, B. (1954) 'Structure formelle des textes et communication', Word 10, 1-27.
Marchal, A. (1988) 'Coproduction: Evidence from EPG data', Speech Communication 7, 287-295.
Miller, J. and Grosjean, F. (1981) 'How the components of speaking rate influence perception of phonetic segments', Journal of Experimental Psychology: Human Perception and Performance 7, 208-215.
Munhall, K. (1985) 'An examination of intra-articulator relative timing', Journal of the Acoustical Society of America 78, 1548-1553.
Munhall, K. and Löfqvist, A. (submitted) 'Gestural aggregation in speech', submitted to Journal of Phonetics.
Munhall, K., Löfqvist, A. and Kelso, J. A. S. (1988) 'Lip-larynx coordination in speech: A perturbation study', Journal of the Acoustical Society of America 80, S109.
Munhall, K. G., Ostry, D. J. and Parush, A. (1985) 'Characteristics of velocity profiles of speech movements', Journal of Experimental Psychology: Human Perception and Performance 11, 457-474.
Nashner, L. and McCollum, G. (1985) 'The organization of human postural movements: A formal basis and experimental synthesis', Behavioral and Brain Sciences 8, 135-172.
Nelson, D. and Marler, P. (1989) 'Categorical perception of a natural stimulus continuum', Science 244, 976-978.
Nittrouer, S., Munhall, K., Kelso, J. A. S., Tuller, B. and Harris, K. S. (1988) 'Patterns of interarticulator phasing and their relation to linguistic structure', Journal of the Acoustical Society of America 84, 1653-1661.
Ohde, R. N. (1984) 'Fundamental frequency as an acoustic correlate of stop consonant voicing', Journal of the Acoustical Society of America 75, 224-230.
Ostry, D. and Munhall, K. (1985) 'Control of rate and duration of speech movements', Journal of the Acoustical Society of America 77, 640-648.
Pattee, H. (1977) 'Dynamic and linguistic modes of complex systems,' International Journal of General Systems 3, 259-266.
Perkell, J. (1986) 'Coarticulation strategies: Preliminary implications of a detailed analysis of lower lip protrusion movements', Speech Communication 5, 47-68.
Port, R. (1979) 'The influence of tempo on stop closure duration as a cue for voicing and place',

Journal of Phonetics 7, 45-56.
Remez, R. (1989) 'When the objects of perception are spoken', Ecological Psychology 1, 161-180.
Rosch, E. and Lloyd, B. (eds.) (1978) Cognition and Categorization, Lawrence Erlbaum, Hillsdale, NJ.
Saltzman, E. (1986) 'Task dynamic coordination of the speech articulators: A preliminary model', Experimental Brain Research, 15, 129-144.
Saltzman, E. and Kelso, J. A. S. (1987) 'Skilled actions: A task dynamic approach', Psychological Review 94, 84-106.
Saltzman, E. and Munhall, K. (in press) 'A dynamical approach to gestural patterning in speech production', Ecological Psychology, in press.
Sawashima, M. and Hirose, H. (1983) 'Laryngeal gestures in speech production', in P. MacNeilage (ed.), The Production of Speech, Springer, New York, pp. 11-38.
Schmidt, R. (1975) 'A schema theory for discrete motor skill learning', Psychological Review 82, 225-260.
Shaiman, S. (1989) 'Kinematic and electromyographic responses to perturbation of the jaw,' Journal of the Acoustical Society of America 86, 78-88.
Shaiman, S. and Abbs, J. (1987) 'Sensorimotor contributions to the temporal coordination of oral and laryngeal movements for speech', SMCL Preprints (Speech Motor Control Laboratories, University of Madison) Spring-Summer 1987, 185-202.
Shapiro, D., Zernicke, R., Gregor, R. and Diestel, J. (1981) 'Evidence for generalized motor programs using gait pattern analysis', Journal of Motor Behavior 13, 33-47.
Shik, M. and Orlovsky, G. (1976) 'Neurophysiology of locomotor automatism', Physiological Reviews 56, 465-501.
Sock, R., Ollila, L., Delattre, C., Zilliox, C. and Zohair, L. (1988) 'Patrons de phases dans le cycle acoustique de détente en français,' Journal Acoustique 1, 339-345.
Soechting, J. and Lacquaniti, F. (1989) 'An assessment of the existence of muscle synergies during load perturbations and intentional movements of the human arm', Experimental Brain Research 74, 535-548.
Stein, P., Mortin, L. and Robertson, G. (1986) 'The forms of a task and their blends', in S. Grillner, P. Stein, D. Stuart, H. Forssberg and R. Herman (eds.), Neurobiology of Vertebrate Locomotion, Macmillan, London, pp. 201-216.
Stetson, R.H. (1951) Motor Phonetics: A Study of Speech Movements in Action, North Holland, Amsterdam.
Summerfield, Q. (1981) 'Articulatory rate and perceptual constancy in phonetic perception', Journal of Experimental Psychology: Human Perception and Performance 7, 1074-1095.
Sussman, H. (1980) 'Methodological problems in evaluating lip/jaw reciprocity as an index of motor equivalence', Journal of Speech and Hearing Research 23, 699-704.
Sussman, H. and Westbury, J. (1981) 'The effects of antagonistic gestures on temporal and amplitude parameters of anticipatory labial coarticulation', Journal of Speech and Hearing Research 46, 16-24.
Terzuolo, C. and Viviani, P. (1979) 'The central representation of learned motor patterns', in R. Talbot and D. R. Humphrey (eds.), Posture and Movement, Raven Press, New York, pp. 113-121.
Tuller, B., Kelso, J. A. S. and Harris, K. S. (1982) 'Interarticulator phasing as an index of temporal regularity in speech', Journal of Experimental Psychology: Human Perception and Performance 8, 460-472.

Tuller, B. and Kelso, J. A. S. (1984) 'The timing of articulatory gestures: Evidence for relational invariants', Journal of the Acoustical Society of America 76, 1030-1036.

Turvey, M. (1977) 'Preliminaries to a theory of action with reference to vision', in R. Shaw and J. Bransford (eds.), Perceiving, Acting, and Knowing: Toward an Ecological Psychology, Lawrence E. Erlbaum, Hillsdale, NJ., pp. 211-265.

van Dijk-Kappers. A. M. L. (1988) 'Comparison of parameter sets for temporal decomposition of speech', IPO Annual Progress Report (Institute for Perception Research, Eindhoven) 23, 24-33.

van Dijk-Kappers, A, M. L. and Marcus, S. (1987) 'Temporal decomposition of speech', IPO Annual Progress Report (Institute for Perception Research, Eindhoven) 22, 41-50.

Viviani, P. and Terzuolo, C. (1980) 'Space-time invariance in learned motor skills', in G. Stelmach and J. Requin (eds.), Tutorials in Motor Behavior, North-Holland, Amsterdam, pp. 525-533.

Yoshioka, H. (1981) 'Laryngeal adjustments in the production of the consonants and devoiced vowels in Japanese', Phonetica 38, 236-251.

Yoshioka, H., Löfqvist, A. and Collier, R. (1982) 'Laryngeal adjustments in Dutch voiceless obstruent production', Annual Bulletin (Research Institute of Logopedics and Phoniatrics, University of Tokyo) 16, 27-35.

Yoshioka, H., Löfqvist, A. and Hirose, H. (1981) 'Laryngeal adjustments in the production of consonant clusters and geminates in American English', Journal of the Acoustical Society of America 70, 1615-1623.

Öhman, S. (1966) 'Coarticulation in VCV utterances: Spectrographic measurements', Journal of the Acoustical Society of America 39, 151-168.

Öhman, S. (1967) 'Numerical model of coarticulation', Journal of the Acoustical Society of America 41, 310-320.

ARTICULATORY PERSPECTIVES OF SPEECH ORGANIZATION

OSAMU FUJIMURA
The Ohio State University, Division of Speech and Hearing Science,
Columbus, OH 43210, U. S. A.

ABSTRACT

Research issues of speech production are discussed assuming that speech is organized based on crucial articulatory gestures, relating the phonological representation of a sentence to a physical utterance with certain information added to the lexicosyntactic specifications. It is argued that recent progress in both phonological theory and experimental methods of recording and processing articulatory gestures substantiates a radical departure from the traditional linguistic view that speech is organized by concatenating phonemic segments.

0. Introduction -- Background and Motivation

Since the foundation of the current acoustic theory of speech production by Fant (1960), research effort has more or less concentrated on understanding speech organization based on concatenation of phonemic segments as a time series. Suprasegmental modulation of such segmental strings together with a smoothing process or the so-called coarticulation constituted the physical implementation. Current phonological theory, in particular the nonlinear phonology, advocates for the abstract representation of speech a basic deviation from such segment concatenation models (see Halle and Vergnaud,1987 for review). In a manner similar to the separation of the intonational "melody" from the segmental temporal stream in early autosegmental accounts, so-called segmental description of speech itself is now being viewed as a composite of multidimensional temporal organizations. Thus, individual phonetic dimensions corresponding to mandible, tongue tip, tongue body, lip, and velum movements, in terms of articulatory organs, form more or less independent time functions under the governance of a certain principle of interdimensional linking (see McCarthy, 1988). This can be viewed as an elaboration of Öhman's (1967) original idea of consonantal perturbation superimposed onto the vocalic time series and the later concepts of phonological projection in connection with the development of the nonlinear theory of phonology (see Halle and Vergnaud, *ibid.*). By setting up different tiers and defining domains for feature specifications independently tier by tier, theoretical difficulties such as long-distance interaction and variable rule application domains may be considered largely resolved. One notable trend in studies of phonological theory is that reference to specific articulators, as opposed to the place of articulation along the vocal tract, for example, is being advocated by theoreticians, based purely on the functional descriptive motivation (see McCarthy, *ibid.*).

In parallel, in recent years, direct observations of articulatory movement patterns in continuous speech have become more and more feasible, using particularly the computer-controlled x-ray microbeam (see Fujimura *et al.*, 1973a; Kiritani *et al.*,

W. J. Hardcastle and A. Marchal (eds.), Speech Production and Speech Modelling, 323–342.

1975; Nadler *et al.*, 1987). The extensive data from utterances under relatively natural conditions have revealed various intriguing deviations from the segment concatenation principle (Fujimura, 1981, 1986, 1987b). Articulatory organs appear, to a certain extent, to freely organize their movement patterns, under physical constraints about interactions among articulators. Interactions among different crucial "segmental" events within each articulatory dimension may affect the temporal course in its own dimension but not necessarily affect events in other dimensions simultaneously, as the segmental model would prescribe. In addition, target values are often not found for articulatory gestures unless they play crucial phonological functions. In fact, according to some emerging ideas of phonology and phonetics, those noncrucial gestures would never appear in the phonological representation (see Browman & Goldstein, 1985, 1986), just as tone values are not specified for each tone bearing unit and the phonetic implementation computes the voice pitch time functions directly out of the sparsely specified representation of tone values (see Pierrehumbert & Beckman, 1988). Recent articulatory studies have revealed the importance of temporal "distortions" of "segmental" patterns in various contexts, shedding new light onto the issue of allophonic variation. Many of the allophonic contrasts previously viewed as discrete oppositions now seem to reveal continuous differences in relative timing of crucial events among different articulators, as will be discussed below. Most of the conditions under which such variation occurs and the reason why such variation must occur still remain to be identified. But it is probably fair to say that we are acquiring a qualitatively new insight into the principles of speech organization. At the same time, it is worth mentioning that the relationship between phonology and phonetics, with respect to speech organization principles, is now one of the main topics of serious study for researchers with both theoretical and experimental orientations.

1. An Overall Description of Speech Utterance

There are many factors that come into the picture when we try to specify the exact time course of speech organization quantitatively. They include the traditional segmental identities such as individual vowels and consonants as well as accent and other prosodic feature specifications (to the extent their features have to be specified within the lexicon), and intonational and configurational feature specifications of the phrasal units in the phonological representation, at different hierarchical and derivational levels of word and phrasal formation (see Kiparsky,1982). Part, or most, of the information comes from the lexicon and from the syntactic description of the sentence to be uttered, but the phrasal organization may not be directly nor completely determined by the syntactic phrasal structure even at the most surface level (see Selkirk,1984). In order to complete the utterance specification, we will need to add information that is not contained within the description of the sentence as given in its syntactic structures. Specifically, utterances vary depending on discourse factors such as focus, contrastive emphasis and other prominence specifications that may be attached to relevant phonological domains.

What Jakobson *et al.* (1963) called expressive features, for example, must be represented appropriately as the input specifications for speech synthesis by rule, in order to produce natural conversational speech. Such "miscellaneous" characteristics of speech can not be ignored not only to succeed in speech technology, but to account for observed speech production and perception phenomena, where quantitative details of observed phenomena concern our description. More importantly, the traditional separation of naturalness from intelligibility in synthetic speech fails to capture the intricate effects of "suprasegmental" factors of speech on speech understanding in conversation. Furthermore, the so-called segmental characteristics of speech, such as formant patterns of vowels pronounced in a sentential utterance, are found to be strongly affected by intonational conditions. For example, the effect of adding prominence to a particular word in a sentence, due to contrastive emphasis, generally tends to cause the jaw to be more widely open, resulting in a considerable change in the formant pattern. These changes are often far beyond the classical account of coarticulation and undershooting. In other words, the suprasegmental modulation of the string of segments, whether phonemes, demisyllables, or syllables, in terms of the classical segment concatenation model, must include manipulation of segmental target gestures themselves. When we study the concrete effects of such "suprasegmental" modulation, we find it difficult to maintain the significance of the concept of a target gesture (or its acoustic consequence) as a completely specified and therefore independent and inherent phonetic gesture of the segment as a whole.

Our goal is to study the organization of speech in such a way that we can identify linguistic factors formally separated from other speech characteristics. In order to arrive at a consistent theoretical framework for speech representation, from this point of view, we need to work out some representational organization, however tentative, to identify and position properly in the overall picture what might be considered gray areas, such as "expressive features" specifying *e. g.* incredulity, along with contrastive or other types of emphases. These phonetically manifest informational elements are clearly linguistic entities to the extent that in many languages, different linguistic devices including lexical items are used to represent (part of) such information. Evidently, for this purpose, it is not sufficient to confine our consideration within the domain of a sentence. There are many readily noticeable interactions among sentences that cause phonetic modifications of sentential utterances. These interactions are often considered discourse effects, and they range from adding lexical adverbs and sentence final particles to subtle intonational adjustments of temporal and voice quality (including pitch) controls. However, we must note that the sentence plays a critical role in the grammatical description of language. Without it, we could hardly formulate any exact syntactic theory. Sentences do define remarkably stable units in this sense. Therefore, in describing utterances in realistic situations, it makes sense to assume the sentence to be the basic domain of description, allowing interactions among sentences and introducing modifications due to external causes as modulations. In this way, we can formulate an approximation theory with successive refinements of the capability of the representation scheme, by studying various effects of the environment of the sentence in question in a realistic discourse situation. By working with a specific situation such as human-machine interaction for a specific domain of tasks, we can assure some feasibility of work without expanding the scope of description beyond reasonable bounds.

2. Components of Input Specification

Figure 1 intends to depict what may be considered an overall picture of the formation process of speech utterance. As the input to the box at the top of the figure containing

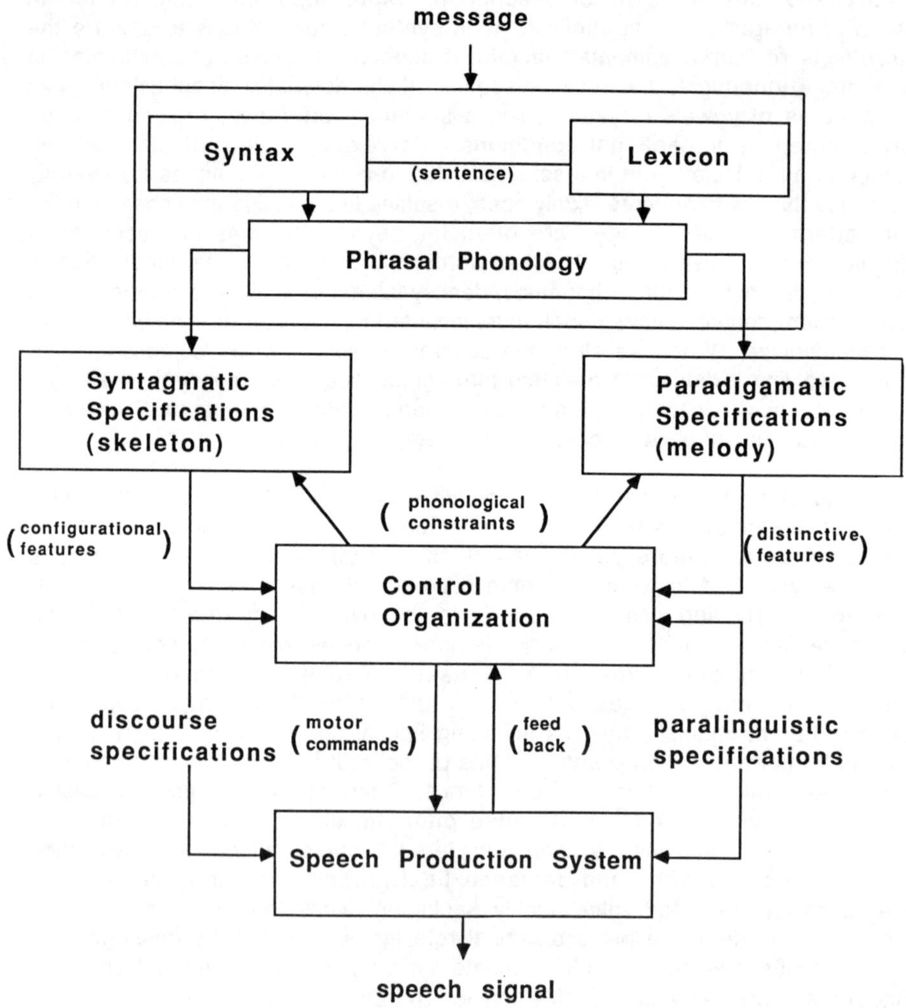

Fig.1 A Model of Speech Utterance

grammatical systems of the given language, what is traditionally considered a linguistic message is specified for selecting a particular sentence to be uttered. The syntactic component of grammar combined with the database describing lexical items defines a sentence using specific words as the ingredients of the sentence. The sound shape of the sentence is formed by a component of grammar called phrasal phonology, using the data provided by the phonological descriptions of entries in the lexicon, based on the organizational structure of the sentence as specified by the syntax of the language.

The phonological representation of the sentence thus formed has two distinct aspects of its linguistic functions: (1) syntagmatic specifications, which provide a skeleton of the phonological structure of the sentence in question, and (2) paradigmatic specifications, which supply many aspects of information concerning the ingredients of the sentential organization. The specifications in the first category identify the organizational framework of the phonological representation, providing in part the abstract temporal organization of the sentence. The second category of information may be assumed to be represented by feature values such as the inherent distinctive feature specifications for consonantal and vocalic aspects of the sound shape of a phrasal unit or its constituents. Within this category, prosodic distinctive features identify the tonal or accentual patterns of the phrase. The so-called intonational patterns may have specifications in both categories: (1) configurational specifications associated with the structural organization or skeleton itself, such as the requirement to have a boundary tone at the end of each phrase, and (2) a paradigmatic choice representing the values of phrase accent, *etc.* All these elements of the phonological representation of the sentence provide input specifications for the control organization of the utterance.

While the input specifications refer to the obviously linguistic components of grammar and database (lexicon including descriptions of ideosyncratic characteristics of certain phrases, see Sproat (1985), the control organization for a speech utterance needs more information. Two categories of such information are identified in Fig. 1. Discourse specifications provide information to be attached to the sentence or its phrasal components, such as contrastive emphasis, focus, presupposition, incredulity, surprise, *etc.* Note that these discourse related elements are sometimes included in the linguistic message at the top of the figure. The use of the definite article can indicate presupposition in English, while the same meaning can be conveyed by the use of the particle 'wa' in Japanese, in certain contexts. In some cases, however, the contrast between two distinct situations, focus *vs.* presupposition for the same word, may be conveyed by different intonation patterns. For example, the same sentence 'Naomi came.' is uttered with different intonation patterns depending on whether it is the answer to 'Who came to the party yesterday?' or 'Did Naomi come to the party yesterday?'. In Japanese under the same circumstances, the distinction is made by the use of the particles 'ga/wa' in 'Naomi ga/wa kimasita', typically in combination with the use of a phrase boundary after 'wa' but not 'ga'. These means of distinction are included in the linguistic message, even though there may well be additional intonational specifications that must be added for a particular situation in natural conversation and manifested in phonetic utterance characteristics.

Certain discourse specifications are in principle difficult to handle within the grammatical representations. Different degrees of prominence attached to a certain word or phrasal unit, for example expressing surprise or excitement, seem to be a matter of numerical control directly affecting parameters of the phonetic implementation process, as opposed to the use of binary distinctions incorporated into grammatical systems. It is not clear, however, that such different components as grammatical *vs.* discourse specifications can be separated consistently in terms of the ordering of rule applications. It seems that the distinction is with respect to symbolic (discrete) *vs.* numerical (continuous) computation rather than with respect to rule application components, even though the latter distinction may also obtain in certain models,. It should be noted that in Fig. 1, there is no specification of logical ordering relating the traditional linguistic process (upper half) to the extragrammatical process (lower half). In this model, rule ordering is assumed to be a matter of theoretical choice, but is confined within specific grammatical components.

Paralinguistic specifications pertain to selections of parameter values such as voice pitch range, articulatory strategies reflecting different styles of speech, *etc.* Note that here again, some of such choices (*viz.* information) can be incorporated into the linguistic message involving choices of lexical items and even choosing different grammars. It varies greatly from language to language which aspects of conversational styles are treated within the lexicogrammatical system and which by phonetic implementation parameters. Expression of politeness or intimacy, for example, can be a matter of selecting a subset of intonational vocabulary (including voice source spectrum control) as opposed to selecting a certain class of sentential particles or pronouns. Choices may be made in the form of specifying parameter values of the system identified as control organization, where phonetic implementation of the discrete symbolic representation of the utterance into numerical time functions of motor commands is achieved. Paralinguistic specifications may select certain parametric settings of the speech production system itself. A clearcut case is the selection of a particular speaker, including the choice of speaker gender. There may well be a certain language- or dialect-dependent set of parameters that are directly specified in the speech production system. If the so-called base of articulation or the phonetic disposition of the given language is to be set as parameter values within a speech synthesizer, for example, the downward arrow from paralinguistic specifications in Fig. 1 would represent such information passage. If the paralinguistic specification pertains to the habit of using particular phrase boundaries under certain semantic environments, the information would produce some bias within the control organization system which would form a specific (phonological) phrase structure for the sentence. A particular phrasal pattern associated with a particular skeleton will be chosen according to the logical interpretation of the expression, of course. In certain cases, however, the skeleton may have to conform to phonological constraints that are imposed by the control organization, in accordance with paralinguistic requirements under the influence of such stylistic or ideosyncratic bias. Therefore, the phonetic manifestation of phrasing distinctions such as

$$2 + (2 * 2) = 6 \textit{ vs. } (2 + 2) * 2 = 8$$

may vary from speaker to speaker, reflecting different biases not only in the choice of a particular physical correlate for the phonological phrase boundary in question, but also in the choice of the phonological phrasing pattern itself.

Interactions between discourse specifications or paralinguistic specifications and the choice of a particular sentence form could be handled by constraints imposed onto any component in the upper parts of this diagram. In this model, it is assumed that the grammar of the given language itself is not interfered with discourse or paralinguistic specifications. Such effects must be represented either in the sentential message itself (at top) or by the operation of the control organization system, which can constrain the choices of syntagmatic or paradigmatic phonological variables in conformity with the wellformedness of the abstract output of phrasal phonology. In other words, the output of phonology must be abstract enough to allow for symbolic manipulations like specifications of unspecified feature values which are sensitive to discourse and paralinguistic factors, without affecting the grammatical identity of the sentence.

3. Phonetic Implementation -- an Overall Picture

Let us assume that the output representation of the phrasal phonology component comprises two sets of specifications for the phonetic implementation. In Fig. 1, these two sets are identified as syntagmatic and paradigmatic specifications, abbreviated as the skeleton and the melody of the sentence, respectively. These represent the sentential message chosen for the utterance in its phonological form.

Skeleton provides a descriptive framework in which the melody information is represented as choices of paradigmatic values, each of the choices playing a specific role in the sentential representation. We may assume a phonological theory such as Selkirk's (1984), where the phonological structure of a sentence is given in the form of a phonological phrase structure, somewhat resembling but distinct from the syntactic (surface) tree. Skeleton must provide, in addition to the organizational information, some specification of the temporal organization of the utterance for the sentence. Phrases at different levels with intonational functions, as discussed in Pierrehumbert and Beckman (1988), as well as their subordinate constituent units such as syllables or demisyllables, are assigned abstract temporal values, such as ordering relations, of crucial physical or physiological events contained therein. The most general elemental unit of temporal organization may be mora, but its function in organizing speech events is expected to vary from language to language.

Melody is specified by variable values more or less like distinctive feature values, but generally we do not expect to have independent segmental units that must be given complete specifications of their contents. Melody is thus represented by sparse specifications in the multidimensional representation of articulatory patterns. A suggestive example of such representational schemes may be found in Browman and Goldstein (1985; in press), where they use the term "gesture score" for the phonological specification as the input for an articulatory signal generating system. A multidimensional organization consists of parallel temporal courses of event sequences in articulatory tiers. Only critical articulatory events are specified in relevant dimensions. The time values of individual events are presumably represented only abstractly. The system of implementation rules must be rather complex, unlike the simple coarticulatory smoothing principle. It computes concrete time functions under the influence of discourse and paralinguistic specifications, as well as feedback signals from the speech production system to conform to phonetic (including perceptual) wellformedness constraints, under each circumstance of the utterance.

4. The Articulatory System

Speech organs form the vocal tract, which defines the acoustic tube as the principal physical system for producing speech sounds. According to the acoustic theory of speech production by Fant (1960), what determines the acoustic properties of the outcoming sound waves, particularly for vocalic sounds, is the shape of this tube with respect to its cross-sectional area as a function of the distance from the glottis, *i. e.* the vocal tract area function. The articulatory system is a physiological system which is related to and responsible for (in terms of the speech production process) the vocal conduit configuration.

Individual articulators must have independent neural control, but positional states of different articulators may be inherently related to each other. For example, lingual gestures involve apical or coronal gestures and tongue dorsum gestures The tongue tip or blade position is affected by the tongue body gesture significantly, as well as by the contraction pattern of inherent muscles. Therefore, the place of the maximal constriction along the vocal conduit varies considerably depending on the vocalic context. The lip gesture can produce either a primarily vertical constriction of the oral orifice or rounding and protrusion, and typically, in many languages, the former is pertinent to consonantal gestures whereas the latter is used for vocalic features. Mandible position is an independent variable, controlled separately from constrictive gestures using the lips or the tongue tip, and is strongly related to prominence control (see Stone, 1981).

Obviously, mandible height must have constraints for producing, for example, open *vs.* close vowels under normal circumstances. Similarly, lip gestures for bilabial closure, in terms of muscular efforts, must make some adjustments, depending on the mandible position determined by the context. These adjustments are likely to be incorporated, in part, within the control organization program, but in part *via* feedback from the speech production system to the control organization affecting motor commands. The latter case may be observed particularly under abnormal circumstances such as bite-block experiments, and in such cases of external disturbance, the control may well respond adaptively to specific conditions affecting the positional effects of the gesture during an utterance. It should be noted, however, that a large part of what may appear to be a feedback effect may only reflect the nonlinearity of the system. For example, a repetition of the same syllable or words may show random variation in terms of jaw height, compensating for variation in lip gesture as in the production of /p/. This apparent compensation is not necessarily the result of sensing (or predicting) the state of lip closing gesture and compensating for its variation by readjusting the jaw gesture (or the other way around). It could well be simply that the jaw raising action is executed normally more than sufficiently for barely making the closure. Due to a saturation effect, then, there it would appear as though there were a trading relation between jaw height and the resultant lip deformation. Even when there is a feedback, for example resulting in readjusted activities of the lip closing effort in response to the (lack of) pressure of contact of the opposing lip, that feedback may be peripheral without involving higher level control. In order to answer these questions, a very careful analysis of timing relations among emg activities and physical movements may provide us with critical information.

For effectively describing the (positional or dynamic) state of each articulator, an appropriate selection of the independent variables is crucial. In general, an articulatory organ such as the tongue and the lips vary not only in position but also in shape, and often there is more than one degree of freedom in describing the phonetically relevant configuration, reflecting complex muscular structures within the organ and muscles attached to it from the outside. Such complex organs need to be decomposed into more than one independently controlled part. Sometimes such independently controlled components may be physically interwoven, and it may be particularly difficult to determine (by electromyography, for example) whether the control is really independent.

In addition, there is the separate problem of how each component can be described, say, in terms of its position, *i. e.* exactly which flesh-point of the structure should be selected as the representative point. Assuming that there is some reason to separate out a part of the organ, such as the tongue blade, from the rest of the same physical object, the tongue body, the center of gravity might be generally a good sample point of the physical portion in specifying its movement characteristics. However, the peripheral parts of the same organ deform in phonetic gestures, particularly under a physical contact (or near contact, as in extreme vowels, see below) with other organs. Acoustic consequences of such constrictions largely depend on the position of peripheral parts rather than the center of gravity (of the tongue tip, the lower lip, the velum, *etc.*). Furthermore, the peripheral deformation causes considerable variability in the relation between the peripheral position and the central position.

In typical microbeam studies, sample positions pertain to nonextreme parts of the articulator. For example, the apical condition is represented by the laterally viewed midsagittal position of the metal pellet placed about one half to one centimeter posterior of the tongue tip, revealing information on the surface position of a relatively central point of the tongue tip rather than on the point of contact for apical stop consonants. Similarly, the lower lip pellet is typically placed at the vermilion border somewhat away from the closure surface. This pellet placement technique is principally motivated by the experimenter's intention to avoid articulatory interference.

While this experimental "artifact" necessitates some caution in interpreting the observed pellet positions in comparison with the acoustic characteristics of the speech signals, it does provide us with additional information often crucial for understanding gestural organizations of speech phenomena. As exemplified by Macchi's (1988) study of lip movements in relation to different (segmental *vs.* suprasegmental) linguistic functions on the one hand and proper labial *vs.* mandibular maneuvers on the other, the central movement of the lower lip as revealed by pellet position, particularly during the closure period, provides us with important information which is not available from acoustic signals. More generally, central position varies temporally more smoothly, without reflecting the highly nonlinear mapping between control time functions and sample-point position or movement. The nonlinearity is largely due to saturation. A more linear mapping relation between the control variable and observed position means a better applicability of the superpositional principle in interpreting the positional data of the articulators. In other words, more than one control factor can be evaluated separately in the analysis of articulatory movement patterns. This of course implies that it is important to record the acoustic signal simultaneously as an independent source of information about, for example, timing of implosion and explosion, which may be crucial for perceptual effects of acoustic discontinuities.

Fig. 2 illustrates the contrast between the continuous articulatory data and the discontinuous acoustic data. The acoustic data are represented by a crude spectrographic display using a limited number of frequency bands. The articulatory data, vertical and horizontal positions of pellets from a microbeam experiment, for the same utterance, are displayed in somewhat comparable form, by converting coordinate values to gray scale.

5. Articulators and Temporal Independence -- "Allophonic" Variation and Other Contextual Effects

There have been observations of qualitatively different phonetic appearances of the same phoneme, depending on the context. From the traditional segmental point of view, these were identified as allophones, often denoting such phonetic variants by different phonetic symbols or by attaching diacritics. Thus, English /l/ is said to be light when it is in syllable-initial position, and dark in syllable-final position. Similarly, English /t/ or /d/ is a flap (more correctly a tap, according to Ladefoged *et al.*,1977) in certain phonological environments (see Kahn, 1976). In some dialects, both phonemes /t/ and /d/ can collapse into the same allophone. In Japanese, (syllable initial) apical consonants are strongly palatalized when followed by tautosyllabic /i/ (see Miyawaki, (1972). As an example of a more subtle phonetic difference which escapes usual phonetic discrimination, some microbeam studies of velum height for nasal consonants in several languages have revealed that generally the velum takes a higher position for a nasal consonant in syllable initial position than in syllable final position (see Fujimura, 1977 for discussion of Japanese and English). Allophonic variation of consonants is typically conditioned by intrasyllabic or ambisyllabic position, and the specific phenomena are often cited as evidence for the significance of syllables as phonological units at some level of phonological representation (*e. g.* Kahn, *ibid*; Fujimura & Lovins, 1978, Borowsky, 1987).

These allophones, as discrete segmental units, obtain as the consequence of changing binary values of distinctive features by context-sensitive phonological rules. In contrast, there is an emerging sense that such allophones are, in essence, extreme and readily noticed cases of a continuum of variation. In addition, there are articulatory observations showing that such variation can be characterized by variable timing relations between characteristic gestures exhibited by different articulators. Fujimura (1986) pointed out the timing relation difference between the stop and flap gestures in American English, and Sproat and Fujimura (1989) give evidence for timing difference between apical and dorsal gestures for English /l/. The latter study demonstrates a detailed gradation of light-dark variation, depending on the type of the boundary involved among other factors, and also reveals strong correlations between rime duration and the articulatory timing shifts among articulators.

In addition to such variation sensitive to syllabic and other syntagmatic conditions, there seem to be a variety of interarticulator timing adjustments or fluctuation that give rise to phonetic observations of deletions and insertions as well as feature

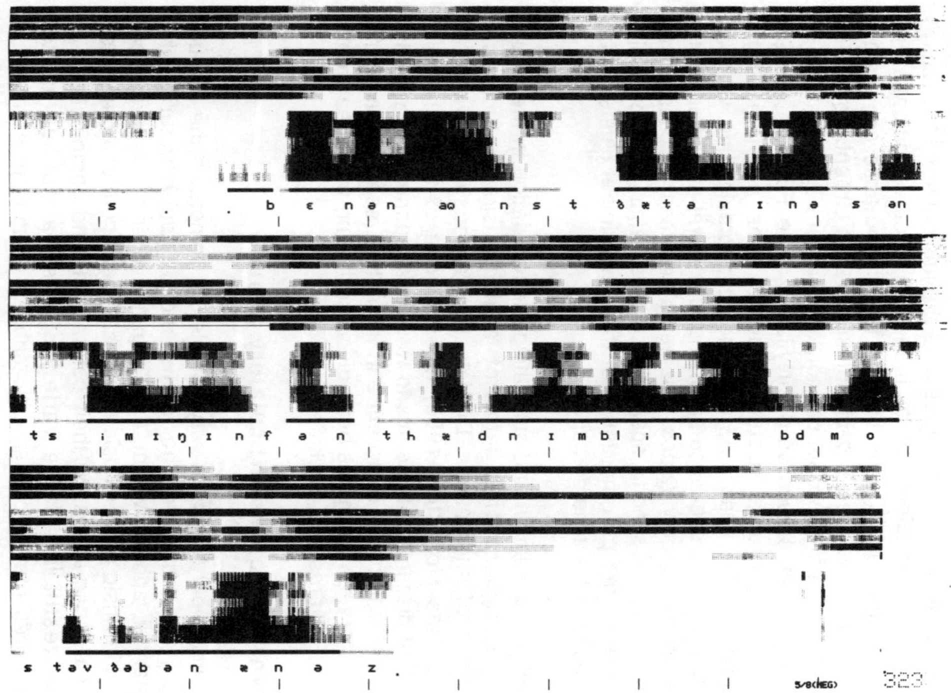

Fig. 2: Comparison of articulatory and acoustic data displays. The upper set of horizontal bars in each of three panels represents pellet coordinate values as follows (from top to bottom): front-back position (dark for front) of lower lip, tongue blade, middle of tongue, rear of tongue; vertical position (dark for up): lower lip, mandible, blade, middle, rear of tongue, velum. The lower half represent a course wide-band spectrogram.

alterations from the segmental point of view. Browman and Goldstein in a recent article (in press) discussed several categories of phenomena of this nature. The basic principle involved is a shift of relative timing among articulators (including laryngeal maneuvers), resulting in salient feature alterations in voicing and nasality, among others, which are interpreted as deletion and insertion of segments. Such phonologically permissible variability in timing of consonantal gestures in general is another aspect of the question of how distinctive features are implemented "correctly" by a general (in part, universal) principle, rather than by explicit specification of segmental sequences within the same demisyllable. The phonotactic constraints discussed extensively in the literature (see *e. g.* Maddieson, 1984; Greenberg, 1978) can be reinterpreted from this point of view: out of unordered feature specifications of units larger than phonemic segments such as demisyllables (see Fujimura, 1979), articulatory events manifesting crucial feature specifications are temporally organized by phonetic implementation rules. As a segmental approximation, we may view this as ordering phonemes into a sequence according to a vowel affinity principle, based on the segments' inherent sonority values. Clements (in press) discussed this issue by introducing a new concept of sonority scale, based on a demisyllabic principle. Browman and Goldstein (1988) interpret the timing of the whole initial cluster to be represented by its central point in the time domain.

In addition to the principle of assimilation and smoothing (coarticulation in the sense of Lindblom, 1983), speech organization in terms of multidimensional articulatory gestures exhibits other types of interaction. Dissimilation, as observed widely in phonologies of languages (see McCarthy, 1988) must be one of the principles. Temporal repulsion, a type of dissimilation, seems to work in a manner similar to the stress clash discussed in metrical theory of accent and intonation (see Liberman and Prince, 1977). Such articulatory repulsion has been discussed using microbeam data by Fujimura (1981) and by Edwards and Beckman (1988).

6. Constraints and Dynamics of the Articulatory System

The physical system of articulatory organs is very complex. It comprises the tongue, the upper and lower lips, the mandible, the larynx and related structures, the structures defining the pharynx, and the velum, each organ involving a large number of independently controlled intrinsic and extrinsic muscles. These controlling muscles generally have independent innervations for their contraction, and combinations of such muscular activities result in a large variety of gestural configurations. Independent phonetic gestures may evoke specific subsets of muscle contraction patterns. Each gesture or neural command may or may not directly correspond to a distinctive feature that functions as the elemental phonological unit or to an independent tier in the nonlinear phonological description.

At least as a first-order approximation, each articulator, such as the tongue tip, the tongue blade, the tongue body, the lower lip, the mandible, the velum, appears to have a well-defined phonetic function such that its representative position (see above) serves as an effective descriptor for an elemental phonetic gesture. Thus gestures such as raising, retroflexion, or protrusion of the tongue tip, raising of the blade (palatalization), tongue body retraction or root advancement, labial constriction with or without lower lip retraction, labial protrusion and rounding, and velum height control, may replace what we considered as segmental articulatory gestures referring to vocal tract configurations.

Obviously, there are many important interactions among these directly controlled gestures using different anatomical structures. Lower lip height cannot be independent of mandibular position, for example. When the back of the tongue is raised, the velum cannot take a very low position regardless of the degree of nasalization, and therefore velum lowering tends to be smaller for velar nasals than for other nasals. Different vowel gestures, due to different tongue body positions, tend to induce different voice fundamental frequencies, other conditions being equal, because of the anatomical linkage between the tongue body and the larynx *via* the hyoid bone (see Honda, 1986). There are many correlations among positions of articulators that still escape our explanation. Some interactions, in relation to phonetic functions in a specific language, are phonologized, in the sense that the anatomically motivated correlation is not necessarily the mechanism for the implementation of the acoustic correlation as the expected phonetic effect. For example, the intrinsic voice fundamental frequency of vowels may be "emulated" by employing an "irrelevant" mechanism of pitch control, in particular the cricothyroid muscle (see Honda, *ibid*). Some of such phenomena may well be a universal phonetic effect, and in that sense what is called motor equivalence (see Abbs *et al.*, 1979) may be observed under certain experimental situations. The feedback paths from the articulatory system up to the control organization in Fig. 1 may reflect such effect-driven phenomena, and this interaction is not necessarily a real-time control but part of a rule formulating procedure, perhaps going up to the level of postlexical phonological organization.

One general point about such interactions in speech production is the nonlinear nature of the mapping from neural control patterns to acoustic characteristics, or the vocal conduit configuration. The most salient and general cause of nonlinearity, apart from the mapping from area function parameters to formant frequencies (see Stevens, in press, Carré and Mrayati, 1989), is the saturation of movement, as discussed above. The saturation may be more complex than the case of lip closure, necessitating an exact three-dimensional modeling of the tongue (see Kiritani *et al.,* 1976). For example, in the case of the [i]-gesture, because of the lateral but not midsagittal support by the lingual contact with the hard palate, the consideration of the midsagittal groove on the tongue surface is critical for relating muscular contraction patterns with acoustic consequences (see Fujimura and Kakita, 1979). Many similar but specifically different situations must exist, and quite possibly these phenomena play crucial roles in determining the acoustic consequences of articulatory gestures.

When we consider the mapping process taking neural commands as the input, the simulation of muscular contraction processes must take into account the rather complex temporal characteristics of the physical system also. Thus, an electromyographic event typically must allow a considerable delay, often 100 msec or even more, before the consequent physical movement appears, when the former reflects the causal control of the latter. In contrast, when the muscle contracts in response to its externally created strain, such a delay may not be observed. The complex of muscles in speech organs forms a network with many feedback routes at different levels, and some loops are not well understood. Furthermore, the muscle contraction process itself exhibits not only nonlinear, but also history dependent characteristics (hysteresis), reflecting the relatively slow and irreversible chemical processes involved.

Independent of these complications, the dynamic properties of the system are critical for modeling the articulatory system. The proper consideration of inertia of subcomponents of the tongue, for example, has not been implemented in working simulation models. When we discuss the temporal characteristics of speech production, this obviously points to a very serious defect of our understanding, and one of our most immediate needs is to design an effective dynamic and three-dimensional model for the tongue and other articulators. Only when we have included in our research methods such a simulation model of appropriate accuracy, will we be in a position to effectively separate out physical phenomena from neurally executed linguistic specifications.

7. Phonetic Phrasing

One crucial aspect of speech organization is phonetic phrasing. Along with the paradigmatic choices of ingredients for utterances, the syntagmatic specifications must include appropriate phrasing of the materials for the utterance of a sentence, as well as the effects of the environment of the sentence. As mentioned above, the phonetic phrase structure reflects in part the underlying syntactic structure, but is independent from it. In fact, it is not a trivial question if there is a hierarchical tree structure, even though it is clear from intonational studies that there must be specifications of different degrees of phrasal separation (see Pierrehumbert and Beckman, 1988; Kubozono,1987 for some relevant discussion). There is some evidence (see Sternberg *et al.*, 1988) that before an utterance starts, the speaker must organize an appropriate phrasal unit in the control organization as a cortical program. If there is any global effect (see Fujimura,1987a for such a model), then the phrasal unit for the execution program must be large enough to contain all interacting components. Otherwise, any change in a part of the utterance material would have to modify other parts that have already been executed. A strictly left-to-right execution model, based on a string of completely specified phonetic units, would not explain such interactions.

There are some preliminary findings (see Fujimura,1987b), using the microbeam observation and iceberg timing measurement, on the temporal control of phrasal organization under the control of contrastive emphasis and semantic phrasing. Figs. 3 and 4 exemplify the findings. The ordinate represents the amount of timing shifts between corresponding articulatory events (icebergs, see Fujimura, 1986) of the two utterances in comparison. By comparing two utterances of the same sentence (in terms of word sequence) but with different phrasing and prominence control, we observe clear discontinuities in the time course of the utterance with respect to the slope of the connecting straight line segments in the figures. Each break point of the continuing line (sometimes accompanying a gap in vertical position) can be interpreted as a boundary between two consecutive phonetic phrases, observing that the local speed of utterance within each such phrase is modified uniformly from one utterance version to another. Parts of the phrasing structures that are common to the two utterances may not reveal any discontinuities in slope of the fitted lines. Empirically, however, such common structures also seem to be affected by the

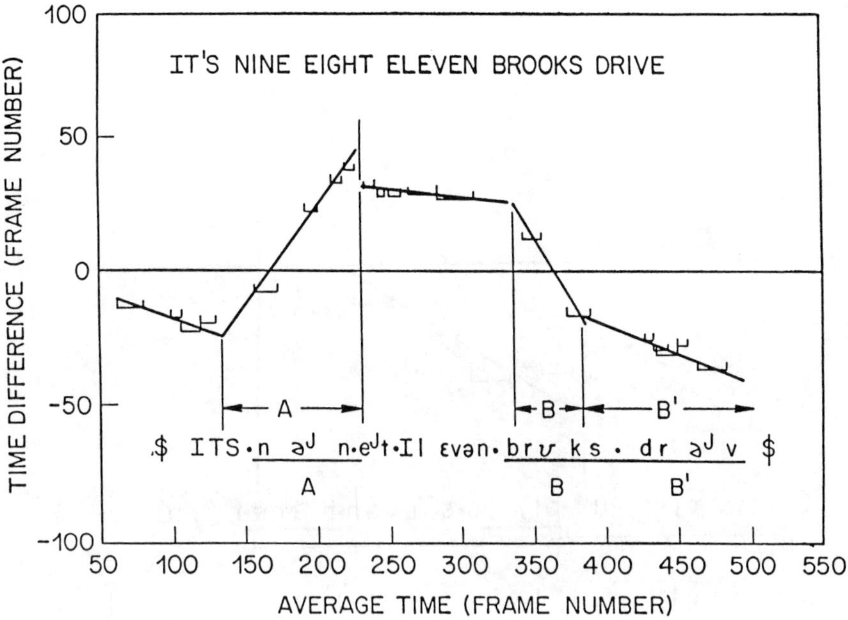

Fig. 3: Relative timing patterns for prominence control.

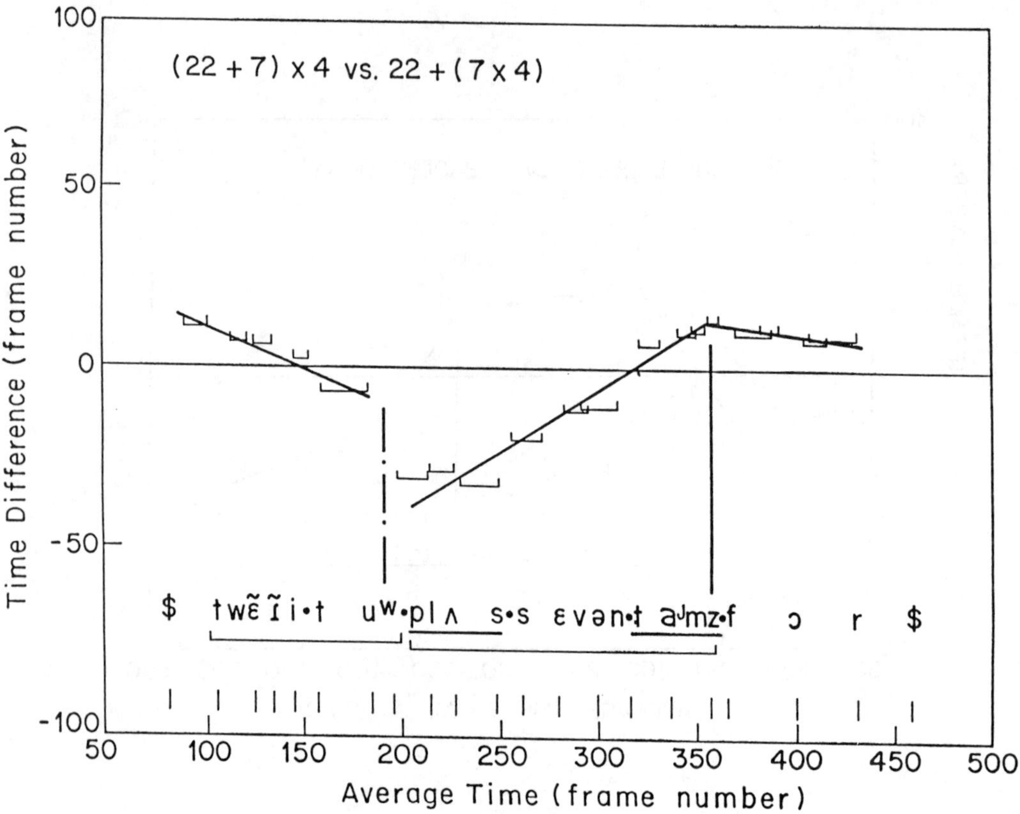

Fig. 4: Relative timing patterns for phrasing control.

perturbation introduced by different "prosodic" controls, resulting in slope differences in practically all phrases in the sentence.

This demonstrates significant interactions among constituent phrases which are not directly affected by the manipulation. The apparent alteration of the direction of the line, resulting in more or less compensated timing differences (which may be identified in the figures as the return to the same height at the end of a set of a few line segments), seems to suggest that phrasal units, larger than those represented by the line segments in the figures, may behave independently from each other in forming their temporal courses. Such quantitative evaluations of "prosodic" control of sentence materials have become feasible using many subjects and well controlled utterance conditions.

8. Concluding Remarks

At the moment, the feasibility of articulatory data collection is due to the x-ray microbeam system. There are other methods of similar and possibly superior observation methods, particularly using magnetic measurements (see Sonoda and Kiritani, 1976; Schönle, 1988; Perkell & Cohen, 1986). Ultrasonic (see Sonies *et al.*,1981, Honda,1985) and dynamic palatography or electropalatography (see Fujimura *et al.*, 1973b, Hardcastle,1974) may serve effectively for such studies. New efforts in computer simulation of the complex articulatory system for studying temporal characteristics of speech are being planned. These experimental and computational methods promise substantive progress in our understanding of speech organization in realistic situations.

In parallel, phonological theory is evolving rapidly. Hopefully these new developments will converge into an integrated understanding of linguistic and behavioral as well as physiological and pathological processes involved in speech production. By the end of the century, we may expect a substantive change of view, hopefully with successful speech technology applications as well, inviting a new social and industrial interest in this area of study.

REFERENCES

Abbs J. H. (1979), `Speech Motor Equivalence: The Need for a Multi-Level Control', *Proc. 9th Int. Cong. Phonetic Sc., Vol II, Copenhagen, Aug. 6-11*, 318-324.

Borowsky, T. J. (1987). *Topics in the Lexical Phonology of English*, Garland, New York.

Browman, C. P. and Goldstein, L. (1985). `Dynamic Modeling of Phonetic Structure', in V. A. Fromkin (ed.), *Phonetic Linguistics -- Essays in Honor of Peter Ladefoged*, Academic Press, New York 35-53.

Browman, C. P. and Goldstein, L. (1986). `Toward an Articulatory Phonology', *Phonology Yearbook 3*, 219-252.

Browman, C. P. and Goldstein, L., (1988), `Some Notes on Syllable Structure in Articulatory Phonology', *Phonetica 45*, 140-155.

Browman, C. P. and Goldstein, L., (in press), `Tiers in Articulatory Phonology, with some Implications for Casual Speech', in Kingston, J. and Beckman M. E. (eds.) *Papers in Laboratory Phonology I: Between the Grammar and the Physics of Speech*, Cambridge Univ. Press, Cambridge.

Carré, R. and Mrayati, M. (1989), `New Concept in Articulatory-Acoustic-Phonetic Relations, Perspectives and Applications', *Proc. Intern. Cong. Acoustics, Speech and Signal Processing, Glasgow*, pp. 231-234.

Clements, G. N. (in press), `The Role of the Sonority Cycle in Core Syllabification', in Kingston, J. and Beckman M. E. (eds.) *Papers in Laboratory Phonology I: Between the Grammar and the Physics of Speech*, Cambridge Univ. Press, Cambridge.

Edwards, J. and Beckman, M. E. (1988), `Articulatory Timing and the Prosodic Interpretation of Syllable Duration', *Phonetica 45*, 156-174.

Fant , G. (1960), *Acoustic Theory of Speech Production*, Mouton, The Hague.

Fujimura, O. (1977), `Recent Findings on Articulatory Processes -- Velum and Tongue Movements as Syllable Features', in Carré, R., Descout, R. and Wajskop, M. (eds.), *Articulatory Modeling and Phonetics*, GALF Groupe de la Communication Parlée, Grenoble, pp. 115-126.

Fujimura, O. (1979), `An Analysis of English Syllables as Cores and Affixes', *Zs. Phonetik, Sprachwiss. u. Kommunikationsforsch. 32*, 471-476.

Fujimura, O. (1981), `Temporal Organization of Articulatory Movements as a Multidimensional Phrasal Structure', *Phonetica 38*, 66-83, corrected version in A. S House (ed.), *Proc. Symp. Acoustic Phonetics and Speech Modeling*, Part 2, Paper 5, Inst. Defense Analysis, Princeton, NJ.

Fujimura, O. (1986), `Relative Invariance of Articulatory Movements: an Iceberg Model', in Perkell, J. S. and Klatt, D. H. (eds.), *Invariance and Variability in Speech Processes*, Lawrence Erlbaum, Hillsdale, NJ, pp. 226-242.

Fujimura, O. (1987a), `A Linear Model of Speech Timing', in Channon, R. and Shockey, L. (eds.), *Ilse Lehiste Puhendusteos*, Foris, Dordrecht.

Fujimura, O. (1987b), `Fundamentals and Applications in Speech Production Research', *Proc. XIth Int. Congr. Phonetic Sc. 6*, Tallinn, Estonia, pp. 10-27.

Fujimura, O. (ed.) (1988), *Articulatory Organization -- from Phonology to Speech Signals* (Special Issue), *Phonetica 45*.

Fujimura, O. and Kakita, Y. (1979), `Remarks on Quantitative Description of the Lingual Articulation', in Öhman, S. and Lindblom, B. (eds.), *Frontiers of Speech Communication Research*, Academic Press, London, pp. 17-24.

Fujimura, O. and Lovins, J. (1978), `Syllables as Concatenative Phonetic Units', in Bell, A. and Hooper, J. B. (eds.), *Syllables and Segments*, North Holland, Amsterdam, pp. 107-120.

Fujimura, O., Ishida, H and Kiritani, S. (1973a), `Computer-Controlled Radiography for Observation of Movements of Articulatory and other Human Organs', *Comp. Biology & Med. 3*, 371-384.

Fujimura, O., Tatsumi, I. F. and Kagaya, R. (1973b), `Computational Processing of Palatographic Patterns', *J. Phonetics 1*, 47-54.

Greenberg, J. H. (1978), `Some Generalizations concerning Initial and Final Consonant Clusters', in Greenberg, J. H. (ed.), *Universals of Human Languages, Vol. 2: Phonology*, Stanford Univ. Press, Stanford, pp. 243-279.

Halle, M. and Vergnaud, J-R. (1987), *An Essay on Stress*, MIT Press, Cambridge MA.

Hardcastle, W. J. (1974), `Instrumental Investigations of Lingual Activity during Speech: a Survey', *Phonetica 29*, 129-157.

Honda, K. (1986), *Biomechanical Interaction between Fundamental Frequency*

Control and Vowel Articulation, Doctoral Diss., U. Tokyo (RILP), Tokyo.

Honda, M. (1985), Dynamical Ultrasonic Measurement System of Tongue Shape, *Trans. Tech. Com. Speech, Acoust. Soc. Japan*, S84-105 (in Japanese).

Jakobson, R., Fant, G. and Halle, M. (1963), *Preliminaries to Speech Analysis*, Cambridge MA, MIT Press, Cambridge MA.

Kahn, D. (1976), *Syllable Based Generalizations in English Phonology*, PhD Diss., MIT, Cambridge, MA.

Kiparsky, P. (1982), `From Cyclic Phonology to Lexical Phonology', in van der Hulst, H. and Smith, N. (eds.), *The Structure of Phonological Representation*, Part I, Dordrecht, Foris, pp. 131-176.

Kiritani, S., Itoh, K. and Fujimura, O. (1975), `Tongue-Pellet Tracking by a Computer Controlled X-Ray Microbeam System', *J. Acoust. Soc. Am. 57*, 1516-1520.

Kiritani, S., Miyawaki, K., Fujimura, O. and Miller, J. E. (1976), `A Computational Model of the Tongue', *Ann. Bull. U. Tokyo RILP 10*, 243-251.

Kubozono, H. (1987), *The Organization of Japanese Prosody*, PhD Dissertation, U. Edinburgh.

Ladefoged, P., Cochra, A. and Disner, S. (1977), Laterals and Trills, *J. Int. Phonetic Assoc. 7*, 46-54.

Liberman M. Y. and Prince, A. (1977), `On Stress and Linguistic Rhythm', *Linguistic Inquiry 8*, 249-336.

Lindblom, B. (1983), `Economy of Speech Gestures', in MacNeilage (ed.), *The Production of Speech*, Springer, New York, pp. 217-246.

Macchi, M. J. (1988), `Labial Articulation Patterns Associated with Segmental Features and Syllable Structure in English', *Phonetica 45*, 109-121.

Maddieson, I. (1984), *Pattern of Sounds*, Cambridge Univ. Press, Cambridge.

McCarthy, J. (1988), `Feature Geometry and Dependency: a Review', *Phonetica 45*, 84-108.

Miyawaki, K. (1972), *A Study of Lingual Articulation by Use of Dynamic Palatography*, MA Thesis, Dept. Linguistics, Univ. of Tokyo.

Nadler, R. D., Abbs, J. H. and Fujimura, O. (1987), `Speech Movement Research Using the New X-Ray Microbeam System', *Proc. 11th Int. Cong. Phonetic Sciences*, Tallinn, Paper Se 11.4.

Öhman, S. (1967), `Numerical Model of Coarticulation', *J. Acoust. Soc. Am. 41*, 310-320.

Perkell, J. S. and Cohen, M. H. (1986), `An Alternating Magnetic Field System for Tracking Multiple Speech Articulatory Movement in the Midsagittal Plane', *Tech. Rept. 512, Res. Lab. Electronics, MIT, Cambridge, MA.*

Pierrehumbert, J. and Beckman, M. E. (1988), *Japanese Tone Structure*, MIT Press, Cambridge MA.

Selkirk, E. O. (1984), *Phonology and Syntax*, MIT Press, Cambridge MA.

Schönle, P. W. (1988), *Electromagnetische Artikulographie, ein neues Verfahren zur klinischen Untersuchung der Sprechmotorik*, Springer, Berlin.

Sonies, B. C., Shawker, T. H., Hall, T. E. Gerber, L. H. and Leighton, S. B. (1981), `Ultrasonic Visualization of Tongue Motion during Speech', *J. Acoust. Soc. Am. 70*, 683-693.

Sonoda, Y. and Kiritani, S. (1976), `Analysis of Tongue Point Movements by a Linear Second-Order System Model', *Ann. Bul. RILP , U. Tokyo 10*, 29-36.

Sproat, R. W. (1985), *On Deriving the Lexicon*, PhD. Diss., MIT, Cambridge, MA.

Sproat, R. W. and Fujimura, O. (1989), `The Articulation of /l/ in English: an Investigation of the Conditioning of Allophony', Tianjin International Conference on Phonetics and Phonology, Tianjin, China.
Sternberg, S., Knoll, R. L., Monsell, S. and Wright, C. E. (1988), `Motor Programs and Hierarchical Organization in the Control of Rapid Speech', *Phonetica 45,* 175-197.
Stevens, K. N. (in press), `On the Quantal Nature of Speech', *J. Phonetics.*
Stone, M. (1981), `Evidence for a Rhythm Pattern in Speech Production: Observations of Jaw Movement', *J. Phonetics 9,* 109-120.

SPEECH MOTOR TIMING

ERIC KELLER
Université du Québec à Montréal
Linguistique, C.P. 8888, Succ. A
MONTRÉAL, QC H3C 3P8, Canada

ABSTRACT. Timing is a fundamental aspect of speech motor control. In a recent, prominent view towards speech timing, temporal organization is seen as arising from the interactions of peripheral articulatory structures and is not considered to be subject to external neurocognitive control. Evidence and arguments countering the main study supporting this position are critically reviewed. New data is presented in favor of neurocognitive control (possibly via the cerebellum) over two specific types of speech timing, rhythm and interarticulatory coordination. It is shown how such a notion of external control can be integrated into a general multifactorial experimental approach towards timing. The multifactorial approach is further illustrated with respect to hypotheses arising from Lindblom's distinction of hyper- and hypospeech.

1.Introduction

In some form or another, most investigations of speech production are concerned with time and timing. At the most elementary semiotic level, speech involves seriation into temporally sequenced elements. At the phonemic level, vowels, fricatives, stops and other segments occupy characteristic periods of time. And at the phonetic output level of analysis, the onsets and offsets of speech gestures are timed according to complex coarticulatory interdependencies--which may change considerably with different repetitions, speakers, speech rates, or dialects. All of speech production appears to be subject to a rather complex and not particularly obvious time scheme.

The objective of the present review is to delineate two contemporary currents of thinking on the motoric aspects of speech timing, and to present a conceptual framework that incorporates and extends the presented concepts. The first current revolves around the notions that timing is inherent in coordinated speech motor action, and that it therefore requires no reference to neural functioning. The second current takes the opposite point of view in proposing that a number of phenomena in speech timing are difficult to explain without reference to absolute time, and that certain neural structures are directly involved in the timing of speech actions. Finally, a proposed conceptual framework circumscribes how complex task objectives, task and system constraints could be seen as they act upon the overall timing portrait of articulated speech.

2. Historical Perspectives: The Quest for Invariance

The traditional approach to the problems of bringing order to the complexity of temporal factors and variability in the observed timing data has been to seek underlying factors of

W. J. Hardcastle and A. Marchal (eds.), Speech Production and Speech Modelling, 343–364.

stability or *invariance*. Inspired by the convenient shorthand of first, graphemic, and later, of phonemic transcriptions, speech research has sought stable physical correlates of transcribed symbols and perceived sound contrasts since the beginnings of phonetic science. In the 1930's, extensive discussions about articulatory invariants accompanied the quest for a satisfactory definition of the phoneme, and the tradition has been pursued down to the most recent day with the search of stable acoustic correlates of features. In the area of speech timing, invariants were first sought in the form of the purported isochrony of speech utterance segments, and later, in the form of stable inter- and intra-articulatory phase relations. This last current shall be addressed below in more detail.

In counterpoint to the search for invariants, a parallel discourse has been carried on in which much importance has been attached to the necessary *variation* associated with motor activity. Nearly all theoretical writings on motor speech systems of the last 20 years have made at least passing mention of the motor equivalence problem. By motor equivalence is meant the motor system's ability to accomplish a given objective by a variety of means. Traditional examples of motor equivalence (see Tolman, 1932; Hull, 1934; Lashley 1942; Hebb, 1949) include the observation that a rat can successfully swim through a maze on the first trial, if it has previously learned to run the maze "in the dry". Although the exact motor actions involved in swimming and running are quite different, the essential task objective of reaching the stimulus in the food box is accomplished in either manner. Motor equivalence is also evident in the organism's behavior within a single task space. A recent prominent example is the frog's wiping reflex (Fukson, Berkinblit & Fel'dman, 1980). Although the wiping limb's initial position may vary quite a bit, the wiping movement is successfully executed. Here successful execution is defined as passing over the spot on the frog's body where tactile stimulation is applied, i.e., where in real life, insects would be landing on the frog's body. Another important example of motor equivalence, this time in speech, stems from the jaw perturbation paradigm (Folkins & Abbs, 1975). In this paradigm, a person is asked to produce a lip closing gesture for [b] or [p], and at certain unpredictable and relatively infrequent moments, the jaw movement is arrested. It is found that the lip closure gesture is satisfactorily executed, even though the system has very little or no time to reorganize itself in these experimental conditions.

Given this type of data to account for, the quest for motor invariants is clearly difficult. It is not evident which aspect of the movement should be considered "stable": so-called target points, certain crucial junctures or intersections of articulator trajectories, underlying temporal or spatial relationships defining such trajectories, or some even more abstract "task space" definition. An interesting proposal for reducing speech timing variation was put forward some time ago by Kelso and colleagues. It is to this proposal that we turn next.

3. Relative Timing within a Dissipative System Approach

In the first half of this decade, Scott Kelso developed a wide-ranging theoretical proposal concerning the motoric aspects of speech. In the given space it is not easy to do full justice to as complex a theoretical approach as this. Nevertheless, the ingeniousness, specificity and testability of the timing-related aspects of the proposal can readily be appreciated. An examination of a much-discussed instantiation of this proposal (Kelso et al., 1986) thus forms an excellent backdrop to the entire rest of the present paper.

The model has two important features with respect to timing. First of all, the timing of coarticulated gestures is seen as being stable over amplitude changes in the movements

(p. 39)[1]. In Kelso et al.'s perspective, timing is not absolute, i.e., it is not counted in milliseconds. In fact, by a simple metric transformation of terms related to the steepness parameter in the non-linear functions defining articulatory gestures, it is possible to model faster and slower movements[2].

The model thus predicts stable phase relationships between events in various task-related articulators. No matter whether speech is fast or slow, the onset of one articulator movement with reference to another will occur at the same point in the first movement's phase. While that point in the phase is stable, its absolute point in time is variable. This rejection of absolute time is entirely intended. Kelso et al. use an interesting metaphor to bring the point across. They say that a candle burns in absolute time, but that its rate of burning, or that the end of its burning has much less to do with absolute time than with the candle's thickness, the ambient temperature, the material the candle is made of, etc. Time and timing is therefore contained in the rate of energy dissipation, not in specific absolute time periods.

It is essentially for the same reason that Kelso et al. also reject the notion that speech timing is driven by, or is produced with reference to, neuronal functioning (e.g., central pattern generators). Since time is given by the properties of the system (is *intrinsic*), time "falls out of the equation", i.e., it does not depend on external neuronal input.

As it turns out, my own reflections, research and reference materials lead me to take a position that with respect to timing is the very opposite of the positions just described. Specifically, evidence contradicts the following two aspects of the hypothesis:

1. In many important details, the relative time hypothesis makes wrong predictions for speech timing.
2. Both normal and pathological speech do show evidence of being regulated to some extent with reference to absolute time.

Furthermore, their proposal is deficient and insufficient in a number of other respects that are crucial to our understanding of speech timing. The following limitations of the hypothesis are particularly bothersome:

1. The task descriptions seem inadequate to the complexities of actual human speech in that humans probably pursue multiple simultaneous goals in task space instead of single sequential goals.
2. The task descriptions do not make any provisions for non-motoric co-determinants of speech timing, such as syntactic pauses, emphatic lengthening, etc.
3. The task descriptions do not incorporate constraints inherent in task execution and system operation.

4. Empirical Examinations of Kelso's Predictions

In support of their theoretical proposal, Kelso et al. provided some data and statistical evaluations that have since been subjected to uncommonly critical scrutiny. Space permits only a sketchy description of the study and its criticisms, and the interested reader is referred to the cited publications for details.

---

[1] Although Kelso's proposal incorporates concepts from Elliot Saltzman's Task Dynamics approach, it is important to note that in his privately authored papers, Eliott Saltzman makes no particular claim concerning relative timing (e.g., Saltzman, 1986). Also, his recent models apparently foresee specific manipulations of timing (K. Munhall, personal communication, 1989).

[2] Kelso et al. furthermore postulate limits on metric tranformations, so that movement transitions are prohibited which would take excessive energy to perform (see also Nelson, 1983, and Nelson et al., 1984, on this issue). This aspect Kelso's proposal has no immediate bearing on the present discussion and will thus be set aside.

The data was first reported in Tuller, Kelso & Harris (1982) and, after reanalysis, in Tuller & Kelso (1984). The present review is based on the second report. Four speakers produced the embedded nonsense utterance /bapab/ at fast and slow speech rates. Half the utterances had the main stress on the first syllable and the other half, on the second syllable. Measurement was for two durations and involved two different movement records. First, the interval between successive onsets for the downward movement associated with the two vowels was measured on the jaw movement record (V1 to V2, let's call this "jaw cycle duration"). Second, an interarticulatory delay was measured between the onset of the jaw cycle (V1) and the onset of the downward movement on the upper lip movement record for the second vowel (V1 to UL, let's call this "interarticulatory latency"). UL occurs roughly in the middle of the jaw cycle.

The interest of choosing these two time measures is that the roughly cyclic upper lip movement is time-offset with respect to the roughly cyclic jaw movement. The jaw being the heavier articulator, it should hold true, according to Kelso et al.'s hypothesis, that the time offset between the lip and jaw cycles work out to a stable proportion of the jaw cycle, no matter the stress pattern nor the speech rate. If true, that would support a hypothesis of a relative timing relationship between the two articulators.

Indeed, the authors reported relatively high Pearson correlation coefficients ($r = .84$ - $.97$) between jaw cycle duration and interarticulatory latency and reported support for the hypothesis. However, several authors (Barry, 1983; Benoit, 1986; Munhall, 1985) have indicated that a certain degree of the apparent proportionality of the two durations is likely due to a statistical artifact linked to the part-whole correlations of the calculation of relative time. In calculating correlations, one must insure that the two sets of samples are based on independent measures, a condition that is violated in this case because the jaw cycle duration and the interarticulatory latency share one measurement point (the "left edge", the point of issue). However, this artifact did not explain the entire correlation. When Munhall (1985) calculated a compensatory factor and measured similar relative timing measures on reiterant [ka] obtained with ultrasound, he still obtained significant correlations.

Munhall indicates that the high correlations may in fact be reflective of the major durational variations that *all* simultaneous time measures of related musculature undergo as syllables expand and contract with different speech rates. A further study performed with 12 French-speaking subjects in our lab (Keller, 1987c) fully confirmed these suspicions. The average *difference* in duration between fast and conversational productions of [ka] comes out to about half of the entire syllable duration for fast [ka] and to about one third of the duration of conversational [ka]. When correlations extend over both fast and conversational measures, nearly all common variance is related to this massive difference in durations. In other words, one could take any measurable time segment of the syllable, and because the syllable undergoes massive contraction from the fast to the conversational speech rate, a substantial correlation between movement records of mechanically linked articulatory musculature would be guaranteed. Such measures are thus no valid test of the relative timing hypothesis. A more potent test would have to examine the prediction of the hypothesis for data inside each durational cell, i.e., separately for fast and slow speech conditions. And when data of this sort are examined, the proposed phase relationship falls apart (Keller, 1987c).

There are a few more wrinkles. Apparently, the study is difficult to replicate (Lubker, 1986, Nittrauer et al., 1988). Also, in a comparison of the variation of absolute durational measures and that of comparable relative measures, relative measures were not more stable (i.e., less variable) than absolute measures (Keller, 1987c). Further, a massive reappraisal of reports in both speech and general motor control has indicated that motor activity performed at different movement rates is in fact far more variable than had at first appeared; stricter statistical measures have turned up non-linear transformations between fast and slow movement rates in many studies that had previously reported linear scaling (Gentner, 1987). Altogether, one may say that over the long run, empirical verification of the relative timing proposal has not fared well, neither in speech nor in general motor control. More sophisticated analyses of speech timing are clearly called for.

5. Relative Time vs. Absolute Time and Central Nervous System Activity in Speech Timing

5.1 SOME EVIDENCE IN FAVOR OF ABSOLUTE SPEECH TIMING

Kelso et al. have argued that speech timing can be seen as independent of either higher cognitive or central nervous system (CNS) activity, if speech timing arises out of movement phase portraits of the participating articulatory gestures. If however speech timing events are related to absolute time, discussion has to expand into the theoretically more complex domains of higher cognitive or CNS involvement[3]. In that case, timing is not a property that arises from, or is inherent in, the activity of the participating musculature, but is conceived to be an activity which at in least part is modulated by a higher cognitive or CNS structure. Because of this, it is important to offer some further motivation for this excursion into non-motor processing.

In one of my studies based on ultrasound measures of tongue dorsum displacement in [ka] I reported that within a single task condition (a single speech rate either in reiterant or in normal speech), out of eight subsyllabic duration measures, it was the consonant cycle duration for the syllable [ka] that showed the least degree of variation (Figure 3A, p. 344, Keller, 1987c). The eight measures also included durations of descending and ascending tongue dorsum movement, time to and from peak velocity in both types of movement, and inter-articulatory delay. Measures for the boundaries of the consonant cycle were taken on the zero-velocity points of the tongue dorsum movement close to the palate, just prior to the descending movement for [ka].

This means that speakers appeared to control their tongue movements in such a way that the syllable onsets occurred at relatively regular time intervals, no matter the variability associated with various subcomponents of the syllable. It is thus possible that speakers made reference to absolute time in some form or another to regulate rhythmic speech output. We have recently reexamined this question with an acoustic analysis of reiterant [CV] syllables, performed to date with two normal subjects as well as with three patients with spino-cerebellar neural disorder (Friedreich's ataxia). Subjects were asked to produce the syllables [pa], [ta], [ka] and [ʔa] repetitively for five seconds or more, and that at two speech rates, a fast speech rate where Ss were asked to produce the series as fast as possible while preserving the consonant-vowel alternation, and at a self-paced "metronome-like" condition at a normal speech rate. Typically there were more than 20 tokens for each syllable and each speaker. High quality acoustic recordings were digitized at 10 kHz and analyzed for evidence of opening and closing gestures, as well as voicing on- and offsets (Figure 1). Averages and standard deviations were computed separately for each speaker for the following periods: opening and closing period (the period of time that the superior vocal tract is open and closed, respectively), voicing and devoicing period (the period of time that voicing and devoicing are in effect, respectively), opening interval (the time between successive onsets of the opening period), voicing interval (the

[3] Strictly speaking, there is no need to refer to neurophysiological (CNS) concepts at all, and all discussion could be handled merely with reference to higher-level cognitive processes. However, both the historical antecedents of the speech timing literature as well as the data introduced below argue powerfully in favor of including reference to neurophysiological and neuroanatomical concepts in the present context.

time between successive onsets of the voicing period), as well as VOT. Data-related estimators of durational variation (log[s.d.]/log[mean]) were calculated for each speaker and each type of syllable, and average estimators were formed for normal and brain-disordered groups[4]. The grouped results reflect the performance of individual subjects in individual conditions. Results are reproduced in Figure 2.

The most interesting aspect is that the opening and voicing intervals, as well as VOT, are the least variable of all of our time measures. This is only true for normal subjects and not for the patients with cerebro-spinal lesion involvement (more on that later). Opening and voicing periods (as well as their counterparts, closing and devoicing periods), by contrast, are far more variable. If this can be taken as a reflection of degree of speech motor control exerted over various speech-timed events, and not some sort of statistical or measurement artifact[5], it means that a greater degree of timing control is exerted over intervals and VOTs than over other aspects of speech segmentation. The low variability of onset-to-onset intervals can be seen as some degree of isochrony, or rhythmicity over successive syllables. Also, the duration of VOT, a well-known measure of co-articulatory timing, appears to be controlled rather closely. This arguments against a speech timing model based on the notion of time arising out of the peripheral interplay of articulators, since it renders plausible a notion of control exerted by the speaker upon specific aspects of speech timing.

Since our estimator is a fairly complex ratio between logged standard deviations and means, it is tempting to think that this result is artifactual. To check against that, means and standard deviations were recomputed on a set of random speaker-speech condition cells (Figure 3). As it turns out, the standard deviations for the two intervals were *smaller* than those for opening, closing, voicing or devoicing periods, even though the two intervals were the longest of any of these measures. In the absence of statistical tests, this appears to be a strong argument in favor of subjets' ability to impose an absolute timing rhythm on repeated syllables.

---

[4] The exact definition of a valid estimator for durational variation is tricky. Simple standard deviations will not do, since longer durations tend to have larger standard deviations than shorter durations, and simple coefficients of variation (s.d./mean) are suspect also, since it is not clear that s.d.s and means are linearly related. Exploratory statistical analyses on the present data base have shown a relatively high linear correlation between log(s.d.) and log(mean) ($r = .577$), and the estimator log(s.d.)/log(mean) was thus chosen as the estimator of durational variation for the present data set. See also Munhall (1989) on this issue.

[5] Significance tests have not yet been performed on these very recent data. A full report with reference to a larger subject group and involving a complete set of statistical analyses is planned for 1990.

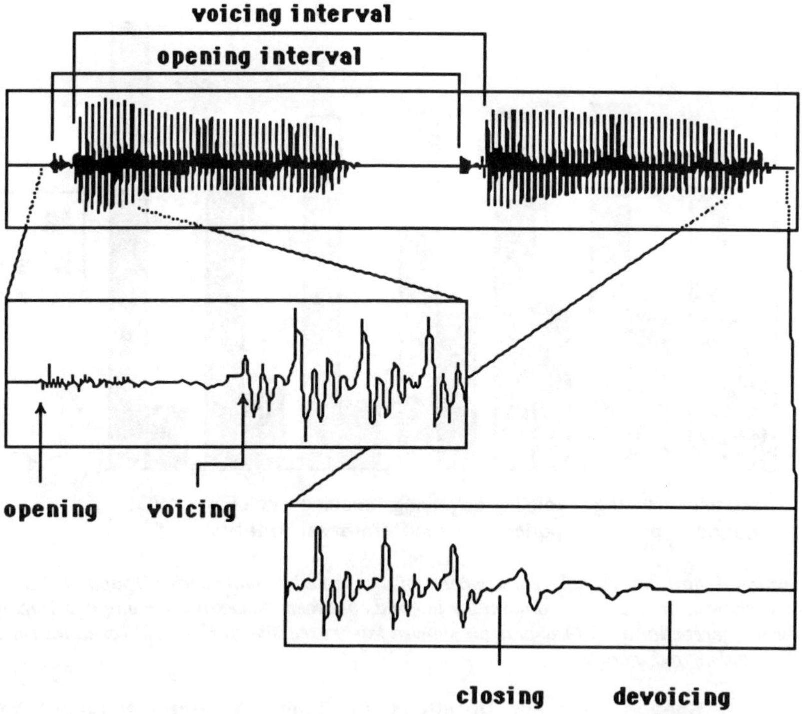

Figure 1: In these CV syllables, opening was considered to occur at the VOT burst onset, closing at the point in the terminating slow oscillating waveform where the disappearance of high-frequency oscillatory components co-occurred with a marked reduction of the amplitude envelope, and voicing on- and offsets were judged with respect to the onset and offset of regular F_0 oscillation. In high-quality acoustic recordings, these criteria were applicable with hardly any exception. Typically, French-speaking subjects' vocal tract opening intervals tend to precede their voicing intervals in both on- and offset of the syllable.

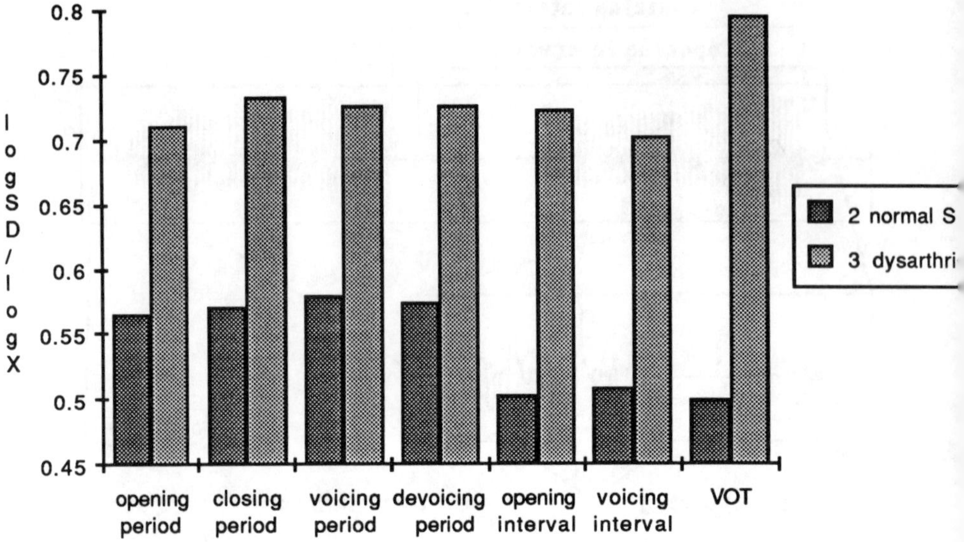

Figure 2. Durational variation on different segments of a repeated syllable series ([papa...], [tata...] etc.). Normal subjects showed the least variation on the intervals between successive opening and voicing onsets and on VOT, while cerebellar dysarthric patients showed fairly excessive variation in on all measures. Data from an ongoing study in our laboratory.

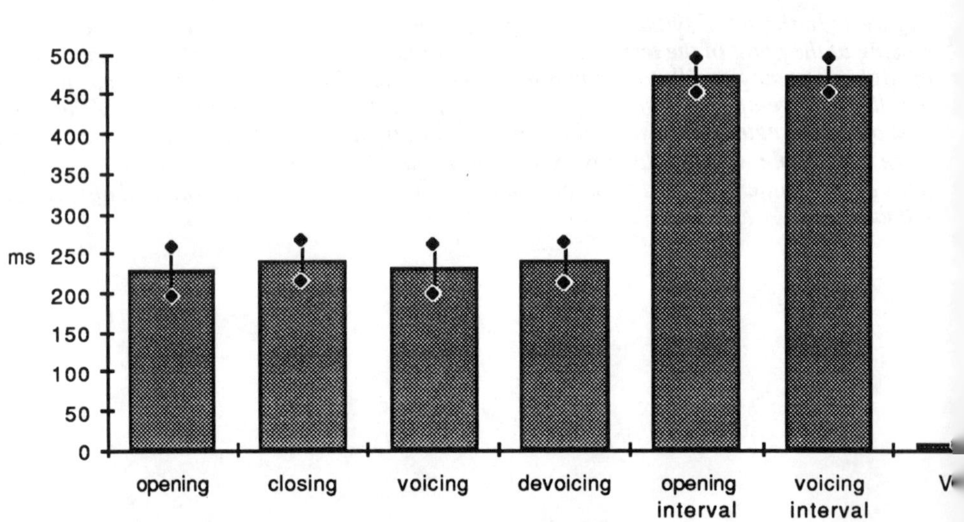

Figure 3. Means and standard deviations of various speech segment durations, taken from a random speaker-s condition. The standard deviations of the opening and voicing intervals are smaller than those for opening, cl voicing or devoicing periods.

Data such as these suggest that speakers can constrain speech utterance segments to occur within specific durations, as measured in absolute time. It could be argued that these results are limited to reiterant speech provoking a frequent counter-argument that reiterant speech is unlike real speech, and thus say nothing about timing in normal speech conditions. Especially when subjects are asked to speak in "metronome fashion", the speech condition seems precariously dissimilar from normal speech. It also is typically argued that past attempts to show isochronicity in connected speech have failed (for a review, see Lehiste, 1977). Even such a tightly controlled type of speech as the production of limericks shows stress group durations varying from 488 to 566 ms (O'Connor, 1965, cited by Lehiste, 1977), and other studies attempting to measure supposedly similar durations in rhythmically produced speech have all failed to provide exact durational evidence in favor of "isochrony"[6] (Shen & Peterson, 1962; O'Connor, 1965, 1968; Bolinger, 1965; Uldall, 1972; Lea, 1974, all cited in Lehiste, 1977).

However, there's another interpretation of isochrony. Allen (1972, 1973, 1975, cited in Lehiste) reports with reference to a considerable review of the literature that the variability on specific speech segments is in fact quite comparable to the diadochokinetic task of finger tapping. Short speech segments are found to have variabilities of about 0.1 CofV, while longer stretches of speech vary with a CofV of about 0.04. These values fall into the same range of values reported above for opening and voicing intervals in normal subjects (calculations not shown here), as well as the variability for a manual tapping task reported by Ivry & Keele, 1989 (see below).

One may conclude that speakers probably use absolute time in the process of acquiring reliable time control over segment durations, and when asked to do so, can use reference to absolute time to impose a metronome-like rhythm on successive syllables in reiterant speech. Further, the degree of variability seen in repeated syllables closely parallels that found in manual tapping and in segment durations in normal speech. It is therefore possible that normal speech processes have access to, and use some judgement of absolute time.

5.2 NEUROGENIC PATHOLOGIES OF SPEECH TIMING

Much recent work on neurogenic disorders of the timing of speech segments provides further support for the notion of CNS involvement in speech timing. Several groups of patients (Broca's aphasics, patients with Parkinsonism and those with cerebellar disorders) suffer from distinct types of timing disorders. However, since the various disorders appear to affect different aspects of motor programming, it is important to distinguish them clearly.

5.2.1. Broca's Aphasia. Motor speech disorders due to lesions at the prerolandic cortical or infracortical level (Broca's aphasia) are typically characterized by excessive slowness (what appears to the ear as slow, labored speech) (see e.g., Goodglass & Kaplan, 1972; Kent & McNeil, 1987; Keller, 1987b). Also, some such patients show excessive slowness in consonant-consonant transitions (Luria, 1966/1962), and have in recent years been shown to suffer from a distinctive impairment in VOTs (Blumstein & et al., 1980; Itoh & et al., 1982; Shewan, Leeper & Booth, 1984 [7]; Ziegler, 1987). In what

---

[6] According to Lehiste (1977) and following Pike (1945), "'isochrony' refers to the phenomenon that in a stress-timed language, such as English, stressed syllables follow each other at approximately equal time intervals".

[7] Shewan et al.'s report is actually somewhat less supportive of a specific VOT impairment than the other reports, since only the linguo-palatal stops were clearly affected in their study.

initially might seem like a paradoxical result, such patients have also been reported to suffer from abnormally short and highly variable vowel durations within single-syllable utterances (Duffy & Gawle, 1984).

When speech processing context, neurophysiological bases, as well as lesional anatomical information is taken into account, it can be hypothesized that such patients suffer from an impairment of the learned coarticulatory basis of speech (Keller, 1987b). Such patients can be seen as being impaired in their ability to program the temporal (phase) and spatial relations that constitute the basis of acquired coarticulatory patterns of speech. This would seem to lead to a speech pattern in which VOTs of voiced and unvoiced plosive consonants are abnormally variable, drawn-out and indistinct, as patients struggle to impose some degree of muscular coordination on the 70-plus muscle groups involved in speech. Furthermore, such a hypothesis is not at odds with Duffy & Gawle's findings. Vocalic segments of speech are the least demanding speech segments when it comes to articulatory coordination, and it may well be that patients try to "gain some time" by shortening vowel durations.

Notice that these disorders are not theoretically informative with respect to the absolute vs. relative time issue, since it is possible to account for the disorder both in terms of a phase portrait and in terms of an absolute timing hypothesis (in Keller, 1987b, I argue in favor of a relative timing conception for this type of cortical processing). However, it does address the issue of central vs. peripheral origin of timing. If these patients do suffer from a syndrome-specific loss of acquired speech timing, it means that timing has to be learned, and that it is learned by cortical structures. Under such a hypothesis, timing is not a necessary consequence of peripheral coarticulatory coordination and does not automatically "fall out of the equation". Coarticulatory coordination has to be acquired by specific CNS structures.

5.2.2. Cerebellar Disorders of Speech. Another group of patients whose deficits are typically associated with timing problems are those with cerebellar lesions. It may well be that their deficit provides a particularly potent argument in favor of CNS regulation of absolute timing. Such patients show extensive timing impairment in so-called diadochokinetic tasks (e.g., finger tapping, reiterant speech); in an ongoing study, we are finding that repeated plosive-vowel syllables are not seriated at the same clockbeat rhythm as found in regular subjects, but are sometimes lengthened and sometimes shortened, leading to between 2.5 and 4 times more variability than that found for similar-aged, normal subjects (Figure 2). These results are in line with earlier work on similar patients (Grémy, Chevrie-Muller & Garde, 1967). A comparison with results obtained on tasks of general motor control shows that these problems generalize to limb motor control. Ivry & Keele (1989) found that cortically lesioned and cerebellar patients were the slowest tappers (together with normal elderly patients) and that cerebellar patients show the greatest variability (Figures 4 and 5).

While this and similar data provide support for the notion of CNS involvement in the control of speech timing, a prominent counter argument is that it is notoriously difficult to disengage motoric from cognitive deficits in motor speech disorders. The noted pathological slowing of speech may be due to the secondary effects of a primary motoric disorganization. However, recent experimental evidence of lesion-induced *perceptual* timing deficits independent of motor impairment has put a kink into this argument. Ivry & Keele (1989) measured the ability to judge whether the second of two pairs of 1-kHz beeps was more or less closely spaced than a first set of beeps. The onsets of the reference pair were always spaced 400 ms apart, while those of the second set were spaced anywhere from 160 to 640 ms apart. They also measured perceptual performance for different degress of loudness varying from 64.9 to 81.1 dB.

Their results are graphed in Figure 6. It is evident that the results on the duration perception task closely parallel those on the CofV of the tapping task in that there is nearly the same order in the average impairment. In order from worst to best performance, the groups are: Cerebellars, corticals and parkinsons, elderly, and young normals. At the same time, impairment on the performance of the *loudness* perception task is quite different, with cortical patients being the most strongly impaired. The strikingly inferior performance of the cerebellar patients for the duration perception task is quite surprising, in that there had been no previous indication in the literature that the cerebellum is involved in duration judgement independent from motor performance[8].

This demonstration of a cognitive impairment of timing judgements suggests an interpretation of cerebellar speech timing disorders independent of motoric lesion effects. It is possible that the slowing be related to a disorder of absolute time perception, rather than to the secondary effects of motoric impairment. If these results can be replicated, they provide an interesting indication that the speech motor control system operates at least in part with reference to absolute time.

[8] The involvement of the cerebellum in the timing of movement and in movement disorders is well-known, as are attempts to account for these effects in terms of neurophysiological models (e.g., Braitenberg, 1967; Kornhuber, 1974).

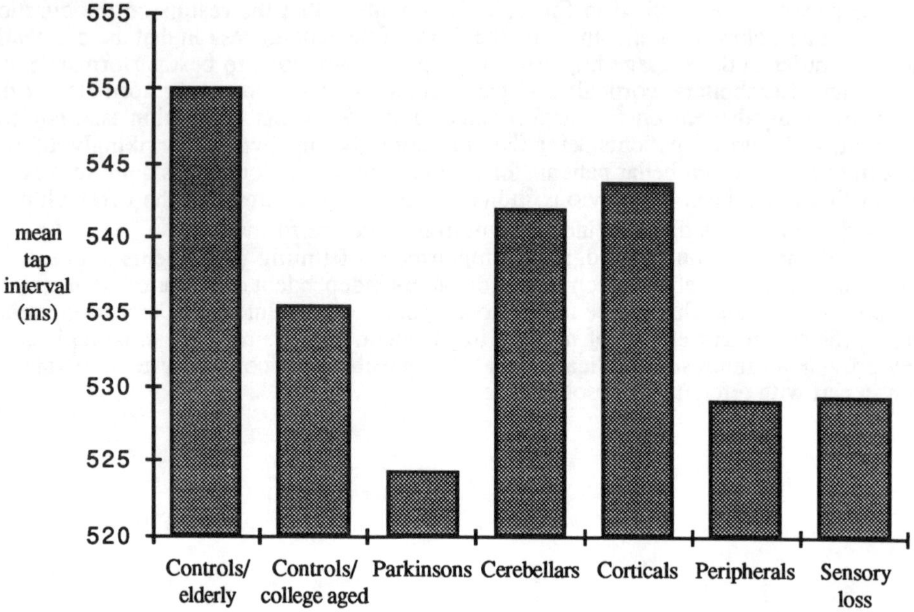

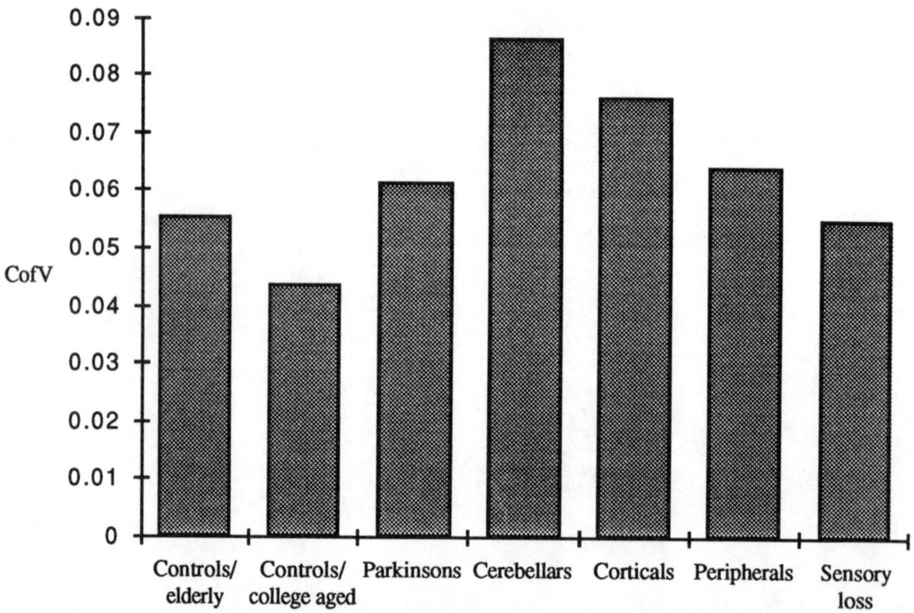

Figures 4 and 5. Mean tap interval and coefficient of variation (s.d./mean) on mean tap interval for simple manual tapping task performed by various subject groups. Data adapted from Ivry & Keele, 1989.

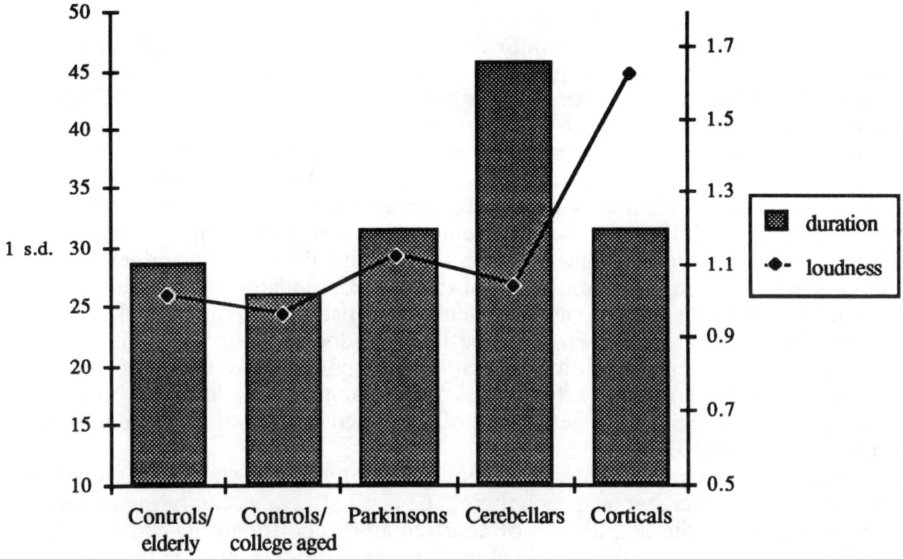

Figure 6. Performance on two perceptual tasks (duration and loudness) in normal subjects and neurologically impaired patients. (Data adapted from Ivry & Keele, 1989).

5.3. CENTRAL PATTERN GENERATORS?

Kelso et al.'s 1986 paper specifically rejected Central Pattern Generators (CPGs) as a source of speech timing. It is thus important to touch briefly upon the present status of CPGs as a hypothetical source of speech timing. CPGs are networks of nerve cells that activate motoneurons serially and rhythmically, and are found in a large number of neural systems; neurons within such networks that fire at specific rates are known as "pacemaker cells". For instance, the spinal cord of decerebrate cats contains a network sophisticated enough to initiate and seriate the actions involved in walking (Shik et al., 1966). These networks are not simple pulse generators, as they are often popularly conceived. Grillner (1986) summarizes evidence that CPG networks modify peripheral motor performance depending on type of action required (e.g., different types of scratch reflexes or locomotor movements). Evidence from locusts indicates that sensory input from the periphery modulates the CPGs rhythmic output pulsations.

In speech, the notion of CPG-induced speech timing attained some popularity at the end of the 1960s, when CPG activity in the 4-5 Hz range in the ventro-lateral nucleus of the thalamus was experimentally linked with parkinsonian tremor (e.g., Hassler et al., 1970). Since the thalamus is part of the cortico-basal-thalamico-cortical loop thought to be crucially involved in the programming of motor action and since the speech of parkinsonian patients is also affected by the tremor, and is furthermore marked by a curious difficulty with slow motor execution of speech (Darley, Aronson & Brown,

1975, p. 190), it was postulated that the rhythmic aspects of normal speech production are related to this central neural pulsation.

However, more recent research into the origins of both normal and pathological origins of tremor throw doubt upon the wisdom of extending to normal speech explanations for Parkinsonian motor abnormality. While it may indeed be that abnormally strong or insufficiently inhibited rhythmic activity in the thalamus contributes to Parkinsonian speech timing abnormalities, it is not likely the case that such rhythmic activity be a direct basis of normal speech timing. The rhythms in question (showing typically a fundamental frequency of 4-5 Hz, but also sometimes prominent frequency components up to 12 Hz) are only with difficulty related to speech timing events. As was seen above, normal speech would seem to require absolute timing judgements over periods that are not necessarily multiples of the 4-5 Hz fundamental frequency period, and both normal subjects and patients with neurogenic pathology (Braun et al., 1985) can be observed to transcend the 4-5 Hz range easily in diadochokinetic tasks. Further, some authors argue that the neurophysiological origin of tremor is unlike CPG activity. It may be related to centrally amplified effects of abnormal muscle spindle firing, and it is not necessarily related to the central generation of a seriated and coordinated neural control event (Freund & Dietz, 1978).

In summary, available evidence from neuropathology of speech supports a notion of differential CNS involvement in speech timing. Cortical structures can be conceived of as being responsible for the acquisition of coarticulatory coordination, while cerebellar structures may provide the basis for absolute durational judgements. On the other hand, while no one questions the existence of CPGs in structures of higher neural organization, or the existence of pacemaker cells in humans (viz., pacemaker cells for cardiac and respiratory rhythms), there is at the present no established link between putative CPGs and speech timing, and there is some functional basis for doubting that CPGs could provide the type of timing information that is needed for speech motor programming tasks.

6. An Alternative Approach: Explaining Variance for Multiple Objectives

At the outset of this article, an approach to timing based on the quest of invariants was reviewed. In view of the considerable difficulty of identifying satisfactory candidates for timing invariants, it is legitimate to ask if the question has been posed the wrong way. Perhaps the speech production system does not attempt to satisfy a simple set of unique articulatory target objectives (identifiable at the surface as a measurable "invariants"), but attempts to satisfy a number of competing objectives, some of which may be of articulatory nature, while others may be of non-motoric origin. The remainder of the paper outlines how research on timing could be undertaken within a multiple objectives approach, and provides an illustration of a multiple simultaneous objective.

The first step is the identification of the success criteria defining a speech production task, as well as of the determinants of variation in the attainment of these objectives. Take a typical communication between a control tower and a pilot. The overarching success criterion of tower-pilot communication is, of course, safe traffic separation. But subsidiary success criteria are: a right balance between rate of speech and clarity of communication, as well as errorfree seriation of standard phrases, correct turn taking, and adequate comprehension of transmitted messages. Each of the subsidiary success criteria may impose its own timing requirements, and some of these requirements could be in mutual competition. Also, there are reasons to assume that speakers often structure the timing patterns of their speech in terms of anticipated perceptual effects on the part of the listener, such as when speaking in noisy conditions. A systematic approach to the examination of speech timing would have to begin by structuring the various success criteria in a systematic and comprehensive manner.

Moreover, it does not appear to be enough to identify *success* criteria. Task and system *constraints* are also likely to have an effect on speech timing. In the case of pilot-tower communication, *task constraints* include: transmission channel limitations, time limitations, and competing activity in the cockpit and the tower. A special type of task constraint may be a set of acquired and language/dialect-specific speech segment[9] durations and segment intervals to which the speaker tends to conform. *System constraints* include limits of speed of cognitive and linguistic organization, level of familiarity with airspace communication procedures, linguistic proficiency, etc.

Depending on the status of any of these success and constraint variables, speech motor execution at either end can range from leisurely and wordy, to very rapid and brief. If the three types of variables surrounding the generation of speech (success criteria, task and system constraints) are lawfully related to speech timing, the duration of various segments of speech should be predictable from the dynamic interaction of goal, task and system factors. In this view, surface variability thus does not "hide" some ill-understood invariants under measurement error and articulator imprecision. On the contrary, surface variability is considered to be the inevitable and theoretically predicted concomitant of a communicative behavior capable of achieving its overall objectives by satisfying a large variety of competing demands and constraints.

6.1. AN EXAMPLE: THE PERCEPTUAL DISTINCTIVENESS/SPEED OF TRANSMISSION TRADEOFF

Let's assume that the timing scheme imposed on a single utterance can at least encapsule the following co-determiners of speech timing:

1. a semantic emphasis encoded through the *lengthening* of a given noun,
2. a syntactic clause marked off by a short *pause* in mid sentence,
3. a rising interrogative intonation leading to a *lengthened* final verb,
4. the language's phonological stress rules reflected by the *lengthening* of specific syllables,
5. the increasing of perceptual distinctiveness of a novel term by the *lengthening* of all syllables,
6. a rapid speech rate reflected in an overall *shortening* of syllables.

Some of these effects are illustrated in the sentence "It's a modem, is that what you mean?", pronounced (1) in normal fashion, (2) with stress on "that", and (3) with a syntactic pause in the middle (Figure 7). It can be seen that an emphatic "that" or the presence of a syntactic pause has the expected timing effects at predictable places in the utterance (Figure 8).

[9] "Speech segment" in the general sense of the word, not in the sense of "phoneme".

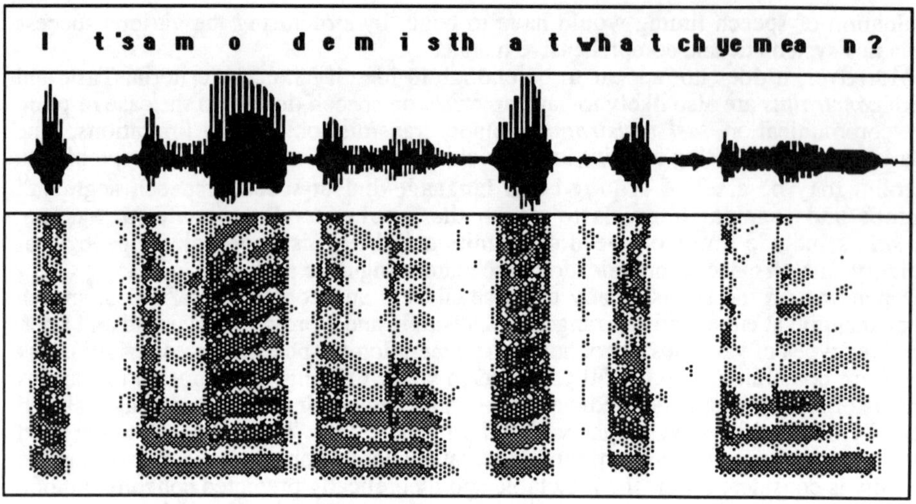

Figure 7. The utterance "it's a modem, is that what you mean" pronounced in normal fashion.

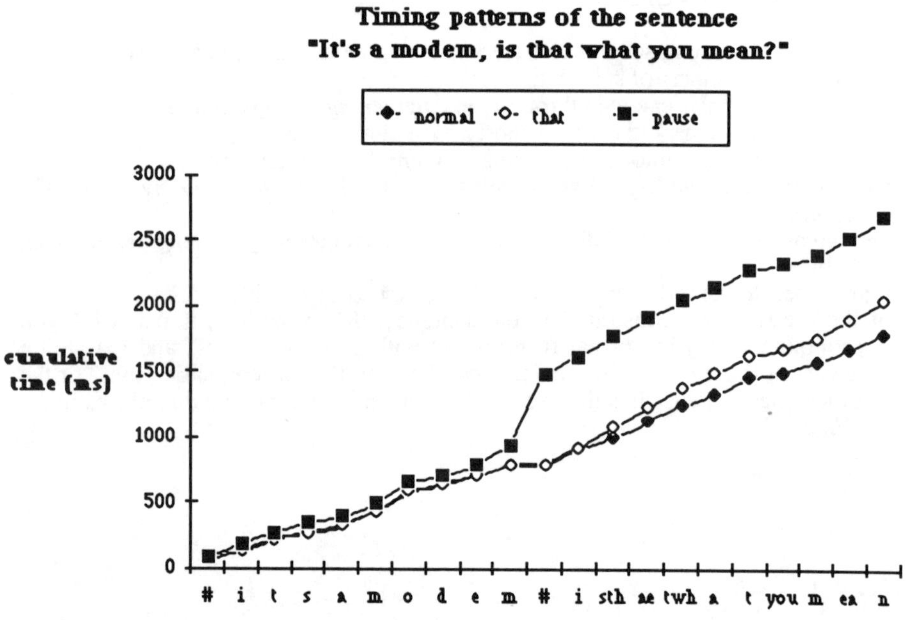

Figure 8. Cumulative time for the utterance "it's a modem, is that what you mean", pronounced (1) in normal fashion, (2) with emphasis on "that", and (3) with a syntactic pause in the middle.

The various postulated timing determiners (independent of task and system constraints) are also sketched in Table I. Notice that the motor and perceptual determiners are portrayed as being potentially in mutual competition. Since it is possible that the same word is subject to competitive demands from lengthening in order to favor perceptual distinctiveness, and from shortening due to a fast speech rate, the two factors are seen as forming a tradeoff relationship.

TABLE I

Hypothetical Communicative, Linguistic and Motor Timing Effects in Speech Production (independent of task and system constraints)

COMMUNICATIVE & LINGUISTIC EFFECTS	MOTOR & PERCEPTUAL EFFECTS
turn taking semantic emphasis prosodic effects syntactic boundaries phonological stress rules	Tradeoff between perceptual distinctiveness and speed of transmission

When the two objectives are in competition for the production of vowels, the classic outcome is "vowel reduction" and "undershoot" (Lindblom, 1963) and can be considered illustrative of "hypospeech" (see Lindblom, this volume). In this speech mode, vowels are produced with what appears to be less expansive tongue and lip gestures when the time available for their execution is reduced. It is of interest to consider some additional data to examine this hypothesis, especially since a recent study has raised some questions concerning the validity of the vowel reduction data.

Gopal and Syrdal (1988), asked five male and five female speakers to produce the well-known series of stimuli containing all vowels of American English "heed", "hid", "had", "hod", "howed", "hood" and "hued" in carrier phrases that were produced at slow, conversational and fast speech rates. Their F1/F2 results are graphed in Figure 9 (*n.b.*, Gopal and Syrdal used bark ratios instead). According to the hypothesis, greater speed of transmission should lead to less extensive tongue movements, that in turn should translate into more central F1/F2 values. The actual results show that nearly all variation occurs along the F2 axis (however pilot data for two French speakers analyzed in our lab showed variation along both F1 and F2 axes). Gopal & Syrdal reject the undershoot hypothesis on the grounds that few of the measured formant differences between fast, conversational and slow articulation reach statistical significance, while those that did could be related to coarticulatory effects. However, the examination of formant differences on specific vowels does not seem like the only satisfactory test of the hypothesis, since both male and female speakers did show consistently more central F1/F2 intersects in the faster speech rate conditions. If a non-parametric chi-square test were used to compare results congruent with the reduction hypothesis with results that are not congruent, their results would clearly support the hypothesis (as do our own data).

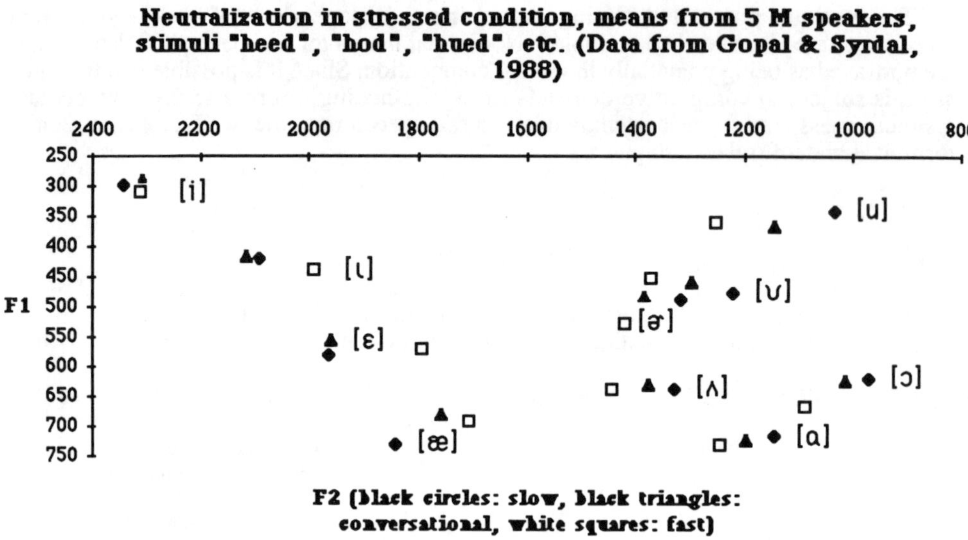

Figure 9: F1/F2 intersects from Gopal & Syrdal's (1988) data.

Nevertheless, more direct tests of the hypothesis would be desirable. The first study relevant in this context is Keller (1989). Here two temporal variables were examined for their ability to predict various segments of the syllable. The first variable was an estimator of speech rate (one estimator for each of 12 subjects and 4 conditions, where conditions were fast and conversational speech rate in reiterant and sentence-embedded [ka]), and the other estimator was an interarticulatory delay parameter (again one estimator per subject and condition). The interarticulatory delay parameter can be postulated to incorporate perceptual distinctiveness information, since it is strongly related to VOT. The study showed that syllable onset duration was reliably predicted by a *combination* of the two independent variables, while the mid and final section of the syllable was primarily predicted by the rate parameter. This appears therefore to be an excellent example of the competitive nature of the two task objectives of speech rate and perceptual distinctiveness.

Using the same database, another test of the tradeoff hypothesis was possible. If there is competition within the system between the demands for speed of production and clarity of articulation, it can be postulated that under conditions of conversational speech, both demands can be accommodated, while under conditions of fast speech, the two demands enter into competition. In terms of displacement and time in a [ka] transition, only minor correlations between time and displacement would be expected without competition, while with competition, one constraint would be expected to clearly dominate over the other, i.e., time would be expected to become a strong predictor of displacement. We therefore expect less correlation between displacement and duration in the conversational condition but a higher correlation in the fast speech rate condition. The results[10] (Figures 10 and 11) fully bear out the prediction with an r of .306 for conversational [ka] (df. 252, $p<.0001$) and a more than twice as large r of .761 for fast [ka] (df. 593, $p<.0001$)[11]. In

[10] Log time was used to permit working with linear correlations.

[11] Alternatively, it could be argued from this correlation that the same underlying variable controls both displacement and duration. This alternative interpretation would support notion that in cases of competing objectives, spatial and temporal variables are controlled by the same underlying variable.

the main, the tradeoff hypothesis thus receives good support from the examined data. The overall schema of timing determiners in Table I is thus supported, as is the notion of the pursuit of multiple simultaneous objectives pursued in task space.

The notion of competitive, simultaneous timing objectives is likely to generalize to most other timing-influencing parameters mentioned above. The multivariate examination of the relative explanatory power of various competitive predictors can therefore be expected to provide more satisfactory solution to the old problem of understanding speech timing than the search for some mythical invariants.

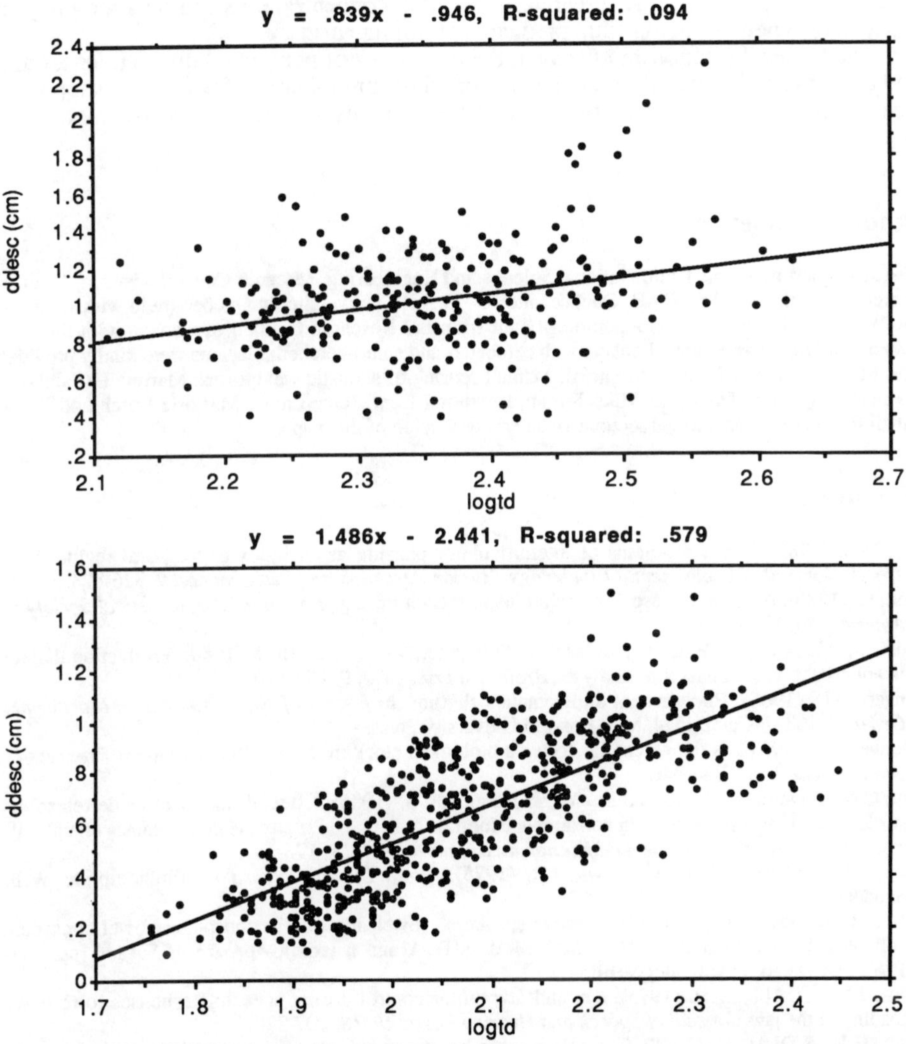

Figures 10 and 11: The relationship between log10 ms time of the descending movement for reiterant [ka] (logtd) and displacement for the same movement (dddesc). Top: Conversational speech rate, bottom: fast speech rate. The linear correlation between the two variables is more than twice as strong at the fast speech rate.

7. Conclusion

The present review of some main issues in speech timing has led to a number of conclusions and proposals that represent departures from accepted phonetic knowledge and analysis practice:

1. Contrary to proposals formulated by Kelso and colleagues some years ago, there is evidence that speech timing uses information of absolute time. Some evidence was presented that this information may be modulated by the cerebellum.

2. Given the theoretical and demonstrated empirical difficulty of identifying timing *invariants*, it is proposed that research in this field concentrate instead on attempting to identify the (sometimes competing) predictors of timing *variance*.

3. The tradeoff relationship between the objectives of perceptual distinctiveness and speed of transmission was found to be supported by a number of of tests. This supports a conceptualization of the pursuit of multiple simultaneous objectives in speech motor task space.

Acknowledgements:

In part supported by grants from: National Science and Engineering Research Council, Canada, Medical Research Council, Canada, FCAR, Quebec, and F.O.D.A.R., Université du Québec (held with J. Rouat, UQAC), and an International Cooperation Grant from the Research Council of Canada with the ICP, University of Grenoble, France. Patients with cerebellar and spino-cerebellar lesions were kindly provided by Dr. M. Botez, Hôtel-Dieu de Montréal. Manual scoring of acoustic waveforms: Martine Laframboise and Patrick Vigneux. Thanks to Mike Jordan, Raymond Kent, Jerry Lame, Marjorie Lorch and Kevin Munhall for feedback on various sections of an earlier version of this paper.

References:

Barry, W.J. (1983). Some problems of interarticulator phasing as an index of temporal regularity in speech. *Journal of Experimental Psychology: Human Percption and Performance, 9*, 826-828.

Benoit, C. (1986). Note on the use of correlations in speech timing. *Journal of the Acoustical Society of America, 80*, 1846-1849.

Blumstein, S.E., Cooper, W.E., Goodglass, H., Statlender, S., & Gottlieb, J. (1980). Production deficits in aphasia: A voice onset time analysis. *Brain and Language, 9*, 153-170.

Bolinger, D.L. (1965). Pitch accent and sentence rhythm. In *Forms of English: Accent, Morpheme, Order*. P. 163 ff. Cambridge, MA: Harvard University Press.

Braitenberg, V. (1967). Is the cerebellar cortex a biological clock in the millisecond range? *Progress in Brain Research, 25*, 334-346.

Braun, C.M.J., Duguay, M., Robitaille, M., & Daignault, S. (1985). Effets d'une thérapie de relaxation sur les symptômes physiques, la performance motrice et le niveau d'anxiété de personnes atteintes de sclérose en plaques. *Revue de modification du comportement, 15*, 176-187.

Darley, F.L., Aronson, A.E., & Brown, J.R. (1975). *Motor Speech Disorders*. Philadelphia: W.B. Saunders.

Duffy, J.R., & Gawle, C.A. (1984). Apraxic speakers' vowel duration in consonant-vowel-consonant syllables. In J.C. Rosenbek, M.R. McNeil, & A.E. Aronson (eds.), *Apraxia of Speech* (pp. 167-196). San Diego, CA: College-Hill.

Folkins, J.W., & Abbs, J.H. (1975). Lip and jaw motor control during speech: Responses to resistive loading of the jaw. *Journal of Speech and Hearing Research, 18*, 207-220.

Freund, H.J., & Dietz, V. (1978). The relationship between physiological and pathological tremor. In J.E. Desmedt (ed.), *Progress in Clinical Neurophysiology, Volume 5* (pp. 66-89). Basel: Karger.

Fukson, O.I., Berkinblit, M.B., & Fel'dman, A.G. (1980). The spinal frog takes into account the scheme of its body during the wiping reflex. *Science, 209*, 1261-1263.
Gentner, D.R. (1987). Timing of skilled motor performance: Tests of the proportional duration model. *Psychological Review, 94*, 255-276.
Goodglass, H., & Kaplan, E. (1972). *The assessment of aphasia and related desorders*. Philadelphia: Lea & Febiger.
Gopal, H.S., & Syrdal, A.K. (1988). Effects of speaking rate on temporal and spectral characteristics of American English vowels. *Speech Communication Group Working Papers, VI*, 162-180. (Research Laboratory of Electronics, Massachusetts Institute of Technology).
Grémy, F., Chevrie-Muller, C., & Garde, E. (1967). Étude phoniatrique clinique et instrumentale des dysarthries. *Revue Neurologique, 116*, 401-426.
Grillner, S. (1986). The transfer of terminology from one field to another may be hazardous and counterproductive for interdisciplinary interation. *Journal of Phonetics, 14*, 109-110.
Hassler, R., Mundinger, F., & Riechert, T. (1970). Pathophysiology of tremor at rest derived from the correlation of anatomican and clinical data. *Confinia Neurologica, 32*, 79-87.
Hebb, D.O. (1949). *Organization of Behavior*. New York: John Wiley & Sons.
Hull, C.L. (1934). *Principles of behavior: An introduction to behavior theory*. New York: Appleton-Century.
Itoh, M., Sasanuma, S., Tatsumu, I.F., Murakami, S., Fukusako, Y., & Suzuki, T. (1982). Voice onset time characteristics in apraxia of speech. *Brain and Language, 17*, 193-210.
Ivry, R.B., & Keele, S. (1989). Timing functions of the cerebellum. *Journal of Cognitive Neuroscience, 1*, 136-152.
Keller, E. (1987a). Factors underlying tongue articulation in speech. *Journal of Speech and Hearing Research, 30*, 223-229.
Keller, E. (1987b). The representation of motor processes of speech. In E. Keller & M. Gopnik (eds.), *Motor and Sensory Processes of Language*. Hillsdale, NJ: Lawrence Erlbaum.
Keller, E. (1987c). The variation of absolute and relative measures of speech activity. *Journal of Phonetics, 15*, 335-347.
Keller, E. (1989). Predictors of subsyllabic durations in speech motor control. *Journal of the Acoustical Society of America, 85*, 322-326.
Kelso, J.A.S., Saltzman, E.L., & Tuller, B. (1986). The dynamical perspective on speech production: Data and theory. *Journal of Phonetics, 14*, 29-59.
Kent, R.D., & McNeil, M.R. (1987). Relative timing of sentence repetition in apraxia of speech and conduction aphasia. In J.H. Ryalls (ed.), *Phonetic Approaches to Speech Production in Aphasia and Related Disorders* (pp. 181-220). San Diego, CA: College-Hill.
Kornhuber, H.H. (1974). Cerebral cortex, cerebellum and basal ganglia: An introduction to their motor functions. In F.O. Schmitt && F.G. Wordern (Eds.), *The Neurosciences, vol. 3* (pp. 267-280). Cambridge, MA: The MIT Press.
Lashley, K.S. (1942). The problem of cerebral organization in vision. In Klüver, H. (ed.), Visual mechanisms. *Biological Symposia, 7*, 301-322.
Lea, W.A. (1974). *Prosodic aids to speech recognition: IV. A general strategy for prosodically-guided speech understanding*. Univac Report No. PX10791. St. Paul, MN: Sperry Univac, DSD.
Lehiste, I. (1977). Isochrony reconsidered. *Journal of Phonetics, 5*, 253-263.
Lindblom, B.E.F. (1963). Spectrographic study of vowel reduction. *Journal of the Acoustical Society of America, 35*, 1773-1781.
Lubker, J. (1986). Articulatory timing and the concept of phase. *Journal of Phonetics, 14*, 133-137.
Luria, A.R. (1966). *Higher cortical function in man* (B. Haigh, Trans.). New York: Basic Books. (Original work published in 1962).
Munhall, K.G. (1985). An examination of intra-articulatory relative timing. *Journal of the Acoustical Society of America, 78*, 1548-1553.
Munhall, K.G. (1989). Articulatory variability. In Paula Storer (Ed.), *Acquired apraxia of speech in aphasic adults*. London: Taylor & Francis.
Nelson, W.L. (1983). Physical principles for economies of skilled movements. *Biological Cybernetics, 46*, 135-147.

Nelson, W.L., Perkell, J.S., & Westbury, J.R. (1984). Mandible movments during increasingly rapid articulations of single syllables: Preliminary observations. *Journal of the Acoustical Society of America, 75*, 945-951.

Nittrauer, S., Munhall, K., Kelso, J.A.S., Tuller, B., & Harris, K.S. (1988). Patterns of interarticulator phasing and their relation to linguistic structure. *Journal of the Acoustical Society of America.* 84, 1653-1661.

O'Connor, J.D. (1965). The perception of time intervals. *Progress Report 2*, Phonetics Laboratory, University College, London, 11-15.

O'Connor, J.D. (1968). The duration of the foot in relation to the number of component sound-segments. *Progress Report 3*, Phonetics Laboratory, University College, London, 1-6.

Pike, K.L. (1945). *The Intonation of American English.* Ann Arbor, MI: University of Michigan Press.

Saltzman, E. (1986). Task dynamic coordination of the speech articulators: A preliminary model. *Experimental Brain Research Series* (Springer-Verlag), 15, 130-144.

Shen, Y., & Peterson, G.G. (1962). Isochonism in English. University of Buffalo, Studies in Linguistics. *Occasional Papers, 9*, 1-36.

Shewan, C.M., Leeper, H.A., & Booth, J.C. (1984). An analysis of voice onset time (VOT) in aphasic and normal subjects. In J.C. Rosenbek, M.R. McNeil, & A.E. Aronson (eds.), *Apraxia of Speech* (pp. 197-220). San Diego, CA: College-Hill.

Shik, M.L., Severin, F.V., & Orlovskii, G.N. (1966). Control of walking and running by means of electrical stimulation of the mid-brain. *Biophysics, 11*, 756-765.

Tolman, E.C. (1932). *Purposive behavior in animals and men.* New York: Century.

Tuller, B., & Kelso, J.A.S. (1984). The timing of articulatory gestures: Evidence for relational invariants. *Journal of the Acoustical Society of America, 71*, 1030-1036.

Tuller, B., Kelso, J.A.S., & Harris, K.S. (1982). Interarticulator phasing as an index of temporal regularity in pseech. *Journal of Experimental Psychology: Human Perception and Performance, 8*, 460-472.

Uldall, E.T. (1972). Relative durations of syllables in two-syllable rhythmic feet in R.P. in connected speech. *Work in Progress, 5*, 110-111. Edinburgh University Department of Linguistics.

Ziegler, W. (1987). Phonetic realization of phonological contrast in aphasic patients. In J.H. Ryalls (Ed.), *Phonetic Approaches to Speech Production in Aphasia and Related Disorders* (pp. 163-179). San Diego, CA: College-Hill.

THE ACOUSTIC AND PHYSIOLOGIC CHARACTERISTICS OF NEUROLOGICALLY IMPAIRED SPEECH MOVEMENTS

R. D. KENT
Department of Communicative Disorders
University of Wisconsin-Madison
1975 Willow Drive
Madison, Wisconsin 53706 U.S.A.

ABSTRACT. The study of neurologic disorders of spoken language is of course important to the improvement of clinical services for those with communicative impairments. But, in addition, the study of these disorders complements the study of normal speech to reach an understanding of the neural regulation of spoken language. Ideally, data on both normal and disordered speech will be embraced by common theoretical frameworks, so as to permit an effective consideration of these two sources of knowledge. Two such frameworks are discussed. One is an information processing perspective on utterance formulation and production. The other is a model of the neural regulation of voluntary movements. These frameworks are considered in relation to the disorders of conduction aphasia, apraxia of speech, and the dysarthrias associated with Parkinson's disease, cerebellar disease, and amyotrophic lateral sclerosis. Finally, points of connection between the two theoretical frameworks are discussed.

1. The Neurologic Disorders of Speech and Language Production

The central goal of this paper is to consider what neurological disorders of speech tell us about the neural regulation of speech. The neurological disorders of greatest currency are aphasia, a language impairment, and dysarthria, a speech impairment. However, speech and language are not always clearly differentiated in clinical diagnosis and nosology. Nor are they always clearly differentiated in human communication. The lack of differentiation is illustrated by the characteristics of spoken language commonly used to distinguish fluent and nonfluent aphasic patients (Benson, 1967; Geschwind, 1966; Howes, 1964; Howes and Geschwind, 1964; Kerschensteiner, Poeck and Brunner, 1972).

As Kerschensteiner et al. pointed out, the fluency-nonfluency dimension gave to aphasia classification an empirical basis more compelling than older clinico-

W. J. Hardcastle and A. Marchal (eds.), Speech Production and Speech Modelling, 365–401.

anatomical subgroupings. (However, a clinico-anatomic substrate for this dimension was described by Benson, 1967, who reported that aphasic patients with anterior lesions had "a mechanical speech difficulty" consisting of low rate, dysarthria, effort, low phrase length, and poor melody.) Kerschensteiner et al. determined that the rank order of variables useful in discriminating aphasic patients with respect to their spontaneous speech was: phrase length, pauses, prosody, rate of speaking, effort, articulation, word choice, verbal paraphasias, literal paraphasias, and perseveration.

The behavorial classification of aphasic patients according to the fluency-nonfluency dimension is particularly interesting because this dimension is rooted largely in speech behavior rather than formal language operations related to, for example, syntax and semantics. Benson (1967) was quoted earlier to the effect that anterior patients have "a mechanical speech difficulty." His remark is testimony to the tenuous separation of speech and language. More recent research emphasizes this problem. For example, evidence that patients with aphasia have impairments at a phonetic or motoric level has been presented for conduction aphasia by Kent and McNeil (1987), and for Broca's aphasia by Itoh and Sasanuma (1987) and Keller (1987).

It appears, then, that speech and language functions are interwoven in aphasia classification based on spoken language. This conclusion has several implications, but one in particular is pertinent to this conference: aphasic syndromes are a source of information on the neurological impairment of spoken language, and therefore on the neurological representation of speech. Another implication is that clinical classification frequently mixes speech and language characteristics.

More generally, speech needs to be viewed within the superordinate behavior of language. Arduini (1987) spoke to this issue when he wrote, "Physical speech...is not a self-supporting or self-sufficient functional property, it is only the outwardly sign of an internal process by which the brain can talk with its 'conscious self'" (p. 15). He points out that the reason why the faculty of speech was accompanied by an enormous increase in brain mass is that speech did more than recruit a pre-existing pool of neurons for oral motor control. It added to these motorneurons but also added neurons for sensory and associative functions.

1.1 THE MOTOR SPEECH DISORDERS

The neurological disorders of speech per se include two broad classes, the dysarthrias and apraxia of speech (also known as verbal apraxia). These two classes are often considered together as the motor speech disorders. Dysarthrias are speech impairments typically associated with the clinical abnormalities of weakness,

Table 1 (cont.)

5. Distinctive feature analyses of substitution errors indicate a variable relationship between error and target sounds, with most errors involving one or two features.
6. Sequential errors (prepositioning, postpositioning, metathesis) make up less than one-tenth of the total errors.
7. Automatic and reactive speech production is less affected than volitional-purposive speech.
8. Articulatory errors tend to be greater for imitative responses than for spontaneous speech. In addition, repetitive responses can have long latencies.
9. Articulation errors increase as word length increases.
10. Articulatory errors during oral reading are more likely for words that carry greater propositionality or greater linguistic or psychological weight.

---

2. A Model of Utterance Formulation and Production

It sometimes has been argued that neurologic disorders of speech and language should help us to understand the neurologic organization of these behaviors if only because disorders can tell us where things can go wrong. In practice, however, the neurologic disorders are surrounded by their own controversies of interpretation. A disordered system is not easily understood without recourse to an intact, functional system. What is needed is a theory that can accomodate data from both normal and disordered systems and generate hypotheses for both kinds of systems.

Speech is a goal-oriented action and it can be understood completely only in the context of language behavior. Ultimately, speech needs to be considered in a general context of utterance formulation and production. Fig. 1 is a modified version of a a model of utterance production proposed by Bock (1982). Some general features of the model will be summarized below.

The General Cognitive Arena is where thought is formulated as nonlinguistic propositions or relations. It may be regarded as the sphere of conscious awareness and intentional action. Included within this arena is a referential component which translates or codes the nonlinguistic representation of thought into a format which can be used by the linguistic system. This translation or coding is accomplished by operations called schematization and framing. Schematization organizes nonverbal

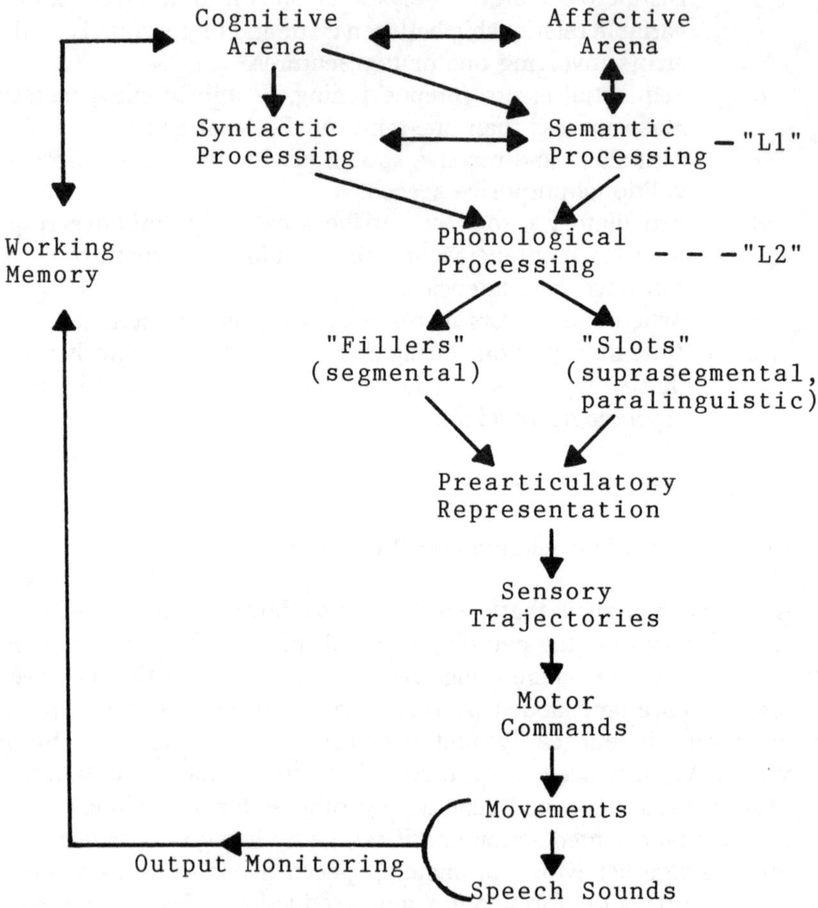

Fig. 1. Information processing model of utterance formulation and production (adapted from Bock, 1982).

material into an already established pattern, for example, by orientation to a "Plan" schema (Lichtenstein and Brewer, 1980), which relates the various components of goal-directed events. Such a schema might consist of the specified or inferred goal of some actor, together with a sequence of subgoals through which actions are determined to achieve the primary goal. Framing (Chafe, 1977) follows schematization when sufficient details are available to formulate a verbal expression suitable to the communicative goals and constraints. Framing is accomplished by the selection of participants and the assignment of these participants to roles.

Bock summarized the Referential Arena as producing a relational structure that "can be neutrally described as a proposition or set of propositions that structure and segment nonlinguistic conceptual patterns" (p. 4). The components and relations that constitute these propositions are processed in the semantic arena.

Bock's original model has been modified to include an Affective Arena which is closely coupled to, but separate from, the General Cognitive Arena. Affective states are determined by cognitive operations, various internal (visceral) states, and by associational mechanisms based on semantics. The Affective Arena acts directly on motor mechanisms of linguistic expression so that emotional factors may facilitate or interfere with such expressions (Kent, 1984). Moreover, because the Affective Arena interacts with the General Cognitive Arena, affective reactions can influence cognitive processes. A developmental perspective on this issue was offered by Bloom and Beckwith (in press), who proposed that the different cognitive requirements for language and emotional expression compete for the attentional resources of a child who is learning language. Eventually, the two means of expression become integrated, so that spoken language can carry strong emotional undertones. Vandamme (1987), in a review of the ways in which cognition and emotion interact, noted that emotion is, in part, a change in excitation that is related to a change in situation. More generally, the change in excitation is related to changed circumstances of any kind, including verbal information.

Syntactic and semantic processing are highly interactive, as indicated by the double-headed arrows in Figure 1. Semantic processing identifies lexical concepts that match propositions formulated in the General Cognitive Arena. The term lexical concept (as opposed to words) is well advised because evidence indicates that lexicalization proceeds in two stages, a relative abstract, prephonological stage, and a second stage of phonological realization (Kempen and Huijbers, 1983). Semantic processing in the model might be likened to the L1 (prephonological) stage, while phonological processing corresponds to L2 operations.

Because semantic processing is not concurrently linked to phonological representation, semantic and syntactic processing may interact before phonological decisions are complete. In this way, semantic and syntactic information is available for phonological processing. Bock reviews a number of studies that point to a central property of language formulation: Semantic (lexical) decisions often carry syntactic privileges and syntactic decisions in turn influence lexical choices for the remainder of the utterance. This interactivity accords with the view that sentence production involves the utterance of the first few words (especially a noun phrase in English) even while the remainder of the utterance is being formulated (Bock, 1982; Kempen and Huijbers, 1982; Lindsley, 1975, 1976). In other words, speakers begin to speak an utterance before the utterance is completely formulated. Thus, the motor mechanisms of language expression are engaged even as language formulation proceeds.

Mutual adjustments are required to coordinate lexical selection and syntax. As Bock (1982) remarked, mismatches between syntactic and lexical processing can arise because lexical information is retrieved in parallel but with different time courses for different words. Eventually, both words and syntax must be reconciled to yield serial output, and the need for adjustments between them becomes greater as the discrepancy increases between the size of the utterance planning unit and the size of the articulatory unit. Studies suggest that the former is roughly the size of the clause (Ford and Holmes, 1978; Garrett, 1980) whereas the latter is essentially phrasal (Lindsley, 1975; Gee and Grosjean, 1984).

The potency of sentential constituent structure on speech production processes is demonstrated by the syntactic determination of dysfluencies (Bernstein, 1981; Bloodstein, 1974). Bernstein's results showed that the disfluencies of both stuttering children and nonstuttering controls occurred primarily in association with the first noun phrase of the utterance and with the conjunction and. The stuttering children differed from the normal controls in having one additional locus of disfluency, the verb phrase.

The Phonological Processor uses information from syntactic and semantic processing to construct the phonological representation of an utterance. It is at this point that the lexical concept becomes phonologically evident as a word. In addition, it would appear that the "tip of the tongue" phenomenon (Brown and McNeill, 1966) arises here when a lexical concept can be accessed before its phonological form is realized and the speaker is obliged to search for the phonological representation.

The two-stage hypothesis of lexicalization may explain certain language abnormalities in the aphasias. As a first example, it is noteworthy that neologistic utterances in jargon aphasia may be preceded by a hesitation pause (Butterworth, 1979). This result detracts somewhat from the common clinical description of these disorders as "fluent" aphasias and illustrates the sensitivity of the output utterance to formulation impairments. As a second example, O'Connell (1981) reported an analysis of neologistic jargon aphasia in which the neologisms were considered to have a complex origin:

1. Some correctly retrieved words were subjected to complex, simultaneous paraphasias.
2. Incorrectly retrieved words were subjected to further disruption in phonologic realization.
3. Recombinations of units from prior utterances were selected to replace target words which either (a) were not retrieved, or (b) were partially or wholly retrieved but could not be correctly specified phonologically because of competition from unsuppressed previously activated elements.

A third example is that effective procedures to facilitate word retrieval vary with type of aphasia (Goodglass, 1983) and it appears that facilitating procedures may be categorized roughly as semantic versus phonological, corresponding, it would seem, to the L1 and L2 stages of lexicalization.

The output of the Phonological Processor is converted to a prearticulatory representation in a Phonetic Coder. The exact form of the prearticulatory representation is open to question, and several candidates have been proposed. The choice has important consequences, among them being implications for the nature of the motor control processes that govern articulation. The more detailed the prearticulatory representation, the smaller the burden placed on motor regulation to account for such properties as allophonic variation. The dominant approach to phonological representation has been in terms of abstract elements such as features or phonemes. However, a motoric representation also has been proposed. Indeed, Ziegler and colleagues (Ziegler and von Cramon, 1987; also see Ploog, 1987) concluded from studies of apraxia of speech that motor goal specifications are an integral aspect of lexical processing. Proposals also have been advanced for an "articulatory phonology" in which aspects of articulation are incorporated into linguistic representation (Browman and Goldstein, 1986; Munhall, 1989).

An idea that is attracting much attention is that the phonological output is effectively divided into two kinds of representation, called "slots and fillers" (Shattuck-Hufnagel, 1983) or "frames and contents" (MacNeilage, Studdert-Kennedy and Lindblom, 1984). As these concepts are incorporated in Fig. 1, they are viewed as follows. Slots are syllable markers and therefore embody information about speaking rate and prosody (including stress). Fillers are phonetic segments that are subsequently assigned to slots in the prearticulatory representation. The independent specification of slots and fillers accounts for most sequencing errors in normal speech. Such errors reflect a number of interesting lawful properties, among which is a general tendency for the interchange of phonetic elements between similarly stressed syllables in an utterance. The slots and fillers model seems to account for some errors in apraxia of speech and conduction aphasia (Kent and McNeil, 1987). Furthermore, this notion has been proposed to account for stuttering as a dissynchrony in the neural signals corresponding to an affective-prosodic channel and a segmental channel (Perkins, Curlee, Kent and Hixon, submitted).

It is assumed that in fluent adult speech, the syntactic, semantic and phonological processing are ordinarily automatic processes, which is to say that they do not require capacity in working memory. Automatic, or capacity-free, processing contrasts with controlled, or capacity-demanding, processing (Schneider and Shiffrin, 1977; Shiffrin and Schneider, 1977). Thus, in Figure 1, connecting arrows are drawn between working memory and both the General Cognitive Arena and the Phonetic Coding, but not between working memory and the other components. However, speakers are flexible in the deployment of automatic and controlled processing. Under difficult situations, a speaker may rely more on controlled processing. It is likely that speakers with a neurological impairment of speech may be more reliant on controlled processing than are normal speakers. Cheng (1985) has questioned the legitimacy of the automaticity versus control distinction and has argued instead that task components are restructured into new perceptual, cognitive or motor units. It is not critical to the present use of the model in Figure 1 whether automaticity versus control or restructuring is the correct conceptualization. The only aspect of the automaticity concept that is especially attractive here is the idea that automatic processes are opaque to monitoring or awareness. But opacity is not necessarily excluded from the restructuring hypothesis, since a restructuring could render some task components less identifiable at the level at which they originally learned. Should the restructuring hypothesis be confirmed, the major change in Figure 1 is to revise the selected connections to Working Memory.

More will be said about motor control mechanisms in a later section. But one final aspect of the utterance formulation and production model of Fig. 1 need to be described. Specifically, provision must be made for self-monitoring, which confirms correct production of an utterance or generates error information from which a corrected utterance can be prepared (Bock, 1982, Levelt, 1983). Speaking is an error prone process even for normal talkers, and effective communication requires self-monitoring to correct mistakes made in verbal formulation and phonetic assembly. Speakers do not seem to have access to semantic, syntactic or phonological processing as an utterance is formulated (Bock, 1982; Levelt, 1983). Thus, self-monitoring is essentially output monitoring insofar as the output (motor or auditory pattern) is compared against the speaker's intention. As would be expected from the concept of automaticity, the normally automatic processes are opaque to self-monitoring.

According to Levelt (1983), monitoring performs two functions. The first is a twofold matching function in which (1) parsed aspects of inner and outer speech are compared with the speaker's intention to determine if the actual utterance corresponds to the intended expression, and (b) the parsed information for inner and outer speech is compared with production standards to detect speech errors and syntactic flaws as well as to monitor rate, loudness and other prosodic features. The second function of monitoring is to generate instructions for correction or adjustment. Whenever an error or nonstandard production is detected, the monitor must alert the formulator and supply corrective information. When the error is serious, the speaker usually stops and restarts. Levelt notes an important characteristic of this restarting: "[It is] not neutral with respect to the interrupted utterance, it usually reinstalls some of the parsed properties of the original utterance" (p. 50). The supply of context for repair of the error aids the listener in solving a "continuation problem," viz., how to relate the speaker's repair to the flawed utterance.

Obviously, the processes represented in Fig. 1 are complex. One may ask if they need to be considered at all to understand the movements of speech. Why not proceed simply with the study of movements in nonsense syllables to avoid the complexities of verbal formulation? To a degree, we can. But ultimately our concern is with spoken language. Nonsense syllables and other contrivances of the laboratory researcher simplify the investigation of speech movements but they also raise serious questions about the ways in which the laboratory data can be generalized to actual speaking situations. The potential limitations in this regard are pointedly described by Bladon, Henton and Pickering (1984), who noted that nearly all of the data in classical acoustic phonetics (and we may add physiological phonetics as well) pertain to "laboratory" or "citation style" speech.

2.1 SENSORY TRAJECTORIES AND MOTOR COMMANDS FOR SPEECH

Perkell (1980) proposed that underlying speech production is an inventory of features that match auditory and orosensory patterns. The auditory patterns permit the detection of distinctive acoustic cues that are quantal in character. Quantal theory maintains that the articulatory-acoustic relationship has stable and instable components that govern the selection of sounds in natural languages (Stevens, 1972). A related conception of distinctive regions and modes has been described by Mrayati, Carre and Guerin (1988). The orosensory patterns matched to the auditory patterns are a complex combination of proprioceptive, tactile, and aerodynamic afference. (Table 2 summarizes the qualities of oral sensation that are available for various oral functions.) The patterning of the sensory information defines the physiological goals of the speech production mechanism, that is, the goals from which acoustic consequences can be predicted.

Table 2. Summary of oral sensation: qualities, receptors and cranial nerve innervation. Drawn from Kent, Martin and Sufit (in press).

Sensory quality	Receptor	Nerve
Jaw position	Jaw muscle spindles	V
	TMJ receptors	
Jaw bite force	Periodontal receptors	V
	TMJ receptors	
	Jaw muscle spindles	
Vibration	Mechanoreceptors (RA)	V
2-pt discrimination	Mechanoreceptors (esp. SA)	V
Touch pressure	Mechanoreceptors	V
Kinetic cutaneous sensitivity	Mechanoreceptors (esp. RA)	V
Oral stereognosis	Mechanoreceptors	V
Texture sensation	Mechanoreceptors	V
Barometric sensation	Mechanoreceptors	V,X
Perioral reflex	Mechanoreceptors	V
Jaw stretch reflex	Mechanoreceptors	V
	Muscle spindles	
Tongue proprioception	Muscle spindles	XII?
Thermal sensation	Warm receptors	V
	Cold receptors	
Pain	Nociceptors	V
Gustation	Taste buds, ant. tongue	VII
	" " palate	VII
	" " post. tongue	IX
	" " larynx	X
	" " pharynx	IX,X
Olfaction	Olfactory receptors	I

Table 2 (cont.)

Abbreviations: TMJ--temporomandibular joint; SA--slowly adapting; RA--rapidly adapting.

Some ideas similar to Perkell's were advanced by Neilson and Neilson (1987) in their "adaptive model theory." In this theory, sensory feedback is used to monitor ongoing movement and correct the regulatory process if necessary. Afference is used in adaptive control insofar as it establishes, confirms and modifies as required the relationship between motor commands and their multimodal consequences. Information that is gained through sensory feedback is the means by which motor commands are derived from a planned sensory trajectory. A critical aspect of the internal model is a representation of the acoustic signal. As Osamu Fujimura has reminded us, the vocal tract is an acoustic concept. The motor control of this system is therefore rooted in its acoustic capabilities.

Neilson and Neilson's adaptive model theory is more general than Perkell's model in that it is not defined specifically in terms of distinctive features. It therefore has a wider application, such as in the imitation of arbitrary sounds, including nonspeech sounds. This is not to deny that features are useful in accounting for speech production, but rather to acknowledge the versatility and multifunctionality of the speech production system, going beyond, but including, conventional linguistic functions.

A fundamental concept is that sensory trajectories are derived from the Prearticulatory Representation of the utterance. Several candidates have been proposed for this representation, including syllables, phonemes, allophones and phonetic features. The preference in this paper is for two coordinated representations corresponding to slots and fillers. Motor commands are derived from the sensory trajectories. The literature mentions several possibilities for the motor realization of speech, including spatial targets, motor commands, acoustic templates and coordinative structures. Perhaps it has been difficult to choose among them because all, or nearly all of them, are possible. The sensory trajectories should include specifications at many sensory levels, including auditory, proprioceptive, tactile and aerodynamic. Thus, there is no contradiction in asserting that the control of speech is based on plurimodal objectives. Nor is speech the only oral function with such a property. Recent conceptions of taste also emphasize the plurimodality of this sensation. Faurion (1987) suggested that the sensory image in taste is minimally gustatory and thermal but may be accompanied by a second complementary image involving gustatory, thermal and mechanical modalities in a cortical primary projection.

Motor commands are generated in accord with the sensory trajectories. The actual sensory consequences can be compared with the sensory trajectories to verify that the commands were appropriate. Because the system is highly practiced, the sensory trajectories are sufficient for accurate motor control even when target sounds are produced under conditions of binaural masking, bilateral TMJ anesthetization, topical anesthetization of the oral mucosa, and fixation of the jaw at various positions (Kelso and Tuller, 1983). It is interesting to recall here the comment of Dubner, Sessle and Storey (1978, p. 74): "The precise motor function associated with speech requires sensory feedback from intraoral structures as well as from auditory cues." This remark may seem difficult to reconcile with the relatively minor speech disturbances associated with short-term disruptions of afference (Kent, Martin and Sufit, in press). However, the inconsistency disappears if one recognizes that the remarkable precision of sensory feedback in the oral and perioral tissues is able to lay down the neural substrate for adaptive control of speech and other oral activities. The neural representation of the tongue in the cerebral cortex is consistent with this functional picture. The following conclusions are drawn from Picard and Olivier's (1983) mapping of sensory cortical representation in 100 patients undergoing craniotomy: (1) The cortical representation is disportionately large; (2) Motor and sensory responses overlap in the cortical region just behind the post-central gyrus; and (3) Although the sensory representation is bilateral, the strongest sensory representation appears to be on the hemisphere that is dominant for speech. In both anesthetized and unanesthetized primates, the majority of neurons in the primary face motor cortex respond to tactile stimulation, but a much smaller proportion of neurons in this area respond to deep orofacial afferent inputs (Huang, Hiraba and Sessle, 1989). Huang et al. concluded that superficial orofacial mechanoreceptors supply the predominant excitatory afferent input to this region of the cortex.

Movement abnormalities in the motor speech disorders are usually confounded with other factors such as voluntary compensation, effects of medication or other management, and prosodic alterations. The last of these deserves particular comment with regard to the issue of speaking rate. With some notable exceptions, such as dysarthria in Parkinson's disease, the motor speech disorders typically are associated with a slow rate of utterance, which, in severe disorders, can take the form of greatly prolonged movements and steady states. The slow speaking rate, whether compensatory or a frank expression of the motor disorder, might in itself lead to certain movement characteristics. Generally, slow speaking rate is associated with a greater variability in the timing of movement components (Kent and Forner,

1980; McNeil, Liss, Chin-Hsing and Kent, submitted; Munhall, 1989). Indeed, McNeil et al. went so far as to propose a slow "register" for speech production. Wieneke, Janssen and Belderbos (1987) concluded that a slow speaking rate generally reduces the overlap or coarticulation in the movement pattern. Considering these possibilities, one should interpret the variability of movement patterns in a speech disorder in terms of the speaking rate for which the data were collected.

3. Neurologically Disordered Speech: Theoretical Implications

The remainder of the paper considers the utility of a general information-processing model and a model for the neural control of movement for the following disorders: conduction aphasia, apraxia of speech, hypokinetic dysarthria (as commonly associated with Parkinson's disease), ataxic dysarthria (as commonly associated with cerebellar disease and with multiple sclerosis), and flaccid-spastic dysarthria (typical of amyotrophic lateral sclerosis). But first, each disorder and its effects on speech will be briefly summarized.

3.1 SUMMARY OF PRIMARY CLINICAL FEATURES

3.1.1. **Conduction aphasia.** The usual clinical description of this disorder is that it is primarily a disorder of repetition, although it is associated with frequent phonemic paraphasias on nearly all tests that require a verbal response. There are several accounts for this disorder, ranging from a disconnectist syndrome (Geschwind, 1965), to a deficit in verbal short-term memory (Warrington, Logue and Pratt, 1971; Warrington and Shallice, 1972), to a dysfunction in "pre-articulatory programming" (Kohn, 1984), to a motor encoding difficulty (Yamadori and Ikumura, 1975), to a phonological disruption (Friedrick, Glenn and Marin, 1984), to a multifaceted syndrome (Shallice and Warrington, 1977; Blumstein, Cooper, Goodglass, Statlender and Gottlieb, 1980). Kent and McNeil (1987) concluded that this disorder is not sufficiently characterized at a phonemic or phonetic level but also has associated motoric disruptions. The prosodic aspects of speech were relatively spared. From this description, it is suggested that conduction aphasia is primarily a phonetic coding disorder, in which (a) phonetic sequencing is mismanaged, (b) sensory trajectories are not correctly specified, and (c) primarily because of the disturbance in (b) above, motor control is subject to subtle errors that are not easily corrected in repeated trials.

The "sensory" quality of this disorder is revealed in descriptions such as: an impairment of auditory-phonologic transcoding (McCarthy and Warrington, 1984); a disruption in auditory verbal short-term memory (Warrington, Logue and Pratt, 1971; Warrington and Shallice, 1972); or a deficit in motor encoding because of a deterioration in the acoustic image of a target word (Yamadori and Ikumura, 1975). Conduction aphasia seems to be, in part, a disorder in the preparation and deployment of sensory trajectories. The relative sparing of prosodic elements (giving to the disorder a "fluent" spontaneous speech despite the numerous phonetic errors) is an indication of a selectivity for the "fillers" rather than "slots" channel.

3.1.2. **Apraxia of Speech.** This disorder is known by several names (verbal apraxia, speech motor programming disorder, little Broca's aphasia, etc.) and even its nosologic existence has been questioned. Its essence from a perceptual point of view is best grasped by a careful reading of the attributes listed in Table 1. A serviceable definition is that apraxia of speech is an impairment of the capacity to form vocal tract configurations and to move from one sequence to another for volitional speech production, in the absence of motor impairments for other responses of the same musculature (Rosenbek, Kent and LaPointe, 1984). A particularly interesting feature of this disorder is that its effects essentially are confined to speech. It is not a language disorder, nor does it affect the movements for nonspeech tasks. Clinical examination often entails the use of tasks that increase the difficulty of utterance formulation and production. The nature of these tasks can be readily imagined from an inspection of Table 1. The clinician manipulates variables such as phonetic sequence, word familiarity and length, and grammatical complexity. The effects of these variables need to be understood through a comprehensive model of spoken language production. Fig. 1 is offered in that spirit.

As noted in an early section of this paper, the first characterizations of apraxia of speech emphasized the frequent phonemic substitutions in most dyspractic speakers. This emphasis encouraged some to view it as a phonological impairment. Darley's (1969) description (cited by Rosenbek et al., 1984) of the disorder as a motor programming impairment is closer to the truth if only because it enlarges our vision to look beyond phonology. Several recent reports converge on a central finding of motor impairment. Ziegler and von Cramon (1986) identify the basic impairment as one of disturbed phase relations of individual speech movements. This description is compatible with other reports (Blumstein and Baum, 1987; Kent and Rosenbek, 1983; Edwards and Miller, 1989). The difficulties in motor phasing may in themselves account for phonological errors, although some frank phonological errors may occur as well (Blumstein and Baum, 1987; Kent and

Rosenbek, 1983; Hardcastle, 1987; Rosenbek, Kent and LaPointe, 1984). Apraxia of speech is in part an impairment in the generation of properly timed (phased) motor commands. These timing errors occur at both subphonetic and phonetic levels of perceptual consequence.

3.1.3. **Hypokinetic dysarthria.** This is the term given by Darley, Aronson and Brown to the dysarthria frequently associated with Parkinson's disease. The lesion in this disease appears to be a depletion of dopaminergic neurons principally in the substantia nigra and to a lesser extent in other nuclei such as the locus ceruleus. The motor abnormalities in Parkinson's disease include tremor (usually a resting tremor), rigidity, akinesia, and bradykinesia.

Perceptual, acoustic and physiologic characteristics reported for this dysarthria are listed in Table 3. The term hypokinetic has been applied to this speech disorder in recognition of reduced range of movement. This reduction often is thought to result from rigidity, which is easily demonstrated in the limb musculature. Rigidity is more difficult to assess in many of the muscles of speech production, but many writers ascribe abnormalities in speech movements to this condition. For example, bowing of the vocal folds, which is probably strongly linked to many of the abnormal vocal qualities in Parkinson subjects, has been related to rigidity of the laryngeal musculature (Hanson, Gerratt and Ward, 1984).

Similarly, rigidity may underlie abnormal movements of the articulators, including restricted movement (Logemann, Fisher, Boshes and Blonsky, 1978). Although rigidity and tremor are the most commonly cited neurologic conditions related to dysarthria, Netsell, Daniel and Celesia (1975) concluded that acceleration phenomena and weakness also contribute to the speech abnormalities. Marsden and Parkes, cited by Gawel (1981), reported that akinesia and dysphonia appeared without rigidity or tremor at the peak of L-Dopa action. Similarly, DeLong noted that rigidity and akinesia can be independently modified in pharmacological intervention. Kent and Netsell (1979) suggested that rigidity may be most likely to appear in muscles that maintain tone or posture, such as certain lip muscles (Leanderson, Meyerson and Persson, 1972), the laryngeal muscles (Hanson et al., 1984) and muscles effecting velopharyngeal closure (Kent and Netsell, 1979). The term posture is intended to include long-term activity states, such as vocal fold adjustments for voiced passages or velopharyngeal settings for nonnasal phonetic strings. Miller and De Long (1988) have offered hypotheses to account for the various motor abnormalities in Parkinson's disease. The basal ganglia and the supplementary motor area figure prominently in these hypotheses.

Table 3. Ten most deviant speech dimensionsnoted in 32 patients with Parkinson's disease(Darley et al., 1969a,b) listed with probable pathophysiology (from various sources noted in text).

Speech Dimension	Probable Pathophysiology
Monopitch	Bowed vocal folds, rigidity of laryngeal muscles, especially cricothyroid
Reduced stress	Rigidity of respiratory and laryngeal muscles, fast rate
Monoloudness/ reduced loudness	Reduced inspiratory and expiratory volume, rigidity of respiratory muscles
Imprecise consonants	Rigidity of vocal tract musculature, reduced movements, inadequate oral closures
Inappropriate silences	Rigidity of musculature, initiation difficulty
Short rushes of speech	Rigidity of musculature, fast rate
Harsh voice quality	Rigidity of laryngeal muscles, hypokinesia, tremor
Breathy voice	Bowed vocal folds, rigidity of laryngeal muscles
Pitch level	Bowed vocal folds, rigidity of laryngeal muscles
Variable rate	Combinations of above

A further possible impairment is indicated by the work of Schneider, Diamond and Markham (1986). Their research on proprioception and tactile localization in Parkinson subjects indicated impairments of jaw proprioception and tactile localization on the tongue, gums and teeth. Possibly, the basal ganglia damage in Parkinson's disease effects sensory as well as motor functions. Lidsky, Manetto and Schneider (1985) proposed that the basal ganglia act as a sensory analyzer for motor systems, putting sensory data into a form that can be used to guide movements. With respect to the model presented earlier, then, the basal ganglia, may participate in early stages of voluntary movement in which sensory trajectories are determined. The basal ganglia also may participate in the evaluation of sensory feedback associated with an ongoing movement.

3.1.4. **Ataxic dysarthria.** Ataxia (literally, without skill) is a general descriptor for a variety of movement disturbances associated with cerebellar lesions. The chief symptoms of cerebellar disease are abnormal postural fixation (hypotonia), disturbed skilled voluntary acts (dyssynergy), disturbed equilibrium (truncal ataxia, ataxia major), and impaired gait and postural reflexes.

The dysarthria was described by Brown, Darley and Aronson, 1970) to consist of three major clusters of related dimensions. The first cluster, termed articulatory inaccuracy, was formed by the speech dimensions of imprecise consonants, irregular articulatory breakdown, and distorted vowels. This cluster was associated with the neurologic defects of inaccuracy of individual movements and dysrhythmia of repetitive movements.
The second cluster, prosodic excess, included the dimensions of excess and equal stress, prolonged phonemes, prolonged intervals, and slow rate. The neurologic defect for this cluster was thought to be slowness of individual and repetitive movements. The third cluster, phonatory-prosodic insufficiency, consisted of the dimensions monopitch, monoloudness and harsh voice. Hypotonia was suggested as the neurologic defect. The cluster names are unfortunate; for example, it is puzzling that prosody could be at once excessive and insufficient. However, most of the deviant speech dimensions are consistent with acoustic and physiologic descriptions of the disorder (as reviewed in the following).

The cerebellum has complex and vast connections to cortical and subcortical structures. It is implicated in the processing of both sensory and motor information, and although the metaphor may not be completely apt, it is instructive to recall Ito's (1984) description of the cerebellum as a special purpose computer adapted to the monitoring and revision of motor commands, to give refinement and accuracy to movement. Pathophysiologic studies of cerebellar disease often implicate the Purkinje cells (Rogers, 1988). These neurons assume particular importance because they are the ultimate target of cerebellar afferents, they dominate information processing in the cerebellum, and they are the final output of the olivocerebellar system. The Purkinje cells are also highly susceptible to damage from vascular, hyperthermic or other insults and from aging processes.

Acoustic and x-ray analyses of ataxic dysarthria have shown that the movements of speech lack precision in direction, velocity and extent (Hirose, 1986; Kent and Netsell, 1975; Kent, Netsell and Abbs, 1979). Interestingly, disturbances in the coordination of speech movements are not easily demonstrated (Hirose, 1986; Kent and Netsell, 1975), perhaps indicating that the basic plan of movement is structured outside the cerebellum and that the role of the cerebellum is to provide the fine control for precise movements.

A similar dysarthria has been observed in association with multiple sclerosis. A rare and particularly interesting form of this dysarthria is paroxysmal dysarthria, in which the speech disorder occurs for a brief interval and then quickly remits. We (Netsell and Kent, 1976) have studied a subject with paroxysmal dysarthria and these data provide an unusual opportunity to study normal and impaired speech movements in the same subject within the same hour. The speech disturbances of the patient with paroxysmal dysarthria were judged to be highly similar to those of patients with cerebellar disease. In both patient groups, the ordering of articulatory movements was not disturbed but individual movements were poorly regulated in respect to rate, range and direction. These results were interpreted to mean that the primary role of the cerebellum in the regulation of acquired motor skill is to supplement and revise as necessary cortically generated motor instructions.

3.1.5. **Spastic-flaccid dysarthria.** This form of dysarthria occurs in several neurological diseases, but this paper will focus on amyotrophic lateral sclerosis (ALS). ALS is diagnosed by the occurrence of lower motor neruon and upper motor neuron signs in the extremities with an absence of sensory abnormalities. The lower motor neuron signs include weakness, atrophy, and fatiguability and fasciculations. The upper motor neuron signs include spasticity, increased tone, exaggerated deep-tendon reflexes and cramps.

Perceptual characteristics of the dysarthria in ALS have been described with rating scales by Darley, Aronson and Brown (1969a,b) and by Carrow, Rivera, Mauldin and Shamblin (1974). Darley et al identified the seven most deviant speech dimensions as (1) imprecise consonants, (2) hypernasality, (3) harsh voice quality, (4) slow rate, (5) monopitch, (6) short phrases, and (7) vowel distortions. Carrow et al. reported that the typical abnormalities and their percentage of occurrence were: harshness (80% of the subjects), hypernasality (75%), breathiness (65%), voice tremor (63%), and imprecise consonants (58%). In comparing the speech characteristics of patients who were judged to have pseudobulbar symptoms with those who were not judged to have such symptoms, Carrow et al. observed that the group with pseudobulbar symptoms tended to be more deviant in three dimensions--strained-strangled voice, slow rate, and imprecise consonants. The dysarthria is termed mixed spastic-flaccid in recognition of the supposed upper motor neuron effects (spasticity) and lower motor neuron effects (flaccidity) on speech.

ALS is a degenerative disease and during its course, many patients lose functional voice and speech. Although the disease has extensive effects on the speech production system, some components are more severely affected than others (Kent, Weismer, Kent, et al., submitted; Kent, Kent, Weismer et al, submitted). Studies in our laboratory indicate that articulatory features particularly vulnerable to impairment are anterior tongue function for consonants, velopharyngeal function and phonatory function. Acoustic analyses of words used in a word-identification test of intelligibility revealed that the slope of formant trajectories is related to intelligibility and may be one index of increasing severity.

It has been suggested that impairment of the intrinsic tongue musculature contributes to the dysarthria in ALS (Shawker and Sonies, 1984). Our own studies are consistent with this view, particularly if one assumes that the intrinsic fibers are responsible for bending of the tongue (as in formation of the liquid consonants). The tongue resembles other nonjointed muscular organs such as the tentacles of an octopus and the trunk of an elephant (Kier and Smith, 1985; Smith and Kier, 1989). The term muscular hydrostat refers to these muscular organs that behave as fluid-filled bags. Under physiological pressures, these organs are incompressible, and the dynamics of their deformation are understood in terms of compensatory changes in their three-dimensional shapes. Bending of such a hydrostat, such as occurs in bending of the lingual apex, is accomplished with two simultaneous adjustments: a decrease in length on one side and a resistance to change in diameter. If diameter change is not resisted, the hydrostat will shorten on one aspect but will not bend. Loss of function of the intrinsic muscles of the tongue may result in inadequate bending and, therefore, in imprecision in the production of lingual consonants. Function of the extrinsic tongue musculature is less impaired in ALS, as indicated by the preservation of the front versus back contrast for vowels.

3.2 RECONSIDERATION OF A MODEL FOR THE NEURAL CONTROL OF SPEECH

These neurological disorders can be considered in relation to models of movement control. One such model appeared initially in Allen and Tsukahara (1974) but has been modified or reprinted by several others, including Hirose (1982), Eccles (1982), Netsell (1982), Kubota (1984), and Barlow and Farley (1989). The current modification is shown in Fig. 2. The premotor and supplementary motor areas are thought to have a major role in the planning and programming of action. The motor program is defined in a system involving the thalamus, premotor cortex, supplementary motor area, lateral cerebellum and motor cortex.

Fig. 2 is, of course, a simplification of a highly complex and still poorly known process. A rather different picture can be drawn to reflect alternative recent conceptions of neural connections. What is depicted in Fig. 2 as a unitary basal ganglia-thalamocortical circuit probably should be presented as segmental circuits comprising at least a "motor" loop and an "association" loop, if not several such loops (De Long and Alexander, 1987). A parallel organization of such segregated

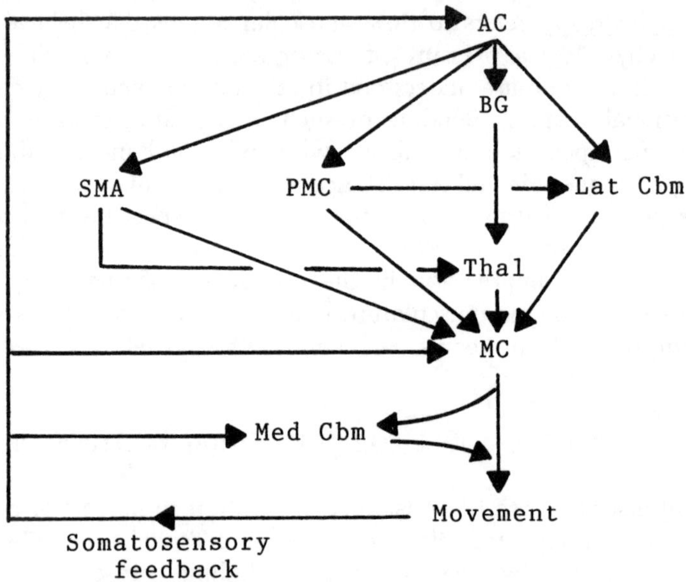

Fig. 2. Diagram of neural control of voluntary movement, adapted from Kubota (1984) and others. Abbreviations: AC, association cortex; BG, basal ganglia; SMA, supplementary motor area; PMC, premotor cortex; Lat Cbm, lateral cerebellum; Thal, thalamus; MC, motor cortex; Med Cbm, medial cerebellum.

circuits has been proposed by Alexander, De Long and Strick (1986). A simplified diagram of such an organization is given in Fig. 3. Alexander et al. (1986) and Miller and De Long (1988) proposed that a restricted portion of the striatum (principally the putamen) receives partially overlapping projections from functionally related areas of the cortex, specifically from motor cortex, premotor cortex, supplementary motor area, primary somatosensory cortex, and the superior parietal lobule. The somatotopic organization of these projections is maintained in the projections from the putamen to the globus pallidus and the substantia nigra. The motor loop is completed through projections to the oral portion of the ventrolateral thalamic nucleus, which in turn projects to the supplementary motor area.

The alternative model in Fig. 3 is compatible with Eccles' hypothesis that the supplementary motor areas (SMA) of both hemispheres are the site of the initial neuronal event in all voluntary movements. Fig. 3 shows more extensive connections between the SMA and other structures than does Fig. 2. As Eccles' conception relates to the ideas developed in the present paper, the SMA may be in a privileged position to initiate both major aspects of movement planning, viz., planning of sensory trajectories and the control of selection and seriation. Jonas (1981) concluded from a review of speech disturbances associated with lesions of the supplementary motor area that this region of the cortex has several roles in speech. Among these roles are: (1) facilitating the initiation of propositional speech; (2) supressing nonpropositional ("automatic") speech; (3) pacing of propositional and nonpropositional speech; (4) controlling some aspects of articulation; (5) controlling phonation. Interestingly, very similar functions have been attributed to the normal dominant hemisphere of the thalamus in language processing (Jonas, 1982).

Evidence favoring the pathway diagram of Fig. 3 is too extensive to be reviewed here, but some general features summarized by De Long and Alexander (1987) and Miller and De Long (1988) should be noted. First, studies of the putamen in primates have shown instruction-dependent changes in discharge. These changes appear to be related to "motor set" and are similar to changes recorded in motor cortex, premotor cortex and the supplementary motor area. Second, many movement-related neurons in the putamen respond to natural somatosensory stimulation. Third, many putamen neurons that discharge to active movements of the wrist also discharge to passive movements of the joint that are similar in amplitude and velocity. Fourth, neuronal activity in primates points to a role of the basal ganglia in controlling the direction of movement and the scaling of movement amplitude or velocity. Fifth, patients with Parkinson's disease often rely on visual guidance of limb movements, as though proproceptive signals were deficient. It is also becoming clear that the slow movements in Parkinson's disease may be related

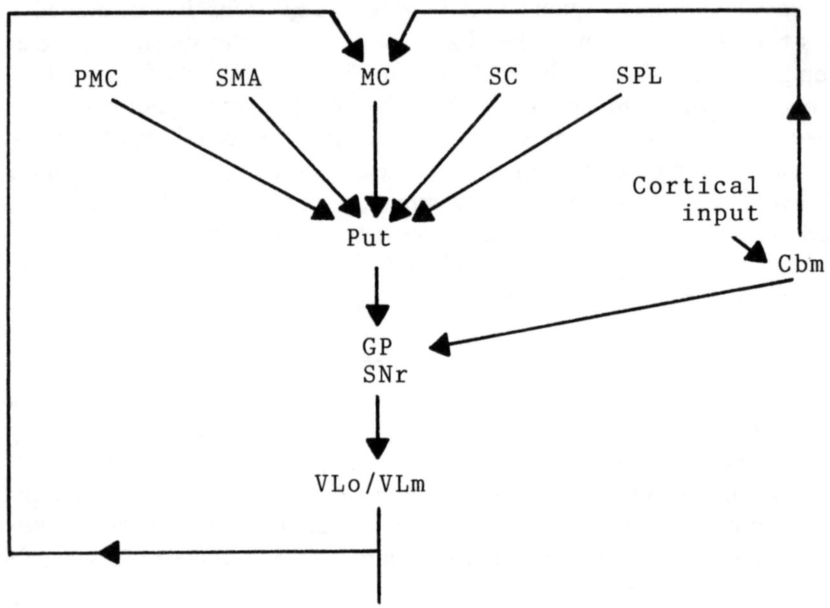

Fig. 3. Diagram of neural control of movement, based on various sources listed in text, especially Alexander, De Long and Strick (1986). Abbreviations: PMC, premotor cortex; SMA, supplementary motor area; MC, motor cortex; SC, somatosensory cortex; SPL, superior parietal lobule; Put, putamen; Cbm, cerebellum; GP, globus pallidus; SNr, substantia nigra pars reticularis; VLo, pars oralis of ventrolateral thalamus; VLm, pars medialis of ventrolateral thalamus.

to abnormal inputs to the motor cortex (Dick, Cowan, Day et al., 1984). This feature may implicate basal ganglia disturbances that produce deviant input to the cortex. Movement abnormalities in Parkinson's disease may be particularly evident for movements of large amplitude or complex movement sequences (Day, Rothwell and Marsden, 1987; Forrest, Weismer and Turner, 1989; Hallet and Khosbin, 1980; Sanes and Evarts, 1985).

As discussed earlier, cerebellar participation in acquired skilled movements appears to be primarily one of supplementing and revising motor instructions prepared elsewhere in the motor circuit. The same may not be true of motor skills that are being learned, in which case the cerebellum could contribute to essential aspects of movement preparation. Furthermore, the effects of cerebellar lesions depends on both the site and extent of the lesion of the cerebellum and the sensorimotor pathways that are affected by the cerebellar lesion and any accompanying lesions elsewhere in the CNS. Ito (1984) summarized cerebellar function in terms of three major principles: the control is open-loop, multivariable and based on adaptive learning. Normally, the cerebellum exercises a continual revisory control that imparts precision to provisional motor instructions issued by the motor cortex. The characteristics of ataxic dysarthria reflect the impairment of cerebellar revisory function.

Conduction aphasia is suggested to reflect disorders in both phonetic sequence management and the specification of sensory trajectories (particularly the auditory component). The lesion is thought to be cortical, affecting either the supramarginal-postcentral gyrus (center lesion) or the arcuate fasciculus (disconnection). The great difficulty in repetition in conduction aphasia may indicate a fragility in the associations between phonetic sequence and auditory representation.

Fig. 3 also is compatible with clinicoanatomic data on apraxia of speech. Although the neural lesion typically is thought to be prerolandic, in the vicinity of the third frontal convolution, subcortical lesions also have been associated with the disorder. Variability in lesion site is not surprising if one considers the pathways of control in Fig. 3. A lesion in several points along the "motor" loop might interfere with aspects of movement selection, sequencing or programming. Indeed, dyspractic disturbances might result from a variety of disruptions in sensorimotor processing. These disruptions include: deficiencies in sensory trajectories (especially important in developmental forms of apraxia), degradation of phonological or motoric representations of an intended utterance, and instabilities in the temporal phasing of movements in a sequence.

Many of the characteristics of apraxia of speech listed in Table 1 resemble characteristics of errors, especially sequencing errors, in normal speech production. To some degree, then, the errors in apraxic speech represent an exaggeration of the errors found in normal speech production, and, moreover, the errors are similarly affected by factors such as complexity, frequency of occurrence, and syllable position. The emerging picture of apraxia of speech is one of a combination of disturbances at various levels, minimally the phonological and motoric levels. The multi-level character of the errors might be explained by a spreading activation model of the kind described by Dell (1986). The activation in this model runs both downward (e.g., from morphemes to phonemes) and upward (e.g., from phonemes to morphemes) and therefore can result in complex loops of positive feedback. On the one hand, this property accounts for output biases that have been described in sequencing errors in normal speech, including lexical bias (the tendency to create existing morphemes or words), syllable bias (the tendency to create existing syllables), frequency bias (the tendency to select sounds in proportion to their frequency of occurrence in the language), contingent frequency bias (the tendency to select frequently occurring combinations of units). On the other hand, this property of the model could account for the complex character of errors in apraxia of speech (and perhaps also the errors in aphasic neologistic jargon or the errors in conduction aphasia). The spreading activation theory allows for simultaneous, interacting activity at various levels of word production. Accordingly, it can deal quite naturally with errors that occur at different levels.

Indeed, one possible point of theoretical convergence between information-processing models of utterance production and neural models of behavioral regulation is a parallel processing of redundant and complementary sources of information. For example, word retrieval for production may rely on semantic, phonological or articulatory information. The different sources of information may be essentially simultaneous in their availability. Similarly, the neural activity for tasks such as word production may be distributed over several sites in the brain, and the interconnections among these sites may facilitate neural processing. In neural feedback circuits, the identification of "origins" of activity is difficult, because each neuron is at once a source and receiver of information (Arduini, 1987). However, several lines of evidence point to the supplementary motor area as playing some kind of leading role in voluntary movement. Through its connections with association cortex, the supplementary motor area may initiate a "motor" loop that parallels activity in an "association" loop. The simultaneous activity in these loops insures the contextual determination of motor performance. For speech, this determination is an intended acoustic signal, hence, intelligible communication.

Although the idea of a series of processing stages has been popular in accounts of the production of spoken language, this kind of model does not seem particularly apt. A more suitable model is one that permits some degree of parallel processing among components. Spreading activation models satisfy this requirement.

Dell's (1986) spreading activation model is particularly interesting because it was tested against data for errors in normal speech. Generally, the model performed quite well. Furthermore, the model seems applicable to the kinds of errors that have been described for conduction aphasia, apraxia of speech and other neurologic speech and language disorders.

A modified version of a network for word retrieval and production is given in Fig. 4. For the purposes of this discussion, the word is taken as the superordinate node. The syllabic representation is defined in terms of an onset (an initial null consonant, single consonant or consonant cluster), nucleus (vowel or diphthong) and coda (a final null consonant, single consonant or consonant cluster). This syllable structure accounts for data from word games (Treiman, 1983) and speech errors (Dell, 1986). As Dell observed, the syllable may not be a unit of retrieval but rather a schema of retrieval. That is, syllable errors per se may be infrequent, but errors that reflect syllable structure are common. The syllable is defined in terms of its segmental constituents (phonemes in Fig. 4), which are in turn defined in terms of phonetic features. The phonetic features are associated with sensory trajectories, which are combined ("compounded" may be a better term) into a movement pattern.

A general assumption of the spreading activation theory is that representations of an utterance are constructed simultaneously at various levels, within the limits of coordinated information operating across levels (e.g. a lexical concept must be selected before its phonological structure can be assigned). A loose yoking across levels permits coordinated representations but allows higher levels of processing to stay ahead of lower ones. Dell treats activation as a real variable defined in terms of spreading (distribution of non-zero levels of activation to all connected nodes), summation (integration or addition of activation at each node), and decay (a passive loss of activation at each node). Bidirectional connections permit excitatory top-down and excitatory bottom-up signals.

In Dell's model, the slipping units that create speech errors are primarily phonemes, consonant clusters, and vowel- consonant units (rimes). Thus, syllables have a primary break between onset and rime and a secondary break between nucleus and coda. Speaking rate interacts with spreading activation such that a slow rate increases the amount of spreading activation. The increased activation increases the similarity between interacting phonemes. But because the decay rate is assumed to be constant, a slow rate also decreases the number of errors of misordering.

These characteristics apply reasonably well to the kinds of errors that occur in apraxia of speech and conduction aphasia. The relatively small proportion of sequencing errors in apraxic speech could be a consequence of the typical slow

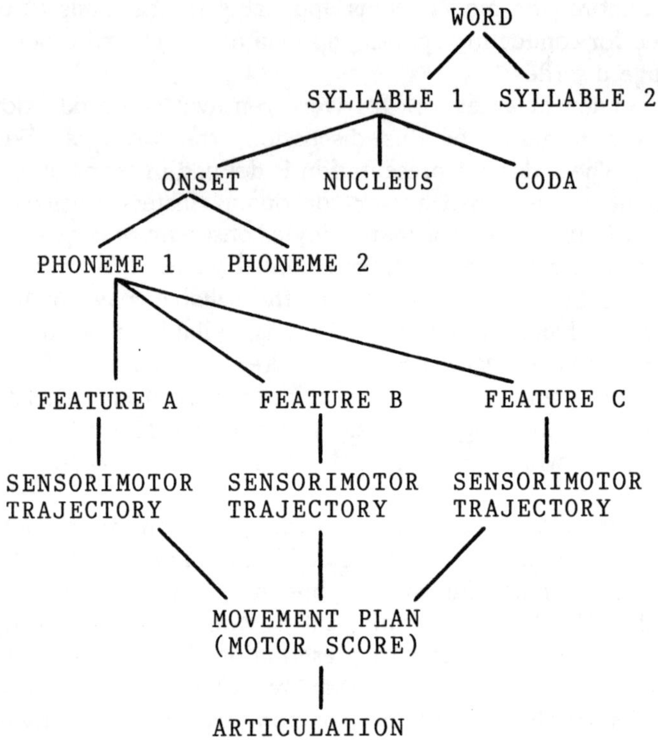

Fig. 4. A portion of a spreading activation network by which a word is encoded phonologically and motorically. Morphemes are expressed as syllables, which have a structure defined in terms of onset, nucleus and coda. The syllable portions are analyzed into phonemes, which have feature specifications. Features are transformed into sensorimotor (SM) trajectories, which are then compiled into a movement plan.

speaking rate. However, the slow rate also increases the opportunity for substitutions because of the spreading activation of features. Substitutions tend to involve one or two feature errors. Repeated phonemes tend to produce errors because of spreading activation between syllables that share phonemes. Apparently phonological and apparently motoric errors occur together in these disorders because of the simultaneously developing representations and their interconnections.

As applied to dysarthria, at least the typical dysarthria with slow speaking rate, the spreading activation model would predict that sequencing errors would be infrequent but that substitution errors would be rather likely. Of course, added to the substitution errors generated through spreading activation effects would be the abnormalities produced by the motoric deficit. In some disorders, such as cerebral palsy and Parkinson's disease, the deficit may be in part an inadequate specification of the sensory trajectories on which movements are based. In other disorders, such as amyotrophic lateral sclerosis, the sensory trajectory may be essentially intact, but the lesions of the upper and lower motoneurons impair the ability of the muscles to effect the intended action.

The theoretical frameworks summarized in this paper appear to have the potential to guide the study of both normal and disordered speech. A particularly important and timely goal is to consider the data on speech production--including normal correct performance, error sequences produced by normal talkers, and abnormal patterns produced by neurologically impaired talkers--in a single framework. Some compatible developments in theories of information processing, spreading activation and neural control augur such a unifying theory.

ACKNOWLEDGMENTS. This work was supported in part by Public Health Service Research grant from the National Institute on Deafness and Other Communication Disorders (NS22458).

REFERENCES

Allen, G.I., and Tsukahara, N. (1974). Cerebrocerebllar communication system. Physiol. Rev., **54**, 957-1006.

Alexander, G.E., DeLong, M.R., and Strick, P.L. (1986). Parallel organization of functionally segregated circuits liking basal ganglia and cortex. Ann. Rev. Neurosci., **9**, 357-381.

Arduini, A. (1987). Principles of Theoretical Neurophysiology. Berlin: Springer-Verlag.

Barlow, S.M., and Farley, G. R. (1989), Neurophysiology of speech. In D.P. Kuehn, M.L. Lemme and J. M. Baumgartner (Eds.), Neural Bases of Speech, Hearing, and Language. Boston: College-Hill.

Benson, D.F. (1967). Fluency in aphasia: Correlation with radioactive scan localization. Cortex, **3**, 373-394.

Bernstein, N.E. (1981). Are there constraints on childhood disfluency? J. Fluency Dis., **6**, 341-350.

Bladon, R.A.W., Henton, C.G., and Pickering, J.B. (1984). Towards an auditory theory of speaker normalization. Lang. Commun., **4**, 59-69.

Bloodstein, O. (1974). The rules of early stuttering. J. Speech Hearing Dis., **39**, 379-394.

Bloom, L., and Beckwith, R. (in press). Talking with feeling: Integrating affective and linguistic expression in early language development. To appear in Cognition and Emotion.

Blumstein, S.E., and Baum, S. (1987). Consonant production deficits in aphasia. In J.H. Ryalls (Ed.), Phonetic Approaches to Speech Production in Aphasia and Related Disorders, pp. 3-21. Boston: College-Hill Press.

Blumstien, S.E., Cooper, W.E., Goodglass, H., Statlender, S., and Gottlieb, J. (1980). Production deficits in aphasia: A voice-onset-time analysis. Brain Lang., **9**, 153-170.

Bock, J.K. (1982). Toward a cognitive psychology of syntax: Information processing contributions to sentence formulation. Psychol. Rev., **89**, 1-47.

Browman, C., and Goldstein, L. (1986). Towards an articulatory phonology. Phonology Yearbook, **3**, 219-252.

Brown, J.R., Darley, F.L., and Aronson, A.E. (1970). Ataxic dysarthria. Intern. J. Neurol., **7**, 302-318.

Brown, R., and McNeill, D. (1966). The "tip of the tongue" phenomenon. J. Verb. Learn. Verb. Behav., **5**, 325-327.

Buckingham, H.W., Avakian-Whitaker, H. and Whitaker, H.A. (1978). Alliteration and assonance in neologistic jargon aphasia. Cortex, **14**, 365-380.

Butterworth, B. (1979). Hesitation and the production of verbal paraphasias and neologisms in jargon aphasia. Brain Lang., **8**, 133-161.

Carrow, E., Rivera, V., Mauldin, M., and Shamblin, L. (1974). Deviant speech characteristics in motor neuron disease. Arch. Otolaryngol., **100**, 212-218.

Chafe, W.L. (1977). Creativity in verbalization and its implications for the nature of stored knowledge. In R.O. Freedle (Ed.), Discourse Production and Comprehension (Vol. 1). Norwood, NJ: Ablex.

Cheng, P. W. (1985). Restructuring versus automaticity: Alternative accounts of skill acquisition. Psychol. Review. **92**, 414-423.

Darley, F.L. (1982). Aphasia. New York: Saunders.

Darley, F.L., Aronson, A.E., and Brown, J.R. (1969a). Differential diagnostic patterns of dysarthria. Journal Speech Hearing Res., **12**, 246-269.

Darley, F.L., Aronson, A.E., and Brown, J.R. (1969b). Clusters of deviant speech dimensions in the dysarthrias. J. Speech Hearing Res., **12**, 462-496.

Darley, F., Aronson, A.E., and Brown, J.R. (1975). Motor Speech Disorders. Philadelphia: Saunders.

Day, B.L., Rothwell, J.C., and Marsden, C.D. (1987). Facets of akinesia in Parkinson's disease. In A. Struppler and A. Weindl (Eds.), Clinical Aspects of Sensory Motor Integration; pp. 212-215. Berlin: Springer-Verlag.

Dell, G.S. (1976). A spreading-activation theory of retrieval in sentence production. Psychol. Rev., **93**, 283-321.

DeLong, M.R., and Alexander, G.E. (1987). The basal ganglia and sensorimotor integration. In A. Struppler and A. Weindl (Eds.), Clinical Aspects of Sensory Motor Integration, pp. 203-211. Berlin: Springer-Verlag.

Dick, J.P.R., Cowan, J.M.A., Day, B.L., Berardelli, A., Kachi, T., Rothwell, J.C., and Marsden, C.D. (1984). The cortico-motoneurone connexion is normal in Parkinson's disease. Nature, **310**, 407-409.

Dubner, R., Sessle, B.J., and Storey, A.T. (1978). The Neural Basis of Oral and Facial Function. New York: Plenum.

Eccles, J.C. (1982). The initiation of voluntary movements by the supplementary motor area. Arch. Psychiat. Nervenkr., **231**, 423-441.

Edwards, S., and Miller, N. (1989). Using EPG to investigate speech errors and motor agility in a dyspraxic patient. Clin. Ling. Phon., **3**, 111-126.

Faurion, A. (1987). Physiology of the sweet taste. In H. Autrum, D. Ottoson, E.R. Perl, R.F. Schmidt, H. Shimazu and W.D. Willis (Eds.), Progress in Sensory Physiology (Vol. 8); pp. 130-201. Berlin: Springer-Verlag.

Ford, M., and Holmes, V.M. (1978). Planning units and syntax in sentence production. Cognition, **6**, 35-53.

Forrest, K., Weismer, G., and Turner, G.S. (1987). Kinematic, acoustic, and perceptual analyses of connected speech produced by Parkinsonian and normal geriatric adults. J. Acoust. Soc. Am., **85**, 2608-2622.

Friedrick, F.J., Glenn, C.G., and Marin, O.S.M. (1984). Interruption of phonological coding in conduction aphasia. Brain Lang., **22**, 266-291.

Garrett, M.F. (1980). Levels of processing in sentence production. In B. Butterworth (Ed.), Language Production (Vol. 1); pp. 177-220. London: Academic Press.

Gawel, M.J. (1981). The effects of various drugs on speech. Brit. J. Dis. Commun. **16**, 51-57.

Gee, J.P., and Grosjean, F. (1983). Performance structures: a psycholinguistic and linguistic appraisal. Cognitive Psychology, **15**, 411-458.

Geschwind, N. (1965). Disconnection syndromes in animals and man. Brain, **88**, 237-274, 585-644.

Geschwind, N. (1966). Discussion to paper by Wepman, J.M., and Jones, L.V. in E.C. Carterette (Ed.), Brain Funcion (Vol. III), pp. 156-171. Berkeley and Los Angeles, CA: University of California Press.

Goodglass, H. (1983). Word retrieval for production. In M. Studdert-Kennedy (Ed.), Psychobiology of Language; pp. 129-138. Cambridge, MA: MIT Press.

Hallett, M., and Khosbin, S. (1980). A physiological mechanism of bradykinesia. Brain, **103**, 301-314.

Hanson, D.G., Gerratt, B.R., and Ward, P.H. (1984). Cinegraphic observations of laryngeal function in Parkinson's disease. Laryngoscope, **94**, 348-353.

Hardcastle, W.J. (1987). Electropalatographic study of articulation disorders in verbal dyspraxia. In J.H. Ryalls (Ed.), Phonetic Approaches to Speech Producion in Aphasia and Related Disorders, pp. 113-136. Boston: College-Hill Press.

Hirose, H. (1986). Pathophysiology of motor speech disorders. Folia Phoniat. **38**, 61-88.

Howes, D. (1964). Application of the word-frequency concept to aphasia. In A.V.S. de Reuck and M. O'Connor (Eds.), Disorders of Language. London: J. and A. Churchill.

Howes, D., and Geschwind, N. (1964). Quantitative studies of aphasic language. Ass. Res. Nerv. Ment. Dis., **42**, 229-244.

Huang, C.-S., Hiraba, H., and Sessle, B.J. (1989). Input-output relationships of the primary face motor cortex in the monkey (Macaca fascicularis). J. Neurophysiol., **61**, 350-362.

Ito, M. (1970). Neurophysiological aspects of the cerebellar motor control system. Int. J. Neurol., **7**, 162-176.

Ito, M. (1984). The Cerebellum and Neural Control. New York:Raven.

Itoh, M., and Sasanuma, S. (1987). Articulatory velocities of aphasic patients. In J.H. Ryalls (Ed.), Phonetic Approaches to Speech Production in Aphasia and Related Disorders; pp. 137-161. Boston: College-Hill Press.

Jonas, S. (1981). The supplementary motor area and speech emission. J. Commun. Dis., **14**, 349-373.

Jonas, S. (1982). The thalamus and aphasia, including transcortical aphasia: A review. J. Commun. Dis., **15**, 31-41.

Keller, E. (1987). Factors underlying tongue articulation in speech. J. Speech Hearing Res., **30**, 223-229.

Kelso, J.A.S., and Tuller, B. (1983). "Compensatory articulation" under conditions of reduced afferent information: A dynamic formulation. J. Speech Hearing Res., **26**, 217-224.

Kempen, G., and Huijbers, P. (1983). The lexicalization process in sentence production and naming: Indirect election of words. Cognit., **14**, 185-209.

Kent, R.D. (1984). Brain mechanisms of speech and language with special reference to emotional interactions. In R.C. Naremore (Ed.), Language Science: Recent Advances; pp. 282-334. San Diego, CA: College-Hill Press.

Kent, R.D., and Adams, S.G. (in press). The concept and measurement of coordination in speech disorders. In S.A. Wallace (Ed.), Perspectives on the Coordination of Movement. Amsterdam: North Holland.

Kent, R.D., and Forner, L.L. (1980). Speech segment durations in sentence recitations by children and adults. J. Phonetics, **8**, 157-168.

Kent, R.D., Kent, J.F., Weismer, G., Martin, R.E., Sufit, R.L., Rosenbek, J.C., and Brooks, B.R. (submitted). Relationships between speech intelligibility and the slope of second-formant transitions in dysarthric subjects.

Kent, R.D., Kent, J.F., Weismer, G., Sufit, R.L., Rosenbek, J.C., Martin, R.E., and Brooks, B.R. (submitted). Impairment of speech intelligibility in amyotrophic lateral sclerosis.

Kent, R.D., Martin, R.E., and Sufit, R.L. (in press). Oral sensation: A review and clinical prospective. To appear in H. Winitz (Ed.), Human Communication and Its Disorders.

Kent, R.D., and McNeil, M.R. (1987). Relative timing of sentence repetition in apraxia of speech and conduction aphasia. In J.R. Ryalls (Ed.), Phonetic Approaches to Speech Production in Aphasia and Related Disorders; pp. 181-220. Boston: College-Hill Press.

Kent, R.D., and Netsell, R. (1975). A case study of an ataxic dysarthric: Cineradiographic and spectrographic observations. J. Speech Hearing Dis., **1**, 115-134.

Kent, R.D. and Netsell, R. (1979). Acoustic-phonetic features of Parkinsonian dysarthria. Paper presented at the Annual Convention of the American Speech-Language-Hearing Associaiton, November 16-19, Atlanta, Georgia.

Kent, R.D., Netsell, R., and Abbs, J.H. (1979). Acoustic characteristics of dysarthria associated with cerebellar disease. J. Speech Hearing Res., **26**, 231-249.

Kent, R.D., and Rosenbek, J.C. (1983). Acoustic patterns of apraxia of speech. J. Speech Hearing. Res., **26**, 231-249.

Kerschensteiner, M., Poeck, K., and Brunner, E. (1972). The fluency-non fluency dimension in the classification of aphasic speech. Cortex, **8**, 233-247.

Kier, W.M., and Smith, K.K. (1985). Tongues, tentacles, and trunks: The biomechanics of movement in muscular-hydrostats. Zool. J. Linn. Soc, **83**, 307-324.

Kohn, S.E. (1984). The nature of the phonological disorder in conduction aphasia. Brain Lang., **23**, 97-115.

Kubota, K. (1984). An introduction to "voluntary movement and the brain. Adv. Neurol. Sci, **28**, 3.

Leanderson, R., Meyerson, B.A., and Persson, A. (1972). Lip muscle function in parkinsonian dysarthria. Acta Otolaryngol., **74**, 271-278.

Levelt, W.J.M. (1983). Monitoring and self-repair in speech. Cognition, **14**, 41-104.

Lichtenstein, E.H., and Brewer, W.F. (1980). Memory for goal-directed events. Cognit. Psychol., **12**, 412-445.

Lidsky, T.I., Manetto, C., and Schneider, J.S. (1985). A consideration of sensory factors involved in motor functions of the basal ganglia. Brain Res. Rev., **9**, 133-146.

Lindsley, J.R. (1975). Producing simple utterances: How far ahead do we plan? Cognit. Psychol., **7**, 1-19.

Lindsley, J.R. (1976). Producing simple utterances: Details of the planning process. J. Psycholing. Res., **5**, 331-354.

Logemann, J., Fisher, H.G., Boshes, B., and Blonsky, E.R. (1978). Frequency and concurrence of vocal tract dysfunctions in the speech of a large sample of Parkinson patients. J. Speech Hearing Dis., **42**, 47-57.

McCarthy, R., and Warrington, E.K. (1984). A two-route model of speech production: Evidence from aphasia. Brain, **107**, 463-485.

McNeil, M.R., Liss, J.M., Chin-Hsing, T., and Kent, R..D. (submitted). Effects of speech rate on the absolute and relative timing of apraxic and conduction sentence production.

Miller, W.C., and De Long, M.R. (1988). Parkinsonian symptomatology: An anatomical and physiological analysis. In J.J. Joseph (Ed.), Central Determinants of Age-Related Declines in Motor Function. Ann. New York Acad. Sci., **515**, 287-302.

Mrayati, M., Carre, R., and Guerin, B. (1988). Distinctive regions and modes: A new theory of speech production. Speech Commun., **7**, 257-286.

Munhall, K. (1989). In P. Square-Storer (Ed.), Acquired Apraxia of Speech in Aphasic Adults: Theoretical and Clinical Issues. London: Taylor and Francis.

Neilson, M.D., and Neilson, P.D. (1987). Speech motor control and stuttering: A computational model of adaptive sensory-motor processing. Speech Commun., **6**, 325-333.

Netsell, R. (1982). Speech control and selected neurologic disorders. In S. Grillner, B. Lindblom, J. Lubker and A. Persson (Eds.), Speech Motor Control. New York: Pergamon Press.

Netsell, R., Daniel, B., and Celesia, G. (1975). Acceleration and weakness in parkinsonian dysarthria. J. Speech Hearing Dis., **40**, 170-178.

Netsell, R., and Kent, R.D. (1976). Paroxysmal ataxic dysarthria. J. Speech Hearing Dis., **41**, 93-109.

O'Connell, P.F. (1981). Neologistic jargon aphasia: A case report. Brain Lang., **12**, 292-302.

Perkell, J.S. (1980). Phonetic features and the physiology of speech production. In B. Butterworth (Ed.), Language Production (Vol. 1); pp. 337-372. New York: Academic Press.

Perkins, W.H., Curlee, R.F., Kent, R.D. and Hixon, T.J. (submitted). A theory of neural function in stuttered speech.

Ploog, D. (Ed.) (1987), Annual Report of the Max Planck Institute for Psychiatry;; pp. 70-71. Munich: Max-Planck-Institut fur Psychiatrie.

Picard, C.O., and Olivier, A. (1983). Sensory cortical tongue. Representation in man. J. Neurosurg., **59**, 781-789.

Putnam, A.H.B. (1988). Review of research in dysarthria. In H. Winitz (Ed.), Human Communication and Its Disorders, A Review, 1988; pp. 107-223. Norwood, NJ: Ablex Publishing Corp.

Rogers, J. (1988). The neurobiology of cerebellar senescence. In J.A. Joseph (Ed.), Central Determinants of Age-Related Declines in Motor Function. Ann. New York Acad. Sci., **515**, 251-268.

Rosenbek, J.C., Kent, R.D., and LaPointe, L.L. (1983). Apraxia of speech: An overview and some perspectives. In J.C. Rosenbek, M.R. McNeil and A. Aronson (Eds.), Apraxia of Speech: Physiology, Acoustics, Linguistics and Management. San Diego, CA: College-Hill Press.

Sanes, J.N., and Evarts, E.V. (1985). Psychomotor performance in Parkinson's disease. In P.J. Delwaide and A. Agnoli (Eds.), Clinical Neurophysiology in Parkinsonism, pp. 117-132. New York: Elsevier.

Schneider, J.S., Diamond, S.G., and Markham, C.H. (1986). Deficits in orofacial sensorimotor function in Parkinson's Disease. Ann. Neurol., **19**, 275-282.

Schneider, W., and Shiffrin, R.M. (1977). Automatic and controlled information processing in vision. In D. LaBerge and S.J. Samuels (Eds.), Basic Processes in Reading: Perception and Comprehension. Hillsdale, NJ: Erlbaum.

Shallice, T., and Warrington, E.K. (1977). Auditory-verbal short-term memory impairment and conduction aphasia. Brain Lang., **4**, 479-491.

Shawker. T.H., and Sonies, B.C. (1984). Tongue movement during speech: A real time ultrasound evaluation. J. Clin. Ultrasound, **12**, 25-33.

Shiffrin, R.M., and Schneider, W. (1977). Controlled and automatic human information processing: II. Perceptual learning, automatic attending, and a general theory. Psychol. Rev., **8**, 127-190.

Shuttuck-Hufnagel, S. (1979). Speech errors as evidence for a serial-ordering mechanism in sentence production. In W. Cooper and E. Walker (Eds.), Sentence Processing. Hillsdale, NJ: Erlbaum.

Smith, K.K., and Kier, W.M. (1989). Trunks, tongues, and tentacles: Moving with skeletons of muscle. Am. Scientist, **77**, 29-35.

Stevens, K.N. (1972). The quantal nature of speech: Evidence from articulatory-acoustic data. In E.E. David and P.B. Denes (Eds.), Human Communication: A Unified View. New York: McGraw-Hill.

Treiman, R. (1983). The structure of spoken syllables: Evidence from novel word games. Cognit., **15**, 49-74.

Vandamme, F. (1987). Emotion and Cognition. Commun. Cognit., **20**, 131-140.

Warrington, E.K., Logue, V., and Pratt, R.T.C. (1971). The anatomical localization of selective impairment of auditory verbal short term memory. Neuropsychologia, **9**, 377-387.

Warrington, E.K., and Shallice, T. (1972). Neuropsychological evidence of visual storage in short-term memory tasks. Quart. J. Exp. Psychol., **24**, 30-40.

Wieneke, G., Jannsen, P., and Belderbos, H. (1987). The influence of speaking rate on the duration of jaw movements. J. Phonetics, **15**, 111-126.

Yamadori, A., and Ikumura, G. (1975). Central (or conduction) aphasia in a Japanese patient. Cortex, **11**, 73-82.

Ziegler, W., and von Cramon, D. (1986). Timing deficits in apraxia of speech. Eur. Arch. Psychiat. Neurol. Sci., **236**, 44-49.

EXPLAINING PHONETIC VARIATION: A SKETCH OF THE H&H THEORY

B LINDBLOM
Departments of Linguistics
University of Texas at Austin
Austin 78712-1196 Texas and
University of Stockholm
Stockholm S-10691 Sweden

ABSTRACT. The H&H theory is developed from evidence showing that speaking and listening are shaped by biologically general processes. Speech production is adaptive. Speakers can, and typically do, tune their performance according to communicative and situational demands, controlling the interplay between production-oriented factors on the one hand, and output-oriented constraints on the other. For the ideal speaker, H&H claims that such adaptations reflect his tacit awareness of the listener's access to sources of information independent of the signal and his judgement of the short-term demands for explicit signal information. Hence speakers are expected to vary their output along a continuum of **hyper-** and **hypospeech**. The theory suggests that the lack of invariance that speech signals commonly exhibit (Perkell and Klatt 1986) is a direct consequence of this adaptive organization (cf MacNeilage 1970). Accordingly, in the H&H program the quest for phonetic invariance is replaced by another research task: Explicating the notion of **sufficient discriminability** and defining the class of speech signals that meet that criterion.

1. Introduction

1.1 EXPLAINING PHONETIC VARIATION

One of the classical topics in phonetic theory is speech sounds in context. The modifications that phonetic segments undergo in the speech of a single individual are known to be extensive and to give rise to what is called the invariance problem: The difficulty of giving a physical phonetic definition of a given linguistic category that is constant and always free of context.

W. J. Hardcastle and A. Marchal (eds.), Speech Production and Speech Modelling, 403–439.

The main theme of the present chapter is: How do we make sense of this pervasive intra-speaker phonetic variation? My account is not new. It uses many well-known facts and traditional ideas. It may nevertheless differ in significant ways from answers currently obtainable from competing frameworks such as the Motor Theory, the Quantal Theory, the Direct Realism and Action Theory approaches, and 'Icebergs' (Fujimura 1989) just to mention a few (see Section 5.1).1)

The ultimate goal of the H&H program is to work out a quantitative and testable theory. At present I shall offer no more than a verbal sketch with a few hints about quantification. For short, I shall refer to it as the H&H theory, H&H referring to 'hyper'- and 'hypo'-articulation.

1.2 THE H&H ARGUMENT

To anticipate where the present review will lead us, I shall state the main argument of the paper right away.

We start out from two basic observations about perception and two about production:

a. Speech perception involves **discrimination** among items stored in the listener's lexicon. Lexical access is thus a function of the distinctiveness (rather than invariance) of the acoustic stimulus.

b. The process of discrimination is facilitated by processes not in the signal and whose contributions show short-term variations. Accordingly lexical access is assumed to be driven also by 'knowledge', that is by **signal-complementary processes.**

c. Speech motor control is future-oriented. It is purpose-driven and prospectively organized. Key words: plasticity and **output-oriented** control. Cf 'hyperspeech'.

d. As the output constraints on a movement become less severe, it tends to default to some low-cost form of behavior. Key words: economy and **system-oriented** control. Cf 'hypospeech'.

Combining those four observations we deduce that:

---

1) The H&H theory has grown out of my collaboration with Peter MacNeilage and Michael Studdert-Kennedy (Lindblom and MacNeilage 1986, Lindblom, MacNeilage and Studdert-Kennedy in preparation). Since language and speech are products of evolution, a minimum requirement on any theorizing in our area is that it be evolutionarily plausible. That means that theories of speech processes and sound patterns must be consistent with current Neo-Darwinian models of biological and cultural variation and selection. In line with such goals, H&H theory makes an attempt to address the systematic nature of intra-speaker phonetic variation.

e. The amount of explicit signal information minimally required for successful lexical access will vary between and within utterances. Cf the predictability of the word "nine" in the following examples (Lieberman 1963):
(i) The next word is ____ .
(ii) A stitch in time saves _____ .
f. In the **ideal case**, the speaker estimates the running contribution that signal-complementary processes will make during the course of an utterance, and dynamically tunes the production of its elements to the short-term demands for either output-oriented control (hyperspeech) or system-oriented control (hypospeech). What he/she needs to control is - not that linguistic units are actualized in terms of **physical invariants** (higher-order or whatever) - but that their signal attributes possess **sufficient contrast**, that is discriminative power that is sufficient for lexical access.
These considerations suggest the H&H working hypothesis:
g. During speech development and adult language use, real speakers adapt to the above-mentioned conditions. Speakers develop a 'feel' for the 'survival value' of phonetic forms through a process not unlike **natural selection.** Hence their speech will tend to vary systematically along an H&H dimension.

2. What Do Listeners Do?

2.1 SIGNAL-DERIVED INFORMATION

In this section a short, selective review will be presented of research on the initial stages of speech signal processing.

A recent theme issue of the Journal of Phonetics (Greenberg 1988) provides a useful overview of what is currently known about the auditory representation of speech signals. It exemplifies the portrayal of signals up to the level of the ventral cochlear nucleus, a projection site of the auditory nerve. It appears that "the cochlea acts as a faithful translator of incoming waveforms" (Geisler 1988:34). However, certain stimulus features are enhanced. And there is evidence for onset detectors, fundamental frequency tracking mechanisms and automatic-gain-control properties of the transduction from the inner hair cells to nerve fibers. Moreover, frequency analysis is non-linear in a way that makes spectral information resistant to noise. With particular interest we note that at these early stages of auditory processing there does not yet seem to be any sorting of stimuli into phonetic categories (Geisler, p 34), no processing in any way special to speech.

Great importance is generally attributed to the **dynamic aspects** of speech signals (Bladon 1985, Remez, Rubin, Pisoni and Carrell 1981, Strange 1989). Note though the experimental work indicating that, under certain conditions, dynamic events may cause blurring rather than enhancement of stimulus cues (Lacerda 1987). Despite widespread agreement on the role of dynamic attributes, the fact is that we do not yet have a great deal of information on precisely what the relevant dynamic aspects might be. Few speech perception models and speech recognition algorithms make explicit proposals about specific dynamic parameters. Klatt (1987) mentions a study by Rabiner (1987) who found that a smoothed estimate of spectral change had a dramatic effet on the performance of many speech recognition systems. Error scores were approximately halved.

A mechanism that derives invariants from the signal was recently presented by Miller (1989) who proposes an F0-dependent version of the so-called **formant-ratio-theory** to account for cross-speaker vowel quality constancy effects. Invariant quality is a function of F0 and fixed logarithmic distances between formants.

Fowler (1986:4) states that "...perception must be direct and, in particular, unmediated by cognitive processes of inference or hypothesis testing, which introduce the possibility of error." This view reflects her attempt to apply the program of **direct perception** (Gibson 1972, 1979) to speech. The phenomenon of cross-speaker vowel color constancy that Miller addresses appears compatible with a direct realist account. Leaving complications such as "non-uniform scaling" (Fant 1973) aside for the moment, we could see Miller's model as an explicit specification of a mechanism that extracts what Gibsonians would call higher-order invariants directly from the signal.

Mechanisms of a similar sort form part of the **feature-based lexical access** scheme proposed by Stevens (1986). Here relationally defined stimulus attributes, offering more stability and invariance than the raw spectral data, are assumed to be extracted from the signal.

The phenomenon of durational contrast - invoked by Diehl and co-workers in developing a **theory of auditory enhancement** (Diehl and Kluender 1989; Diehl, Kluender, Walsh and Parker in press) - highlights an interesting aspect of auditory processing. In one of their experiments they address the widely observed interplay between closure duration and preceding vowel duration as a cue for signaling voicing contrasts in medial stops (Kluender, Diehl and Wright 1988). Two series of VCV-stimuli were used each varying only in the duration of the first vowel: The first series was derived by editing a natural /apa/. It ranged perceptually from /aba/ to /apa/. The second series replaced the vowels by square-wave segments. Although their temporal and amplitude-envelope characteristics were those of the speech stimuli, these stimuli were not at all speech-like. For both series the task was to classify a given stimulus as one of the end-point stimuli. When plotted as a function of the duration of the medial silent interval, the results showed that a long initial vowel duration produced more /b/ responses than the short initial vowel condition. Significantly, a long initial square-wave segment analogously favored more short-gap responses. In other words, the longer initial segment durations made the silent gaps sound shorter in both cases. The authors see their parallel speech and non-speech results as arising from durational contrast, a general signal-processing phenomenon, and suggest that this effect explains the voicing-dependent vowel duration effect traditionally attributed to production factors.

In a recent formulation of the **Motor Theory** (MT), Liberman and Mattingly (1985:7) state: " ... adaptations of the motor system for controlling the organs of the vocal tract took precedence in the evolution of speech. These adaptations made it possible, not only to produce phonetic gestures, but also to coarticulate them so that they could be produced rapidly. A perceiving system specialized to take account of the complex acoustic consequences, developed concomitantly." How this specialized mechanism works remains to be specified. Nevertheless, the current version of MT assumes that phonetic invariance is gestural and counts on its being derivable from the signal: "...the invariant source of the phonetic percept is somewhere in the processes by which the sounds of speech are produced" (p 21).

2.2 SIGNAL-COMPLEMENTARY PROCESSES

Studying speech signal processing physiologically, psycho-acoustically or by numerical modeling, will be highly relevant to obtain more sophisticated formulations of the invariance problem. How far a better understanding of signal-derived processes will take us towards a solution of that problem is not known. Assuming, as the direct realists do, that "the information is in the signal" is a methodologically important move since it forces the investigator to squeeze all he/she can out of the stimulus (Ohala 1986).

Nevertheless, there is a great deal of evidence indicating that speech percepts are also influenced by the current dynamic state of the processing system, that is by signal-complementary processes.2) That conclusion is sometimes vividly illustrated in introductory classes on auditory analysis and phonetic transcription. When students know a language and are thoroughly familiar with the transcription conventions they show a high degree of interperson agreement. The segments and their phonetic values are by and large "heard correctly" (=according to the conventions taught). However, as soon as the speech sample comes from an unknown language subjects tend to differ more widely both with respect to segmentation and quality judgements. Interestingly, the transcription errors generally make good sense from a physical-acoustic viewpoint and in terms of phonetic similarity.

A similar situation arises when the trained phonetician compares his own transcriptions with spectrograms. Although he knows the language and has considerable experience from listening analytically, he is nevertheless continually surprised by the spectrographic patterns that typically show omissions and contextual modifications in excess of his expectations.

What is going on here? It seems that, if we know a certain language, we cannot help imposing that knowledge on the signal. Physically ambiguous information is disambiguated and incomplete stimulus information is restored. It appears as if the signal-complementary processes modulate the input and shape the percept in a most tangible way. And the process is highly automatic.

Introspectively, there is something utterly compelling about it. As listeners we do not in any way have the impression that disambiguation and perceptual filling in are products of "cognitive processes of inference or hypothesis testing". The compelling nature of the process is illustrated by examples of the following sort: 1. What is your homework assignment? - 2. Less**on** five. - 3. How many came? - 4. Less **than** five. Although the pronunciation of 2 and 4 may both contain physically identical **less'n**, there is no ambiguity. In either case listeners perceive the ambiguous syllabic nasal as intended and are unaware of any signal overlap.

We here enter a huge literature including topics such as intelligibility of filtered speech in noise, effect of increasing access to signal context, perceiving continuity in pulsating signals in noise, phonemic restoration and fluent restoration in shadowing. There is also the phenomenon of **amodal completion** as when subjects are asked to identify an utterance on the basis of visual and reduced

---

2) Cf Lashley (1951:112): "... the input is never into a quiescent or static system, but always into a system which is already actively excited and organized." Also Miller, Galanter and Pribram (1960).

auditory cues. Consider experiments using acoustic stimuli generated from a natural utterance by intensity- and F0-modulation of a steady-state complex buzz spectrum. When these stimuli are presented in synchrony with the video recording of the speaker some of the segments reported are demonstrably physically absent (Risberg 1979).

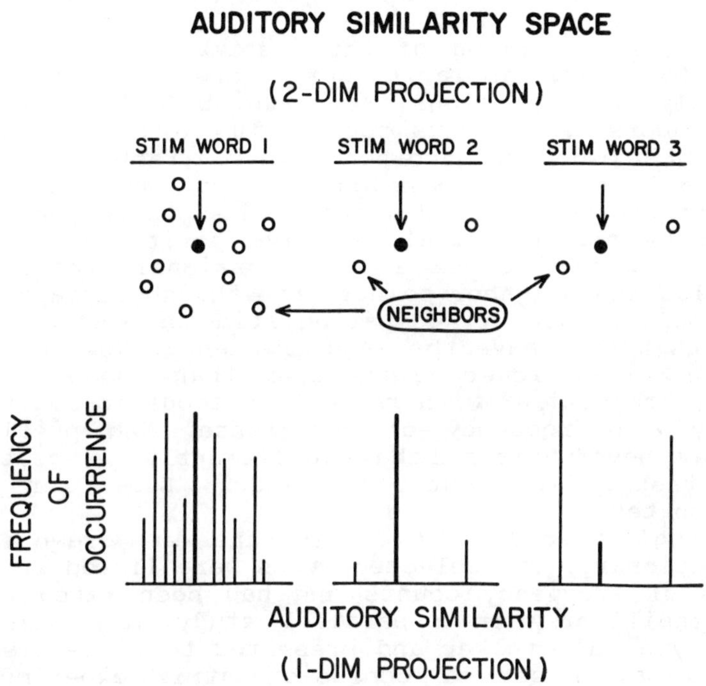

Figure 1. Auditory similarity and frequency of occurrence as determinants of word recognition. Adapted from Luce (1986).

It is impossible to do justice to all the experimental paradigms that elucidate the nature of signal-complementary processes. Let us focus on a particular topic: the **word frequency effect** in auditory word recognition. A recent thesis by Luce (1986) takes a new look at this phenomenon. A model, the **Neighborhood Activation Model** (NAM) is developed and tested. The literature indicates that high-frequency words exhibit a number of processing advantages. The higher the frequency of a word the higher its probability of being correctly recognized. That is true in general but NAM predicts departures from this basic rule. It views word identification as the task of discriminating among lexical items in memory and makes this discrimination a function of the number and nature of the elements that the input signal activates. The recognition process is assumed to be influenced not only by the frequency of the stimulus word itself but also by how auditorily similar it is to other words in the "neighborhood", by how many such neighbors it has and by what the frequencies of these neighbors are. Figure 1 is our attempt to capture those ideas.

At the top the locations of three stimulus words in a hypothetical "auditory similarity space" are indicated. For simplicity only two dimensions are drawn. Unfilled circles refer to neighbors, solid dots to stimulus words. Stimulus word 1 is in a crowded neighborhood. In contrast, stimulus words 2 and 3 have only two neighbors each. The auditory similarity structures of 2 and 3 are assumed to be the same as indicated by the identical relative positions of the points. At the bottom we see a one-dimensional projection of the stimulus and neighbor points onto the abscissas. The ordinate is word frequency. Comparing stimulus words 2 and 3 we notice that they have the same frequency. However the neighbors of 2 have lower frequencies than those of 3. Although they are similar with respect to acoustic-auditory confusability and frequency of occurrence, NAM predicts that they may nevertheless behave differently on certain processing tasks. To test the model Luce ran an identification test.

About 900 English words of CVC structure were drawn from Webster's dictionary. The selected items were listed in the Brown corpus of frequency counts and had been rated with respect to familiarity in a parallel study. The stimuli were spoken by a male talker and presented to listeners at three S/N-ratios. A set of confusion-matrix experiments were also carried out. One test contained the CV sequences formed by by combining 23 initial consonants (including a null consonant) with 15 vowels. Another set was produced from 15 vowels and 22 final consonants (again including a null element). These tests were run at the same three S/N-ratios.

NAM defines the probability of correctly recognizing a stimulus word in the context of its neighbors as:

$$p=S/(S+\Sigma N) \qquad (1)$$

where S is the product of the discriminability and frequency of the stimulus word. ΣN is the sum of terms associated with individual neighbors, each computed in a manner analogous with the calculation of S, that is an individual N is the product of the confusability and frequency of a given neighbor. The discriminability of a given word is derived from the confusion matrices. The probability of correctly recognizing the word "beat", from its acoustic waveform only, is computed as the product of three terms all available from the confusion-experiment data: The probability of responding by /b/ when presented with /b/, the probability of responding by /i/ when presented with /i/ and the probability of responding by /t/ when presented with /t/. Similarly, the confusability of of a neighbor, say "meat", is the product of the three terms: probability of responding by /m/ when presented with /b/, the probability of responding by /i/ when presented with /i/ and the probability of responding by /t/ when presented with /t/.

Let us examine Eq (1), as Luce did, by considering four cases obtained by systematically combining a high and a low value of S with a high and a low value of ΣN. A word with a highly distinctive acoustic pattern and with a high frequency would make S large. A high value of ΣN would be observed in a case where there are many neighbors that could easily be confused with the stimulus word and whose frequencies are high. On the other hand, a low frequency and a high confusability with neighbors makes S low. A low value of ΣN would result when the stimulus word has few neighbors whose frequencies and similarity scores are low.

TABLE I Testing the NAM model. A comparison of identification performance (%) and qualitative predictions (in parentheses) derived from Eq (1).

		value of S	
		low	high
value of ΣN	low	51 (intermed)	64 (high)
	high	38 (low)	55 (intermed)

Table I presents predictions (in parentheses) derived by plugging high and low values for S and ΣN into Eq (1). Also shown are the percent correct identifications observed for the four groups of words meeting the criteria of the matrix cells. The percentages can be seen to vary as expected from the predictions.

Our summary of NAM serves the purpose of demonstrating that factors independent of the signal - in this case word frequency and lexical sound structure - do indeed play a role in perceptual processing. Secondly, how these factors come into play can be studied in a rigorous manner by building quantitative models.

2.3 DIRECT PERCEPTION

It might at first seem that the conclusions reached in the preceding paragraphs are incompatible with the Direct Realism approach. This is not necessarily so. The source of signal-complementary processes is knowledge ('internal representations') of various kinds that interacts with the incoming signal in percept formation. If we assume that that knowledge would become accessible more or less instantly upon stimulation (cf parallel processing), then postulating a role for "information not the signal" need not include "mediation by cognitive processes or hypothesis testing". The signal-complementary information embodied in the lexicon network would start modulating the signal immediately and in a direct manner.

In his Gibson Memorial Lecture, Shepard (1984) presents a theory that attempts to reconcile the direct nature of perception with a role for 'internal representations', notions normally avoided and usually explicitly rejected by Gibsonians. Basic to the ecological approach is the belief in "smart" perceptual processes capable of extracting higher-order invariants in an automatic way. Shepard agrees with Gibson that "the brain has evolved to extract invariants under favorable conditions ..".

But he adds that we must also assume that it has "evolved to serve the organism under less favorable conditions of nighttime, obstructed, and spatially or temporally limited viewing and, even, of structural damage to the brain itself" (p 419). He goes on to develop the idea that, to deal with non-optimal stimulus conditions, brains were driven to internalize significant external conditions, and to develop an ability to build mental models of the world. This process which underlies the ability to remember, to anticipate, and to imagine objects and to plan events in their absence is not separate from perceptual functions. This is where Shepard differs from Gibson.

If, as Shepard suggests, signals are assumed to "resonate with" internalized knowledge, direct perception and signal-complementary processes ('internal representations') are perfectly compatible notions.

3. What Do Speakers Do?

3.1 CLUES FROM BIOLOGY I: ECONOMY

The themes of the next two sections are summarized in Figure 2. It shows a window cleaner and the compensatory articulation of a bite-block vowel (Lindblom 1983). Two important movement characteristics are illustrated here: economy and plasticity. The bottom diagrams illustrate the expected normal behaviors in which the feet and the jaw synergistically participate to facilitate the two tasks, cleaning the pane on the right and producing the vowel /i/. The top diagrams show compensatory behaviors. The feet and jaw assume positions that necessitate extreme displacements of the arm-body and the tongue respectively. The point is this: **Unconstrained, a motor system tends to default to a low-cost form of behavior.**

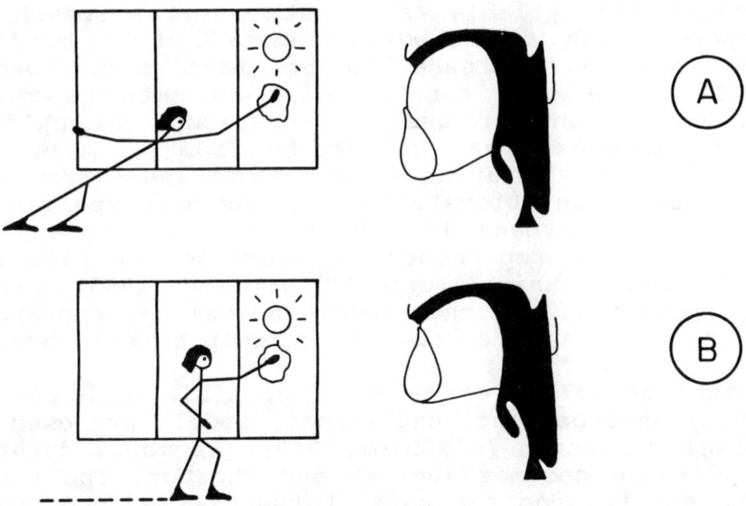

Figure 2. Movement economy and plasticity. A: Compensatory behaviors. B: Normal behaviors.

This trend is pervasive in all kinds of motor behavior including speech. First let us review some non-speech cases. Hoyt and Taylor (1981) observed rates of energy consumption (measured as amount of oxygen consumed to move a certain distance) for a walk, a trot and a gallop in three small horses. The animals were trained to vary their speed within each gait while moving on a motorized treadmill. 'Cost functions' were derived by plotting oxygen consumption against running speed. They form parabolic contours that are concave upward and have clear minima at speeds that depend on the type of gait.

Data on oxygen consumption and speed are presented also for free movement. They indicate that, within each gait, speeds tended to be selected near the minima of the cost functions established in the first experiment. The authors conclude that "the natural gait at any speed indeed entails the smallest possible energy expenditure." Using biomechanical models Milsum (1966) worked out cost functions for breathing and bipedal walking that are similar in form to those reported by Hoyt and Taylor. Rates predicted on the basis of least power expenditure compare favorably with actual rates observed under conditions of rest and exercise.

Nelson, Perkell and Westbury (1984) compared jaw movements during increasingly rapid opening and closing alternations ('wags') and during increasingly rapid repetitions of the syllable /sa/. Both for the speech and the non-speech task the results indicate that mandible excursions tend to decrease as movement times become shorter. They show that for jaw displacements to remain large and independent of gesture duration, it would be necessary to increase peak velocities from 40 mm/s (for movements of long durations) to over 250 mm/s (for movements of short durations). In fact the observed maximum velocities cluster around 100-120 mm/s. They argue that, since peak velocity represents a theoretically plausible measure of "biomechanical effort" (Nelson 1983), their findings support the idea that speech as well as non-speech movements are constrained by a principle of physical "economy".

The results of Nelson et al are compatible with the so-called **duration-dependent undershoot** model proposed to explain vowel reduction (Lindblom 1963): As vowel duration in a CVC syllable becomes shorter and shorter, the extent of the movement towards the vowel target is reduced: Hence both articulatory and acoustic undershoot. However, several studies (Kuehn and Moll 1976, Nord 1986, Engstrand 1988) provide data that refute the idea that presence/absence of undershoot depends only on duration as implied by the original undershoot model. For instance, it is evident that sometimes targets can be reached also at very short durations with no trace of undershoot (e.g. Gay 1978).

A theoretical analysis along the lines of Nelson (1983) and Nelson, Perkell and Westbury (1984) provides the clue to resolving this issue. For simplicity consider a single articulator, say the jaw, which we model biomechanically as a strongly damped spring-mass system. Further let us analyze movements in a CVC syllable as the response of such a system to a sequence of stepwise alternating force values: "close!"-"open!"-"close!" This metaphor helps us understand on the one hand why the system will exhibit undershoot when the commands arrive too fast for the system to follow. On the other hand, it also shows how undershoot behavior could in principle be avoided: Compensation for rapid timing of the force commands can be achieved by increasing movement velocities which in turn is brought about by increasing force amplitudes (cf velocity data in Kuehn and Moll (1976)). However, as pointed out by Nelson et al increasing peak velocities entails greater biomechanical power expenditure. If the speech system operates so as to minimize 'articulatory effort' (peak velocities), we should expect it to undershoot phonetic targets quite often, but not necessarily in every single instance. The key point is: **Speakers have a choice.**

3.2 CLUES FROM BIOLOGY II: PLASTICITY

The second point made in Figure 2 is the output-oriented

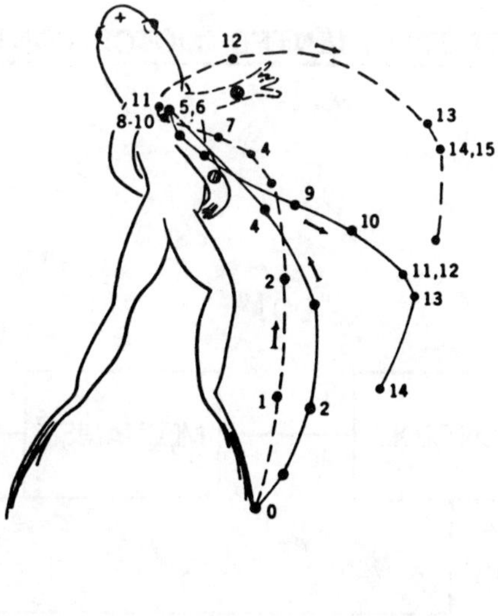

Figure 3. Wiping reflex of spinal frog as a function of foreleg position. (Adapted from Grillner 1982).

nature of motor control, the fact that movements appear to be 'purpose-driven' (Sherrington 1941, Granit 1979). The end products of the compensatory and unconstrained /i/-articulations (or window cleaning) are the same although the movements achieving them are very different. This prospective means-end organization is not special to speech. A lot of evidence shows that it is a perfectly general feature of motor control.

Grillner (1982) draws attention to a study of the wiping reflex in a frog whose spinal cord had been transected at a high cervical level (Fukson, Berkinblit and Feldman 1980). The trajectory of the frog's hindlimb was traced from film frames. When an irritant substance was applied to the forelimb, the frog responded by trying to wipe it away. Significantly, the path of the trajectory varied depending on the position of the foreleg at the moment of stimulation (Figure 3). Although the details of the computational mechanism in the spinal cord remain unknown, the result clearly demonstrates the presence of both adaptive plasticity and goal-orientation in non-speech movements.

Recall also the followng familiar observation. With ease we often recognize a friend's handwriting, whether the letters are small and occur on a piece of paper, or they are large and appear on a blackboard.

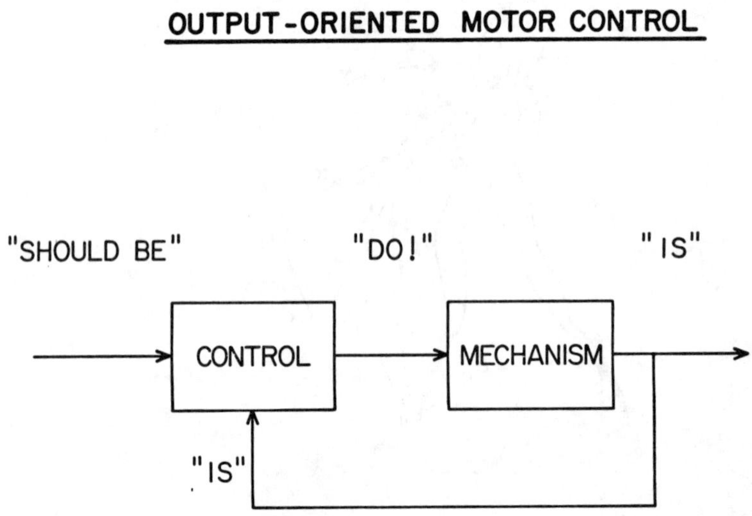

Figure 4. Some components of a feedback system.

Personal characteristics are sometimes preserved under unusual circumstances, as when right-handers are asked to try upside-down mirror writing with the left hand and with eyes closed (Lashley 1951:124).

Experimental observations of this sort suggest that, paradoxically, the execution of a given set of movements is preceded by the construction of a representation of the end-product of those very movements - a representation that also shapes their implementation (Sherrington 1941, Lashley 1951, MacNeilage 1970, Granit 1979).

Figure 4 is an attempt to summarize our discussion so far. On the one hand, we have stated that unconstrained a movement tends to default to a low-cost form of behavior. Accordingly, when an /i/ is produced without a bite-block, a tongue gesture is invoked that deviates little from a neutral configuration ('economy'). However, when a bite-block is in fact present, the system is perfectly capable of compensating ('plasticity').

Common to both situations is the goal, the "SHOULD-BE" value, and an "IS"-signal, the current state of the system as reported to the CONTROL component by various sensory feedback channels. Given "IS" and "SHOULD-BE" the controller computes a course of action and tells the mechanism what to do (the "DO"-signal).

What can we learn from this simplification of the complex phenomena that go on in speech? Recall our discussion of reduction and undershoot. We concluded that within limits speakers appear to have a choice whether to undershoot or not to undershoot. We also noted that avoiding undershoot at short segment durations entails a higher biomechanical cost. The model in Figure 4 offers a way of thinking about how such choices could be implemented.

The control component operates on the basis of error information. It looks at the difference between "SHOULD-BE" and "IS". The current error value can be amplified, or attenuated, by varying the "gain of the feedback loop". When the error is allowed to have a strong influence, the system response will exhibit faster and more accurate target attainment. This corresponds to a compensatory and output-oriented mode. When the role of the error information is attenuated, the "DO"-signal drives the system less forcefully. In this non-compensatory and system-oriented mode target attainment is less efficient and the response is shaped more by system constraints. There is little compensation for the intrinsic dynamic characteristics of individual articulators, should they interfere with the implementation of the "SHOULD-BE"-command.

By conceptualizing speech in this way we obtain a way of describing the mechanism for tuning target attainment according to the demands of the listener and the situation, the notion of "controlling the gain of the feedback loop". This, we suggest, is how speakers move around along the dimension of **hyper-** and **hypospeech**. In global terms, this is how they are able to over- or underarticulate.

4. The H&H Argument Illustrated

We have looked at two sources of information for speech perception: signal-driven and signal-independent processes. We have singled out two significant characteristics of speech motor control: economy and plasticity. We are ready to move on by putting two and two together (cf 1.2, **e-g**).

Let us examine H&H against the background of speech as produced not only in the laboratory but also in its natural, ecological settings. Without aspiring to be exhaustive, in Figure 5 we survey factors that shape the phonetic variation in the speech of a single individual. The simplification that H&H imposes is that phonetic variation is basically one-dimensional: When output constraints dominate, we expect to see hyperforms, whereas with system constraints dominating, hypospeech will be observed.

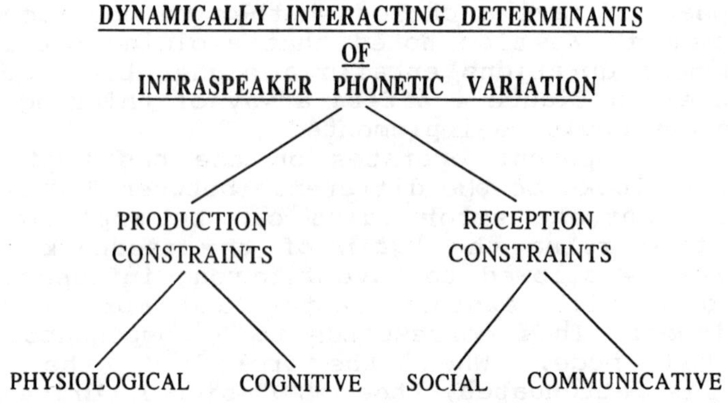

Figure 5: Factors causing the pronunciation of spoken forms to vary.

In keeping with this view Figure 5 distributes the factors influencing intra-speaker variation between production constraints and perception constraints. Under these two nodes we place physiological factors (mostly involuntary modulations such as emotion, disease, etc...), cognitive factors (speaking to oneself, propositional vs automatic styles (Bates 1979:359), etc...), social and communicative factors (channel, listener, situation, degree of formality ...).

These complex phenomena are mentioned to make clear that the assumption about H&H variation being one-dimensional is a deliberate simplification which is likely to be revised in the course of further work. However, for the moment we regard this assumption as heuristic and methodologically justified.

4.1 OUTPUT-ORIENTED CONTROL

It might be suggested that the fact that subjects do well on experimental tasks involving compensatory articulation (e g bite-block speech) is interesting but irrelevant to phonetic theory since bite-blocks create unnatural, non-speech conditions. We favor an alternative view: Good performance on compensatory tasks is a consequence of the fact that motor behavior in general is organized to be inherently compensatory. Sundberg's work on singing (Sundberg 1987) provides some instructive non-speech examples of output-oriented control.

Female opera singers often produce pitches whose frequencies radically exceed the normal values of the first formant. This is particularly true for close vowels. Acoustic theory predicts that when there are no strong harmonics to support the first formant, the intensity of the sound will go down. Sundberg has shown that singers "track" the increases in fundamental frequency by adjusting the frequency of F1. This is done by lowering the mandible so as to raise F1 in step with F0, thus maintaining an output of sufficient intensity.

Another example is the account of the origin and function of the "singer's formant" (Sundberg 1987:118). This is a spectral prominence primarily due to a clustering of the third, fourth and fifth formants. According to Sundberg's measurements and studies of vocal tract models, this cluster seems to be brought about by larynx lowering which has the effect of making the vocal tract longer as well as making certain structures (laryngeal ventricle and sinus piriformis) wider. He measured long-term average (LTA) spectra for three types of signal: orchestra, speech, orchestra + singing. The LTA-spectra of the orchestra and the speech were found to coincide closely, whereas that of the singer exhibited a salient peak in the region of the "singer's formant" rising above the orchestra envelope. Recordings of singing - produced with and without the "singer's formant" and superimposed on noise with the long-term spectral properties of orchestral sound - demonstrate that the "singer's formant" makes the singing more audible in the presence of an orchestra. Sundberg notes that the "singer's formant" is typical of all voiced sounds produced in Western concert-style singing except soprano singing. Why not sopranos? According to Sundberg the higher pitch range of sopranos causes their LTA-spectra to overlap less extensively with the LTA-spectra of orchestral sound. Their audibility may therefore be sufficiently high without a "singer's formant" - a circumstance fully in line with his explanation and a further instance of the notion of output-oriented motor control.

4.2 OUTPUT- VS SYSTEM-ORIENTED CONTROL

So far our story conforms with accounts of speech production that view it as a continual **tug-of-war** between demands on the output on the one hand and system-based constraints on the other (cf the papers by Keller and by Kohler, this volume). Here is another finding from the Sundberg research team that highlights this balance of forces.

In an X-ray investigation Johansson, Sundberg and Willbrand (1983) studied the vowels /i: a: u:/ spoken and sung. The fundamental frequencies of the sung versions were 230, 465 and 940 Hz. Mid-sagittal contours of the tongue body were traced (Sundberg 1987:Fig 5.25). For the lowest pitch the tongue contours of sung and spoken vowels coincide rather closely. At 930 Hz, however, essentially the same tongue configuration was used for all three vowels. It is interesting that this shape comes closest to the most neutral of the vowels /a:/. This result is a good illustration of the rule proposed in Section 3.1: "Unconstrained, a motor system tends to default to a low-cost form of behavior".

4.3 PARTIAL COMPENSATION

Compensations can be seen as cases where output constraints overrule system constraints. It is an interesting fact about most articulatory compensations that, when examined closely, they turn out to be partial, not perfect. Such observations support the correctness of the present tug-of-war scenario. Let us look at four examples of compensatory speech production behavior.

In a study of the duration of open and close vowels (Lindblom 1967) I measured lip and jaw movement in /I'bVb(:)I/-sequences and found evidence supporting the Extent of Movement Hypothesis. This hypothesis was proposed by Eli Fischer-Jorgensen (1964) to explain the durational differences between open and close vowels: I found that the jaw began its opening gesture for the V during the first /b/-segment and returned to a high position around the middle of the second /b/. In test words with open and close vowels, say /I'ba:bI/ and /I'bi:bI/, the jaw contributed to an earlier separation of the lips in the first /b/ and delayed their coming together again for the final /b/.

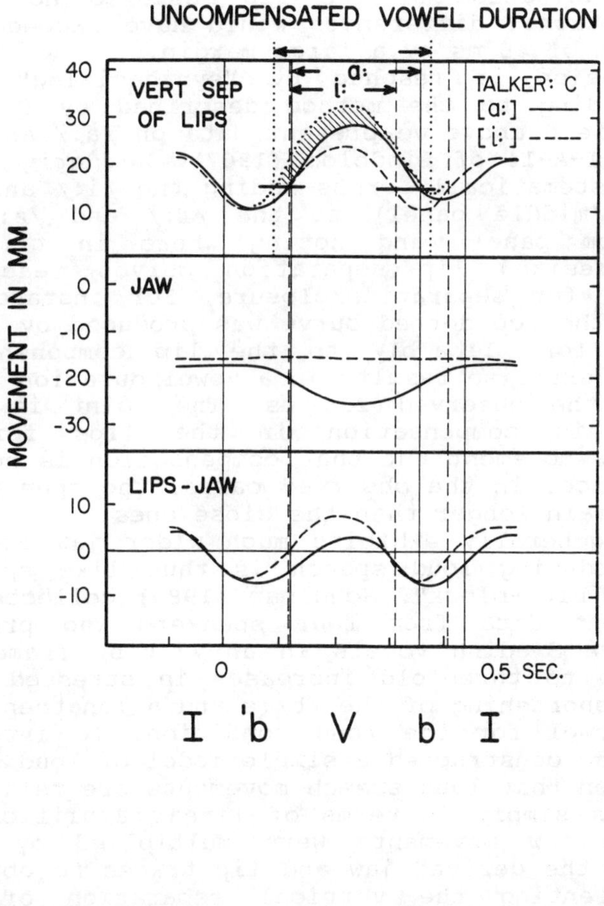

Figure 6: Compensated and uncompensated vowel duration. Vertical lines indicate acoustic segment boundaries. Solid and dashed lines show measurements from which the dotted lines and the dotted curve were derived.

Since open vowels had larger excursions of the jaw than close vowels, the effect on the timing of the vowel segment boundaries was greater for the open vowels. It was thus, everything else being equal, responsible for the longer duration of those segments.

The reason for returning to those findings here is that the condition of "everything else being equal" was not always met. It was violated in an interetsting way. In words with open vowels the lip gestures for the /b/:s (corrected for superimposed jaw movement) were found to actively resist a premature opening of the first /b/ and to be closer to each other for the second /b/. Had this vowel-dependent reorganization of the lip gestures not taken place, the durational differences would have exceeded the observed maximum of 45 ms by a large margin.

This conclusion is reached by "synthesizing" vowel durations according to the method described by Schulman (1989). In Figure 6 above we present data on /a:/ and /i:/ taken from Fig I-A-14 of Lindblom (1967). We derive vowel durations by systematically cross-adding the /i:/ and /a:/ jaw movements (middle panel) to the /i:/ and /a:/ lip gestures (bottom panel) and noting where in time the resulting synthesized lip separation curves reach the criterion value for separation/closure. For instance, in Figure 6 above the top dotted curve was produced by adding the jaw curve for /I'ba:bI/ to the lip component for /I'bi:bI/. This exercise results in a vowel duration of 310 ms instead of the observed 235 ms. The point is this: Clearly there is compensation in the lips for the superimposed jaw movement but that compensation is no more than partial since, in the observed cases, the open vowels nevertheless remain longer than the close ones.

Loud speech generally exhibits much wider jaw openings for vowels. Producing loud speech is thus like speaking with a "natural bite-block". Schulman (1989) collected lip and jaw movement data from four speakers who produced tokens of twelve Swedish vowels in an /i'b_b/ frame. His study showed up to threefold increases in stressed vowel jaw opening, a shortening of the /b/:s and a lengthening of the stressed vowel for the loud condition. To interpret these changes, he constructed a simple model of loud speech on the assumption that loud speech movements are related to normal movements simply in terms of linear amplification. Normal lip and jaw movements were multiplied by scale factors. Adding the derived jaw and lip traces he obtained a curve representing the vertical separation of lips (Schulman 1989, Figure 12). This procedure demonstrates that loud speech involves a more complex transform than mere linear amplification of movements. For instance, for the loud vowels the jaw makes excursions that are 2 to 3 three times larger than those of the normal vowels. However, for the surrounding consonants in loud tokens, the jaw is not thus scaled. It remains in its normal high position. Schulman's simulation shows us why: To keep the effort-dependent increases in vowel duration within bounds and so as not to totally jeopardize the attainment of stop closure. We conclude that the loud-speech transform is goal-directed and involves articulatory manoeuvres that are compensatory. Note that this compensation is partial. The system tolerates some increase of loud vowel durations but shows restraint.

The interaction of jaw movement and segment durations is also the topic of Lindblom, Lubker, Lyberg, Branderud and Holmgren (1987). In this study reiterant speech samples were used modeled on real Swedish mono-, bi- and trisyllabic words: ['bab: 'bab:ab ba'bab: 'bab:abab ba'bab:ab baba'bab:]. The subjects were all unaware of the goal of the experiment. They spoke the test words both normally and with sizeable bite-blocks. The difference in jaw opening between normal and bite-block conditions was of the order of 15-20 mm. The question addressed was this: Would speakers be able to achieve /b/-closures in spite of the fixed and abnormally low jaw position? And if so, what segment durations would they produce under such conditions? It was found that bite-block segment durations reproduced normal vowel and consonant durations within 15 ms and that there was a tendency for the bite-block condition to induce complementary consonant shortenings and vowel lengthenings (cf Schulman's findings). It appears that these "results do not preclude that the reorganization of articulatory gestures takes place with respect to segment duration only in so far as it is necessary to maintain perceptual constancy of stress patterns. Hence we observe **selective** or partial segment duration **compensation**" (p 178).

Gay, Lindblom and Lubker (1981) obtained articulatory data on normal and bite-block steady-state vowels from still lateral X-ray pictures. Normal and bite-block contours were compared in terms of a maxilla-based coordinate system. Compensations were close to perfect near maximum constrictions, but incomplete at points with large cross-sectional areas. Area functions derived from the X-ray tracings were subjected to a theoretical perturbation analysis which showed that the strategy actually followed by the subjects was the acoustically optimal one. Once again the same pattern emerges: Although compensations are invoked that are selective, acoustic equivalence is nevertheless achieved.

4.4 SUFFICIENT CONTRAST: AN ANALYSIS OF COARTICULATION

One influential view of coarticulation is the one that has guided the development of the Motor Theory of Speech Perception. Recall the earlier quotation from Liberman and Mattingly (1985:7): The MT "assumes that adaptations of the motor system for controlling the organs of the vocal tract took precedence in the evolution of speech. These adaptations made it possible, not only to produce phonetic gestures, but also to coarticulate them so that they could be produced rapidly. A perceiving system, specialized to take account of the complex acoustic consequences, developed concomitantly." In other words, coarticulation is seen as an adaptation to a demand for a preferred speaking tempo. "A function of coarticulation is to evade" ... "that each unit would become a syllable, in which case talkers could speak only as fast as they could spell" (p 13). Also, reviewing the multiplicity and variety of cues, Liberman and Mattingly write: "... we should conclude that there is simply no way to define a phonetic category in purely acoustic terms" (p 12).

Another well-known, but entirely different view is that explored by Stevens and co-workers (Stevens and Blumstein 1978, 1981; Blumstein and Stevens 1979, 1981). This research program involves a theoretically motivated search for acoustic properties that remain invariant across talkers and phonetic contexts and whose perceptual role is demonstrated experimentally.

Below I shall discuss two sets of data that bear on theories of coarticulation. In particular, we shall examine the possibility of interpreting coarticulation phenomena in terms of "sufficient contrast" as suggested by the H&H theory.

The first point is a restatement of Öhman's (1966) findings. Öhman's speech samples were symmetrical and asymmetrical VCV sequences. They were produced by generating all possible combinations of /b d g/ and /y ö a o u/ and were spoken by a Swedish subject. His Tables II and IV contain formant frequency measurements for the initial and final vowels and for the points corresponding to the VC and CV boundaries (here called "locus" patterns). Vowel-consonant coarticulation is extensive. The locus patterns and the formant transitions are not unique. "... place information for a given consonant is carried by a rising transition in one vowel context and a falling transition in another" (Liberman, Delattre, Cooper and Gerstman 1954). At the CV boundary formant patterns depend not only on the identity of V_2 but also on V_1. Conversely, at the VC boundary they depend on both V_1 and V_2. At first glance one might feel tempted to agree with Liberman and Mattingly conceding "that there is simply no way to define a phonetic category in purely acoustic terms".

Is such a conclusion really justified? In Figure 7 we plot F2 measured at the CV-boundary along the x-axis, F3 also measured at the CV-boundary along the y-axis and the F2 vowel target on the z-axis. This 3-D view displays three "clouds" that enclose all the V_1bV_2, V_1dV_2 and V_1gV_2

measurements. Although the consonant formant patterns exhibit massive context-dependence, the 3-D space shows three configurations that do not overlap. The point is this: If other dimensions were also considered (spectral dynamics, aspects of the burst spectra etc), we would expect even better separability. Owing to the large coarticulation effects there are no absolute invariants. However, the phonetic correlates of the three stop categories are nevertheless distinct. They meet the condition of "sufficient contrast". For similar results for English stops see Sussman (1989).

"Unconstrained, a motor system tends to default to a low-cost form of behavior". Is the preceding analysis of coarticulation compatible with that rule? If it is true that the constraints on perceptual distinctiveness tolerate some variation in locus patterns - manifested as clouds rather than points in a multi-dimensional representation - what is the source of that variation? Why is it there? H&H theory assumes that locus coarticulation is there - not to serve perception in the first instance (although it sometimes provides valuable cues) - but primarily to facilitate production. If so, in what sense is vowel-consonant coarticulation "a low-cost form of behavior"? An extension of work previously done on vowels (Lindblom and Sundberg 1971) sheds some light on those questions.

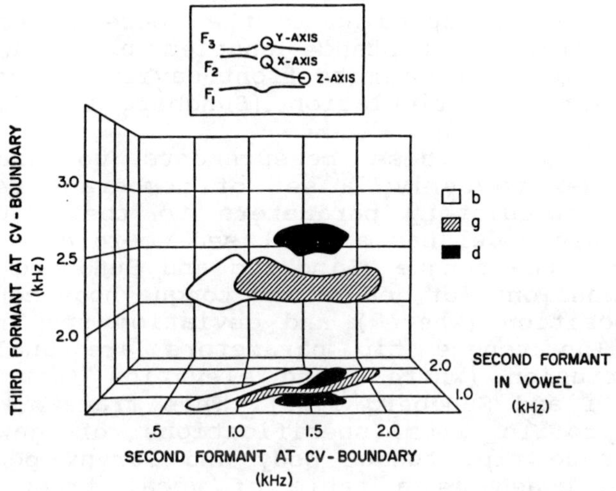

Figure 7: Acoustic separability of stop consonants. Adapted from Öhman (1966).

Two sets of experiments were done in which the following articulatory parameters were controlled: (1) Lip configuration: The subject (JS) held a cylindrical brass tube between his lips so that it made contact with his upper incisors. This allowed us to control the length and the cross-sectional area of the lip section; (2) Jaw opening: A bite-block creating a 7 mm jaw opening was introduced between the subject's molar teeth; (3) Place of apical closure: The place of tongue tip was marked with a small pellet on a false palate.

In the first series of experiments we collected data on coarticulation effects in the formant patterns of Swedish dental and retroflex stops in VCV utterances (cf Öhman 1966). We included naturally spoken VCV utterances as well as tokens produced with the set-up described above. The second series was aimed at determining the acoustically relevant properties of the cavity in front of the apical closure. Using the lip tube, the bite-block and the false palate with the marker, the subject produced VCV utterances in which the place of articulation of the consonant was systematically varied. The subject was instructed to freeze his articulatory movement as he reached the closure. While he thus held his articulation steady, a sine-wave sweep-tone was applied to the front cavity. The frequency at which its amplitude had a maximum was taken to be the resonance frequency of the front cavity system. The introduction of the lip tube made that system similar to a Helmholtz resonator with a neck corresponding to the lip tube and a volume corresponding to the space in front of the occlusion. This circumstance enabled us to use the resonance frequencies to estimate front cavity volumes as a function of point of articulation (Sundberg and Lindblom 1989).

In parallel with these measurements we used our articulatory model to derive a set of nomograms relating variations in articulatory parameters to their acoustic consequences. This model has a jaw-based coordinate system for the lips and the tongue (Lindblom and Sundberg 1971). It uses two dimensions for generating tongue body contours: constriction position (where?) and deviation from neutral (how much?). The tongue tip parameters are analogous: retraction-protrusion (where?) and elevation (how much?) (Lindblom, Pauli and Sundberg 1975). This framework first generates a profile from specifications of jaw, lip parameters, tongue tip, tongue body and larynx position. The profile is coded as a table of vocal tract cross-distances. The distances are converted into cross-sectional areas following two methods. Posterior to the closure we used our previously established rules (Lindblom and

Sundberg 1971). In front of the closure we derived areas from the volume data gathered in the sweep-tone experiment.

To investigate the patterns of coarticulation observed in the spectrographic VCV measurements, we adopted an analysis-by-synthesis procedure addressing the following question: What model configurations show formant patterns identical with those measured in the VCV words? Recall that in one of the VCV experiments we controlled articulation with the aid of the lip tube, a 7 mm jaw opening and indications of place of apical closure. With the model we had obtained parallel information simulating those conditions. For any given place of articulation the parameters free to vary were the retraction-protrusion, constriction position and deviation from neutral. For each place of apical articulation we prepared a set of diagrams showing how F2 and F3 of the model varied as a function of these three parameters. Then we adopted the following graphical procedure for the comparing the measurements with the model: (i) Input the F2 and F3 observed for a given place of apical closure; (ii) Determine the best acoustic match that the model produces for that place; (iii) Read off the articulatory parameter values associated with that match.

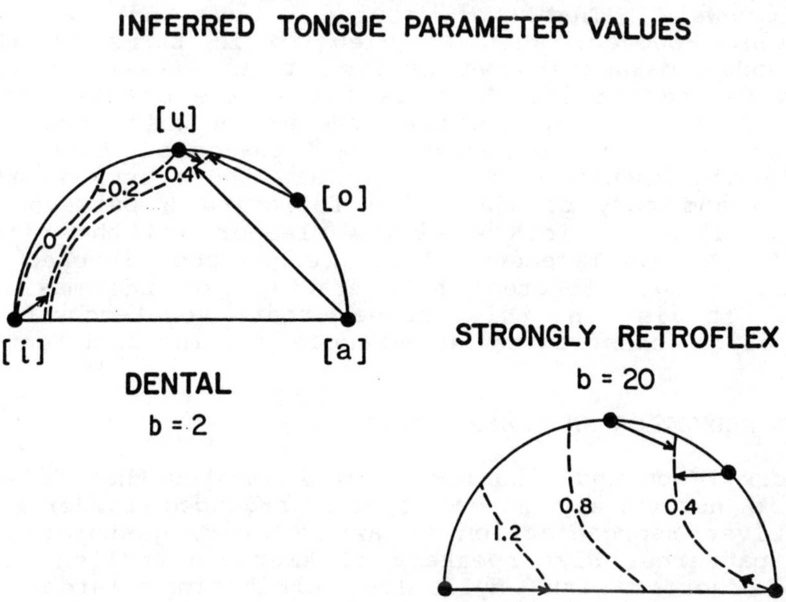

Figure 8: Coarticulation in dental and retroflex stops. Values of tongue body parameters were inferred from articulatory model simulations and matching the derived F-patterns to the observed formant data.

Figure 8 shows the analysis-by-synthesis results for apical stops produced at 2 and 20 mm behind the upper incisors in five contexts: /i-i/, /ɛ-ɛ/, /a-a/, /o-o/ and /u-u/. Formant measurements were made at VC boundaries. The semicircles stylize the tongue body space. A point within the semicircle is specified by two numbers: angle of radius (constriction location) and displacement from origin along a given radius (deviation from neutral). Vowel contexts are shown as solid dots. The arrows indicate the values of the best matches. The dashed wavy lines are **compatibility contours**. They represent families of tongue body shapes, that is those tongue body contours that are compatible with a given place of articulation, **b**, and a given degree of retraction-protrusion. (The more positive the number next to dashed line the greater the retraction).

The point that Figure 8 makes is this: As indicated by the compatibility contours in both semicircles, the model makes available a large range of tongue shapes. However, the arrows do not show arbitrary positions on those contours. They occur in the local neighborhood of the vowel dots.

The significance of that finding is simply that the tongue shapes of apical stops tend to resemble those of the adjacent vowels. That, of course, is not a new observation. However, when we view it in terms of the matched model parameters we arrive at an answer to the questions raised earlier: What is the source of the locus pattern variations? Why are they there? In what sense is vowel-consonant coarticulation "a low-cost form of behavior"? The answer is that they are there because the tongue tip and body do their job following a pattern of synergy we also saw in the window cleaner and the bite-block /i/. The displacement of the tongue body invoked to facilitate apical closures is a fraction of the maximum possible. It is in this sense that vowel-consonant coarticulation constitutes an example of low-cost motor behavior.

4.5 VOWEL REDUCTION IN CLEAR SPEECH

Recent work (Moon and Lindblom 1989) indicates that "clear speech" is not merely normal speech produced louder. It also involves reorganization of articulatory gestures and acoustic patterns. Five speakers of American English read lists of words with syllables exhibiting large F2 transitions: wheel, will, well, wail. To obtain vowel duration variations the syllables occurred as monosyllbles and were embedded in bi- and trisyllabic frames generated by appending -ing and -ingby or -ingham (place name endings): e g well, welling, Wellingby. These lists were read in two ways: First at a rate and loudness spontaneously chosen by the subject (citation forms); Then in response to an explicit instruction to "overarticulate", and "to speak as clearly as possible". Citation-form style repetitions of each vowel in an /h_d/ environment were also recorded for each subject. Vowel durations and F2 and F3 at vowel midpoints were measured for all tokens. Formant frequencies were plotted as a function of vowel duration to see whether there were undershoot effects. All speakers showed undershoot. Its degree varied somewhat depending on vowel, talker and style.

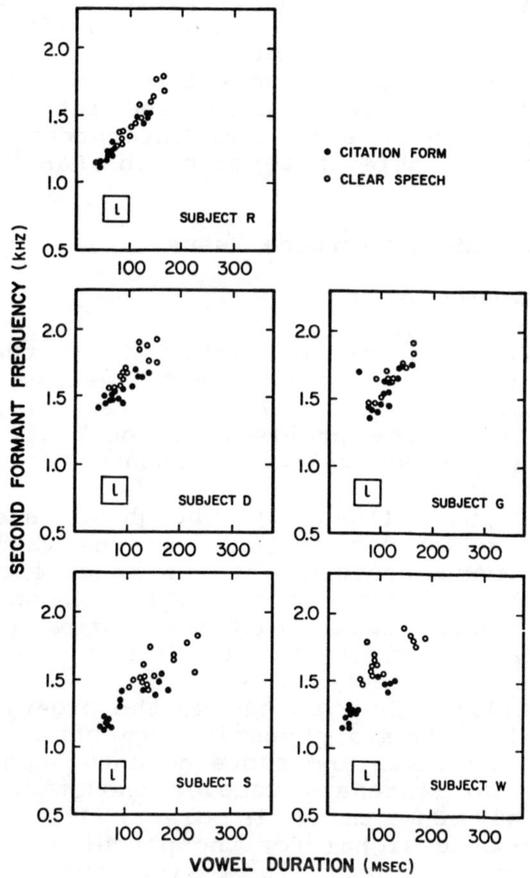

Figure 9: Duration-dependent undershoot in clear and citation-form speech. Measurements from syllable <u>will</u>-.

Both clear and citation-form speech showed undershoot. But clear speech formants were generally closer to the /h_d/ reference values than the corresponding citation-form measurements. (This effect was not simply due to clear speech tokens being longer (which they often were). When exponentials were fitted to the formant-vs-duration plots (cf Lindblom 1963), separate curves were needed to describe the two styles (Moon (in prep), cf also Engstrand and Krull (1989))

Why does clear speech have these properties? Both the existence and the nature of a clear-speech mode make eminently good sense from the H&H viewpoint. So does the preliminary evidence that formant patterns approximate targets more closely in clear than in citation-form speech. Should further analyses establish that clear-speech transforms are indeed communicatively functional - that is, they increase both the audibility and the intelligibility of speech (Moon in prep) - a natural interpretation would immediately become available in terms of the H&H theory.

5. Implications for the Invariance Issue

5.1 IN-PRINCIPLE SOLUTIONS

The Motor Theory (Liberman and Mattingly 1985), the Quantal Theory (Stevens 1989) and the Direct Realism approach (Fowler 1986) can all be said to offer in-principle solutions to the invariance problem. Although different in other respects, MT and DR explicitly assume that phonetic invariance is gestural.

Liberman and Mattingly (1985:22) "Though we have a great deal to learn before we can account for the variation in instances of the same gesture, it is nonetheless clear that, despite such variation, the gestures have a virtue that the acoustic cues lack: instances of a particular gesture always have certain topological properties not shared by any other gesture."

And Fowler (1986:11): "By hypothesis, the organization of the vocal tract to produce a phonetic segment is invariant over variation in segmental and suprasegmental context." Cf also (p 13): "How do listeners recover phonetic structure from such a signal? One thing is clear; the functional parsing of the acoustic signal for the perceiver is not one into acoustic segments. Does it follow that perceivers impose their own parsing on the signal? There must be a "no" answer to this question for any event theory devised from a direct-realist perspective to be viable. The perceived parsing must be in the signal; the special role of the perceptual system is not to create it, but only to select it."

The QTS is developed from the assumption that acoustic/auditory stability plays a major role in the selection of the phonetic segments and features of the languages of the world. There seems to be a clear link between the functional value attributed to stability in QTS and the experimental search for acoustic invariants that Stevens has been engaged in. In several papers the approach taken builds on the fact that the "work of Fant (1960), Jakobson et al (1963) and Halle et al (1957) in particular, has suggested that spectral analysis of the speech signal results in distinctively different patterns corresponding to individual phonetic features." ... "The notion that spectral analyses reveal distinct patterns for individual phonetic features suggests that there may indeed be an invariance of acoustic patterns which can characterize place of articulation independent of the following vowel" (Blumstein and Stevens 1980:648).

These research programs represent distinct working hypotheses about the resolution of the invariance problem. However, they do have one theme in common: They all share the assumption that the ultimate solution will be found in the signal.

In contrast the H&H theory assumes that, in all instances, speech perception is the product of both signal-driven and signal-independent information, that the contribution made by the signal-independent processes show short-term fluctuations and that speakers adapt to those fluctuations. It says that - whether communicatively successful or not - that adaptive behavior is the reason for the alleged lack of invariance in the speech signal. Hence it predicts that the quest for signal-based definitions of invariance will continue to remain unsuccessful as a matter of principle. In the H&H model the need to solve the invariance issue disappears. But the problem is replaced by another - one that appears equally forbidding but should, in principle, have better prospects of success: That of describing the class of speech signals that satisfy the condition of "sufficient discriminative power".

Let us pause to try to envision a hypothetical situation in the future. Imagine that the invariance problem has been solved and that one of the signal-based theories is the winner. Assume that this theory is successful because it defines the physical correlates of linguistic units in a rigorous and quantitative way that makes them independent of context. No matter what instance of phonetic gesture or acoustic pattern we come up with, this theory will tell how to make the measurement that establishes the one-to-one correspondence between the linguistic category and the signal or gesture attributes.

Given this state of affairs, what will the status of "intra-speaker phonetic variation" be? It would be true, would it not, that once a successful new theory emerges, the notorious variability of speech signals would by definition vanish. That variability would turn into epiphenomena arising from our incorrect 1989 way of looking at speech signals. Nonetheless, as soon as we looked at speech as we do today, all the extensive modifications (coarticulation, reduction etc) would still be there and would appear as systematic as they do now. What will successful signal-based theories of invariance have to say about that variability?

Do we really want our theories to succeed in solving the invariance problem without having anything to say about the structure of intra-speaker phonetic variation? No, the systematicity of phonetic variation is very real and demands an explanation.

5.2 CLUES FROM THE TYPOLOGY OF SOUND SYSTEMS

The reality of phonetic variation comes to the fore also when we begin to think about the relationship between on-line phonetic processes and the phonetic structure of lexicons and phonologies. The following findings make that evident.

The UPSID database (Maddieson 1984) lists the vowel and consonant inventories of 317 languages. 58 phonetic dimensions are used to specify several hundred consonant and vowel segment types. For descriptive purposes these elements can be naturally divided into three categories: Basic, Elaborated and Complex segments. To illustrate this classification consider [d d̤ ɖ ɖ̤]. The middle two symbols are Elaborated segments in that they show a superposition of retroflexion and breathy voice on the more elementary articulation of the Basic [d]. Any segment with a combination of at least two elaborating processes, e. g. retroflexion and breathy voice, is treated as Complex. Classifying all the segments of UPSID in this way we found a very robust pattern that we have called the **Size Principle** (Lindblom and Maddieson 1988): Small systems use only Basic segments, medium-sized systems use both Basic and Elaborated articulations. Large systems show all three kinds. This principle emerges clearly when the number of B, E and C segments that a language uses is plotted as a function of its inventory size. In systems with six to ten consonants the recruitment of B segments grows in step with system size. Then B segments seem to reach a saturation point at which E segments are brought in growing linearly in number. For inventories with more than 25 consonants, C segments are invoked in addition to the other types. They too exhibit a linear growth with system size. The picture is similar for the vowels and highly lawful making quantitative size-dependent predictions of the number of B, E and C segments possible.

It appears likely that a **sufficient contrast** constraint underlies the formation of these regularities. In small systems demands for perceptual distinctiveness are less than in larger systems. Articulatorily complex articulations seem to be brought into play only in so far as intrasystemic competition for contrast calls for them.

Clearly there is a parallel between the use of B, E and C segments in these inventories and the use of H&H forms in on-line speech production: A similar tug-of-war and balancing of production-oriented and listener-oriented forces. What mechanism underlies the selection and shaping of phonetic inventories? We cannot seriously address that question here. Suffice it to say that, if natural speech varies along an H&H continuum, then the raw materials are available for building inventories structured according the Size Principle. Elsewhere (Lindblom, MacNeilage and Studdert-Kennedy in preparation) we propose that a process not unlike the variation-selection mechanism of evolutionary biology provides the appropriate model for understanding also the phonological fossilization of on-line phonetic variation.

What do clues from the typology of sound systems tell us about phonetic variation? As we have just seen, they suggest that it is very real. It builds phonologies.

6. References

Bates E (1979): The Emergence of Symbols, Academic Press:New York.

Bladon A (1985): "Diphthongs: A Case Study of Dynamic Auditory Processing", Speech Communication 4:145-154.

Blumstein S and Stevens K N (1979): "Acoustic Invariance in Speech Production: Evidence from Measurement of the Spectral Characteristics of Stop Consonants", J Acoust Soc Am 72, 43-50.

Blumstein S and Stevens K N (1981): "Phonetic Features and Acoustic Invariance in Speech", Cognition 10, 25-32.

Diehl R L and Kluender K R (1989): "On the Objects of Speech Perception", Ecological Psychology 1(2), 121-144.

Diehl R L, Kluender K R, Walsh M A and Parker E M (in press): "Auditory Enhancement in Speech Perception and Phonology", to appear in Hoffman, R and Palermo, D (eds): Cognition: The State of the Art, LEA:Hillsdale, NJ.

Engstrand O (1988): "Articulatory Correlates of Stress and Speaking Rate in Swedish VCV Utterances", J Acoust Soc Am 83:1863-1875.

Engstrand O and Krull D (1989): "Determinants of Spectral Variation in Spontaneous Speech", pp 88-91 in Proceedings of Speech Research '89, Budapest.

Fant G (1973): Speech Sounds and Features, MIT Press:Cambridge, MA.

Fischer-Jørgensen E (1964): "Sound Duration and Place of Articulation", Zeitschrift für Sprachwissenschaft und Kommunikationsforschung 17:175-207.

Fowler C A (1986): "An Event Approach to the Study of Speech Perception from a Direct-Realist Perspective", J of Phon 14:1, 3-28.

Fujimura O (1989): "Articulatory Perspectives of Speech Organization" lecture presented at the Nato Advanced Institute on Speech Production and Speech Modeling, see this volume.

Fukson O I, Berkinblit A G and Feldman A G (1980): "The Spinal Frog Takes into Account the Scheme of its Body during the Wiping Reflex", Science 209, 1261-1263.

Gay T (1978): "Effect of Speaking Rate on Vowel Formant Movements", J Acoust Soc Am 63(1):223-230.

Gay T, Lindblom B and Lubker J (1981): "Production of Bite-Block Vowels: Acoustic Equivalence by Selective Compensation", J Acoust Soc Am 69(3), 802-810.

Geisler C D (1988): "Representation of Speech Sounds in the Auditory Nerve", J of Phon 16:1, 19-35.

Gibson J J (1972): "Outline of a Theory of Direct Visual Perception", in Royce, J R and Rozeboom, WW (eds): The Psychology of Knowing, Gordon&Breach:New York.

Gibson J J (1979): The Ecological Approach to Visual Perception, Houghton Mifflin:Boston, MA.

Granit R (1979): The Purposive Brain, MIT Press:Cambridge MA.

Greenberg S (1988): Representation of Speech in the Auditory Periphery, J of Phon 16:1-149 (theme issue).

Grillner S (1982): "Possible Analogies in the Control of Innate Motor Acts and the Production of Speech", 217-229 in Grillner S, Lindblom B, Lubker, J and Person, A (eds): Speech Motor Control, Pergamon Press:Oxford.

Hoyt D F and Taylor C R (1981): "Gait and the Energetics of Locomotion in Horses", Nature 292, 239-240.

Johnson A, Sundberg J and Willbrand H (1983): "'Kölning': A Study of Phonation and Articulation in a Type of Swedish Herding Song", 187-202 in Askenfelt A, Felicetti S, Jansson E and Sundberg J (eds): Proc of SMAC 83 (vol 1), Royal Swedish Academy of Music:Stockholm.

Keller E (1989): "Speech Motor Timing", lecture presented at the Nato Advanced Institute on Speech Production and Speech Modeling, see this volume.

Kelso J A S, Saltzman, E L and Tuller, B (1986): "The Dynamical Perspective on Speech Production: Data and Theory", J of Phon 14:1, 29-59.

Klatt D H (1987): "Review of Selected Models of Speech Perception", to be published in Marslen-Wilson W D (ed): Lexical Representation and Process, MIT Press:Cambridge, MA.

Kluender K R, Diehl R L and Wright B A (1988): "Vowel-Length Difference before Voiced and Voiceless Consonants: An Auditory Explanation", J of Phon 16, 153-169.

Kohler K J (1989): "Segmental Reduction in Connected Speech in German: Phonological Facts and Phonetic Explanations", lecture presented at Nato Advanced Institute on Speech Production and Speech Modeling, see this volume.

Kuehn D P and Moll K L (1976): "A Cineradiographic Study of VC and CV Articulatory Velocities", J of Phon 4:303-320.

Lacerda F (1987b): "Effects of Peripheral Auditory Adaptation on the Discrimination of Speech Sounds", dissertation monograph, Perilus VI, Department of Linguistics, Stockholm University.

Lashley K S (1951): "The Problem of Serial Order in Behavior", 112-146 in Jeffress, L A (ed): Cerebral Mechanisms in Behavior, Wiley:New York.

Liberman A M and Mattingly I G (1985): "The Motor Theory of Speech Perception Revised", Cognition 21:1-36.

Liberman A M, Harris K S, Hoffman H S and Griffith B C (1957): "The Discrimination of Speech Sounds within and across Phoneme Boundaries", J of Experimental Psychology 54:358-368.

Lieberman P (1963): "Some Effects of Semantic and Grammatical Context on the Production and Perception of Speech", Language and Speech 6:172-187.

Lindblom B (1963): "Spectrographic Study of Vowel Reduction", J Acoust Soc Am 35:1773-1781.

Lindblom B (1967): "Vowel Duration and a Model of Lip Mandible Coordination", STL-QPSR 4/1967, 1-29, (Department of Speech Communication, RIT, Stockholm).

Lindblom B (1983): "Economy of Speech Gestures", 217-245 in MacNeilage, P.F. (ed): Speech Production, Springer Verlag:New York.

Lindblom B and Sundberg J (1971): "Acoustical Consequences of Lip, Tongue, Jaw and Larynx Movement", J Acoust Soc Am 50(4):1166-1179.

Lindblom B, Pauli S and Sundberg J (1975): "Modeling Coarticulation in Apical Stops", 87-94 in Fant G (ed): Proceedings of the Speech Communication Seminar, Vol. 2, Almqvist&Wiksell:Stockholm.

Lindblom B, and Maddieson I (1988): "Phonetic Universals in Consonant Systems", 62-78 in Hyman L M and Li C N (eds): Language, Speech and Mind, Routledge:London and New York.

Lindblom, B and MacNeilage, P (1986): "Action Theory: Problems and Alternative Approaches", J of Phon 14:1, 117-132.

Lindblom B, Lubker J, Lyberg B, Branderud P and Holmgren K (1987): "The Concept of Target and Speech Timing", 161-182 in Channon R and Shockey L (eds): In Honor of Ilse Lehiste, Foris:Dordrecht, Holland.

Lindblom B and Sundberg J (in prep): Acoustical Consequences of Articulatory Movement.

Lindblom B, MacNeilage P and Studdert-Kennedy M (in prep): Evolution of Spoken Language, Orlando, FL:Academic Press.

Luce P A (1986): Neighborhoods of Words in the Mental Lexicon, Doctoral dissertation, Department of Psychology, Indiana University.

MacNeilage P (1970): "Motor Control of Serial Ordering of Speech", Psychological Review 77:182-196.

Maddieson I (1984): Patterns of Sound, Cambridge University Press:Cambridge.

Miller G A, Galanter, E and Pribram, K (1960): Plans and the Structure of Behavior, Holt, Rinehart & Winston:New York.

Miller J D (1989): "Auditory-Perceptual Interpretation of the Vowel", J Acoust Soc Am 85(5):2114-2133.

Milsum J H (1966): Biological Control Systems Analysis, McGraw-Hill:New York.

Moon S-J and Lindblom, B (1989): "Formant Undershoot in Clear and Citation-Form Speech: A Second Progress Report", 121-123 in STL-QPSR 1/1989, (Dept of Speech Communication, RIT, Stockholm).

Nelson W L (1983): "Physical Principles for Economies of Skilled Movements", Biol Cybernetics 46, 135-147.

Nelson W L, Perkell, J S and Westbury, J R (1984): "Mandible Movements during Increasingly Rapid Articulations of Single Syllables: Preliminary Observations", J Acoust Soc Am 75(3):945-951.

Nord L (1986): "Acoustic Studies of Vowel Reduction in Swedish", 19-36 in STL-QPSR 4/1986, (Dept of Speech Communication, RIT, Stockholm).

Ohala J J (1986): "Against the Direct Realist View of Speech Perception", J of Phonetics 14:1, 75-82.

Öhman S (1966): "Coarticulation in VCV Utterances: Spectrographic Measurements", J Acoust Soc Am 39(1):151-168.

Perkell J and Klatt D (1986): Invariance and Variability in Speech Processes, LEA:Hillsdale, N J.

Rabiner L R (1987): "Use of Spectral Change Information Can Significantly Reduce the Error Rate in Speech Recognition", oral presentation at DARPA meeting at Bolt Beranek and Newman Inc, Cambridge, MA, Nov 1987.

Remez R E, Rubin P E, Pisoni D B and Carrell T D (1981): "Speech Perception without Traditional Speech Cues", Science 212, 947-950.

Risberg A (1979): Bestämning av hörkapacitet och talperceptionsförmåga vid svåra hörselskador, Doctoral dissertation, Royal Institute of Technology, Stockholm.

Schulman R (1989): "Articulatory Dynamics of Loud and Normal Speech", J Acoust Soc Am 85(1):295-312.

Shepard R N (1984): "Ecological Constraints on Internal Representation: Resonant Kinematics of Perceiving, Imagining, Thinking and Dreaming", Psychological Review 91(4), 417-447.

Sherrington C S (1941): Man on His Nature, MacMillan:London.

Stevens K N (1986): "Models of Phonetic Recognition II: A Feature-Based Model of Speech Recognition", 66-67 in Mermelstein P (ed): Proceedings Montreal Satellite Symposium on Speech Recognition, Twelfth International Congress on Acoustics.

Stevens K N (1989): "On the Quantal Nature of Speech", J of Phonetics 17:1/2, 3-45.

Stevens K N and Blumstein S (1978): "Invariant Cues for Place of Articulation in Stop Consonants", J Acoust Soc Am 64, 1358-1368.

Stevens K N and Blumstein S (1981): "The Search for Invariant Acoustic Correlates of Phonetic Features", 1-38 in Eimas P and Miller J (eds): Perspectives on the Studies of Speech, LEA:Hillsdale, N J.

Strange W (1989): "Evolving Theories of Vowel Perception", J Acoust Soc Am 85(5), 2081-2087.

Sundberg J (1987): The Science of the Singing Voice, Northern Illinois University Press:Delkalb, Illinois.

Sundberg J and Lindblom B (1989): "Area Functions for Apical Stops and Some Acoustic Problems", submitted to J Acoust Soc Am.

Sussman H M (1989): "The Representation of Stop Place in Multi-Dimensional Space: A Graphic and Statistical Investigation Consonantal Separability as a Function of Vowel Place", submitted to J Acoust Soc Am.

INDEX